Lecture Notes in Computer Science 14435

Founding Editors

Gerhard Goos
Juris Hartmanis

Editorial Board Members

The series Lecture Notes in Computer Science (LNCS), including its subseries Lecture Notes in Artificial Intelligence (LNAI) and Lecture Notes in Bioinformatics (LNBI), has established itself as a medium for the publication of new developments in computer science and information technology research, teaching, and education.

LNCS enjoys close cooperation with the computer science R & D community, the series counts many renowned academics among its volume editors and paper authors, and collaborates with prestigious societies. Its mission is to serve this international community by providing an invaluable service, mainly focused on the publication of conference and workshop proceedings and postproceedings. LNCS commenced publication in 1973.

Qingshan Liu · Hanzi Wang · Zhanyu Ma ·
Weishi Zheng · Hongbin Zha · Xilin Chen ·
Liang Wang · Rongrong Ji
Editors

Pattern Recognition and Computer Vision

6th Chinese Conference, PRCV 2023
Xiamen, China, October 13–15, 2023
Proceedings, Part XI

Springer

Editors
Qingshan Liu [ID]
Nanjing University of Information Science
and Technology
Nanjing, China

Zhanyu Ma [ID]
Beijing University of Posts
and Telecommunications
Beijing, China

Hongbin Zha [ID]
Peking University
Beijing, China

Liang Wang
Chinese Academy of Sciences
Beijing, China

Hanzi Wang [ID]
Xiamen University
Xiamen, China

Weishi Zheng [ID]
Sun Yat-sen University
Guangzhou, China

Xilin Chen [ID]
Chinese Academy of Sciences
Beijing, China

Rongrong Ji [ID]
Xiamen University
Xiamen, China

ISSN 0302-9743   ISSN 1611-3349 (electronic)
Lecture Notes in Computer Science
ISBN 978-981-99-8551-7   ISBN 978-981-99-8552-4 (eBook)
https://doi.org/10.1007/978-981-99-8552-4

Preface

Welcome to the proceedings of the Sixth Chinese Conference on Pattern Recognition and Computer Vision (PRCV 2023), held in Xiamen, China.

PRCV is formed from the combination of two distinguished conferences: CCPR (Chinese Conference on Pattern Recognition) and CCCV (Chinese Conference on Computer Vision). Both have consistently been the top-tier conference in the fields of pattern recognition and computer vision within China's academic field. Recognizing the intertwined nature of these disciplines and their overlapping communities, the union into PRCV aims to reinforce the prominence of the Chinese academic sector in these foundational areas of artificial intelligence and enhance academic exchanges. Accordingly, PRCV is jointly sponsored by China's leading academic institutions: the Chinese Association for Artificial Intelligence (CAAI), the China Computer Federation (CCF), the Chinese Association of Automation (CAA), and the China Society of Image and Graphics (CSIG).

PRCV's mission is to serve as a comprehensive platform for dialogues among researchers from both academia and industry. While its primary focus is to encourage academic exchange, it also places emphasis on fostering ties between academia and industry. With the objective of keeping abreast of leading academic innovations and showcasing the most recent research breakthroughs, pioneering thoughts, and advanced techniques in pattern recognition and computer vision, esteemed international and domestic experts have been invited to present keynote speeches, introducing the most recent developments in these fields.

PRCV 2023 was hosted by Xiamen University. From our call for papers, we received 1420 full submissions. Each paper underwent rigorous reviews by at least three experts, either from our dedicated Program Committee or from other qualified researchers in the field. After thorough evaluations, 522 papers were selected for the conference, comprising 32 oral presentations and 490 posters, giving an acceptance rate of 37.46%. The proceedings of PRCV 2023 are proudly published by Springer.

Our heartfelt gratitude goes out to our keynote speakers: Zongben Xu from Xi'an Jiaotong University, Yanning Zhang of Northwestern Polytechnical University, Shutao Li of Hunan University, Shi-Min Hu of Tsinghua University, and Tiejun Huang from Peking University.

We give sincere appreciation to all the authors of submitted papers, the members of the Program Committee, the reviewers, and the Organizing Committee. Their combined efforts have been instrumental in the success of this conference. A special acknowledgment goes to our sponsors and the organizers of various special forums; their support made the conference a success. We also express our thanks to Springer for taking on the publication and to the staff of Springer Asia for their meticulous coordination efforts.

We hope these proceedings will be both enlightening and enjoyable for all readers.

October 2023

Qingshan Liu
Hanzi Wang
Zhanyu Ma
Weishi Zheng
Hongbin Zha
Xilin Chen
Liang Wang
Rongrong Ji

Organization

General Chairs

Hongbin Zha	Peking University, China
Xilin Chen	Institute of Computing Technology, Chinese Academy of Sciences, China
Liang Wang	Institute of Automation, Chinese Academy of Sciences, China
Rongrong Ji	Xiamen University, China

Program Chairs

Qingshan Liu	Nanjing University of Information Science and Technology, China
Hanzi Wang	Xiamen University, China
Zhanyu Ma	Beijing University of Posts and Telecommunications, China
Weishi Zheng	Sun Yat-sen University, China

Organizing Committee Chairs

Mingming Cheng	Nankai University, China
Cheng Wang	Xiamen University, China
Yue Gao	Tsinghua University, China
Mingliang Xu	Zhengzhou University, China
Liujuan Cao	Xiamen University, China

Publicity Chairs

Yanyun Qu	Xiamen University, China
Wei Jia	Hefei University of Technology, China

Local Arrangement Chairs

Xiaoshuai Sun   Xiamen University, China
Yan Yan   Xiamen University, China
Longbiao Chen   Xiamen University, China

International Liaison Chairs

Jingyi Yu   ShanghaiTech University, China
Jiwen Lu   Tsinghua University, China

Tutorial Chairs

Xi Li   Zhejiang University, China
Wangmeng Zuo   Harbin Institute of Technology, China
Jie Chen   Peking University, China

Thematic Forum Chairs

Xiaopeng Hong   Harbin Institute of Technology, China
Zhaoxiang Zhang   Institute of Automation, Chinese Academy of
                           Sciences, China
Xinghao Ding   Xiamen University, China

Doctoral Forum Chairs

Shengping Zhang   Harbin Institute of Technology, China
Zhou Zhao   Zhejiang University, China

Publication Chair

Chenglu Wen   Xiamen University, China

Sponsorship Chair

Yiyi Zhou   Xiamen University, China

Exhibition Chairs

Bineng Zhong	Guangxi Normal University, China
Rushi Lan	Guilin University of Electronic Technology, China
Zhiming Luo	Xiamen University, China

Program Committee

Baiying Lei	Shenzhen University, China
Changxin Gao	Huazhong University of Science and Technology, China
Chen Gong	Nanjing University of Science and Technology, China
Chuanxian Ren	Sun Yat-Sen University, China
Dong Liu	University of Science and Technology of China, China
Dong Wang	Dalian University of Technology, China
Haimiao Hu	Beihang University, China
Hang Su	Tsinghua University, China
Hui Yuan	School of Control Science and Engineering, Shandong University, China
Jie Qin	Nanjing University of Aeronautics and Astronautics, China
Jufeng Yang	Nankai University, China
Lifang Wu	Beijing University of Technology, China
Linlin Shen	Shenzhen University, China
Nannan Wang	Xidian University, China
Qianqian Xu	Key Laboratory of Intelligent Information Processing, Institute of Computing Technology, Chinese Academy of Sciences, China
Quan Zhou	Nanjing University of Posts and Telecommunications, China
Si Liu	Beihang University, China
Xi Li	Zhejiang University, China
Xiaojun Wu	Jiangnan University, China
Zhenyu He	Harbin Institute of Technology (Shenzhen), China
Zhonghong Ou	Beijing University of Posts and Telecommunications, China

Contents – Part XI

Low-Level Vision and Image Processing

Efficiently Amalgamated CNN-Transformer Network for Image Super-Resolution Reconstruction

Mengyuan Zheng(✉)🆔, Huaijuan Zang🆔, Xinzhi Liu, Guoan Cheng, and Shu Zhan🆔

Hefei University of Technology, Hefei 230000, Anhui, China
1582431913@qq.com

Abstract. Currently, heavy and sophisticated neural network models are designed to improve image super-resolution reconstruction accuracy. However, the model requires high computation resources and is difficult to deploy on mobile devices. Therefore, designing an efficient lightweight image super-resolution network is urgent. We propose to apply the self-attention mechanism in transformer to compensate for the limitations of convolutional neural networks in global representation. In particular, we apply the split depth-wise transposed orthogonal transformer encoder, which splits the input tensor into feature subsets, and depth-wise separable convolution and self-attention across channel dimensions to implicitly increase the receptive field and encode multi-scale features. Experimental results confirmed that our proposed lightweight image super-resolution network (ACTNet) based on an amalgamate CNN-Transformer can achieve better performance than state-of-the-art methods.

Keywords: Super-resolution · Lightweight · Split depth-wise

1 Introduction

Single image super-resolution (SISR) is intended to restore a high-resolution (SR) image with more distinct detail that is visually satisfactory from a corresponding low-resolution (LR) image while preserving the image content. Therefore, the image super-resolution algorithm is widely applied to various advanced computer vision tasks, such as medical image analysis [1], remote sensing imaging [3], and security monitoring [5]. However, SISR was born with the notorious ill-posed problem, a low-resolution image may have countless corresponding high-resolution images, so the previous methods based on manual design have two defects. First, it is difficult to establish a correctly defined mapping from low-dimensional space to high-dimensional space. Second, the efficiency of establishing complex high-dimensional mapping under a large number of original data is hard to meet realistic requirements. Fortunately, with the benefit of the development of efficient computation hardware and the progress of complex algorithms, deep learning has been fully explored. Therefore, researchers apply

Q. Liu et al. (Eds.): PRCV 2023, LNCS 14435, pp. 3–13, 2024.
https://doi.org/10.1007/978-981-99-8552-4_1

learning-based methods to image super-resolution reconstruction. The experimental results confirm that compared with the traditional task-learning algorithm that uses specific domain knowledge to select useful manual features, the deep learning methods can automatically learn the information hierarchy representation and achieve better results both quantitatively and qualitatively.

Since convolutional neural networks made waves on SISR missions, various image super-resolution methods [6,16] based on convolutional networks erupt. After decades of development, researchers continue to design more sophisticated structures to pursue higher accuracy, which inevitably brings unbearable computational complexity. The deployment of mobile terminals with limited resources, such as mobile phones and tablets, is a fatal challenge. Currently, most existing methods typically utilize well-designed and efficient convolution variants to achieve a trade-off between speed and accuracy on resource-constrained mobile devices. Despite these lightweight models are easy to train and effective at encoding local feature details, they do not take advantage of the global interactions between pixels.

In recent years, researchers have applied Transformer, a mainstream framework for natural language processing (NLP), to computer visual tasks [20]. ViT [25] pioneered the visual Transformer. Benefit from the self-attention mechanism, which compensates for CNN's difficulty in modeling global interrelationships. However, the global modeling capability of the Transformer at the cost of increased reasoning speed, and its computational complexity is quadratic to the input, thus creating a gap that is difficult to bridge when directly applying Transformer to mobile devices.

Most of the existing methods pursue to design lightweight network structures for the mobile deployment of image super-resolution. However, due to the limited receptive field of CNN, it is difficult to model the long-distance dependence relationship between image pixels. Although more block structures can be superimposed, it is at the cost of an unbearable computational amount. The emergence of visual transformers is good medicine for CNN. The global modeling capability of the self-attention mechanism can well save CNN's local limitations. Therefore, this paper considers aggregating CNN and Transformer to design a simple and efficient image super-resolution network and make it possible to deploy on mobile terminals. The main contributions of this paper are as follows:

- Aggregating the local modeling capability of convolution operation and the global modeling capability of self-attention, a lightweight image super-resolution reconstruction network, named ACTNet, is designed to be friendly to mobile devices. It is very effective in terms of model size, parameters, and addition operations. The problem of local limitation of convolution operation is solved without significantly increasing the amount of computation.
- Convolution operation and self-attention mechanism are combined in a cascaded way, and the computational complexity of the self-attention mechanism is reduced from an orthogonal perspective. In particular, the depth-wise segmentation mechanism splits the input tensor into feature subsets, implicitly

increases the sensitivity field, and encodes multi-scale features through depth-wise convolution.

– Sufficient experimental results confirm that the proposed ACTNet balances computational complexity and performance well and is superior to the previous state-of-the-art methods in terms of experimental metrics.

2   Related Work

2.1   CNN for SISR

SRCNN [26] is the first work to apply convolutional networks to SISR. It uses 9×9 and two 5×5 convolutional filters to learn the LR to HR nonlinear mapping and uses the MSE loss function to regularize the network structure. Due to SRCNN learning effective representations from mega data in an end-to-end way, it showed tremendous advantages over traditional methods. However, the input of SRCNN is the result of bicubic sampling, if the network structure is to be wider and deeper, the CNN-based super-resolution method faces the problems of calculation and memory. ESPCN [28] uses the subpixel convolution layer, and the interpolation function of the image size enlargement process is implicitly included in the previous convolution layer. It can be automatically learned that the convolution operation is carried out on the size of a low-resolution image, thus greatly reducing the computational complexity. Therefore, the subsequently proposed network structure embeds the upsampler layer at the end. With the development of deep learning, many subsequent works continue to expand the depth and width of the SISR network. VDSR [29] introduced a small convolution kernel into the network structure and extended the depth of the model to 20 layers by using 3×3 convolution kernel stacking layers. RDN [30] uses residual connection and dense connection to solve the problem of information flow loss in the process of gradient propagation, thus increasing the number of network layers to more than 400 layers.

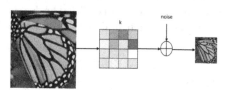

Fig. 1. Degradation process of SISR

2.2   Lightweight SISR

Despite achieving excellent performance, most deep methods face some challenges. First of all, deepening and broadening the network has become an

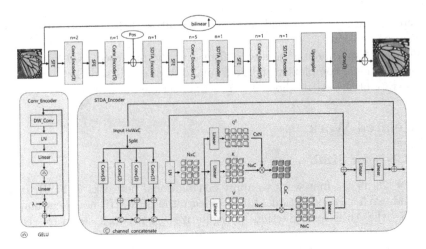

Fig. 2. The network structure of ACTNet

inevitable design trend to improve accuracy, but these methods based on complex computation and memory consumption are difficult to directly deploy in real life. To solve this problem, researchers have focused on designing simple and efficient lightweight networks. The information distillation module proposed by IDN [23] can learn an effective representation of long and short-distance features by combining a strengthening unit and a compression unit, and improve the reasoning speed by using relatively few filters and group convolution. Benefiting from IDN, IMDN [7] proposed a multi-distillation module to realize the optimization between memory and real-time performance. Gradient propagation is carried out by dividing the feature map of a quarter of the channel, while the rest is used as the channel connection to enrich the efficient characterization of the feature map. However, the above methods all take convolutional neural networks as the backbone. The convolutional operation based on the kernel window inevitably causes the extracted feature map to bias to the local receptive field, failing to model the global context information. Secondly, the learned network weight is stable in the reasoning process, which makes the model unable to adapt to the input content.

3   Method Overview

In this section, we will describe the proposed lightweight image super-resolution network in detail. 3.1 will briefly introduce the basic principles of SISR, and 3.2 will introduce the special structure of the proposed ACTNet.

3.1   The Fundamentals of SISR

SISR is committed to restoring missing feature information from low-resolution images. In typical SISR framework, as shown in Fig. 1, We can clearly observe the

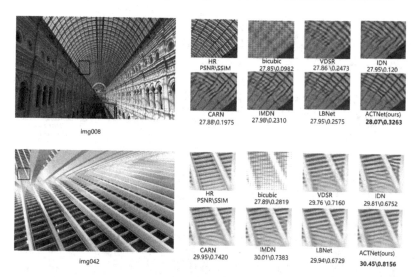

Fig. 3. Comparison of visualization results of Urban100

degradation process of the LR image. Its mathematical formula can be expressed as:

$$y = (k \otimes x) \downarrow_s + n \tag{1}$$

where x represents the corresponding high-resolution image, $\otimes$ represents the convolution operation, k represents the size of the blurred kernel, $\downarrow_s$ represents the downsampling operation with the scale factor of s, n represents the independent noise term, and SISR is actually its inverse process.

3.2 Network Structure

Benefit from [2], the overall network structure of the proposed ACTNet is shown in Fig. 2. We divide the model into four parts: shallow feature extraction module (SFE), convolution layer encoder ($Conv\_Encoder$), depth-wise segmentation transposed orthogonal transformer encoder ($STDA\_Encoder$), and upsampler module. It can be expressed as:

$$I_{SR} = B(I_{LR}) + H_{UP}(H_N(\cdots H_1(I_{LR}))) \tag{2}$$

where I_{LR} represents the input low-resolution image, I_{SR} represents the corresponding high-resolution image, B represents bilinear interpolation operation, H_i represents the aggregate function of the i-th shallow feature extraction module, convolutional layer encoder and depth-wise segmentation transposed orthogonal transformer encoder, and H_{UP} represents the function of the upsampler layer.

Considering the trade-off between computational complexity and network performance, each shallow feature extraction module consists of a 3×3 convolution layer with 64 filters and a standard layer normalization (LayerNorm)

Fig. 4. Comparison of visualization results of Manga109

operation. Note that all the LayerNorm operations in the SFE are located in front of the convolution layer, except for the first SFE. And the shallow feature extraction module is defined relative to the subsequent encoder. And all 3×3 filter size is set to 64 except for the first and last convolution layer.

The convolutional layer encoder consists of depth-wise separable convolution with adaptive kernel size. In this paper, depth-wise separable convolution with a kernel size of 3, 5, 7, 9 is used to gradually increase the receptive field of the deep network, followed by two point-wise convolution layers to fecundate the local representation, and nonlinear feature mapping is carried out by combining standard layer normalization and GELU activation function. A learnable parameter λ is added to enhance the representation ability. Finally, residual connection is used to render the information flow across the network hierarchy and prevent the loss of information flow. Mathematically, it can be expressed as:

$$x_{i+1} = (\lambda * H_{PW}^2(H_{PW}^1(H_{DW}^N(x_i)))) + x_i \tag{3}$$

where x_i denotes the input of the i-th layer convolutional encoder, x_{i+1} denotes the output of the i-th layer convolutional encoder, H_{PW}^1 and H_{PW}^2 denotes the i-th layer two point-wise convolutional functions, H_{DW}^N denotes the depth-wise convolutional function of the size $N \times N$.

The depth-wise segmentation transposed orthogonal transformer encoders have two core components. One is to segment the input feature map to learn the adaptive multi-scale feature representation at different spatial levels. As shown in Fig. 2, suppose we divide the input feature map into four equal parts according to channel dimension and conduct channel connection at different layers after

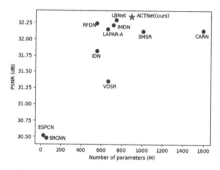

Fig. 5. Calculation complexity and performance comparison

high-dimensional mapping. The learned features are equipped with more flexible spatial receptive field. Moreover, we set different segmentation numbers at different stages to allow efficient and flexible feature encoding. In the experiment, feature subsets of size were set 2, 2, 6, 2 respectively. Second, we reduce the burden of self-attention on network computational complexity from the orthogonal view. Traditional self-attention is calculated by the following formula:

$$H_{Attn} = ((QK^T)/d^{1/2})V \quad (4)$$

Thus, the mathematical expression of the depth-wise segmented transposed orthogonal transformer layer can be written as follows:

$$H_{Attn}^1 = V((Q^T K)/d^{1/2})$$
$$x_{i+1} = (\lambda * H_{PW}^2(H_{PW}^1(H_{Attn}^1(H_{split}(x_i))))) + x_i \quad (5)$$

where H_{Attn}^1 represents the aggregation function of orthogonal self-attention. Different from H_{Attn}, H_{Attn}^1 can effectively reduce the computational complexity of self-attention from the perspective of orthogonal.

We used pixel-shuffle and a convolutional layer with 3 filters to form the upsampler module. Note that, unlike other lightweight SR methods, we stacked two pixel-shuffle operations for the scale factor of 4 instead of using the lighter direct upsampler.

4 Experimental Results and Analysis

4.1 Training Details and Evaluation Metrics

Similar to most existing methods [7,19,21,23], we use DIV2K [4] as our dataset. DIV2K consists of 900 high and low-resolution image pairs, 800 of which serve as the training set and the remaining 100 as the verification set, in which the low-resolution images are obtained by the bicubic downsampling model. The test datasets consist of five benchmark datasets: Set5 [8], Set14 [9], BSD100 [10], Urban100 [11], and Manga109 [12], which contain various scenes such as people,

Table 1. Quantify the comparison results. The best performance is highlighted by red.

Method	Scale	Params (K)	Multi-Adds (G)	Set5 PSNR/SSIM	Set14 PSNR/SSIM	BSD100 PSNR/SSIM	Urban100 PSNR/SSIM	Manga109 PSNR/SSIM
SRCNN [26]	×2	8	52.7	36.66/0.9542	32.42/0.9063	31.36/0.8879	29.50/0.8946	35.60/0.9663
VDSR [29]	×2	666	612.6	37.53/0.9587	33.03/0.9124	31.90/0.8960	30.76/0.9140	37.22/0.9729
LapSRN [27]	×2	251	29.9	37.52/0.9590	33.08/0.9130	31.80/0.8950	30.41/0.9100	37.27/0.9740
IDN [23]	×2	553	144.6.1	37.83/0.9600	33.30/0.9148	32.08/0.8950	31.27/0.9196	38.01/0.9749
CARN [19]	×2	1592	222.8	37.83/0.9600	33.30/0.9148	32.08/0.8985	31.27/0.9196	38.01/0.9749
IMDN [7]	×2	694	158.8	38.00/0.9605	33.63/0.9177	32.19/0.8996	32.17/0.9283	38.88/0.9774
LAPAR-A [18]	×2	548	171.0	38.01/0.9605	33.62/0.9183	32.19/0.8999	32.10/0.9283	38.67/0.9772
RFDN [17]	×2	534	95.0	38.05/0.9606	33.68/0.9184	32.16/0.8994	32.12/0.9278	38.88/0.9773
ACTNet(ours)	×2	642	80.8	38.15/0.9617	33.75/0.9193	32.18/0.8999	32.27/0.9290	38.97/0.9789
SRCNN [26]	×3	57	52.7	32.75/0.9090	29.28/0.8209	28.41/0.7863	26.24/0.7989	30.59/0.9107
VDSR [29]	×3	665	612.6	33.66/0.9213	29.77/0.8314	28.82/0.7976	27.14/0.8279	32.01/0.9310
IDN [23]	×3	553	56.3	34.11/0.9253	29.99/0.8354	28.95/0.8013	27.42/0.8359	32.71/0.9381
CARN [19]	×3	1592	118.8	34.29/0.9255	30.29/0.8407	29.06/0.8034	28.06/0.8493	33.50/0.9440
IMDN [7]	×3	703	71.5	34.36/0.9270	30.32/0.8417	29.09/0.8046	28.17/0.8519	33.61/0.9445
LAPAR-A [18]	×3	544	114.0	34.36/0.9267	30.34/0.8421	29.11/0.8054	28.15/0.8523	33.51/0.9441
RFDN [17]	×3	541	42.2	34.41/0.9273	30.34/0.8420	29.09/0.8050	28.21/0.8525	33.67/0.9449
LBNet [21]	×3	736	68.4	34.47/0.9277	30.38/0.8417	29.13/0.8061	28.42/0.8559	33.82/0.9460
ACTNet(ours)	×3	665	36.8	34.59/0.9283	30.48/0.8431	29.22/0.8066	28.52/0.8564	33.94/0.9473
SRCNN [26]	×4	8	52.7	30.48/0.8626	27.50/0.7513	26.90/0.7101	24.52/0.7221	27.58/0.8555
VDSR [29]	×4	666	612.6	31.35/0.8838	28.01/0.7674	27.29/0.7251	25.18/0.7524	28.83/0.8870
LapSRN [27]	×4	813	149.4	31.54/0.8852	28.019/0.7700	27.32/0.7275	25.21/0.7562	29.09/0.8900
IDN [23]	×4	553	32.3	31.82/0.8903	28.25/0.7730	27.41/0.7297	25.41/0.7632	29.41/0.8942
CARN [19]	×4	1592	90.9	32.13/0.8937	28.60/0.7806	27.58/0.7349	26.07/0.7837	30.47/0.9084
IMDN [7]	×4	715	40.9	32.21/0.8948	28.58/0.7811	27.56/0.7353	26.04/0.7838	30.45/0.9075
LAPAR-A [18]	×4	659	94.0	32.15/0.8944	28.61/0.7818	27.61/0.7366	26.14/0.7871	30.42/0.9074
RFDN [17]	×4	550	23.9	32.24/0.8952	28.61/0.7819	27.57/0.7360	26.11/0.7858	30.58/0.9089
LBNet [21]	×4	742	38.9	32.29/0.8960	28.68/0.7832	27.62/0.7382	26.27/0.7906	30.76/0.9111
ACTNet(ours)	×4	698	20.4	32.34/0.8966	28.74/0.7839	27.70/0.7385	26.31/0.7914	30.80/0.9112

animals, landscapes, buildings, and cartoons. In addition to the convolution of the first layer and the last layer, the number of all convolution channel dimensions is set to 64. We apply Adam [14] optimizer to train our model, and the initial learning rate was set as 2×10^{-4}. We train the model 1000 epochs, and the learning rate decreased to half of the previous epochs after every 200 epochs. We set the batch size to 16 and the input patch size to 64×64. Here, we used the $L1$ norm as the loss function and PSNR and SSIM [13] as the evaluation metrics. And we apply the network weight with the scale factor of 2 as the pre-training weight to train the network structure of ×3 and ×4.

4.2   Experimental Results and Analysis

In this section, we analyze the experimental results quantitatively and qualitatively and compare them with the current 9 state-of-the-art lightweight methods. It includes SRCNN [26], VDSR [29], LapSRN [27], IDN [23], IMDN [7], SMSR [15], RFDN [17], CARN [19], LAPSR-A [18] and LBNet [21]. The comparison results are shown in Table 1. It can be observed from the table that our proposed

ACTNet almost achieves the best performance on the five benchmark datasets regardless of the scale factor. And from Fig. 5, we can clearly observe that we achieved higher performance without significantly increasing the amount of computation, but achieved a good balance between computational complexity and performance.

In addition to the objective analysis of experimental metrics, we visually analyzed the results of different methods from a subjective point of view. It can be observed from Fig. 3 and Fig. 4 that no matter the complex building images in Urban100 or the cartoon images in Manga109, ACTNet has been able to recover results with sharper details and more realistic lines. For example, "Arisa" in Manga109 in Fig. 4, other methods are more or less accompanied by a little blurring and artifacts when restoring hair lines, while our ACTNet restores sharper results and better satisfies the human visual experience.

5   Conclusion

Designing an efficient and lightweight network structure is necessary in the fast-paced era. This paper proposes an efficient and simple image super-resolution model integrating CNN and Transformer, which effectively combines the self-attention mechanism to make up for the shortcomings of CNN in global modeling ability. Moreover, we introduce depth-wise segmentation transposed orthogonal Transformer encoder that splits input tensors into multiple feature subsets and utilizes depth-wise convolution along with self-attention across channel dimensions to implicitly increase the receptive field and encode multi-scale features. And the amount of computation in the self-attention layer is greatly reduced. The experimental results show that the proposed ACTNet surpasses state-of-the-art methods no matter in terms of objective performance metrics or subjective human visual perception.

References

1. Fu, H., Xu, Y., Lin, S., Kee Wong, D.W., Liu, J.: DeepVessel: retinal vessel segmentation via deep learning and conditional random field. In: Ourselin, S., Joskowicz, L., Sabuncu, M.R., Unal, G., Wells, W. (eds.) MICCAI 2016. LNCS, vol. 9901, pp. 132–139. Springer, Cham (2016). https://doi.org/10.1007/978-3-319-46723-8_16
2. Maaz, M., et al.: EdgeNeXt: efficiently amalgamated cNN-transformer architecture for mobile vision applications. arXiv e- prints (2022)
3. Xueyang, F., Zihuang, L., Yue, H., Xinghao, D.: A variational pan-sharpening with local gradient constraints. In: Proceedings of the IEEE Conference on Computer Vision and Pattern Recognition, pp. 10265–10274 (2019)
4. Eirikur, A., Radu, T.: Ntire 2017 challenge on single image super-resolution: dataset and study. In: Proceedings of the IEEE Conference on Computer Vision and Pattern Recognition Workshops, pp. 126–135 (2017)
5. Vijay, P.S., Nicolas, H.Y., Roger, L.K.: An efficient pan-sharpening method via a combined adaptive PCA approach and contourlets. IEEE Trans. Geosci. Remote Sens. 46(5), 1323–1335 (2008)

6. Xintao, W., Liangbin, X., Chao, D., Ying, S.: Real-ESRGAN: Training real- world blind super-resolution with pure synthetic data. In: ICCVW2021, pp. 1905–1914 (2021)

7. Hui, Z., Gao, X., Yang, Y., Wang, X.: Lightweight image super resolution with information multi distillation network. In: Proceedings of the 27th ACM International Conference on Multimedia, pp. 2024–2032 (2019)

8. Bevilacqua, M., Roumy, A., Guillemot, C., Alberi-Morel, M.L.: Low-complexity single-image super-resolution based on nonnegative neighbor embedding (2012)

9. Zeyde, R., Elad, M., Protter, M.: On single image scale-up using sparse-representations. In: Boissonnat, J.-D., et al. (eds.) Curves and Surfaces 2010. LNCS, vol. 6920, pp. 711–730. Springer, Heidelberg (2012). https://doi.org/10.1007/978-3-642-27413-8_47

10. Martin, D., Fowlkes, C., Tal, D., Malik, J.: A database of human segmented natural images and its application to evaluating segmentation algorithms and measuring ecological statistics. In: Proceedings Eighth IEEE International Conference on Computer Vision, ICCV 2001, vol. 2, pp. 416–423. IEEE (2001)

11. Huang, J.B., Singh, A., Ahuja, N.: Single image super-resolution from transformed self-exemplars. In: Proceedings of the IEEE Conference on Computer Vision and Pattern Recognition, pp. 5197–5206 (2015)

12. Yusuke, M., et al.: Sketch-based manga retrieval using manga109 dataset. Multimed. Tools Appl. **76**(20), 21811–21838 (2017)

13. Zhou, W., Alan, C.B., Hamid, R.S., Eero, P.S.: Image quality assessment: from error visibility to structural similarity. IEEE Trans. Image Process. **13**(4), 600–612 (2004)

14. Kingma, D.P., Ba, J.: Adam: a method for stochastic optimization. arXiv preprint arXiv:1412.6980 (2014)

15. Wang, L., et al.: Exploring sparsity in image super-resolution for efficient inference. In: IEEE Conference on Computer Vision and Pattern Recognition, CVPR2021, virtual, 19–25 June 2021, pp. 4917–4926 (2021)

16. Lim, B., Son, S., Kim, H., Nah, S., Mu Lee, K.: Enhanced deep residual networks for single image super-resolution. In: proceedings of the 2017 IEEE Conference on Computer Vision and Pattern Recognition Workshops (CVPRW), pp. 1132–1140 (2017)

17. Liu, J., Tang, J., Wu, G.: Residual feature distillation network for lightweight image super-resolution. In: Bartoli, A., Fusiello, A. (eds.) ECCV 2020. LNCS, vol. 12537, pp. 41–55. Springer, Cham (2020). https://doi.org/10.1007/978-3-030-67070-2_2

18. Wenbo, L., Kun, Z., Lu, Q., Nianjuan, J., Jiangbo, L., Jiaya, J.: LAPAR: Linearly-assembled pixel-adaptive regression network for single image super-resolution and beyond. Adv. Neural. Inf. Process. Syst. **33**, 20343–20355 (2020)

19. Ahn, N., Kang, B., Sohn, K.A.: Fast, accurate, and lightweight super-resolution with cascading residual network. In: Proceedings of the 15th European Conference Computer Vision - ECCV 2018, Part X, Munich, Germany, 8–14 September 2018, pp. 256–272 (2018)

20. Mehta, S., Rastegari, M.: Mobilevit: light-weight, general-purpose, and mobile-friendly vision transformer. In: International Conference on Learning Representations (2022)

21. Gao, G., Wang, Z., Li, J., Li, W., Yu, Y., Zeng, T.: Lightweight bimodal network for single-image super-resolution via symmetric CNN and recursive transformer. In: Proceedings of the Thirty-First International Joint Conference on Artificial Intelligence, IJCAI 2022, Vienna, Austria, 23–29 July 2022, pp. 913–919 (2022)

22. Ma, N., Zhang, X., Zheng, H.T., Sun, J.: ShuffleNet v2: practical guidelines for efficient CNN architecture design. In: The European Conference on Computer Vision (2018)
23. Hui, Z., Wang, X., Gao, X.: Fast and accurate single image super-resolution via information distillation network. In: Proceedings of the IEEE Conference on Computer Vision and Pattern Recognition, pp. 723–731 (2018)
24. Sandler, M., Howard, A., Zhu, M., Zhmoginov, A., Chen, L.C.: MobileNetv 2: inverted residuals and linear bottlenecks. In: Proceedings of the IEEE/CVF Conference on Computer Vision and Pattern Recognition (2018)
25. Dosovitskiy, A., et al.: An image is worth 16x16 words: transformers for image recognition at scale. In: International Conference on Learning Representations(ICLR), Austria (2021)
26. Dong, C., Loy, C.C., He, K., Tang, X.: Learning a deep convolutional network for image super-resolution. In: Fleet, D., Pajdla, T., Schiele, B., Tuytelaars, T. (eds.) ECCV 2014. LNCS, vol. 8692, pp. 184–199. Springer, Cham (2014). https://doi.org/10.1007/978-3-319-10593-2_13
27. Lai, W.S., Huang, J.B., Ahuja, N., Yang, M.H.: Deep Laplacian pyramid networks for fast and accurate super-resolution. In: Proceedings of the IEEE Conference on Computer Vision and Pattern Recognition, pp. 624–632 (2017)
28. Shi, W., Caballero, J., Huszár, F.: Real-time single image and video super-resolution using an efficient sub-pixel convolutional neural network. arXiv e-prints (2016)
29. Kim, J., Lee, J.K., Lee, K.M.: Accurate image super-resolution using very deep convolutional networks. In: Proceedings of the IEEE Conference on Computer Vision and Pattern Recognition, pp. 1646–1654 (2016)
30. Zhang, Y., Tian, Y., Kong, Y., Zhong, B., Fu, Y.: Residual dense network for image super-resolution. In: 2018 IEEE Conference on Computer Vision and Pattern Recognition, Salt Lake City, USA, pp. 2472–2481 (2018)

A Hybrid Model for Video Compression Based on the Fusion of Feature Compression Framework and Multi-object Tracking Network

Yunyu Chen[1], Lichuan Wang[1], and Yuan Zhang[1,2](✉)

[1] College of Big Data and Artificial Intelligence, China Telecom Research Institute, Shanghai, China
chenyy63@chinatelecom.cn, 12031110@zju.edu.cn
[2] College of Information Science and Electronic Engineering, Zhejiang University, Hangzhou, China

Abstract. This paper proposes a video compression algorithm which bases on the combination of feature compression framework and multi-object tracking network. Here, the multi-object tracking network uses joint detection and embedding (JDE) network. The feature compression framework is consisted of extraction network, encoder/decoder and reconstruction network. More specifically, features of the last layer which is before FPN (Feature Pyramid Network) in JDE are first condensed by extraction network to obtain a low-dimensional compact representation. The compact representation allows for losing feature details, whilst capturing several key important cues within the feature at an extremely low bit-rate cost. Then, the condensed features are encoded and decoded by DCT (Discrete Cosine Transform) method. After that, the features are restored by reconstruction network. Therefore, our method combines deep neural network and DCT, so as to constitute a new hybrid compression architecture to ensure robust and efficient compression. The proposed method is evaluated on object tracking task and the experimental results prove that this proposed framework significantly outperforms both image anchor and feature anchor. In particular, it is reported that 46.70% BD-rate saving can be achieved against the image anchor and 92.17% BD-rate saving can be achieved against the feature anchor for the object tracking task.

Keywords: Feature Compression · Video Compression · Object Tracking

1 Introduction

In recent years, some reports [1] show that more than 80% of internet traffic are video contents, which will have an ascending trend in future. Meanwhile, video resolution and fidelity have also made huge steps forward (e.g., 4k, 8k, gigapixel [2], high dynamic range [3], bit-depth [4]) that pose huge challenge to video transmission. Therefore, it is crucial to develop effective video compression systems which can reduce the video size while keeping an acceptable visual quality reflected in the human visual system. In order to achieve this aim, many researchers and engineers have invested many efforts and proposed multiple compression systems. For instance, researchers propose typical block-based video compression standards including H.264/AVC [5], H.265/HEVC [6] and H.266/VVC [7], so as to narrow bandwidth. These three methods rely on handcrafted modules including large quantization parameter and coding unit size to decrease the high level of spatial or temporal redundancy in video sequences. However, these methods produce inevitable defects such as undesirable red artifacts or blurring artifacts near the boundaries during block-wise operation [8], which will have adverse impact on video quality degradation and user experience. Besides, these methods cannot optimize in an end-to-end manner, which will impede the efficiency of optimization.

In order to tackle these problems, researchers pay their attention to learning-based neural networks especially convolutional neural networks (CNN). Different from conventional hand-crafted block-based coding such as H.264, CNN based compression methods enable the usage of non-linear transforms and do not require handcrafting features. Hence, these merits attract many researchers' huge interest. For instance, Schiopu et al. [9] proposed a novel block-wise prediction paradigm based on Convolutional Neural Networks (CNNs) for lossless video coding. Their main contribution is replacing all HEVC-based angular intra-prediction models with CNN-based intra-prediction method. Their experimental results showed that this proposed coding system outperforms the HEVC standard with an average bitrate improvement of around 5%. Zhang et al. [10] surveyed a deep learning-based framework in order to further improve the compression efficiency of HEVC intra frame coding. Their main work was reducing computational complexity by skipping the non-linear mapping layer, and incorporating the residual learning to obtain better predicted residual for CTU encoding. The results proved that the proposed method achieves 3.2% bitrate reduction in average BDBR (Bjentegaard delta bit rate) with only 37% encoding complexity increased. Liu et al. [11] introduced a generic method for helping CNN-filters deal with variable quantization noises, which is introducing a quantization step (Qstep) into the CNN. Their data proved that this method used only one CNN filter and achieved about 3.6% BD-rate reduction for the luminance component of random-access configuration comparing with VVenC anchor.

Although CNN based compression methods have extraordinary performance on video compression, their drawbacks are also obvious such as additional computation, time consuming, heavy computation cost and requirement of powerful hardware. In order to solve these problems and keep the compression performance at the same time, this paper introduces a feature compression framework, which mainly contains feature compression/reconstruction models basing on light-weight convolutional neural networks, and encoding/decoding modules basing on DCT (Discrete Cosine Transform) method [12]. After that, this feature compression framework is inserted between 1/8 Downsample layer and 1/16 Down-sample layer in multi-object tracking network. Here, the multi-object tracking network is chosen as JDE (Joint Detection and Embedding model) [13]. This hybrid model has good performance on video compression. Meanwhile, it can greatly reduce the time and computational cost in encoding and decoding process.

2   Related Works

2.1   JDE (Joint Detection and Embedding Model)

Multiple object tracking (MOT) aims at predicting trajectories of multiple targets in video sequences, which plays a critical role in autonomous driving and smart video analysis. The solution strategy of multiple tracking task can be classified as two methods: SDE (Separate Detection and Embedding) and JDE (Joint Detection and Embedding). For SDE, this method breaks MOT down to two steps [14, 15]: (1) the detection step, in which targets in single video frames are localized; and (2) the association step, where detected targets are assigned and connected to existing trajectories. This method requires at least two compute-intensive components: a detector and an embedding (re-ID) model and its overall inference time is roughly the summation of the two components. For JDE, this idea integrates the detector and the embedding model into a single network. The two tasks thus can share the same set of low-level features, and re-computation is avoided. The backbone of JDE can be Faster R-CNN framework or YOLO framework [13]. In this paper, JDE with YOLO framework proposing by Wang et al. [13] is used as multiple object tracing model because it can greatly reduce inference time and show extraordinary tracking performance.

2.2   DCT (Discrete Cosine Transform) Method

DCT method uses the sum of cosine functions oscillating at different frequencies to represent the sum of sequences of data points. This method is introduced by Nasir Ahmed in 1972 [12], which is a widely used conversion technology in signal processing and data compression. Because the DCT transform is symmetric, the original image information can be restored at the receiving end by using DCT inverse transform. The principles of DCT and inverse DCT transform based on the Eq. (1) and Eq. (2) respectively:

$$F(u, v) = c(u)c(v) \sum_{i=0}^{N-1} \sum_{j=0}^{N-1} f(i,j) cos\left[\frac{(i+0.5)\pi}{N}u\right] cos\left[\frac{(j+0.5)\pi}{N}v\right] \quad (1)$$

$$f(i,j) = \sum_{u=0}^{N-1} \sum_{v=0}^{N-1} c(u)c(v)F(u,v) cos\left[\frac{(i+0.5)\pi}{N}u\right] cos\left[\frac{(j+0.5)\pi}{N}v\right] \quad (2)$$

where, $F(\mu, v)$ is the coefficient after DCT transformation, and $f(i,j)$ is the original signal. Although DCT itself is lossless and symmetric, when combining with quantization method for compression, the image quality restored by inverse DCT will be inferior than that of original image. Hence, DCT compression is lossy and the quality will be more inferior with the rise in quantization coefficients.

3   Methodology

The architecture of the hybrid method is shown in Fig. 1. It mainly contains feature extractor, feature encoder, feature decoder and feature reconstructor:

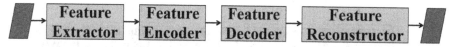

Fig. 1. The architecture of feature compression framework

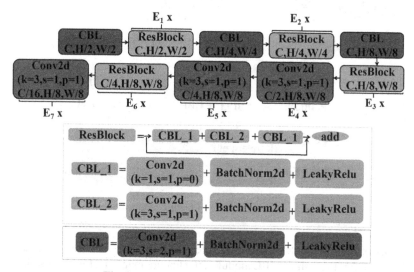

Fig. 2. The architecture of feature extractor

3.1 Feature Extractor

The feature extractor architecture is shown in Fig. 2:

In Fig. 2, the feature extractor is mainly consisted of deep convolutional neural networks. Specifically, three CBL modules are used to down-scale the feature map size and the feature map become its original size of 1/8 after these three CBL modules. Each of these three CBL modules includes 3x3 convolution with stride $= 2$ and LeakyReLU activation function. After each CBL module, some ResBlock modules are inserted, which include two 1x1 convolutions with stride $= 1$ and one 3×3 convolution with stride $= 1$. The purpose of these ResBlock modules is alleviating the problem of network degradation and enhancing feature extraction especially when the size of feature map becomes smaller. E_1, E_2, E_3 and E_6 represent the number of ResBlock modules and their specific values will be defined depending on the degree to which the feature map is compressed. The 3×3 convolution modules with stride $= 1$ in feature extractor are used to reduce the channel of the feature map, so as to further reduce the size of feature map for better compression (E_4, E_5 and E_7 represent the number of the 3x3 convolutions).

3.2 Feature Reconstructor

The feature reconstructor architecture is shown in Fig. 3:

In Fig. 3, the CBL modules are used to restore the channel of the feature map to the original size, but not change the width or height of the feature map (D_1 represents

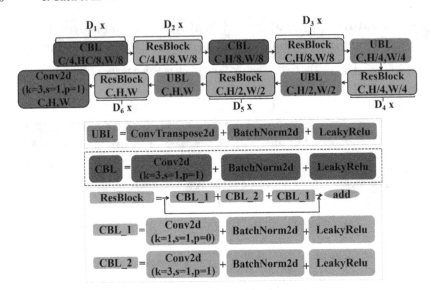

Fig. 3. The architecture of feature reconstructor

the number of the initial CBL modules). Three UBL modules are used to up-scale the feature map. After these three UBL modules, the feature map is restored to original size. Some number of ResBlock modules including two 1x1 convolutions with stride = 1 and one 3 × 3 convolution with stride = 1 have obvious good effect on enhancing feature restoration especially when the size of feature map is very much small. D_2, D_3, D_4, D_5 and D_6 represent the number of ResBlock modules and their specific values will be defined depending on the degree to which the feature map is compressed. The last 3 × 3 convolution with stride = 3 is used to fuse the final feature and output it.

3.3 Feature Encoder and Decoder

The feature encoder and decoder are constructed according to DCT method [12]. The specific flow chart is shown in Fig. 4:

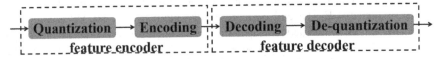

Fig. 4. The flow chart of feature encoder and decoder

In Fig. 4, the quantization module is used to obtain dct_coefficient by performing DCT on each extracted feature which is output by Feature Extractor module and to quantize dct_coefficient by using quantization matrix which is set as a hyper parameter. After the quantization module, the quantized coefficient is obtained and is fed to encoding module which mainly uses binary arithmetic coding (BAC) [5]. After encoding module, a bit stream is obtained and fed to decoding module which is the inverse encoding process for restoring quantized coefficient. The restored quantized coefficient is a little different from the original quantized coefficient because the decoding module has bias. Then, the restored quantized coefficient is transferred to de-quantization module to restore dct_coefficient by using inverse quantization matrix and to recovery the extracted feature by performing inverse DCT on the restored dct_coefficient. The final reconstructed extracted feature is different from the original extracted feature because of quantization and the degree of difference depending on the quantization matrix.

4 Experiment

4.1 The Architecture of the Hybrid Model

This hybrid model is constructed on the combination of feature compression framework and JDE. Specifically, the feature compression framework is inserted after the 1/8 Down-sample layer of JDE, as shown in Fig. 5:

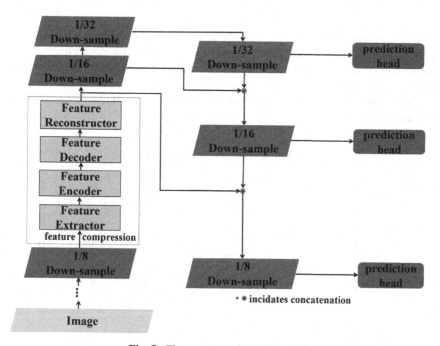

Fig. 5. The structure of hybrid model

In Fig. 5, the feature compression network is used to compress and restore the feature which is output by 1/8 Down-sample layer in JDE.

4.2 Training Details

According to MPEG Technical requirements N193 [16], there are 6 standard points for comparison, as shown in Table 1. The six points represent different compression parameters.

Table 1. The 6 standard points from MPEG

Point No	Bitrate (kbps)	MOTA
1	4756.92	51.10%
2	2148.92	49.90%
3	1002.07	46.80%
4	472.07	40.00%
5	207.59	32.70%
6	87.30	26.10%

In order to surpass the compression performance of these 6 standard points in Table 1, the JDE combining with three different feature compression networks are trained. These three feature compression networks are produced by changing the values of E and D which have been described in Feature Extractor and Feature Reconstructor. The details are listed in Table 2 and Table 3, Table 4 shows the training information of software and hardware:

Table 2. The specific values of E

Point No	E_1	E_2	E_3	E_4	E_5	E_6	E_7
1	0	0	0	1	0	0	0
2	1	2	4	0	1	0	0
3	1	2	4	0	1	0	0
4	1	2	4	0	1	0	0
5	1	2	4	0	1	4	1
6	1	2	4	0	1	4	1

In Table 2 and Table 3, for the first point, the first feature compression model is used. For the second, third and fourth points, the second feature compression model is used. For the fifth and sixth points, the third feature compression model is used.

The three different feature compression models combining with the same JDE are warmed up by loading the best parameters given by authors [13]. Each of the three feature compression networks with the same JDE is trained for 120 epochs with batch size equaling 16 on JDE training datasets [17–22].

Table 3. The specific values of D

Point No	D_1	D_2	D_3	D_4	D_5	D_6
1	0	0	0	0	0	0
2	0	0	0	0	0	0
3	0	0	0	0	0	0
4	0	0	0	0	0	0
5	1	1	1	1	1	1
6	1	1	1	1	1	1

Table 4. The training information of software and hardware

Nvidia Driver	CUDA	GPU	Ubuntu	Python	Pytorch
515.48.07	11.3	Tesla T4, 64GB	18.04 LTS	3.8	1.12

4.3 Evaluation Results

After completing training process, the hybrid model is evaluated on TVD video sequences [23, 24] on CPU. The experimental results are listed in Table 5:

In Table 5, TVD-01, TVD-02 and TVD-03 are three video sequences in TVD datasets. Overall values are the average of TVD-01, TVD-02 and TVD-03 values. QP means quantization points, kbps means kilobits per second. The smaller kbps means deeper compression and more inferior quality. MOTA means Multiple Object Tracking Accuracy, which is a performance measurement for the object tracking task. The MOTA accounts for all object configuration errors made by the tracker, false positives, misses (true negative), mismatches, overall frames [25]. Then, these experimental results are compared with image anchor and feature anchor given by MEPG [16]. Results are shown in Table 6, Table 7, Table 8 and Fig. 6:

In Table 6, comparing with image anchor, the proposed method in this paper has better performance in TVD-01 sequences and a little worse performance in TVD-02 sequences and TVD-03 sequences, but the overall performance is better. In Table 7, comparing with feature anchor, this proposed method is significantly better in TVD-01 sequences and TVD-02 sequences, and have the same performance in TVD-03 sequences. For overall performance, this method also has obvious advantage in BD-rate savings. In Table 8 and Fig. 6, they show the comparison results for overall performance. Comparing with image anchor, this method has worse performance if value of kbps is more than 1000. But when the kbps becomes smaller, the performance of this method becomes better. In other words, bigger kbps means that the quality of video is closer to lossless and the degree of loss in video quality will increase with the decrease of kbps. It is obvious that this method has better performance compared with image anchor in lossy compression. Comparing with feature anchor, this proposed method has excellent performance in both lossless and lossy compression.

Table 5. The experimental results of the hybrid model

Sequence	QP	Bitrate (kbps)	MOTA
Overall	R0	3950.03	50.2%
	R1	1145.60	49.0%
	R2	659.10	48.4%
	R3	386.56	45.0%
	R4	185.10	38.8%
	R5	79.28	27.0%
TVD-01	R0	3887.83	41.6%
	R1	1103.46	41.2%
	R2	613.68	40.7%
	R3	349.95	36.0%
	R4	167.61	33.6%
	R5	69.45	16.9%
TVD-02	R0	3905.28	53.7%
	R1	1138.87	46.1%
	R2	658.36	48.5%
	R3	387.80	43.5%
	R4	195.84	48.5%
	R5	83.31	40.7%
TVD-03	R0	4056.99	60.9%
	R1	1194.47	59.3%
	R2	705.26	58.2%
	R3	427.93	56.8%
	R4	191.86	44.3%
	R5	85.08	38.5%

Table 6. The BD-rate savings of the hybrid model when compared with image anchor

Sequences	BD-rate MOTA	BD-rate Pareto MOTA	BD-MOTA
TVD-01	−79.81%	−79.81%	0.12
TVD-02	−100.00%	55.45%	−0.03
TVD-03	90.07%	90.07	−0.05
Overall	−46.70%	−46.70%	0.04

Table 7. The BD-rate savings of the hybrid model when compared with feature anchor

Sequences	BD-rate MOTA	BD-rate Pareto MOTA	BD-MOTA
TVD-01	−97.47%	−97.45%	0.26
TVD-02	−100.00%	−93.64%	0.07
TVD-03	−56.19%	−56.19%	−0.01
Overall	−92.17%	−92.17%	0.14

Table 8. The Bitrate and MOTA comparison for overall performance

image anchor		feature anchor		hybrid model	
Bitrate(kbps)	MOTA	Bitrate(kbps)	MOTA	Bitrate(kbps)	MOTA
4756.92	51.10%	8254.63	49.52%	3950.03	50.2%
2148.92	49.90%	6075.86	47.92%	1145.60	49.0%
1002.07	46.80%	4393.07	45.62%	659.10	48.4%
472.07	40.00%	2520.69	38.47%	386.56	45.0%
207.59	32.70%	1664.72	33.00%	185.10	38.8%
87.30	26.10%	1106.31	26.30%	79.28	27.0%

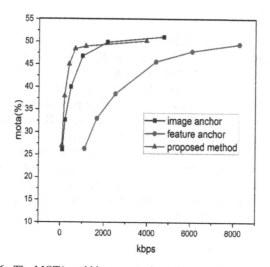

Fig. 6. The MOTA and kbps comparison (for overall performance)

In addition, encoding and decoding time are also crucial measurements for video compression. The time of this proposed model is evaluated on CPU and results are listed in Table 9:

Table 9. The encoding and decoding time of the hybrid model for object tracking task (overall)

Sequence	QP	Encoding Time(s)	Decoding Time(s)	Total Time(s)
Overall	R0	20.58	12.95	33.53
	R1	16.94	11.99	28.93
	R2	13.77	9.25	23.02
	R3	16.35	12.61	28.96
	R4	13.02	8.65	21.67
	R5	15.74	10.90	26.64

In Table 9, the encoding and decoding time of this proposed model are just a few seconds, which is very fast. Through referring relevant previous inputs given by MPEG, the encoding and decoding time of image anchor and feature anchor last for several hours or days, which is very much time-consuming. Hence, in comparison with operation time, this proposed method has huge advantage. Considering the good compression performance, it can be concluded that this hybrid model finds a balance between compression quality and time. In other words, it greatly accelerates the compression process at the cost of slightly sacrificing image quality in comparison with the standard methods given by MPEG.

5   Conclusions

This paper proposes a hybrid model which can greatly reduce compression time while not excessively damaging compression quality. This hybrid model combines feature compression framework with JDE and the feature compression framework mainly uses light-weight CNN-based network to extract or reconstruct feature. The experimental results verify that this method offers significant BD performance gains compared to standard methods given by MEPG. Specifically, the overall BD-rate savings of the proposed compression system over image anchor and feature anchor for object tracking are 46.70% and 92.17% respectively. Moreover, benefiting from using DCT as the core codec in feature compression framework, this hybrid system just needs a few seconds for encoding or decoding, which has outstanding advantage in comparison with standard methods given by MPEG. Therefore, based on the merits of this proposed method, it has potential application value in the future.

References

1. Networking, C.V.: Cisco Global Cloud Index: Forecast and Methodology, 2015–2020, white paper, Cisco Public, San Jose, p. 2016 (2016)
2. Brady, D.J., Gehm, M.E., Stack, R.A., Marks, D.L., et al.: Multiscale gigapixel photography. Nature **486**(7403), 386–389 (2012)
3. Kang, S.B., Uyttendaele, M., Winder, S., Szeliski, R.: High dynamic range video. ACM Trans. Graph (TOG) **22**(3), 319–325 (2003)

4. Winken, M., Marpe, D., Schwarz, H., Wiegand, T.: Bit-depth scalable video coding. In: Proceedings of the IEEE International Conference on Image Processing, vol. 1, pp. I–5. IEEE (2007)
5. Wiegand, T., Sullivan, G.J., Bjontegaard, G., Luthra, A.: Overview of the H. 264/AVC video coding standard. IEEE Trans. Circuits Syst. Video Technol. 13(7), 560–576 (2003)
6. Sullivan, G.J., Ohm, J.-R., Han, W.-J., Wiegand, T.: Overview of the high efficiency video coding (HEVC) standard. IEEE Trans. Circuits Syst. Video Technol. 22(12), 1649–1668 (2012)
7. Bross, B., Chen, J., Liu, S., Wang, Y.-K.: Versatile video coding (draft 5). In: Joint Video Experts Team (JVET) of ITU-T SG, vol. 16, pp. 3–12 (2019)
8. Lim, H., Park, H.: A ringing-artifact reduction method for block-DCT-based image resizing. IEEE Trans. Circuits Syst. Video Technol. 21(7), 879–889 (2011)
9. Schiopu, I., Huang, H., Munteanu, A.: CNN-based intra-prediction for lossless HEVC. IEEE Trans. Circuits Syst. Video Technol. 30(7), 1816–1828 (2019)
10. Zhang, Z.T., Yeh, C.H., Kang, L.W., Lin, M.H.: Efficient CTU-based intra frame coding for HEVC based on deep learning. In: Proceedings of the Asia-Pacific Signal and Information Processing Association Annual Summit and Conference, pp. 661–664. IEEE (2017)
11. Liu, C., Sunyz, H., Kattoz, J., Zeng, X., Fan, Y.: A QP-adaptive mechanism for CNN-based filter in video coding. In: 2022 IEEE International Symposium on Circuits and Systems (2022).https://doi.org/10.1109/ISCAS48785.2022.9937233
12. Ahmed, N., Natarajan, T., Rao, K.R.: Discrete cosine transform. IEEE Trans. Comput. 100(1), 90–93 (1974). https://doi.org/10.1109/T-C.1974.223784
13. Wang, Z., Zheng, L., Liu, Y., Li, Y., Wang, S.: Towards real-time multi-object tracking. In: Vedaldi, A., Bischof, H., Brox, T., Frahm, J.M. (eds.) Computer Vision – ECCV 2020, ECCV 2020, vol. 12356. Springer, Cham (2020). https://doi.org/10.1007/978-3-030-58621-8_7
14. Milan, A., Leal-Taix´e, L., Reid, I., Roth, S., Schindler, K.: Mot16: a benchmark for multi-object tracking. arXiv preprint arXiv:1603.00831 (2016)
15. Yu, F., Li, W., Li, Q., Liu, Y., Shi, X., Yan, J.: Poi: multiple object tracking with high performance detection and appearance feature. In: Hua, G., Jégou, H. (eds.) ECCV 2016. LNCS, vol. 9914, pp. 36–42. Springer, Cham (2016). https://doi.org/10.1007/978-3-319-48881-3_3
16. Curcio, I.: Evaluation Framework for Video Coding for Machines, N193 (2022)
17. Dollár, P., Wojek, C., Schiele, B., Perona, P.: Pedestrian detection: a benchmark. In: CVPR (2009)
18. Zhang, S., Benenson, R., Schiele, B.: CityPersons: a diverse dataset for pedestrian detection. In: CVPR (2017)
19. Xiao, T., Li, S., Wang, B., Lin, L., Wang, X.: Joint detection and identification feature learning for person search. In: CVPR (2017)
20. Zheng, L., Zhang, H., Sun, S., Chandraker, M., Yang, Y., Tian, Q.: Person re-identification in the wild. In: Proceedings of the IEEE Conference on Computer Vision and Pattern Recognition, pp. 1367–1376 (2017)
21. Ess, A., Leibe, B., Schindler, K., Van Gool, L.: A mobile vision system for robust multi-person tracking. In: IEEE Conference on Computer Vision and Pattern Recognition (2008)
22. Milan, A., Leal-Taixé, L., Reid, I., Roth, S., Schindler, K.: MOT16: A benchmark for multi-object tracking. arXiv preprint arXiv:1603.00831 (2016)
23. Gao, W., Xu, X., Qin, M., Liu, S.: An open dataset for video coding for machines standardization. In: 2022 IEEE International Conference on Image Processing (ICIP) (2022). https://doi.org/10.1109/ICIP46576.2022.9897525
24. TVD Datasets Homepage. https://multimedia.tencent.com/resources/tvd
25. Bernardin, K., Elbs, A., Stiefelhagen, R. Multiple object tracking performance metrics and evaluation in a smart room environment. In: Sixth IEEE International Workshop on Visual Surveillance in Conjunction with ECCV (2008)

Robust Degradation Representation via Efficient Diffusion Model for Blind Super-Resolution

Fangchen Ye, Yubo Zhou, Longyu Cheng, and Yanyun Qu[✉]

School of Informatics, Xiamen University, Xiamen, China
yyqu@xmu.edu.cn

Abstract. Blind super-resolution (SR) is a challenging low-level vision task, dedicated to recovering corrupted details in low-resolution (LR) images with complex unknown degradations. The mainstream blind SR methods mainly adopt the paradigm of capturing the robust degradation representation from the LR images as condition and then perform deep feature reconstruction. However, the manifold degradation factors make it challenging to achieve flexible estimation. In this paper, we propose a residual-guided diffusion degradation representation scheme (Diff-BSR) for blind SR. Specifically, we leverage the powerful generative capability of the diffusion model (DM) to implicitly model the diverse degradations representation, which helps to resist to the disturbance of varied input. Meanwhile, to reduce the expensive computational complexity and training costs, we design a lightweight degradation extractor in the residual domain. It transforms the target residual distribution in a low-dimension feature space. As a result, Diff-BSR requires only about 60 sampling steps and a much smaller scale denoising network. Moreover, we designed the Degradation-Aware Multihead Self-Attention mechanism to effectively fuse the discriminative representations with the intermediate features of the network for robustness enhancement. Extensive experiments on mainstream blind SR benchmarks show that Diff-BSR achieves SOTA or comparable performance compared to existing methods.

Keywords: Blind SR · Diffusion model · Degradation representation

1 Introduction

Image super-resolution is a fundamental research direction in the field of computer vision. In order to enhance the robustness of SR models in handling real-world complex degraded images, blind SR aims to restore high-resolution (HR) images from low-resolution (LR) images that suffer from various unknown degradations, which is a challenging and ill-posed problem. Some approaches utilize the synthesis ability of Generative Adversarial Networks (GANs) to restore

Supplementary Information The online version contains supplementary material available at https://doi.org/10.1007/978-981-99-8552-4_3.

damaged texture details [6,29,35]. However, their unstable training process has been criticized, and the generated results often contain unrealistic textures. Another category of methods [12,28] involves designing additional modules to extract image-level degradation representations and fuse them into the restoration process. Unfortunately, we have found that the existing methods have weak representation capability in degradation extraction, and the corresponding feature fusion mechanisms are too simplistic.

Recently, denoising diffusion probabilistic models (DDPM) [9,24] emerge as a new trend for various synthesis tasks [4,7,10]. In principle, DDPM follows a Markov chain to convert the target data into the latent variable by gradually adding noise (diffusion process), and then predicts the noise in each step to reconstruct the original data by a neural network (reverse process). DDPM can be trained easily by optimizing a variant of the variational lower bound, which avoids the unstable training of GANs. Besides, the emergency of Transformer [17, 33] has pushed the further development of strong feature modelling capacities by capturing the global long-range dependencies among pixels.

Therefore, we propose an efficient DM framework for Blind SR, namely Diff-BSR. To mimic the flexible and diverse degradation representation, we integrate the DM into the degradation representation extraction procedure. It is worth noting that conventional DMs require repeated function evaluations in the high-dimensional space of RGB images for training and inference, and their impressive generative capability relies on accurate noise map predictions for hundreds or thousands of times. To utilize the DM in a hardware-friendly manner, we perform the diffusion process only on degradation representations mined in the latent feature space. Regarding the latter, we have designed the Degradation-Aware Multihead Self-Attention mechanism (DA-MSA) to linearly map the degradation representation into degradation *query*, which is then combined with *key* and *value* generated from intermediate features to perform self-attention computation, allowing effective capturing of spatial degradation information within a large receptive field. The main contributions can be summarized as follows:

1. We have designed an efficient DM to implicitly predict low-dimensional degradation representations, resulting in a significant improvement in representation capability.
2. Based on the powerful global modeling capability of the attention calculation, we designed the DA-MSA mechanism to fully leverage the degradation representations.
3. Experimental results on popular benchmarks demonstrate the effectiveness of our Diff-BSR, achieving SOTA or comparable results in terms of visual quality and evaluation metrics.

2   Related Work

Blind SR. There are implicit and explicit ways to construct datasets for blind SR. Implicit methods [8,18,27,32] focus on collecting real-world LR samples and employ unsupervised training frameworks, such as GANs or self-supervised contrastive learning, to train models in the absence of groud truth (GT) labels. Explicit methods, on the other hand, emphasize parameterizing the degradation process of real images by designing an image degradation pipeline [29,35]. To expand the degradation space of LR images, a high-order combination of multiple degradations (blur, noise, JPEG compression) is used to replace the original simple bicubic downsampling operation. In this paper, we focus on the explicit degradation ways and aim to design a supervised framework for blind SR.

Among blind SR methods using synthetic datasets, GAN and their variants [6,16,29,35] have gained widespread use due to their exceptional generative capabilities. However, GAN is known for its unstable training and unrealistic texture generation. Besides, another approach involves extracting degradation representations to assist the main network in detail restoration. DAN [12] predicts the blur kernel from the LR input, and then simply concatenates it with the feature maps before feeding it into the backbone network. DASR [28] extracts a linear degradation vector from the LR, and then performs linear scaling on the feature maps within the backbone network.

Diffusion Models. Nowadays, DMs [9,19,24] have exhibited remarkable generative capabilities in various computer vision tasks, including image restoration, image editing, image generation, etc [4,7,10]. In the field of SR, SR3 [22] first utilizes a U-Net as a denoising backbone and performs reverse diffusion through an iterative denoising process. Building upon SR3, SR3+ [21] incorporates a self-supervised training strategy combined with noise-conditioning augmentation to address the challenges of complex parameterized degradation in SR problems. SRDiff [15] employs a pre-trained LR encoder to extract the conditional LR information before the diffusion process while incorporating a residual prediction strategy, significantly reducing the required number of diffusion steps and accelerating the convergence speed of the DM. Later, ResDiff [23] utilizes discrete wavelet transform operations to guide the DM in predicting high-frequency details in the frequency domain, complemented with a Convolutional Neural Network (CNN) to recover the main low-frequency components in the image.

The aforementioned pixel-level DMs suffer from high hardware requirements and training costs. Therefore, the latent DM methods propose to encode images into a latent feature space as a preprocessing step, reducing the number of sampling steps and improving inference efficiency. LSGM [26] incorporates DM into the latent space through a variational autoencoder (VAE), where DM only needs to learn a prior distribution. LDM [20] employs a pre-trained image autoencoder to generate a low-dimensional latent space. The DM is then trained to generate latent codes, which are subsequently decoded to produce pixel space outputs. In this work, we mainly adopt DM to model the diverse degradation representation in latent residual feature space.

3   Methods

Overall Framework. Our Diff-BSR aims to leverage the powerful generation capabilities of DMs to capture high-quality degradation representations for

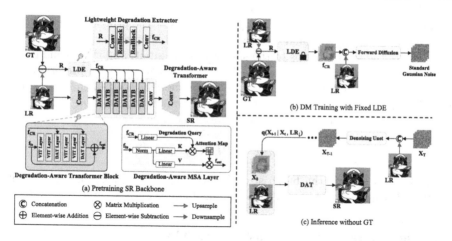

Fig. 1. The proposed framework of Diff-BSR.

assisting blind SR tasks. The overall framework of Diff-BSR is shown in Fig. 1. The whole procedure consists of three stages: SR backbone pretraining, DM training, and fine-tune. The SR backbone includes a lightweight degradation extractor (LDE) and Degradation-Aware Transformer (DAT). LDE is designed to fully mine the degradation information between GT and LR images, which can be regarded as an upper bound on performance. DAT is used for deep feature reconstruction conditional with the degradation representation of LDE. In the SR backbone pretraining stage, we introduce LR and GT data to jointly train the LDE and DAT. Then, we train the DM to mimic the pre-trained LDE in predicting the degradation representation in the absence of GT. Afterward, we fine-tune DAT while keeping the diffusion process fixed, which is omitted in the figure. Finally, the LR image and the degradation representation generated by the latent DM are input to DAT to obtain the SR results.

3.1 Lightweight Degradation Extractor (LDE)

As shown in Fig. 1(a), the LDE is composed of several residual blocks and bicubic downsampling operations, which is lightweight with negligible computation. Unlike the existing degradation extraction methods, we introduce GT data and calculate the residual map R with upsampled LR image through element-wise subtraction, which serves as the input to LDE. On one hand, the residual map primarily captures the missing high-frequency components of the original image, encompassing rich textures details.

On the other hand, SRDiff [15] has demonstrated that residual prediction strategy can accelerate the convergence and reduce the number of sampling steps in the diffusion process. An intuitive inference is that degradation representations extracted from the residual map can also reduce the optimization difficulty of prediction in the subsequent diffusion process. Considering the above two reasons, we extract degradation representations from the residual space.

Furthermore, the downsampling operations follow the principle of latent DM to generate a low-dimensional degradation space. The first downsampling operation scales the feature maps to the size of the LR input, and the second downsampling operation scale is set to 4, achieving a balance between preserving the texture details of feature maps and improving the efficiency of the subsequent DM. For a given pair of LR image $\mathbf{x}$ and GT image $\mathbf{y}$, the ideal degradation representation obtained by LDE can be formulated as:

$$R = \mathbf{y} - \mathbf{x}_{\uparrow bic}; \ f_{CR} = LDE(R), f_{CR} \in \mathbb{R}^{b \times c \times \frac{h}{4} \times \frac{w}{4}}, \tag{1}$$

where b, c, h, w denote the batch size, the number of channels, and the height, width of LR images. "$\uparrow bic$" refers to the bicubic upsample operation.

3.2 Degradation-Aware Transformer (DAT)

The DAT follows the design paradigm of common ViT models [17,33], which includes a Conv layer for shallow feature extraction, multiple Degradation-Aware Transformer Blocks (DATB) for deep feature extraction, and a final Conv layer for image reconstruction, as shown in Fig. 1(a).

To preserve the carefully designed structure of ViT, we designe the DATB as a stack of several conventional ViT layers and DAT layers. For the DAT layer, we integrate the degradation-aware representation f_{CR} into the Multihead Self-Attention calculation as Degradation-Aware Multihead Self-Attention (DA-MSA), which aims to enhance the robustness for processing diverse degradation types.

Specifically, we first use 1×1 Conv to increase the dimension of f_{CR}, and then use PixelShuffle operation to align the size of f_{CR} and LR features. During the deep feature extraction process, the DA-MSA layer has two inputs: the aligned degradation representation f_{CR} and the input feature f_{in}. Following the standard MSA mechanism, we partition both inputs into $\frac{hw}{M^2}$ local windows with a size of $M \times M$. Within each window, we first use the linear transformation to map f_{CR} to degradation queries Q_D, and transform f_{in} to K and V. Then, we perform standard attention computation, which can be represented as follows:

$$Q_D = f_{CR} \cdot P_Q, K = f_{in} \cdot P_K, V = f_{in} \cdot P_V, \ Q_D, K, V \in \mathbb{R}^{M^2 \times d}, \tag{2a}$$

$$Attention(Q_D, K, V) = SoftMax(Q_D K^T / \sqrt{d} + B)V, \tag{2b}$$

where P_Q, P_K, P_V denote the projection matrices, d refers to the embedding dimension, and B refers to the learnable relative positional encoding. In summary, as a variant of MSA, we leverage the powerful global modeling capabilities of DA-MSA to effectively integrate spatial degradation representation with intermediate features within a large receptive field. In addition, as an advanced and high-performance degradation-aware architecture, DA-MSA exhibits great versatility for other feature fusion scenarios.

3.3   Diffusion Model Training and Inference

After the SR backbone pretraining stage, we train the DM to mimic the output of LDE in generating the same compact representation in the absence of GT. We first utilize the pretrained LDE to extract f_{CR} from the paired training sample as the target distribution of DM.

Forward Diffusion Process. Subsequently, we apply the forward diffusion process on X_0 (f_{CR}), to obtain X_T by sampling for T steps, which follows a standard Gaussian distribution, as shown in Fig. 1(b). By utilizing the reparameterization method, the sampling at any time step $t, t \in \{1, 2, \ldots T\}$ can be summarized as:

$$q(X_t \mid X_0) = \mathcal{N}(X_t; \sqrt{\overline{\alpha}_t}X_0, (1 - \overline{\alpha}_t)\mathbf{I}), \tag{3a}$$

$$X_t(X_0, \epsilon) = \sqrt{\overline{\alpha}_t}X_0 + \sqrt{1 - \overline{\alpha}_t}\epsilon, \ \epsilon \sim \mathcal{N}(\mathbf{0}, \mathbf{I}) \tag{3b}$$

where $\overline{\alpha}_t = \prod_{i=1}^{t} \alpha_i$, $\alpha_i = 1 - \beta_i$ and β_i is the predefined scale factor, $\mathcal{N}$ refers to Gaussian distribution. In DMs, U-Net $\epsilon_\theta(X_t, cond, t)$ is commonly employed to estimate the noise ϵ, where $cond$ is the conditional information used to guide the diffusion process. Since the size of the predicted ϵ is much smaller than conventional DMs, we have streamlined the U-Net architecture, resulting in approximately 30% parameter reduction and 50% computation savings. During the training process, X_t is first generated according to Eq. (3) with a randomly sampled time step t and $\epsilon \sim \mathcal{N}(\mathbf{0}, \mathbf{I})$. We follow the widely-used DDPM [9] algorithm to optimize the parameter θ of U-Net ϵ_θ by:

$$\nabla_\theta \|\epsilon - \epsilon_\theta(\sqrt{\overline{\alpha}_t}X_0 + \sqrt{1 - \overline{\alpha}_t}\epsilon, \mathbf{x}_{\downarrow bic}, t)\|_2^2, \tag{4}$$

where $cond$ is set to $\mathbf{x}_{\downarrow bic}$ with the same size as f_{CR}. Both are then concatenated along the channel dimension and input into the U-Net.

Reverse Diffusion Process. In the reverse process of DM, we first sample a random variable $X_T \sim \mathcal{N}(X_T; \mathbf{0}, \mathbf{I})$ with the same dimensions as the f_{CR}. Then, we run T steps of denoising iterations to obtain X_0 ($\hat{f}_{CR}$). For any sampling step t, the denoising process during training can be written as the following Gaussian distribution form, the noise ϵ is first generated by $\epsilon_\theta(X_t, \mathbf{x}_{\downarrow bic}, t)$, and then the mean and variance are calculated for noise removal to obtain X_{t-1}:

$$p(X_{t-1} \mid X_t, X_0) = \mathcal{N}(X_{t-1}; \mu_t(X_t, X_0), \sigma_t^2 \mathbf{I}), \tag{5}$$

where $\mu_t(X_t, X_0) = \frac{1}{\sqrt{\alpha_t}}(X_t - \frac{\beta_t}{\sqrt{1 - \overline{\alpha}_t}}\epsilon)$ and $\sigma_t^2 = \frac{1 - \overline{\alpha}_{t-1}}{1 - \overline{\alpha}_t}\beta_t$.

Due to the inevitable slight discrepancy between the predicted $\hat{f}_{CR}$ and f_{CR}, at the end of the training process, we fix the diffusion process and fine-tune the SR backbone DAT. The final inference process, as shown in Fig. 1(c), can be expressed as $\hat{y} = DAT_\phi(\mathbf{x}, \hat{f}_{CR})$, where ϕ denotes DAT parameters.

Loss Functions. In the SR backbone pre-training stage, we optimize the parameters of DAT and LDE by minimizing the L_1 pixel loss. In addition, we use the contrastive loss following CSD [31] to further improve the degradation extraction ability of LDE. The total loss can be summarized as:

$$L_{CL} = \sum_i^N \sum_j^M \lambda_j \frac{d(\varphi_j(f_{CR}^{(i)}), \varphi_j(f_{Pos}^{(i)}))}{\sum_k^K d(\varphi_j(f_{CR}^{(i)}), \varphi_j(f_{Neg}^{(k)}))}, \tag{6a}$$

$$\mathcal{L}_{total} = \|\hat{\mathbf{y}} - \mathbf{y}\|_1 + \lambda_c L_{CL}, \tag{6b}$$

where λ_c is the hyperparameter, N is the number of training images. M is the total number of hidden layers. K is the number of negative samples. Intuitively, we randomly sample two patches in f_{CR} as positive samples, and set f_{CR} generated from different images as negative samples. φ_j is the intermediate features from the jth layer of the pretrained VGG19. λ_j is the learnable balancing weight for each layer. $d(*, *)$ is the L_1-distance between two samples. The loss function in the DM training stage has been given by Eq. (4). During the fine-tune stage, only L_1 loss is used. Overall, the Diff-BSR procedure is described in Algorithm 1.

Algorithm 1. Efficient Diffusion Model for Blind SR (Diff-BSR)

Data: LR and GT image dataset: D_{lr}, D_{gt}
Input: Initialized parameters of DAT ϕ, LDE ψ and U-Net θ
Output: Optimized parameters DAT ϕ and U-Net θ

1: /*Stage1: SR Backbone Pretraining*/
2: **while** *not done* **do**
3:      Sample LR-GT batch from D_{lr}, D_{gt};
4:      Generate f_{CR} defined in Eq. (1) and predict SR result $\hat{\mathbf{y}} = DAT_\phi(\mathbf{x}, f_{CR})$;
5:      Calculate the training loss $\mathcal{L}_{total}$ defined in Eq. (6) and update ϕ, ψ;
6: **end while**
7: /*Stage2: Diffusion Model Training*/
8: **while** not done **do**
9:      Sample LR-GT batch from D_{lr}, D_{gt};
10:     Generate f_{CR} defined in Eq. (1) as X_0 with fixed LDE parameters ψ;
11:     Optimize U-Net parameters θ with Eq. (3)(4);
12: **end while**
13: /*Stage3: Fine-tune DAT*/
14: **while** *not done* **do**
15:     Sample LR-GT batch from D_{lr}, D_{gt};
16:     Generate $\hat{f}_{CR}$ using Eq. (5) with fixed θ and predict $\hat{\mathbf{y}} = DAT_\phi(\mathbf{x}, \hat{f}_{CR})$;
17:     Fine-tune DAT parameters ϕ using L_1 loss;
18: **end while**

---

4   Experiments

4.1   Training and Testing Datasets

DIV2K [1], Flickr2K [25] and OutdoorSceneTraining [30] are adopted to train the SR model. The HR patch size is set to 256×256 and we follow BSRGAN [35] to generate the LR samples. The data augmentation is implemented by random flipping and 90° rotation. Besides, the trained models are evaluated on four public benchmark datasets: Set5 [3], Set14 [34], BSD100 [2], and Urban100 [11]. Different from the training samples, we use a mixed degradation pipeline of BSRGAN and Real-ESRGAN [29], denoted as *bsrgan-plus*, to generate the diverse test LR samples. In addition, we conduct no-reference image quality evaluation on three real-world datasets, including RealSR [5], DRealSR [32] and DPED-iphone [13].

4.2   Implementation and Training Details

Implementation Details. ADAM [14] optimizer is used during training by setting $\beta_1 = 0.9$, $\beta_2 = 0.999$. We set the minibatch size to 16. In the three-stage training process, the learning rates are set as follows: 10^{-4}, 10^{-4} and 2×10^{-5} and halved at every 2×10^5 iterations. The total number of iterations is set as 300K, 200K, and 50K, respectively. We set the hyper-parameters $\lambda_c = 0.01$, and keep the rest of the settings in L_{CL} consistent with CSD [31]. The diffusion configuration is provided in Table 1. For more details, please refer to the supplementary material. We implemented our model with PyTorch framework using two NVIDIA 3090 GPUs. The overall training time for Diff-BSR is approximately 4 days.

Table 1. Detailed configuration of training the DM for ×4 Diff-BSR. U-Net scale is indicated by "number of parameters (M)/computational complexity (GFlops)".

	SR3+ [21]	SRDiff [15]	Diff-BSR
Schedule	linear	cosine	cosine
Time Step	2000	100	60
Start β_0	10^{-6}	8×10^{-3}	10^{-6}
End β_T	10^{-2}	2×10^{-2}	10^{-2}
U-Net Scale	132M/56.5G	109M/36.4G	93M/27.8G

Evaluation Criterion. For synthetic datasets, we evaluate the performance using the widely used peak signal-to-noise ratio (PSNR) and structure similarity index (SSIM). For real-world datasets, since there are usually no available GT images, we provide the well-known NIQE score as the evaluation index.

Table 2. Quantitative comparison with mainstream blind SR baselines for scaling factor ×2,×4. The best and second-best results are **highlighted** and underlined respectively.

Type	Method	Scale	Set5 PSNR/SSIM	Set14 PSNR/SSIM	BSD100 PSNR/SSIM	Urban100 PSNR/SSIM
w/o DM	DAN [12]	×2	25.27 / 0.6278	22.79 / 0.5083	23.46 / 0.4923	20.93 / 0.4793
	DASR [28]	×2	25.31 / 0.6312	22.81 / 0.5110	23.49 / 0.4958	20.94 / 0.4819
	BSRGAN [35]	×2	27.65 / **0.7799**	24.59 / 0.6475	24.88 / 0.5967	**22.76** / 0.6391
	Real-ESRGAN+ [29]	×2	26.73 / 0.7771	23.65 / 0.6299	24.11 / 0.5860	21.66 / 0.6148
	SwinIR-GAN [17]	×2	27.07 / 0.7793	23.76 / 0.6364	23.83 / 0.5717	21.54 / 0.6195
	FeMaSR [6]	×2	26.46 / 0.7470	23.38 / 0.5982	23.83 / 0.5599	21.90 / 0.5956
w/ DM	SR3+[21]	×2	26.51 / 0.7371	23.39 / 0.5981	23.88 / 0.5367	21.41 / 0.5740
	SRDiff [15]	×2	26.58 / 0.7377	23.41 / 0.5983	23.90 / 0.5368	21.40 / 0.5743
	ResDiff [23]	×2	27.34 / 0.7782	24.10 / 0.6428	24.52 / 0.5912	22.51 / 0.6300
	Diff-BSR (Ours)	×2	**27.71** / 0.7791	**24.68** / **0.6508**	**24.90** / **0.5970**	22.60 / **0.6395**
w/o DM	DAN [12]	×4	20.85 / 0.5319	21.44 / 0.4937	22.52 / 0.4818	20.20 / 0.4757
	DASR [28]	×4	20.87 / 0.5336	21.43 / 0.4953	22.49 / 0.4818	20.18 / 0.4752
	BSRGAN [35]	×4	21.65 / 0.5576	**22.14** / 0.5163	22.60 / 0.5036	20.62 / **0.5335**
	Real-ESRGAN+ [29]	×4	21.33 / 0.5446	21.56 / **0.5284**	22.48 / 0.5041	19.98 / 0.5282
	SwinIR-GAN [17]	×4	20.94 / 0.5130	21.60 / 0.5043	22.23 / 0.4925	20.11 / 0.5303
	FeMaSR [6]	×4	20.47 / 0.4865	21.22 / 0.4806	22.11 / 0.4830	20.25 / 0.5243
w/ DM	SR3+[21]	×4	21.15 / 0.5419	21.29 / 0.5156	22.18 / 0.4998	19.56 / 0.5241
	SRDiff [15]	×4	21.20 / 0.5422	21.32 / 0.5158	22.19 / 0.4999	19.59 / 0.5243
	ResDiff[23]	×4	21.52 / 0.5498	21.96 / 0.5170	22.47 / 0.5036	20.48 / 0.5322
	Diff-BSR (Ours)	×4	**21.89** / **0.5590**	21.91 / 0.5245	**22.70** / **0.5053**	**20.63** / 0.5311

4.3 Comparison with Existing Blind SR Methods

We compare the proposed Diff-BSR with several recent DM-based methods: SR3+ [21], SRDiff [15], ResDiff [23], and non-DM blind SR methods: DAN [12], DASR [28], BSRGAN [35], Real-ESRGAN [29], SwinIR-GAN [17], FeMaSR [6]. Table 2 shows that our Diff-BSR achieve SOTA or comparable results to recent methods on mainstream ×2, ×4 benchmarks. The quantitative and qualitative results on ×4 real-world dataset are shown in Fig. 2 and Table 3. We can observe that Diff-BSR is more favorable and can recover more texture details.

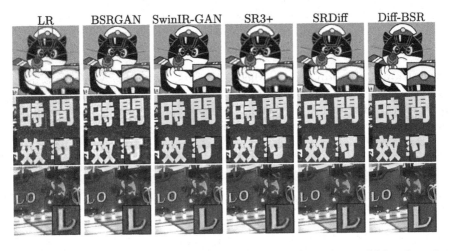

Fig. 2. Visual comparison with existing Blind SR baselines on RealSR [5] benchmarks.

4.4 Ablation Study

To demonstrate the effectiveness of each core component in our Diff-BSR, we removed the degradation extraction and the diffusion process, and only used the SR backbone DAT as the baseline. As shown in Table 4, plain-structured DAT performs poorly in case 1. In case 2, we introduced the degradation vector following DASR [28]. While this approach has provided some improvement in performance, it does not fully utilize the degradation information. As shown in case 3, we employed the DA-MSA mechanism to capture spatial degradation representation, resulting in a significant performance improvement. The PSNR was increased by approximately 0.4 dB. In case 4 and 5, we followed the standard training procedure of Diff-BSR, which demonstrated the effectiveness of incorporating GT and contrastive learning loss into the degradation extraction process.

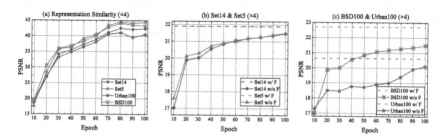

Fig. 3. Analysis of the effectiveness of $\hat{f}_{CR}$ generated by DM.

Table 3. Comparison of the NIQE↓ score on real-world benchmarks for ×4 scaling.

Benchmark	Method				
	DASR	BSRGAN	Real-ESRGAN+	SwinIR-GAN	Diff-BSR
RealSR [5]	8.1916	5.7348	4.7831	<u>4.7645</u>	**4.7230**
DRealSR [32]	9.1442	6.1356	4.8455	**4.7053**	<u>4.7562</u>
DPED-iphone [13]	6.9880	5.9900	5.2628	<u>4.9468</u>	**4.9468**

Table 4. Ablation studies on ×4 blind SR benchmarks (PSNR) for Diff-BSR.

Index	Method					Benchmark		
	Baseline	LDE	DA-MSA	DM	L_{CL}	Set5	Set14	BSD100
1	✓					20.88	20.79	21.54
2	✓	✓				21.12	21.10	21.83
3	✓	✓	✓			21.56	21.59	22.34
4	✓	✓	✓	✓		21.83	21.90	22.67
5	✓	✓	✓	✓	✓	**21.89**	**21.91**	**22.70**

Effectiveness of Diffusion Model. In addition to the fewer sampling steps and smaller U-Net scale shown in Table 1, Diff-BSR demonstrates faster convergence compared to pixel-level DM, as illustrated in Fig. 4 (approximately 200K iterations). To further validate the effectiveness of the DM, we computed the PSNR metric of the $\hat{f}_{CR}$ predicted by the DM, and the f_{CR} extracted by the fixed-parameter LDE during the DM training stage. The results are shown in Fig. 3, which demonstrates that the PSNR of the two reach approximately 40 after convergence, indicating a high degree of structural similarity. We also conducted experiments without fine-tune, directly combining DAT with DM. In Fig. 3(b)(c), we can clearly observe the rapid performance recovery.

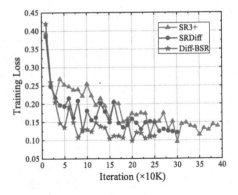

Fig. 4. Comparison of convergence speed of DM-based SR methods.

5 Conclusion

In this paper, we propose an efficient diffusion framework for blind super-resolution, called Diff-BSR. We have devised a strategy that leverages a latent DM to implicitly predict the degradation representation extracted from the residual domain, resulting in a significant enhancement in representation capability as well as fast inference. Additionally, the DA-MSA mechanism is designed to fuse spatial degradation information with intermediate features within a large receptive field. Quantitative and qualitive results on benchmarks demonstrate the effectiveness of Diff-BSR. Although the visual quality of the restored images still remains relatively poor for extremely degraded samples, we believe that our method has such potential for fully mining the image degradation representation and can inspire the research of this direction in the future.

Acknowledgment. This work is supported by the National Key Research and Development Program of China No. 2020AAA0108301; National Natural Science Foundation of China under Grants No. 62176224; CCF-Lenovo Blue Ocean Research Fund.

References

1. Agustsson, E., Timofte, R.: NTIRE 2017 challenge on single image super-resolution: dataset and study. In: CVPRW, pp. 126–135 (2017)
2. Arbelaez, P., Maire, M., Fowlkes, C., Malik, J.: Contour detection and hierarchical image segmentation. TPAMI **33**(5), 898–916 (2010)
3. Bevilacqua, M., Roumy, A., Guillemot, C., Alberi-Morel, M.L.: Low-complexity single-image super-resolution based on nonnegative neighbor embedding (2012)

4. Blattmann, A., et al.: Align your latents: high-resolution video synthesis with latent diffusion models. In: CVPR, pp. 22563–22575 (2023)

5. Cai, J., Zeng, H., Yong, H., Cao, Z., Zhang, L.: Toward real-world single image super-resolution: a new benchmark and a new model. In: ICCV, pp. 3086–3095 (2019)

6. Chen, C., et al.: Real-world blind super-resolution via feature matching with implicit high-resolution priors. In: ACMMM, pp. 1329–1338 (2022)

7. Dhariwal, P., Nichol, A.: Diffusion models beat GANs on image synthesis. In: NeurIPS, vol. 34, pp. 8780–8794 (2021)

8. Fritsche, M., Gu, S., Timofte, R.: Frequency separation for real-world super-resolution. In: ICCVW, pp. 3599–3608 (2019)

9. Ho, J., Jain, A., Abbeel, P.: Denoising diffusion probabilistic models. In: NeurIPS, vol. 33, pp. 6840–6851 (2020)

10. Ho, J., Salimans, T., Gritsenko, A., Chan, W., Norouzi, M., Fleet, D.J.: Video diffusion models. arXiv:2204.03458 (2022)

11. Huang, J.B., Singh, A., Ahuja, N.: Single image super-resolution from transformed self-exemplars. In: CVPR, pp. 5197–5206 (2015)

12. Huang, Y., Li, S., Wang, L., Tan, T., et al.: Unfolding the alternating optimization for blind super resolution. In: NeurIPS, vol. 33, pp. 5632–5643 (2020)

13. Ignatov, A., Kobyshev, N., Timofte, R., Vanhoey, K., Van Gool, L.: DSLR-quality photos on mobile devices with deep convolutional networks. In: ICCV, pp. 3277–3285 (2017)

14. Kingma, D.P., Ba, J.: Adam: a method for stochastic optimization. In: Bengio, Y., LeCun, Y. (eds.) ICLR (2015)

15. Li, H., et al.: SRDiff: single image super-resolution with diffusion probabilistic models. Neurocomputing **479**, 47–59 (2022)

16. Li, X., Zuo, W., Loy, C.C.: Learning generative structure prior for blind text image super-resolution. In: CVPR, pp. 10103–10113 (2023)

17. Liang, J., Cao, J., Sun, G., Zhang, K., Van Gool, L., Timofte, R.: SwinIR: image restoration using swin transformer. In: ICCV, pp. 1833–1844 (2021)

18. Maeda, S.: Unpaired image super-resolution using pseudo-supervision. In: CVPR, pp. 291–300 (2020)

19. Nichol, A.Q., Dhariwal, P.: Improved denoising diffusion probabilistic models. In: ICML, pp. 8162–8171 (2021)

20. Rombach, R., Blattmann, A., Lorenz, D., Esser, P., Ommer, B.: High-resolution image synthesis with latent diffusion models. In: CVPR, pp. 10684–10695 (2022)

21. Sahak, H., Watson, D., Saharia, C., Fleet, D.J.: Denoising diffusion probabilistic models for robust image super-resolution in the wild. CoRR abs/2302.07864 (2023)

22. Saharia, C., Ho, J., Chan, W., Salimans, T., Fleet, D.J., Norouzi, M.: Image super-resolution via iterative refinement. TPAMI **45**(4), 4713–4726 (2023)

23. Shang, S., Shan, Z., Liu, G., Zhang, J.: ResDiff: combining CNN and diffusion model for image super-resolution. arXiv preprint arXiv:2303.08714 (2023)

24. Sohl-Dickstein, J., Weiss, E., Maheswaranathan, N., Ganguli, S.: Deep unsupervised learning using nonequilibrium thermodynamics. In: ICML, pp. 2256–2265 (2015)

25. Timofte, R., Agustsson, E., Van Gool, L., Yang, M.H., Zhang, L.: NTIRE 2017 challenge on single image super-resolution: methods and results. In: CVPRW, pp. 114–125 (2017)

26. Vahdat, A., Kreis, K., Kautz, J.: Score-based generative modeling in latent space. In: NeurIPS, vol. 34, pp. 11287–11302 (2021)

27. Wan, Z., et al.: Bringing old photos back to life. In: CVPR, pp. 2747–2757 (2020)

28. Wang, L., et al.: Unsupervised degradation representation learning for blind super-resolution. In: CVPR, pp. 10581–10590 (2021)

29. Wang, X., Xie, L., Dong, C., Shan, Y.: Real-ESRGAN: training real-world blind super-resolution with pure synthetic data. In: ICCVW, pp. 1905–1914 (2021)

30. Wang, X., Yu, K., Dong, C., Loy, C.C.: Recovering realistic texture in image super-resolution by deep spatial feature transform. CoRR abs/1804.02815 (2018)

31. Wang, Y., et al.: Towards compact single image super-resolution via contrastive self-distillation. In: Zhou, Z. (ed.) IJCAI, pp. 1122–1128 (2021)

32. Wei, P., et al.: Component divide-and-conquer for real-world image super-resolution. In: Vedaldi, A., Bischof, H., Brox, T., Frahm, J.-M. (eds.) ECCV 2020. LNCS, vol. 12353, pp. 101–117. Springer, Cham (2020). https://doi.org/10.1007/978-3-030-58598-3_7

33. Zamir, S.W., Arora, A., Khan, S., Hayat, M., Khan, F.S., Yang, M.: Restormer: efficient transformer for high-resolution image restoration. In: CVPR, pp. 5718–5729 (2022)

34. Zeyde, R., Elad, M., Protter, M.: On single image scale-up using sparse-representations. In: ICCV, pp. 711–730 (2010)

35. Zhang, K., Liang, J., Van Gool, L., Timofte, R.: Designing a practical degradation model for deep blind image super-resolution. In: ICCV, pp. 4791–4800 (2021)

MemDNet: Memorizing More Exogenous Information to Dehaze Natural Hazy Image

Guangfa Wang and Xiaokang Yu[✉]

College of Computer Science and Technology, Qingdao University, Qingdao, China
{2020020635,xyu}@qdu.edu.cn

Abstract. Model trained on synthetic datasets are difficult to obtain good results in the real world due to domain shift. Previous works have been devoted to mining the potential of existing natural datasets to obtain better results, but the number of images in the dataset is ultimately limited. In this paper, we propose MemDNet, a two-branch dehazing network that incorporates an atmospheric scattering model. Our approach utilizes a residual network and a U-shaped network to estimate the corresponding parameters. To alleviate the issue of insufficient natural hazy image pairs, we propose a generalized memory branch that incorporates information from exogenous images. This allows the network to memorize the fine details of the exogenous images, thereby enhancing its ability to recover image colors. Extensive experimental results demonstrate that our proposed memory branch is general and effective, and our method outperforms the state-of-the-art methods on real datasets.

Keywords: Natural image dehazing · Atmospheric scattering model · Memory branch

1 Introduction

The suspended particles in the air absorb and scatter light, which is more serious under hazy conditions and can cause distortion, blurring and color bias in the images received by the receiving device. However, advanced vision tasks often require high-quality clear images, so obtaining clear images from hazy images has received a lot of attention from researchers.

The process with hazy image generation was modeled by Narasimhan et al. as the following equation [1,2]:

$$I(x) = J(x)t(x) + A(1 - t(x)) \tag{1}$$

where I(x) is the hazy image received by the camera, J(x) represents the corresponding clear image, t and A denote the transmission map and atmospheric value, respectively, and x is the pixel position. This is clearly a ill-posed problem: there is only one input I(x), but the solution J(x), t(x), A is required.

© The Author(s), under exclusive license to Springer Nature Singapore Pte Ltd. 2024
Q. Liu et al. (Eds.): PRCV 2023, LNCS 14435, pp. 39–49, 2024.
https://doi.org/10.1007/978-981-99-8552-4_4

To solve this problem, early researchers have used a series of statistical methods to obtain the common features of hazy images or clear images to form a relatively common prior, and use these priors to obtain clear image, including: dark channel prior (DCP) [3], color attenuation prior (CAP) [4], haze-line (HL) [5], etc. However, due to the shortcomings of these priors, they can only be applied to a limited number of scenes, such as DCP cannot work effectively in the sky area.

Since the hand-designed prior can not be applied to each scenario, researchers have introduced convolutional neural networks (CNNs) to the dehazing task, using the powerful fitting ability of CNNs to replace the prior to find valid features, which has proven to be effective. Early CNNs, such as ADN [6], DehazeNet [7], MSCNN [8], used CNNs to estimate the parameters needed for atmospheric scattering model, but this strategy was soon eliminated due to the drawbacks of atmospheric scattering models and the problem of insufficient arithmetic power. Later researchers have tended to discard atmospheric scattering models and use an end-to-end method to obtain clear images, proposing GDN [9], MSBDN [10], FFANet [11], etc. However, these models can only obtain good results on synthetic datasets and cannot work effectively on natural image, and have a high number of parameters or require high video memory.

The recent research trend is to lightweight the network and exploit the potential of the network or dataset as much as possible by some means, such as ARCRNet [12], C2PNet [13] by contrast learning to obtain good results, RIDIP [14], PSD [15] by introducing some priors to obtain better results for visual perception, D4 [16], RDN [17] and other methods to obtain clear images by exploiting the feature of generative adversarial networks that do not require paired image training.

These methods are constantly exploring to obtain good results using existing natural datasets, but the results remain unsatisfactory because the amount of data is really too small. In this paper, we propose MemDNet, which incorporates the atmospheric scattering model into the network in order to improve interpretability. Using two sub-networks TNet, ANet to estimate the corresponding parameters, we use the attention mechanism, dense connection, and deformable convolution to improve the network's ability to extract features. On top of this network, we improve a generic memory branch from a training perspective to explore the impact of exogenous images on the network performance.

The contributions of this paper are as follows:

- Based on dense connection, we propose a simple and effective dehazing module to guarantee the diversity of deep network features as well as feature reusability.
- From a training perspective, we propose a simple memory branch that enables the network to remember more artificially given exogenous details, and we demonstrate its effectiveness and generality through numerous experiments.
- We propose an effective MemDNet that combines atmospheric scattering models and CNNs to improve interpretability and demonstrate its superiority experimentally.

2   Proposed Method

To improve interpretability, we reincorporate the atmospheric scattering model into the CNNs, we transform Eq. (1) to obtain Eq. (2):

$$J(x) = \frac{I(x) - A}{t(x)} + A \qquad (2)$$

We define I(x) as I, $\frac{1}{t(x)}$ as T and define J(x) as J to obtain the following equation:

$$J = (I - A)T + A \qquad (3)$$

We propose an MemDNet which consists of two sub-networks to estimate A and T, respectively, where ANet consists of residual module, attention mechanism module and deformable convolution, and TNet consists of dense module, enhanced module and deformable convolution, and its structure is shown in Fig. 1. The first convolution changes the channel from 3 to 16 and the last deformable convolution changes the channel from 16 to 3. ANet does not involve changes in channel and feature size, the TNet encoding stage multiplies the feature channel and divides the size, and the decoding stage recovers accordingly.

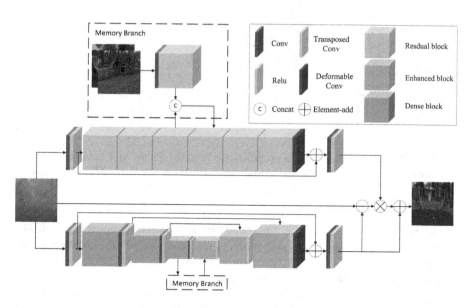

Fig. 1. The architecture of the proposed MemDNet

2.1   Dense Block

MemDNet is a relatively lightweight network, and in order to ensure the diversity of deep features, we choose dense connection as the main way to process features in the encoder stage, which includes three convolutional layers, three relu functions and two attention mechanisms, and our proposed dense module is experimentally proven to be effective and necessary, it is schematically shown in Fig. 2(a).

2.2 Enhanced Block

In FFANet [11], the authors prove the effectiveness of its base module, but through our experiments, we find that its performance can be further improved completely, so in this paper, we propose a new enhanced module which includes three convolutional layers, two relu functions, and two attention mechanisms, and through the subsequent ablation experiments, we prove that it has stronger enhancement ability, it is schematically shown in Fig. 2(a).

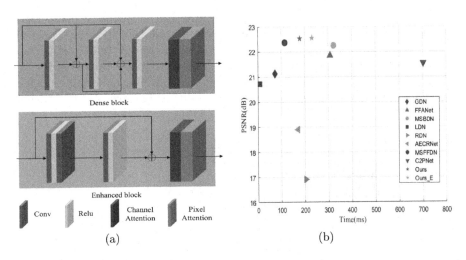

Fig. 2. (a) The architecture of the proposed module. (b) PSNR and Time trade-off of the state-of-the-art dehazing methods and our method on the O-HAZE [20] dataset.

2.3 Memory Branch

CNNs are data-driven, and the "imagination" capability of the network likely depends on the amount of data it receives and the size of the network itself. Our goal is to enable relatively lightweight networks to have similar capability to larger networks. Therefore, we sacrifice some SSIM performance by artificially feeding unpaired and external images to the network, allowing the network to directly memorize fragmented local information. This compensates for its limited "imagination" capability.

As shown in Fig. 1, the memory branch is simply composed of convolutional layer and residual block. After processing the images, they are concatenated with the original network's features and feed into the subsequent network, enabling plug-and-play functionality, which is very convenient. Through our experiments, we find that feeding two images simultaneously produces the best results.

Table 1. Quantitative Evaluations with the state-of-the-art methods on the real-world datasets.Red represents the best, blue represents the second best, green represents the third best. The unit of Params is MB.

Method	Venue&Year	I-HAZE		O-HAZE		Dense-HAZE		NH-HAZE		Params
		PSNR	SSIM	PSNR	SSIM	PSNR	SSIM	PSNR	SSIM	
DCP	CVPR2009	13.67	0.714	17.63	0.842	14.26	0.570	15.01	0.680	-
ADN	ICCV2017	19.69	0.853	20.15	0.849	15.68	0.477	16.73	0.637	0.002
GDN	ICCV2019	18.98	0.833	21.13	0.781	16.01	0.587	17.33	0.738	0.960
FFANet	AAAI2020	19.78	0.899	21.85	0.785	18.42	0.682	18.84	0.702	4.680
MSBDN	CVPR2020	21.10	0.831	22.25	0.792	15.99	0.505	19.38	0.741	31.35
LDN	TIP2021	19.85	0.859	20.72	0.845	15.85	0.487	17.17	0.679	0.030
RDN	TIP2021	13.97	0.755	16.92	0.808	12.23	0.422	12.46	0.546	68.56
AECRNet	CVPR2021	19.23	0.719	18.90	0.649	15.43	0.388	18.57	0.701	2.610
MSFDN	ACCV2022	22.98	0.820	22.37	0.836	15.85	0.520	19.83	0.731	7.670
C2PNet	CVPR2023	22.44	0.864	21.54	0.868	16.48	0.595	19.34	0.759	7.170
Ours		22.51	0.891	22.53	0.843	17.30	0.482	19.42	0.751	1.200
Ours_E		22.54	0.813	22.55	0.759	17.93	0.520	19.44	0.698	1.710
Ours_P		22.46	0.902	23.68	0.860	18.61	0.526	19.13	0.790	1.710

3 Experiments

3.1 Experimental Settings

Datasets. We are targeting the natural image dehazing task, so we choose four natural image datasets: I-HAZE [21], O-HAZE [20], NH-HAZE [22] and Dense-HAZE [23], all from the NTIRE dehazing challenge, with around 50 pairs of images in each dataset. We selected 5 images in each dataset as the validation set, 5 images as the testing set and the rest as the training set, each of which was broadened to 15,000 pairs by cropping and rotation, the size of each image is 256×256.

Implementation Details. The batchsize is 8, and the initial learning rate is 1e-4, multiplied by 0.1 every 25 epochs. We use Adam [25] as the optimizer with default parameters. Loss function is MSE. Each dataset is trained with 100 epochs. All experiments are on a single NVIDIA 3090 GPU.

Competitors and Evaluation Metrics. We compare our method with the prior-based methods (e.g., DCP [3], HL [5]) and learning-based methods (e.g., ADN [6], FFA-Net [11], MSBDN [10], ADN [18], RDN [17], AECR-Net [12], MSFDN [19], and C2PNet [13]). We utilize the peak signal-to-noise ratio (PSNR) and structural similarity (SSIM) to evaluate the performance.

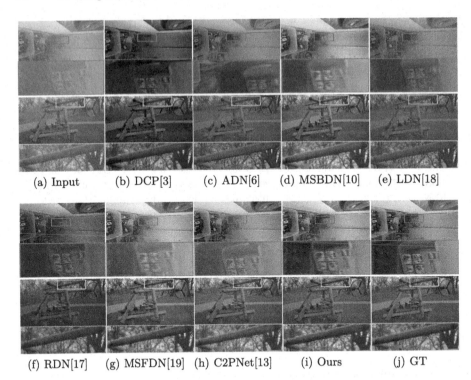

(a) Input   (b) DCP[3]   (c) ADN[6]   (d) MSBDN[10]   (e) LDN[18]

(f) RDN[17]   (g) MSFDN[19]   (h) C2PNet[13]   (i) Ours   (j) GT

Fig. 3. Visual result comparison on I-HAZE [21], O-HAZE [20] dataset.

3.2 Comparison with SOTAs

As shown in Table 1, Ours means no memory branch is inserted and no pre-trained model is loaded, Ours_E means a memory branch is inserted, and Ours_P means a pre-trained model is loaded on top of Ours_E. It can be seen that our method has relatively better results compared to other SOTAs. In particular, on the O-HAZE [20] dataset, the three models are ahead of the other SOTAs across the board in PSNR, but lacking in SSIM, which is in line with our design intent.

The visualisation results on the natural dataset are shown in Fig. 3, where we select an image from I-HAZE [21] and O-HAZE [20], respectively, the exact details are shown in the green window and the yellow window for comparison, compared to the latest C2PNet [13], our approach looks relatively dark and visually not as good as C2PNet, but you can see that our method reconstructs the image more closely to ground true. The other methods have some other drawbacks, such as the results of both DCP [3] and RDN [17] are darker, ADN [6] and LDN [18], being lightweight networks, both discard visual effects accordingly, and the red ground of the second result of both MSBDN [10] and MSFDN [19] appears whitish.

To verify the generalisation performance of our method, we choose two more classical natural images in the GT-free image, as shown in Fig. 4, and our method still reconstructed the images well.

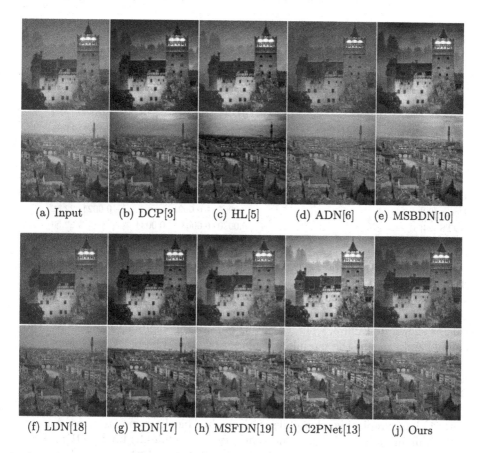

(a) Input   (b) DCP[3]   (c) HL[5]   (d) ADN[6]   (e) MSBDN[10]

(f) LDN[18]   (g) RDN[17]   (h) MSFDN[19]   (i) C2PNet[13]   (j) Ours

Fig. 4. Visual result comparison on natural images.

3.3   Ablation Study

Our memory branch introduces exogenous images, so it is important to consider whether different exogenous images will have an impact on the final results. We are considering the case of adding an external image, trained on both the Dense-HAZE [23] and I-HAZE [21] datasets, all in the same configuration as in Sect. 3.1.

During training, we input some images from the remaining 4 datasets as exogenous images into the network to get the final weights. Then at testing time, the rest of the images from each dataset were input to get the final results. The final results were then tested for significance and as shown in Table 2, the differences in results between most of the datasets were not significant, while SOTS was chosen as the best in terms of mean PSNR and maximum PSNR, so we choose RESIDE [24] as the exogenous source for the training-time exogenous images as well as SOTS among them as the test-time exogenous images.

Table 2. ✓ indicates a significant difference in the PSNR obtained between different datasets and the opposite for ×. Row-Col indicates the difference of mean PSNR between two datasets.

	NH	O-HAZE	I-HAZE	SOTS	Row-Col(MeanPSNR)				MaxPSNR
NH [22]	-	×	✓	✓	-	−0.0044	−0.0167	−0.0304	17.82
O-HAZE [20]		-	×	✓	0.0044	-	−0.0123	−0.0260	17.83
I-HAZE [21]			-	×	0.0167	0.0123	-	−0.0137	17.83
SOTS [24]				-	0.0304	0.0260	0.0137	-	17.84
	NH	OHAZE	Dense	SOTS	Row-Col(MeanPSNR)				
NH [22]	-	×	×	×	-	−0.0026	−0.0005	-0.0029	22.52
O-HAZE [20]		-	×	×	0.0026	-	0.0021	−0.0003	22.52
Dense [23]			-	×	0.0005	−0.0021	-	−0.0024	22.52
SOTS [24]				-	0.0030	0.0003	0.0024	-	22.53

In this article we propose two modules: Dense Blcok (DB), Enhanced Block (EB), use Dense Connection (DC), Skip Connection (SC) too, and to demonstrate the necessity of each part we have done the corresponding experiments, the dataset used for the relevant experiments in Tables 3 and 4 are O-HAZE [20], as shown in Table 3, where × means remove this module, ✓ means use this module and - means the module is replaced by a residual block.

As shown in Table 3, the metrics only reach their optimal values when the network includes all proposed modules and operations. This indicates the necessity of each proposed module and operation in our approach.

To prove the effectiveness of our proposed module, we define the following network: **1) base:** replacing all modules with residual blocks, removing dense connection within modules and skip connection between modules;**2) base+DB:** replacing the residual block of down branch with DB; **3) base+DB+EB:** replacing the DB of the decoder part with EB on top of 2; **4) base+DB+EB+DC:** adding dense connection within modules; **5) Ours:** add skip connection between modules.

Table 3. Ablation experiments on the necessity of each module.

DB	EB	DC	SC	PSNR	SSIM
-	✓	✓	✓	21.94	0.777
✓	-	✓	✓	21.69	0.754
✓	✓	×	✓	22.00	0.758
✓	✓	✓	×	20.85	0.755
✓	✓	✓	✓	22.42	0.757

Table 4. Ablation experiments on the effectiveness of each module.

Model	PSNR	SSIM
base	19.17	0.641
base+DB	21.70	0.720
base+DB+EB	22.16	0.723
base+DB+EB+DC	22.39	0.733
Ours	22.42	0.757

Effectiveness of DB. Since our approach is a relatively lightweight network, it is necessary to use dense connection to reuse features, as shown in Table 4, and the relatively large gains obtained by applying DB to the network is very much in line with our intention in designing the module.

Effectiveness of EB. The effect of the attention mechanism on the results has been well illustrated in FFANet [11], and we continued this idea by further enhancing the dehazing capability of the enhancement module, as shown in Table 4, where an improvement of 0.46dB PSNR was obtained with the addition of EB.

Effectiveness of Connection. The role of connection in lightweight network has been consistently demonstrated in experiments, ensuring that deep network can converge and that features between different levels can be linked and exploited. As shown in Table 4, the addition of two connections gives a 0.26dB PSNR and 0.034 SSIM improvement.

Table 5. Manual selection or random selection of external images.

	O-HAZE	I-HAZE	NH	SOTS
Manual	18.21	18.14	18.06	18.61
	0.506	0.503	0.500	0.508
Random	17.83	17.83	17.82	17.83
	0.524	0.524	0.523	0.524

Table 6. The effect of the number of exogenous images on the final result.

	PSNR	SSIM
One Picture	17.76	0.588
Two Pictures	17.93	0.520
Three Pictures	17.64	0.518
Four Pictures	17.24	0.519

Both the training and testing sets in Tables 5, 6 and 7 are Dense-HAZE [23], and the other configurations are the same as in the previous section, Table 5 considers the case of one exogenous image.

After our experiments, the way the exogenous images are selected affects the results. Hand-selected images refer to the selection of images that are similar to the colour and environment of the testing set, as shown in Table 5, Hand-selected images have obvious advantages in terms of PSNR, but considering that memory branch will destroy the image structure to some extent and, we choose to select exogenous images in a random way from the perspective of image structure and generality.

The number of images feed into the network via the memory branch has an effect on the results, based on the random selection of the external source images, and we consider four scenarios to determine the best results with two images feed via the memory branch, the result is shown in Table 6.

Table 7. Analysis of the generality of memory branch.

	ADN [6]	MSBDN [10]	LDN [18]	MSFDN [19]
PSNR	15.88(0.200↑)	16.04(0.050↑)	15.99(0.140↑)	16.12(0.270↑)
SSIM	0.474(−0.003↓)	0.513(0.008↑)	0.454(−0.033↓)	0.480(−0.040↓)

To demonstrate the generality of the proposed memory branch, we apply the memory branch to some other SOTAs, and in line with our design intent, most methods obtain a gain in PSNR with a corresponding partial loss of SSIM, as shown in Table 7.

4   Conclusion

In this paper, we propose a relatively lightweight network, MemDNet, which consists of dense block and enhanced block that we propose. In addition, to enable the network to better process natural hazy image, we propose a generic memory branch to allow the network to remember some fragmented details. Through extensive experiments, we demonstrate the effectiveness of the proposed method on four natural datasets compared to other SOTAs.

References

1. Narasimhan, S.G., Nayar, S.K.: Vision and the atmosphere. Int. J. Comput. Vis. **48**(3), 233–254 (2002)
2. Cantor, A.: Optics of the atmosphere–scattering by molecules and particles. IEEE J. Quantum Electron. 698–699 (1978)
3. He, K., Sun, J., Tang, X.: Single image haze removal using dark channel prior. IEEE Trans. Pattern Anal. Mach. Intell. **33**(12), 2341–2353 (2011)
4. Zhu, Q., Mai, J., Shao, L.: A fast single image haze removal algorithm using color attenuation prior. IEEE Trans. Image Process. **24**(11), 3522–3533 (2015)
5. Berman, D., Treibitz, T., Avidan, S.: Non-local image dehazing. In: 2016 IEEE Conference on Computer Vision and Pattern Recognition (CVPR) (2016)
6. Li, B., Peng, X., Wang, Z., Xu, J., Dan, F.: AOD-Net: all-in-one dehazing network. In: 2017 IEEE International Conference on Computer Vision (ICCV) (2017)
7. Cai, B., Xu, X., Jia, K., Qing, C., Tao, D.: Dehazenet: an end-to-end system for single image haze removal. IEEE Trans. Image Process. **25**(11), 5187–5198 (2016)
8. Ren, W., Liu, S., Zhang, H., Pan, J., Cao, X., Yang, M.-H.: Single image dehazing via multi-scale convolutional neural networks. In: Leibe, B., Matas, J., Sebe, N., Welling, M. (eds.) ECCV 2016. LNCS, vol. 9906, pp. 154–169. Springer, Cham (2016). https://doi.org/10.1007/978-3-319-46475-6_10
9. Liu, X., Ma, Y., Shi, Z., Chen, J.: Griddehazenet: attention-based multi-scale network for image dehazing (2019)
10. Dong, H., et al.: Multi-scale boosted dehazing network with dense feature fusion. In: Proceedings of the IEEE/CVF Conference on Computer Vision and Pattern Recognition, pp. 2157–2167 (2020)

11. Qin, X., Wang, Z., Bai, Y., Xie, X., Jia, H.: FFA-Net: feature fusion attention network for single image dehazing. In: Proceedings of the AAAI Conference on Artificial Intelligence, vol. 34, pp. 11908–11915 (2020)
12. Wu, H., Qu, Y., Lin, S., et al.: Contrastive learning for compact single image dehazing. In: Proceedings of the IEEE/CVF Conference on Computer Vision and Pattern Recognition (CVPR), pp. 10551–10560 (2021)
13. Zheng, Y., Zhan, J., He, S., et al.: Curricular contrastive regularization for physics-aware single image dehazing. In: Proceedings of the IEEE/CVF Conference on Computer Vision and Pattern Recognition (CVPR), pp. 5785–5794 (2023)
14. Wu, R., Duan, Z., Guo, C., et al.: RIDCP: revitalizing real image dehazing via high-quality codebook priors. In: Proceedings of the IEEE/CVF Conference on Computer Vision and Pattern Recognition (2023)
15. Chen, Z., Wang, Y., Yang, Y., et al.: PSD: principled synthetic-to-real dehazing guided by physical priors. In: Computer Vision and Pattern Recognition (2021)
16. Yang, Y., Wang, C., Liu, R., et al.: Self-augmented unpaired image dehazing via density and depth decomposition. In: Proceedings of the IEEE/CVF Conference on Computer Vision and Pattern Recognition (CVPR), pp. 2037–2046 (2022)
17. Zhao, S., Zhang, L., Shen, Y., et al.: RefineDNet: a weakly supervised refinement framework for single image dehazing. IEEE Trans. Image Process. **30**, 3391–3404 (2021)
18. Ullah, H., Muhammad, K., Irfan, M., et al.: Light-DehazeNet: a novel lightweight CNN architecture for single image dehazing. IEEE Trans. Image Process. **30**, 8968–8982 (2021)
19. Wang, G., Yu, X.: MSF2DN: multi scale feature fusion dehazing network with dense connection. In: Proceedings of the Asian Conference on Computer Vision (ACCV), pp. 2950–2966 (2022)
20. Ancuti, C.O., Ancuti, C., Timofte, R., et al.: O-HAZE: a dehazing benchmark with real hazy and haze-free outdoor images. In: 2018 IEEE/CVF Conference on Computer Vision and Pattern Recognition Workshops (CVPRW) (2018)
21. Ancuti, C.O., Ancuti, C., Timofte, R., et al.: I-HAZE: a dehazing benchmark with real hazy and haze-free indoor images (2018)
22. Ancuti, C.O., Ancuti, C., Timofte, R.: NH-HAZE: an image dehazing benchmark with non-homogeneous hazy and haze-free images. In: 2020 IEEE/CVF Conference on Computer Vision and Pattern Recognition Workshops (CVPRW) (2020)
23. Ancuti, C.O., Ancuti, C., Sbert, M., et al.: Dense Haze: a benchmark for image dehazing with dense-haze and haze-free images. arXiv (2019)
24. Li, B., Ren, W., Fu, D., et al.: RESIDE: A Benchmark for Single Image Dehazing (2017)
25. Kingma, D.P., Ba, J.: Adam: a method for stochastic optimization. In: Bengio, Y., LeCun, Y. (eds.) 3rd International Conference on Learning Representations, ICLR 2015, San Diego, CA, USA, 7–9 May 2015, Conference Track Proceedings (2015)

Technical Quality-Assisted Image Aesthetics Quality Assessment

Xiangfei Sheng[1], Leida Li[1]([✉]), Pengfei Chen[1], Jinjian Wu[1], Liwu Xu[2], Yuzhe Yang[2], and Yaqian Li[2]

[1] School of Artificial Intelligence, Xidian University, Xi'an, China
ldli@xidian.edu.cn
[2] OPPO Research Institute, Shanghai, China

Abstract. Image aesthetics assessment (IAA) aims at predicting the perceived aesthetic quality of images. Intuitively, the technical quality of an image has significant impact on its aesthetic quality, e.g., an image with noticeable distortions is not likely to have very high aesthetic quality. However, this characteristic has rarely been considered when designing modern IAA models. Motivated by this, this paper presents a new Technical Quality-assisted multi-task deep network for image Aesthetic Quality assessment, dubbed TQ4AQ. Specifically, we first extract theme-aware general aesthetic features based on the attention mechanism. Meantime, hand-crafted technical quality features are extracted from aesthetic images. Then the general aesthetic features are utilized to predict the technical quality features and the aesthetic quality simultaneously, based on which technical quality features are integrated. By this means, the aesthetic features are empowered the capability of understanding technical distortions, and more comprehensive aesthetic feature representations are obtained for IAA. Extensive experiments demonstrate the advantage of the proposed TQ4AQ model over the state-of-the-arts.

Keywords: Image Aesthetics Assessment · Technical Quality · Multi-task Learning

1 Introduction

With the surge of mobile Internet, the image data on social media and mobile devices has exploded. While people commonly pursue images of high technical quality, the demand for high aesthetics is also increasing. Image aesthetics assessment (IAA) aims at predicting the perceived beauty of images in a computationally efficient way. IAA has extensive applications in a variety of fields, such as photo recommendation [23], image enhancement [26] and AI painting [19], etc.

Technical quality and aesthetic quality are two innate aspects of image quality. Technical quality assessment measures the severity of image distortions, which mainly involves low-level features. By contrast, aesthetic quality is mainly

Q. Liu et al. (Eds.): PRCV 2023, LNCS 14435, pp. 50–62, 2024.
https://doi.org/10.1007/978-981-99-8552-4_5

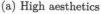

(a) High aesthetics   (b) Low aesthetics

Fig. 1. An exemplar explanation of the relationship between aesthetic quality and technical quality. Technical distortions deteriorate the aesthetic quality of the left images.

related to photography attributes, e.g., color harmony, light usage and composition, which mainly involve high-level features. Intuitively, the technical quality of an image has significant impact on its aesthetic quality, and an example is shown in Fig. 1. Images with high aesthetics typically have excellent technical quality, whereas distortions tend to degrade the aesthetic quality of the image. However, in the literature, technical quality assessment (TQA) and aesthetic quality assessment (AQA) are typically regarded as two independent tasks. The relationship between technical quality and aesthetic quality, particularly how technical quality influences the perception of aesthetic quality is largely underexplored.

Motivated by the above facts, this paper presents a technical quality-assisted image aesthetic quality assessment method (dubbed TQ4AQ), which models the relationship between technical quality and aesthetic quality, so that the aesthetic quality of an image can be evaluated better by integrating technical quality features. To this end, we first extract theme-aware general aesthetic features and hand-crafted technical quality features. Then the aesthetic features are employed to predict the technical quality features and the image aesthetic quality simultaneously through a multi-task learning network with feature fusion. By introducing the prediction of technical quality features as well as the feature fusion, the proposed aesthetic model is empowered the capability of understanding technical distortions when generating the aesthetic scores, hence the impact of technical quality on aesthetic quality can be integrated.

The contributions of this paper can be summarized into the following aspects:

- We propose a technical quality-assisted image aesthetic quality assessment model, dubbed **TQ4AQ**, where technical quality features are integrated to boost the performance of image aesthetics assessment.
- We propose a prediction-based technical quality feature fusion approach. Prediction of technical quality features is treated as an auxiliary task, which enhances the representation ability of general aesthetic features.
- We conduct extensive experiments and comparisons on three public IAA databases. The experimental results demonstrate the advantage of the proposed **TQ4AQ** model over the state-of-the-arts.

2 Related Work

2.1 Technical Quality Assessment

Traditional image technical quality assessment models are mainly based on hand-crafted features. Mittal *et al.* [15] proposed a Natural Scene Statistics (NSS)-based model in the spatial domain. Gu *et al.* [2] proposed a method using the free energy theory and Human Visual System (HVS)-inspired features. Xue *et al.* [27] proposed a model that utilizes the joint statistics of the gradient magnitude (GM) and the Laplacian of Gaussian (LOG). Recently, deep learning-based image quality metrics have shown better evaluation ability than hand-crafted feature-based metrics. In [10], Liu *et al.* proposed a rank-based image quality model based on massive synthetically generated distortions. Zhu *et al.* [31] proposed the MetaIQA model based on deep meta-learning, which learns the prior knowledge shared by diversified distortions. Sun *et al.* [22] proposed the graph representation learning based-GraphIQA, in which each distortion is represented as a graph to model the relationship between perceptual image quality and distortion types and levels. More recently, Zhang *et al.* [29] proposed a continual learning method for technical quality assessment, where a model learns continually from a stream of IQA datasets, building on what was learned from previously seen data.

2.2 Aesthetic Quality Assessment

Early IAA models utilized hand-crafted features and shallow classifiers, which require a considerable amount of domain knowledge and engineering skills. For instance, Ke *et al.* [5] extracted aesthetic features with high-level semantics from photographic knowledge. While hand-crafted feature-based IAA models have achieved notable advances, their representation power is usually limited in face of the highly abstract nature of image aesthetics. Recently, a number of deep aesthetics assessment models have been proposed. Lu *et al.* [11] designed a deep multi-patch aggregation network by aggregating patches from an image. Li *et al.* [9] proposed a personality-assisted model to achieve both generic and personalized image aesthetics assessment. She *et al.* [20] proposed a hierarchical layout-aware Graph Convolutional Network to capture the layout information for IAA.

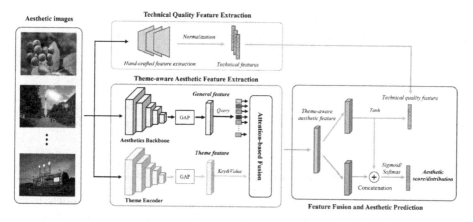

Fig. 2. Framework of the proposed technical quality-assisted image aesthetic quality model (TQ4AQ).

Technical quality and aesthetic quality are two innate aspects of image quality. The technical quality of an image has significant impact on its aesthetic quality. In this work, we make an attempt to integrate image technical quality features to achieve better aesthetic quality assessment.

3 Proposed Method

In the proposed IAA model, image technical quality features are explored to assist the aesthetic quality evaluation through a multi-task network. The overall structure is shown in Fig. 2, which consists of three basic modules, including *theme-aware aesthetic feature extraction, technical quality feature extraction,* and *feature fusion and aesthetic prediction.*

3.1 Theme-Aware Aesthetic Feature Extraction

We first extract general aesthetic features (f_g) using a vanilla backbone network. Considering that people typically adopt different standards for judging the aesthetics of images with different themes [6], we also introduce an attention-based theme fusion strategy to enhance the aesthetic features. Specifically, the theme features (f_t) are extracted with a scene recognition model pre-trained on the *Places* database [30], which contains 10 million images labeled with semantic scene categories. In implementation, we freeze the parameters of the theme backbone during training. With the general aesthetic feature f_g and the theme feature f_t, the cross-attention mechanism [25] is adopted to explore the relationship between them. To this end, we first generate a set of query (Q), key(K) and value (V) features by linear transformations as $Q = w_q f_t$, $K = w_k f_g$ and

$V = w_v f_g$, where w_q, w_k and w_v are learnable transformation parameters. Then the attention-based theme-aware aesthetic feature $\mathcal{F}$ is obtained by:

$$\mathcal{F} = f_g \oplus \text{softmax} \left(\frac{QK^T}{\sqrt{d_k}} \right) V, \tag{1}$$

where $\oplus$ denotes the concatenation, T and d_k denote the transpose and dimension of the theme feature f_t. The theme-aware feature is beneficial for the aesthetics assessment.

3.2   Technical Quality Feature Extraction

In the literature, a great number of technical quality assessment models have been proposed with encouraging performances. For simplicity, we extract hand-crafted technical quality features. In this work, we try different feature extraction approaches, including BRISQUE [15], GM-LOG [27] and NFERM [2], which are all popular image technical quality models. In BRISQUE [15], 36 features were extracted based on the Mean-Subtracted Contrast-Normalized (MSCN) coefficients in the spatial domain. In GM-LOG [27], 40 features were extracted based on the joint statistics of two normalized local contrast features, including the gradient magnitude (GM) and the Laplacian of Gaussian (LOG) response. In NFERM [2], 23 features were extracted following the brain-inspired free-energy theory. In implementation, technical quality features are first extracted using different feature extractors $\mathcal{E}$. Then they are z-score normalized to obtain the final quality features:

$$f_q = Norm\left(\mathcal{E}(I)\right), \tag{2}$$

where I denotes the input image, and $Norm$ denotes the normalization operation. It is worth noting that the technical quality features are not unique in the proposed method, and an ablation study will be conducted to demonstrate their effectiveness in the experiment section.

3.3   Feature Fusion and Aesthetic Prediction

After obtaining the theme-aware aesthetic features $\mathcal{F}$ and the technical quality features f_q, we employ the latter to assist the aesthetic prediction. The underlying idea is to empower the aesthetic model the ability to understand image technical quality, so that the impact of technical distortions on image aesthetic quality can be integrated. To this end, we design a prediction-based technical quality feature fusion module under the multi-task learning framework. Specifically, the theme-aware aesthetic features $\mathcal{F}$ are adopted to predict the technical quality features f_q and the image aesthetic score/distribution simultaneously. In implementation, we first add two Fully Connected (FC) layers, which both contain 512 nodes. Then, we can obtain the new features for the technical quality feature prediction and aesthetics assessment, which are denoted by d_q and

d_a respectively. In addition, the feature d_q is further concatenated to the aesthetic feature d_a for obtaining more discriminative aesthetic feature representation, which is subsequently used to predict the aesthetic quality. Formally, the predicted technical quality features and the aesthetic score/distribution can be obtained by

$$f_{q_i} = MLP_{\theta_q}(d_q), \; p_i = MLP_{\theta_a}(d_a \oplus d_q), \tag{3}$$

where θ_q and θ_a represent the parameters of two multi-layer perceptions respectively. The two MLPs both consist of a linear layer followed by a nonlinear activation, but the former uses the *tanh* activation function and the latter uses the *Sigmoid* function (for score regression) or *Softmax* function (for distribution prediction).

The Mean Square Error (MSE) loss is adopted to measure the difference between the predicted technical quality features and the ground truth features, which is defined as

$$L_q = \frac{1}{n} \sum_{i=1}^{n} (\hat{f}_{q_i} - f_{q_i})^2. \tag{4}$$

The Earth Mover's Distance (EMD) loss is used to measure the distance between the predicted aesthetic distribution $\hat{p}$ and the ground-truth distribution p, which is defined as

$$L_a = \left(\frac{1}{N} \sum_{k=1}^{N} |CDF_p(k) - CDF_{\hat{p}}(k)| \right)^{\frac{1}{r}}, \tag{5}$$

where N is the number of the images, and CDF is the cumulative distribution function. In implementation, r is set to 2 to penalize the Euclidean distance between the two CDF_s, which facilitates easier optimization with gradient descent. Kindly note that when predicting the aesthetic score instead of aesthetic distribution, we also use the MSE loss as L_a.

Finally, the total loss is calculated as

$$L_{total} = \frac{\lambda_1 L_q + \lambda_2 L_a}{\lambda_1 + \lambda_2}, \tag{6}$$

where λ_1 and λ_2 are learnable weights to balance the trade-off between technical quality feature prediction and aesthetic prediction.

4   Experimental Results

4.1   Databases and Settings

We evaluate the performance of the proposed TQ4AQ model on three public IAA databases, including AVA [17], AADB [6] and EVA [3]. AVA is the most

Table 1. Experimental results on AVA dataset.

Method	Backbone	Image Size	Classification Accuracy↑	Score Regression			Distribution	
				PLCC↑	SRCC↑	MSE↓	EMD1↓	EMD2↓
Random Split								
NIMA [24]	Inception-v2	299 × 299	81.5%	0.636	0.612	-	0.050	-
AFDC [1]	ResNet-50	320 × 320	83.0%	0.671	0.649	0.271	0.045	-
Split from [9]								
PA-IAA [9]	Inception-v3	299 × 299	83.7%	-	0.677	-	0.047	-
HLA-GCN [20]	ResNet-50	300 × 300	84.1%	0.678	0.656	0.264	0.045	0.065
TQ4AQ	ResNet-50	224 × 224	84.3%	0.685	0.663	0.263	0.045	0.064
TQ4AQ	Swin-T	224 × 224	**85.2%**	**0.736**	**0.720**	**0.225**	**0.041**	**0.059**
Split from [17]								
DMA-Net [11]	AlexNet	227 × 227	75.4%	-	-	-	-	-
MNA-CNN [13]	VGG16	224 × 224	77.1%	-	-	-	-	-
Zeng et al. [28]	ResNet-101	384 × 384	80.8%	0.720	0.719	0.275	-	0.065
APM [16]	ResNet-101	Resize(500)	80.3%	-	0.709	0.279	-	0.061
A-Lamp [12]	VGG16	224 × 224	82.5%	-	-	-	-	-
MP_{ada} [21]	ResNet-18	224 × 224	**83.0%**	-	-	-	-	-
MUSIQ [4]	VIT	Full resolution	81.5%	0.738	0.726	**0.242**	-	-
TQ4AQ	ResNet-50	224 × 224	80.5%	0.713	0.705	0.277	0.046	0.065
TQ4AQ	Swin-T	224 × 224	81.9%	**0.753**	**0.744**	0.247	**0.043**	**0.062**

popular IAA dataset, which contains 255,530 images with aesthetic distributions annotated on a scale 1–10. In the literature, there are two commonly used training-test data splits, i.e., the official split and the split used in [9]. The official split is beneficial for the aesthetic classification task, while the latter is beneficial for score regression. For completeness, we test our model under both splits. AADB contains 10,000 images with aesthetic scores annotated on a scale 1–5. Following the standard split [6], 8,500 images are used for model training, 500 images are used for validation, and the rest 1,000 images are used for testing. EVA contains 4,070 images, which provides the labels for aesthetic score on a scale 1–10. Considering that the EVA dataset itself does not provide standard data split, we perform ten-fold cross-validation in our experiment. Since it is relatively new and small in size, very few image aesthetic assessment models report performance on EVA dataset. Therefore, we are not able to compare the performance of the proposed metric with other IAA metrics on EVA. Nevertheless, to demonstrate the effectiveness of the proposed method, we still show the performances using different backbone networks on EVA database.

We adopt backbone networks pre-trained on the ImageNet [7] to build the proposed model. Images are resized to 224 × 224 before feeding to the network. The whole model is optimized using the Adam optimizer with a weight decay of 1e-5. All experiments are implemented in PyTorch on a computer with RTX 3090 GPU. For aesthetic distribution prediction, we employ the EMD to measure model performance. For aesthetic score regression, we use the Spearman Rank

Correlation Coefficient (SRCC), Pearson Linear Correlation Coefficient (PLCC) and Mean Squared Error (MSE). For binary classification, we adopt the accuracy.

4.2   Comparison with the State-of-the-Arts

Table 1 and Table 2 summarize the results of the state-of-the-art models on AVA and AADB datasets respectively. For AVA, EMD with $r = 1$ and $r = 2$ are both reported, which are denoted by EMD1 and EMD2 respectively. Since most of the existing models only report SRCC results on AADB, we only compare the SRCC values on this dataset. It is observed from Table 1 that the proposed TQ4AQ achieves very encouraging performances in terms of all three aesthetic tasks, especially when the Swin Transformer backbone is adopted. Table 2 shows similar results, and TQ4AQ delivers the best result. These results demonstrate the advantage of integrating technical quality features in image aesthetics assessment. Kindly note that in AADB, most of the existing models only reported SRCC results, so we only compare the SRCC values.

Table 2. Experimental results on AADB dataset.

Method	Backbone	SRCC
RA-DCNN [6]	AlexNet	0.678
DCNN [14]	ResNet-50	0.689
AAL [18]	Resnet-50	0.704
NIMA [24]	ResNet-50	0.708
Zeng *et al.* [28]	ResNet-101	0.726
HIAA [8]	ResNet-50	0.739
TQ4AQ	ResNet-50	0.741
TQ4AQ	Swin-T	**0.763**

4.3   Ablation Experiments

Analysis of Model Components. Theme-aware aesthetic feature extraction and technical quality feature fusion are two crucial components of the proposed model. Here, we conduct an ablation experiment to analyze their relative contributions to the overall model performance based on the AVA dataset, and the results are summarized in Table 3. It is observed from the table that both components contribute to the model performance, while technical quality features play a more significant role, which in turn demonstrates the advantage of the proposed technical quality fusion strategy for IAA.

Table 3. Evaluation of model components. TA: theme-aware feature extraction, TQ: technical quality feature fusion.

Model Component	ResNet-50		Swin-T	
	PLCC	SRCC	PLCC	SRCC
Baseline	0.685	0.676	0.730	0.721
Baseline + TA	0.698	0.685	0.743	0.734
Baseline + TQ	0.705	0.696	0.747	0.736
Baseline + TA + TQ	**0.713**	**0.705**	**0.753**	**0.744**

Impact of Technical Quality Features. In our experiments, we mainly use the technical quality features extracted by the BRISQUE model [15]. As aforementioned, the technical quality features are not unique, meaning that different feature extraction approaches can be used. To further investigate the impact of different technical quality features on the proposed model, we test different technical quality feature extraction approaches as well as their combinations, including BRISQUE [15], GM-LOG [27] and NFERM [2], which are all representative image technical quality models. Table 4 summarizes the experimental results (ResNet-50 backbone) based on the AVA dataset.

It is known from Table 4 that among the three kinds of features, BRISQUE features deliver slightly better results than the other two. The performance of the proposed model can be further improved slightly by combining two or all three features. These results indicate that the proposed model is not sensitive to the selection of technical quality features, and the common image quality features can be used. For simplicity and fast computation, we only use the BRISQUE features in our experiments.

Table 4. Performance of the proposed TQ4AQ model when using different technical quality features.

Feature Group	Feature Dim.	PLCC	SRCC
Baseline(ResNet-50)	-	0.685	0.676
+ BRISQUE	36	0.713	0.705
+ GM-LOG	40	0.711	0.703
+ NFERM	23	0.709	0.700
+ BRISQUE + GM-LOG	76	0.716	0.709
+ BRISQUE + NFERM	59	0.717	0.706
+ GM-LOG + NFERM	63	0.716	0.708
+ ALL	**99**	**0.718**	**0.709**

Table 5. Performance evaluation on different backbones.

Model	AVA		AADB		EVA	
	PLCC	SRCC	PLCC	SRCC	PLCC	SRCC
EfficientNet-b0	0.668	0.660	0.698	0.691	0.727	0.707
TQ4AQ	0.687	0.676	0.728	0.724	0.744	0.730
Gain	**1.9%**	**1.6%**	**3.0%**	**3.3%**	**1.7%**	**2.3%**
MobileNet-v3	0.666	0.655	0.710	0.700	0.729	0.712
TQ4AQ	0.682	0.671	0.729	0.724	0.746	0.726
Gain	**1.6%**	**1.6%**	**1.9%**	**2.4%**	**1.7%**	**1.4%**
ResNet-50	0.685	0.676	0.722	0.718	0.745	0.728
TQ4AQ	0.713	0.705	0.745	0.741	0.759	0.745
Gain	**2.8%**	**2.9%**	**2.3%**	**2.3%**	**2.3%**	**2.3%**
DeiT	0.683	0.674	0.725	0.721	0.751	0.732
TQ4AQ	0.704	0.696	0.747	0.739	0.773	0.756
Gain	**2.1%**	**2.2%**	**2.2%**	**1.8%**	**2.2%**	**2.4%**
Swin Transformer	0.730	0.721	0.743	0.739	0.799	0.782
TQ4AQ	0.753	0.744	0.769	0.763	0.813	0.799
Gain	**2.3%**	**2.3%**	**2.6%**	**2.4%**	**1.4%**	**1.7%**
Average Gain	**2.1%**	**2.1%**	**2.4%**	**2.4%**	**1.7%**	**2.1%**

Evaluation on Different Backbone Networks. To demonstrate the universality of the proposed technical quality feature fusion strategy, we further test the proposed method using different backbone networks, ranging from lightweight EfficientNet to the up-to-date Swin Transformer. The results are listed in Table 5, where we report the baseline and performance gains obtained by different backbones. It is observed that the proposed technical quality fusion consistently improves the model performance on all tested backbone networks, mostly with a sizable gain higher than 2%.

Use as a Plug-and-Play Module. We further test the performance of the proposed technical quality fusion as a plug-and-play module to boost the existing IAA models. To this end, we integrate it into two popular IAA models, i.e., NIMA [24] and HLA-GCN [20], and the results are listed in Table 6. It is observed that after integrating the technical quality features, the performance of both models can be further improved. This further confirms the effectiveness of technical quality in aesthetic quality assessment.

Table 6. Evaluation of the proposed technical quality feature fusion as a plug-and-play module on the AVA dataset.

Method	Backbone	PLCC	SRCC
NIMA [24]	VGG16	0.610	0.592
HLA-GCN [20]	ResNet-50	0.678	0.656
	ResNet-101	0.687	0.665
NIMA + Quality	VGG16	0.633 (**+0.023**)	0.610 (**+0.018**)
HLA-GCN + Quality	ResNet-50	0.684 (**+0.006**)	0.663 (**+0.007**)
	ResNet-101	0.696 (**+0.009**)	0.674 (**+0.009**)

5   Conclusion

In this paper, we have presented a new technical quality-assisted image aesthetic quality assessment model. Particularly, we investigated the relationship between image technical quality and aesthetic quality, based on which we have designed a multi-task deep network to integrate technical quality features to achieve more comprehensive image aesthetics assessment. A prediction-based technical quality feature fusion strategy is proposed to enhance the representation ability of aesthetic features. We have conducted extensive experiments and comparisons on public IAA databases, and the results confirm that integrating technical quality features consistently improves the performance of aesthetics assessment models. Image aesthetics is a highly abstract task and diversified factors could be considered to design more advanced IAA models, e.g., emotion and multimodal clues.

Acknowledgments. This work was supported in part by the National Natural Science Foundation of China under Grants 62171340, 62301378 and 61991451, and the OPPO Research Fund.

References

1. Chen, Q., et al.: Adaptive fractional dilated convolution network for image aesthetics assessment. In: Proceedings of the IEEE/CVF Conference on Computer Vision and Pattern Recognition, pp. 14114–14123 (2020)
2. Gu, K., Zhai, G., Yang, X., Zhang, W.: Using free energy principle for blind image quality assessment. IEEE Trans. Multimedia **17**(1), 50–63 (2015)
3. Kang, C., Valenzise, G., Dufaux, F.: EVA: an explainable visual aesthetics dataset. In: Joint Workshop on Aesthetic and Technical Quality Assessment of Multimedia and Media Analytics for Societal Trends, pp. 5–13 (2020)
4. Ke, J., Wang, Q., Wang, Y., Milanfar, P., Yang, F.: MUSIQ: multi-scale image quality transformer. In: Proceedings of the IEEE/CVF International Conference on Computer Vision, pp. 5148–5157 (2021)
5. Ke, Y., Tang, X., Jing, F.: The design of high-level features for photo quality assessment. In: Proceedings of the IEEE Conference on Computer Vision and Pattern Recognition, pp. 419–426 (2006)
6. Kong, S., Shen, X., Lin, Z., Mech, R., Fowlkes, C.: Photo aesthetics ranking network with attributes and content adaptation. In: Leibe, B., Matas, J., Sebe, N., Welling, M. (eds.) ECCV 2016. LNCS, vol. 9905, pp. 662–679. Springer, Cham (2016). https://doi.org/10.1007/978-3-319-46448-0_40
7. Krizhevsky, A., Sutskever, I., Hinton, G.E.: Imagenet classification with deep convolutional neural networks. Commun. ACM **60**(6), 84–90 (2017)
8. Li, L., Duan, J., Yang, Y., Xu, L., Li, Y., Guo, Y.: Psychology inspired model for hierarchical image aesthetic attribute prediction. In: Proceedings of the IEEE Conference on Multimedia and Expo, pp. 1–6 (2022)
9. Li, L., Zhu, H., Zhao, S., Ding, G., Lin, W.: Personality-assisted multi-task learning for generic and personalized image aesthetics assessment. IEEE Trans. Image Process. **29**, 3898–3910 (2020)

10. Liu, X., Van De Weijer, J., Bagdanov, A.D.: Rankiqa: learning from rankings for no-reference image quality assessment. In: Proceedings of the IEEE International Conference on Computer Vision, pp. 1040–1049 (2017)

11. Lu, X., Lin, Z., Shen, X., Mech, R., Wang, J.Z.: Deep multi-patch aggregation network for image style, aesthetics, and quality estimation. In: Proceedings of the IEEE International Conference on Computer Vision, pp. 990–998 (2015)

12. Ma, S., Liu, J., Wen Chen, C.: A-lamp: adaptive layout-aware multi-patch deep convolutional neural network for photo aesthetic assessment. In: Proceedings of the IEEE Conference on Computer Vision and Pattern Recognition, pp. 4535–4544 (2017)

13. Mai, L., Jin, H., Liu, F.: Composition-preserving deep photo aesthetics assessment. In: Proceedings of the IEEE Conference on Computer Vision and Pattern Recognition, pp. 497–506 (2016)

14. Malu, G., Bapi, R.S., Indurkhya, B.: Learning photography aesthetics with deep CNNs. arXiv preprint arXiv:1707.03981 (2017)

15. Mittal, A., Moorthy, A.K., Bovik, A.C.: No-reference image quality assessment in the spatial domain. IEEE Trans. Image Process. **21**(12), 4695–4708 (2012)

16. Murray, N., Gordo, A.: A deep architecture for unified aesthetic prediction. arXiv preprint arXiv:1708.04890 (2017)

17. Murray, N., Marchesotti, L., Perronnin, F.: AVA: a large-scale database for aesthetic visual analysis. In: Proceedings of the IEEE Conference on Computer Vision and Pattern Recognition, pp. 2408–2415 (2012)

18. Pan, B., Wang, S., Jiang, Q.: Image aesthetic assessment assisted by attributes through adversarial learning. In: Proceedings of the AAAI Conference on Artificial Intelligence, vol. 33, pp. 679–686 (2019)

19. Ramesh, A., Dhariwal, P., Nichol, A., Chu, C., Chen, M.: Hierarchical text-conditional image generation with clip latents (2022)

20. She, D., Lai, Y.K., Yi, G., Xu, K.: Hierarchical layout-aware graph convolutional network for unified aesthetics assessment. In: Proceedings of the IEEE/CVF Conference on Computer Vision and Pattern Recognition, pp. 8475–8484 (2021)

21. Sheng, K., Dong, W., Ma, C., Mei, X., Huang, F., Hu, B.G.: Attention-based multi-patch aggregation for image aesthetic assessment. In: Proceedings of the 26th ACM International Conference on Multimedia, pp. 879–886 (2018)

22. Sun, S., Yu, T., Xu, J., Zhou, W., Chen, Z.: GraphiQA: learning distortion graph representations for blind image quality assessment. IEEE Trans. Multimedia 1 (2022). https://doi.org/10.1109/TMM.2022.3152942

23. Sun, W.T., Chao, T.H., Kuo, Y.H., Hsu, W.H.: Photo filter recommendation by category-aware aesthetic learning. IEEE Trans. Multimedia **19**(8), 1870–1880 (2017)

24. Talebi, H., Milanfar, P.: NIMA: neural image assessment. IEEE Trans. Image Process. **27**(8), 3998–4011 (2018)

25. Vaswani, A., et al.: Attention is all you need. In: Advances in Neural Information Processing Systems, vol. 30. Curran Associates, Inc. (2017)

26. Wang, W., Shen, J., Ling, H.: A deep network solution for attention and aesthetics aware photo cropping. IEEE Trans. Pattern Anal. Mach. Intell. **41**(7), 1531–1544 (2018)

27. Xue, W., Mou, X., Zhang, L., Bovik, A.C., Feng, X.: Blind image quality assessment using joint statistics of gradient magnitude and laplacian features. IEEE Trans. Image Process. **23**(11), 4850–4862 (2014)

28. Zeng, H., Cao, Z., Zhang, L., Bovik, A.C.: A unified probabilistic formulation of image aesthetic assessment. IEEE Trans. Image Process. **29**, 1548–1561 (2019)

29. Zhang, W., Li, D., Ma, C., Zhai, G., Yang, X., Ma, K.: Continual learning for blind image quality assessment. IEEE Trans. Pattern Anal. Mach. Intell. **45**(3), 2864–2878 (2023)
30. Zhou, B., Lapedriza, A., Khosla, A., Oliva, A., Torralba, A.: Places: a 10 million image database for scene recognition. IEEE Trans. Pattern Anal. Mach. Intell. **40**(6), 1452–1464 (2017)
31. Zhu, H., Li, L., Wu, J., Dong, W., Shi, G.: MetaiQA: deep meta-learning for no-reference image quality assessment. In: Proceedings of the IEEE/CVF Conference on Computer Vision and Pattern Recognition, pp. 14143–14152 (2020)

Self-supervised Low-Light Image Enhancement via Histogram Equalization Prior

Feng Zhang[1], Yuanjie Shao[2], Yishi Sun[1], Changxin Gao[1], and Nong Sang[1(✉)]

[1] National Key Laboratory of Multispectral Information Intelligent Processing Technology, School of Artificial Intelligence and Automation, Huazhong University of Science and Technology, Wuhan, China
{fengzhangaia,yishisun,cgao,nsang}@hust.edu.cn
[2] School of Electronic Information and Communication, Huazhong University of Science and Technology, Wuhan, China
shaoyuanjie@hust.edu.cn

Abstract. Deep learning-based methods for low-light image enhancement have achieved remarkable success. However, the requirement of enormous paired real data limits the generality of these models. Although there have been a few attempts in training low-light image enhancement model in the self-supervised manner with only low-light images, these approaches suffer from inefficient prior information or improper brightness. In this paper, we present a novel self-supervised method named HEPNet to train an effective low-light image enhancement model with only low-light images. Our method drives the self-supervised learning of the network through an effective image prior termed histogram equalization prior (HEP). This prior is a feature space information of the histogram equalized images. It is based on an interesting observation that the feature maps of histogram equalized images and the reference images are similar. Specifically, we utilize a mapping function to generate the histogram equalization prior, and then integrate it into the model through a spatial feature transform (SFT) layer. Guided by the histogram equalization prior, our method can recover finer details in real-world low-light scenarios. Extensive experiments demonstrate that our method performs favorably against the state-of-the-art unsupervised low-light image enhancement algorithms and even matches the state-of-the-art supervised algorithms.

Keywords: Low-Light Image Enhancement · Self-Supervised Learning · Histogram Equalization Prior

1 Introduction

Images captured under low-light conditions often suffer from poor visibility, unexpected noise, and color distortion. In order to take high-quality images

Supplementary Information The online version contains supplementary material available at https://doi.org/10.1007/978-981-99-8552-4_6.

Q. Liu et al. (Eds.): PRCV 2023, LNCS 14435, pp. 63–75, 2024.
https://doi.org/10.1007/978-981-99-8552-4_6

in low-light conditions, several operations including setting long exposures, high ISO, and flash are commonly applied. However, solely turning up the brightness of dark regions will inevitably amplify image degradation. To further mitigate the degradation caused by low-light conditions, several traditional methods have been proposed. Histogram Equalization (HE) [15] rearranges the pixels of the low-light image to improve the dynamic range of the image. Retinex-based methods [18,19] decompose the low-light images into illumination and reflection maps and obtain the intensified image by fusing the enhanced reflection map and illumination map. Dehazing-based methods [3,11] regard the inverted low-light image as a haze image and improve visibility by applying dehazing. Although these methods can improve brightness, especially for dark pixels, they barely consider realistic lighting factors, often making the enhanced results visually tenuous and inconsistent with the actual scene.

Recently, Deep Convolutional Neural Networks (CNNs) set the state-of-the-art performance in low-light image enhancement [21]. Compared with traditional methods, the CNNs learn better feature representations to obtain enhanced results with superior visual quality, which benefit from the large dataset and powerful computational ability. However, most CNN-based methods require training examples with paired reference images, whereas it is extremely challenging to simultaneously capture low-light images and normal-light images of the same visual scene. To eliminate the reliance on paired training data, several unsupervised deep learning-based methods [5,7,9,13,14] have been proposed. These algorithms are able to restore images with decent illumination and contrast in real world scenarios. However, they heavily rely on carefully selected multi-exposure training data or unpaired training data, which makes the workload of data collection still burdensome. Although some self-supervised approaches [12,25] are proposed to reduce the burden, these approaches suffer from inefficient prior information or improper brightness. Therefore, it is of great interest to seek a novel strategy to provide more valid information.

In this paper, we propose a self-supervised low-light image enhancement algorithm HEPNet based on an effective prior termed histogram equalization prior (HEP). Our method takes only low-light images as input, and eliminates the requirement of any paired or unpaired reference images. This work is motivated by an interesting observation: the feature maps of histogram equalized image and the ground truth are similar. Intuitively, the feature maps of histogram equalized image can directly provide abundant texture and structure information [4]. We show statistically and empirically that this generic property of the histogram equalization theory holds for many low-light images, more details are shown in next section.

Following [2], we design a self-supervised model on the basis of Retinex theory to decompose the low-light images into illumination maps and reflectance maps, and the reflectance maps can be regarded as restored images. We further introduce a feature mapping network to generate the proposed histogram equalization prior, and integrate it into the network through a spatial feature transform (SFT) layer [20]. Furthermore, we propose a histogram equalization prior loss to guide the training process, and introduce an illumination smoothness loss to suppress the texture and structure information in the illumination maps. Extensive experiments demonstrate that our method performs favorably against the state-of-the-art unsupervised low-light enhancement algorithms and even matches the state-of-the-art supervised algorithms.

In summary, the main contributions of this work are as follows:

1. We build a novel self-supervised low-light image enhancement framework HEPNet on the basis of the Retinex theory, possessing more effective training and faster convergence speed with limited computational cost.
2. We propose an effective prior termed histogram equalization prior (HEP) for low-light image enhancement. It delivers effective structure and luminance information to drive the self-supervised learning.
3. We conduct extensively experiments on some benchmark datasets to demonstrate the notability and generality of our proposed prior and framework.

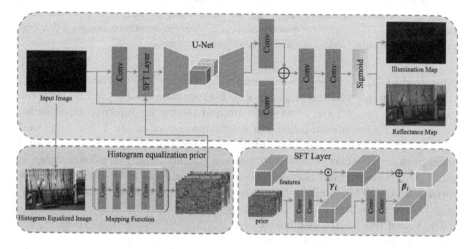

Fig. 1. Overview of the proposed framework. The input low-light image is first fed into a convolution layer to extract the initial feature map, then the histogram equalization prior (HEP) extracted from the histogram equalized image by a 5-layer mapping function and integrated into the initial feature map by a spatial feature transform (SFT) layer, and finally the reflectance and illumination maps are obtained through a U-Net and several convolution layers. The reflectance map is considered as the restored image contaminated by noise.

2   Methodology

2.1   Histogram Equalization Prior

For low-light image enhancement, self-supervised learning-based approach is difficult to implement. The main reason is that extracting texture and color information from low-light images without the help of paired or unpaired reference images is extremely challenging. To address this issue, we propose an effective prior information based on histogram equalization to drive the self-supervised learning.

Traditional histogram equalization can make the dark images visible by stretching the dynamic range of dark images via manipulating the corresponding histogram. However, it is not flexible enough for visual property adjustment in local regions and leads to undesirable local appearances, e.g., under/over-exposure and color bias. Encouraging pixels to match the histogram equalized image captures unpleasant local impressions.

(a) Input   (b) Feature map of a   (c) HE   (d) Feature map of c   (e) Reference image   (f) Feature map of e

Fig. 2. The visual examples on histogram equalization prior. First column: input low-light image. Second column: feature map of input image. Third column: histogram equalized image. Forth column: feature map of histogram equalized image. Fifth column: reference image. Sixth column: feature map of reference image.

Inspired by [8], we observe some interesting phenomena on the visualization of the feature maps, which is shown in Fig. 2. From the visual representation, the input low-light image barely has structure and luminance information in sight, consequently, its feature map has scarce semantic information. When processed with histogram equalization, the enhanced image suffers from color bias and underexposure. However, the feature map of the histogram equalized image is abundant in semantic information, which are even similar to the feature map of reference image. This similarity inspires that the information in feature space might be quite valuable.

The visual results are insufficient to explain the validity of the proposed prior. Hence, we attempted to explain it statistically. We selected 500 paired images from the LOL [2] dataset, and calculate the cosine similarity between the feature maps of histogram equalized images and reference images. Figure 3 is the histogram of cosine similarities over all 500 low-light images. We can observe that over 90% of the cosine similarities are concentrated above 0.8. Compared with the cosine similarities between the feature maps of input low-light images and reference images, the cosine similarities

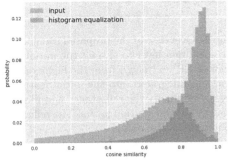

Fig. 3. Histogram of the cosine similarities. Green: cosine similarities between the feature maps of input low-light images and reference images. Blue: cosine similarities between the feature maps of histogram equalized images and reference images. (Color figure online)

between the feature maps of histogram equalized images and reference images have been substantially improved. The statistic provides a strong support to the histogram equalization prior, and indicates that we can adopt this prior instead of reference image to guide the training process.

Based on the above analysis, we attempt to utilize the feature map as a prior information for the nearual network, and also utilize the feature map instead of the histogram equalized image as the regularization term to guide the deep network training. However, since the feature map is still derived from the histogram equalized image, it may cause the learned enhanced image to converge to the histogram equalized image. Therefore, motivated by the Deep Image Prior (DIP) [17], which demonstrates that the deep neural network offers high impedance to noise and low impedance to signal, we adopt the same learning strategy that restrict the number of iterations in the optimization process to a certain number of iterations. By restricting the number of iterations, the learned enhanced image can be effectively suppressed to converge to the histogram equalized image.

Following the above phenomena, we have to extract the histogram equalization prior from the histogram equalized image through a mapping function $\mathcal{M}$. In this study, we use a 5-layer neural network to implement this mapping function, which enables end-to-end optimization through the network. By incorporating histogram equalization prior as part of the network, which provides fine-grained guidance for the intermediate feature map, the network can handle the low-light images in a self-supervised manner.

To make the model better characterize the properties of histogram equalization prior, we utilize Spatial Feature Transform (SFT) [20] layer to learn a modulation parameter pair (γ, β) based on the histogram equalization prior. The learned parameter pair adaptively influences the outputs by applying an affine transformation spatially to intermediate feature maps in networks. After obtaining learned parameter pair from histogram equalization prior, the transformation is carried out by scaling and shifting feature maps of a specific layer, the formula is as follow:

$$(\gamma, \beta) = Conv(\mathcal{M}(H(I))), \quad SFT(F|\gamma, \beta) = \gamma \odot F + \beta \tag{1}$$

where $H(\cdot)$ stands for histogram equalization operation, $\mathcal{M}$ is the mapping function, $Conv$ is the convolution layers to extract modulation parameter pair (γ, β), I is the input low-light image, F stands for feature maps extract from input low-light image, $\odot$ represents element-wise multiplication, i.e., Hadamard product.

2.2 Architecture

The framework improves the brightness of the low-light images on the basis of Retinex theory. According to the Retinex theory, the images can be decomposed into reflectance maps and illumination maps. Mathematically, a degraded low-light image can be naturally modeled as follows:

$$I = R \circ L + N \tag{2}$$

where I stands for input image, R stands for reflectance map, L is the illumination map, N represents the noise component, ∘ represents element-wise multiplication.

By taking simple algebra steps [23], we can have the following formula:

$$I = R \circ L + N = R \circ L + \tilde{N} \circ L = (R + \tilde{N}) \circ L = \tilde{R} \circ L \qquad (3)$$

where $\tilde{N}$ stands for the degradation having the illumination decoupled, $\tilde{R}$ represents for the polluted reflectance map.

According to the above theory, the reflectance maps can be regarded as the restored images with noise. Therefore, we design a novel network to decompose the low-light images into reflectance and illumination maps. Since the reflectance map is a restored image with noise, we have to constrain the noise contained in the reflectance map. The Total variation minimization (TV) loss [1] is often used as smoothness prior for various image restoration tasks, and we can utilize it to minimizes the gradient of the reflectance map to suppress noise.

The overall framework is shown in Fig. 1. We first fed the input low-light image into a convolution layer to extract the initial feature map. Secondly, we extract the histogram equalization prior with a mapping function, and then integrate it into the initial feature map through a spatial feature transform layer (SFT) [20]. Finally, the fused features are fed into a U-Net, and the reflectance and illumination maps are obtained through several convolution layers and a sigmoid layer.

Briefly, our proposed architecture consists of three components: a retinex network which takes the low-light image and its bright channel as input and decomposes into an illumination map and a reflectance map, a mapping function that extracted the histogram equalization prior from the histogram equalized image, and a spatial feature transform layer, which integrate the histogram equalization prior into intermediate feature maps. Given that the SFT layer is proposed by [20], the detailed structure is recommended to refer to the original article.

2.3   Loss Function

Our framework is based on self-supervised learning, yet it is difficult to recover low-light images with this manner, so we need to obtain some reference information to guide the training process. Benefiting from the proposed histogram equalization prior, we can constrain the restored image by the feature space information of histogram equalized image. However, it is not practicable to directly constrain the image space with the feature space, so we refer to the perceptual loss [8], and constrain the feature space between restored image and histogram equalized image. We define an MSE loss between the feature maps of the output image and the histogram equalized image, which we call the histogram equalization prior loss. The loss function can be formulated as follows:

$$\mathcal{L}_{hep} = \parallel F(H(\tilde{R})) - F(H(I)) \parallel_2^2 \qquad (4)$$

where $F(\cdot)$ denotes the feature map extracted from a VGG-19 model pre-trained on ImageNet, $H(\cdot)$ represent the histogram equalization operation.

Table 1. Quantitative comparisons on the LOL test set in terms of PSNR, SSIM, and LPIPS. The RT (running time), FLOPs, and model size (Params) are also represented. T, SL, UL and SSL represent the traditional method, supervised learning method, unsupervised learning method, and self-supervised learning method, respectively.

Learning	Method	PSNR↑	SSIM↑	LPIPS↓	RT	Params	FLOPs
T	HE [15]	15.56	0.540	0.419	0.0021 s	/	/
	LIME [6]	17.18	0.639	0.354	0.4914 s	/	/
SL	Retinex-Net [2]	16.77	0.585	0.419	0.2912 s	0.445M	136.02G
	KinD [24]	20.38	0.882	0.199	0.3872 s	8.016M	127.68G
	KinD++ [23]	21.32	0.891	0.207	0.5486 s	8.275M	2.53T
	LLFlow [22]	25.00	0.919	0.164	0.3345 s	17.42M	1.05T
	MAXIM [16]	23.41	0.908	0.149	0.2176 s	14.12M	216G
UL	Zero-DCE [5]	14.86	0.687	0.331	0.0064 s	0.082M	19.02G
	EnlightenGAN [7]	17.48	0.768	0.327	0.0178 s	8.637M	61.03G
	SCI [14]	14.78	0.646	0.327	0.0017 s	0.0003M	0.238G
SSL	Self-Sup [25]	19.13	0.778	0.363	0.2655 s	**0.486M**	126.73G
	DUNP [12]	15.49	0.654	0.405	1.8147 s	2.208M	**65.66G**
	Ours	**20.03**	**0.822**	**0.306**	**0.0199 s**	0.487M	76.79G

Based on Retinex theory, the network decomposes the images into illumination and reflectance maps, while the decomposed maps should reproduce the input image. Thus, we introduce the reconstruction loss to constrain this process. The formula is as follows:

$$\mathcal{L}_{recon} = \parallel \tilde{R} \circ L - I \parallel_1 \tag{5}$$

Since the reflectance map is a restored image contaminated with noise, we need to constrain the noise contained in the reflectance map. Hence, we adopt Total variation minimization (TV) [1] to minimizes the gradient of the reflectance map so that the noise is suppressed. The TV loss is formulated as:

$$\mathcal{L}_{tv} = \parallel \nabla R \parallel_1 \tag{6}$$

As the reflectance map should preserve more texture and color details. In other words, the illumination map should be smooth in textural information while still preserving the structural boundaries. To make the illumination map smooth in textural information, we modify the illumination smoothness loss proposed in [23]. Different from the previous loss, our illumination smoothness loss only takes the low-light input image as the reference. This term constrains the relative structure boundaries of the illumination map to be consistent with the input image, which can reduce the risk of over-smoothing on the structure boundaries. The illumination smoothness loss is formulated as:

$$\mathcal{L}_{is} = \parallel \frac{\nabla L}{max(\mid \nabla I \mid, \epsilon)} \parallel_1 \tag{7}$$

where $|\cdot|$ means the absolute value operator, ϵ is a small positive constant for avoiding zero denominators, ∇ denotes the gradient including ∇h (horizontal) and ∇v (vertical).

As a result, the loss function is as follows:

$$\mathcal{L} = \mathcal{L}_{recon} + \lambda_{hep}\mathcal{L}_{hep} + \lambda_{is}\mathcal{L}_{is} + \lambda_{tv}\mathcal{L}_{tv} \tag{8}$$

In our experiment, these parameters are set to $\lambda_{hep} = \lambda_{is} = 0.1$, $\lambda_{tv} = 0.2$, $\epsilon = 0.01$. Due to the careful settings of these loss terms, the proposed framework can perform sufficiently well.

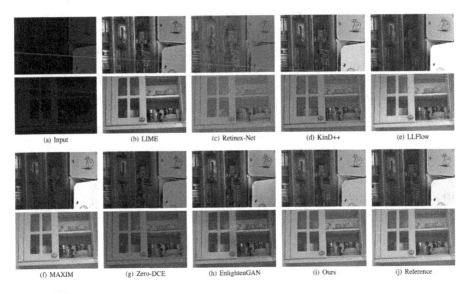

Fig. 4. Visual comparison with other state-of-the-art methods on LOL dataset [2]. Best viewed in color and by zooming in. (Color figure online)

3 Experimental Validation

3.1 Implementation Details

We train and evaluate our method on the representative low-light image dataset LOL [2], which includes 500 low/normal-light paired images. These images are taken in both indoor and outdoor scenes. LOL is currently the most abundant dataset available for low-light image enhancement. We use Adam [10] optimizer to perform optimization with the weight decay equal to 0.0001. The batch size is set to 16 and the patch size is 64×64, and the learning rate is set to 10^{-4}.

3.2   Quantitative Evaluation

We have also quantitatively compared our method to the state-of-the-art methods. As shown in Table 1, our proposed HEPNet achieves the best performance among all unsupervised methods on the LOL dataset, although it only takes low-light images as input. It demonstrates that the proposed HEPNet possesses favorable capability among all unsupervised methods, and its performance is even close to some state-of-the-art supervised methods. The promising PSNR and SSIM values reveal that our method HEPNet is capable of recovering color and luminance information without any reference image, while preserving more structural details. We can see from Table 1 that our method HEPNet obtains the best LPIPS scores among all unsupervised methods. It illustrates that the low-light images enhanced by our proposed HEPNet are more consistent with human perception. Furthermore, we also report the running time (RT), model size (Params), and FLOPs of all methods in Table 1, with the running time averaged on 485 images of size $600 \times 400 \times 3$. Evidently, our proposed method HEPNet is relatively lightweight while yielding excellent performance.

Fig. 5. Visual comparison with the state-of-the-art methods on real-world dataset VV (no reference images). Best viewed in color and by zooming in. (Color figure online)

3.3   Qualitative Evaluation

We first visually evaluate our proposed HEPNet on the classical low-light image datasets: LOL dataset [2], and compare it with the state-of-the-art approaches with available codes, including LIME [6], Retinex-Net [2], KinD++ [23], Zero-DCE [5], EnlightenGAN [7], MAXIM [16], and LLFlow [22]. We train all models on the LOL training set and then evaluate them on the LOL test set. Figure 4 shows a representative result for visual comparison. From the result, we can see that our method HEPNet is quite competitive with the state-of-the-art methods, while it is a self-supervised manner that does not rely on any reference image. The visual comparison results illustrate the effectiveness of our proposed method HEPNet. More visual comparison results are shown in the supplementary material.

3.4   Generalization Ability on Real-World Images

To further demonstrate the generality of the proposed method, we evaluate our proposed HEPNet and the state-of-the-art methods in a cross-dataset manner. We train all methods on the LOL training set and then test in some real-world low-light datasets VV[1](24 images). We trained all models on the LOL training set and then evaluated on these datasets. Figure 5 show the results of some challenging images on the real-world dataset. From the results, we can observe that our proposed HEPNet can enhance dark regions and simultaneously preserve the color. The results are visually pleasing without obvious noise and color casts. It demonstrates that our method has great generality in real-world images with more naturalistic quality.

4   Ablation Studies

4.1   Comparison with Other Prior Information

A major advantage of our method over existing self-supervised methods is that histogram equalization prior can better extract the structure and luminance information through the feature space. To demonstrate its validity, we need to compare it with other prior information. However, since there is no similar prior information available for comparison in previous methods, we employ the feature

Table 2. Ablation study of the contribution of histogram equalization prior in terms of PSNR, SSIM and LPIPS.

Prior	PSNR↑	SSIM↑	LPIPS↓
Input	7.77	0.221	0.405
Gamma Correction	16.38	0.685	0.363
Multi-Scale Retinex	15.88	0.622	0.385
Histogram Equalization	20.03	0.822	0.306

Table 3. Ablation study of the contribution of histogram equalization prior loss in terms of PSNR, SSIM and LPIPS.

Loss	PSNR↑	SSIM↑	LPIPS↓
Input	7.77	0.221	0.405
with $\mathcal{L}_{L1}$	18.11	0.687	0.368
with $\mathcal{L}_{MSE}$	18.51	0.705	0.368
with $\mathcal{L}_{SSIM}$	18.29	0.681	0.356
with $\mathcal{L}_{max}$	18.39	0.745	0.380
with $\mathcal{L}_{hep}$	20.03	0.822	0.306

---

[1] https://sites.google.com/site/vonikakis/datasets.

maps of different low-light image enhancement operations as prior information for comparative assessment. In this comparison, we adopt the feature maps of Gamma Correction (GC) and Multi-Scale Retinex (MSR) as prior. Table 2 shows the results of the comparison, and it is obvious that histogram equalization prior provides more better feature space information. The inefficacy of other prior information is due to the fact that all these methods require carefully selected parameters, while histogram equalization does not require any parameters and thus works more generally for different sorts of low-light images.

4.2   The Effectiveness of Histogram Equalization Prior Loss

In order to build a strong constraint, we further propose the histogram equalization prior loss, which improves the performance of the network by constraining the variance between low-light image and histogram equalized image in feature space. To verify the effectiveness of such strategy, we have evaluated different loss functions with the histogram equalized image: L1 loss $\mathcal{L}_{L1}$, MSE loss $\mathcal{L}_{MSE}$, and SSIM loss $\mathcal{L}_{SSIM}$, and max information entropy loss $\mathcal{L}_{max}$ [25]. The formula of max information entropy loss is as follow:

$$\mathcal{L}_{max} = \| \max_{c \in R, G, B}(R^c) - H(\max_{c \in R, G, B}(I^c)) \|_1 \tag{9}$$

where $H(\cdot)$ stands for histogram equalization operation, $max(\cdot)$ stands for the maximum channel operation, R^c represents the reflectance map, I^c represents the input low-light image.

The comparison results evaluated on LOL dataset are reported in Table 3. Using the L1 loss $\mathcal{L}_{L1}$ or the MSE loss $\mathcal{L}_{MSE}$ achieves similar SSIM and LPIPS scores. Nevertheless, for the PSNR values, the estimation from the MSE loss $\mathcal{L}_{MSE}$ exceeds those from the L1 loss $\mathcal{L}_{L1}$ with 0.40dB. The SSIM loss $\mathcal{L}_{SSIM}$ improves the LPIPS score, but it failed in PSNR and SSIM by a large margin. The max information entropy loss $\mathcal{L}_{max}$ surpasses the SSIM loss $\mathcal{L}_{SSIM}$ in PSNR and SSIM score, but the SSIM loss $\mathcal{L}_{SSIM}$ outperformed by 0.024 in LPIPS. The model constraint by the histogram equalization prior loss $\mathcal{L}_{hep}$ has a huge improvement in all metrics comparing with other loss functions.

5   Conclusions

In this paper, we present a novel self-supervised framework for low-light image enhancement. Inspired by Retinex theory, the network brightens the image by decomposing the images into reflectance and illumination maps. We also propose an effective prior termed histogram equalization prior to drive the self-supervised learning, which is an extension of histogram equalization that investigates the spatial correlation between feature maps. Benefiting from the abundant information of the histogram equalization prior, the reflectance maps generated by the network simultaneously improve brightness and preserve texture and color information. Both qualitative and quantitative experiments demonstrate the superiority of our model over state-of-the-art unsupervised methods.

References

1. Chan, S.H., Khoshabeh, R., Gibson, K.B., Gill, P.E., Nguyen, T.Q.: An augmented lagrangian method for total variation video restoration. IEEE Trans. Image Process. **20**(11), 3097–3111 (2011)

2. Wei, C., Wang, W., Yang, W., Liu, J.: Deep retinex decomposition for low-light enhancement. In: British Machine Vision Conference. British Machine Vision Association (2018)

3. Dong, X., et al.: Fast efficient algorithm for enhancement of low lighting video. In: 2011 IEEE International Conference on Multimedia and Expo, pp. 1–6. IEEE (2011)

4. Geirhos, R., Rubisch, P., Michaelis, C., Bethge, M., Wichmann, F.A., Brendel, W.: Imagenet-trained CNNs are biased towards texture; increasing shape bias improves accuracy and robustness. arXiv preprint arXiv:1811.12231 (2018)

5. Guo, C., et al.: Zero-reference deep curve estimation for low-light image enhancement. In: Proceedings of the IEEE/CVF Conference on Computer Vision and Pattern Recognition, pp. 1780–1789 (2020)

6. Guo, X., Li, Y., Ling, H.: Lime: low-light image enhancement via illumination map estimation. IEEE Trans. Image Process. **26**(2), 982–993 (2016)

7. Jiang, Y., et al.: Enlightengan: deep light enhancement without paired supervision. IEEE Trans. Image Process. **30**, 2340–2349 (2021)

8. Johnson, J., Alahi, A., Fei-Fei, L.: Perceptual losses for real-time style transfer and super-resolution. In: Leibe, B., Matas, J., Sebe, N., Welling, M. (eds.) ECCV 2016. LNCS, vol. 9906, pp. 694–711. Springer, Cham (2016). https://doi.org/10.1007/978-3-319-46475-6_43

9. Kandula, P., Suin, M., Rajagopalan, A.: Illumination-adaptive unpaired low-light enhancement. IEEE Trans. Circuits Syst. Video Technol. (2023)

10. Kingma, D.P., Ba, J.: Adam: a method for stochastic optimization. arXiv preprint arXiv:1412.6980 (2014)

11. Li, L., Wang, R., Wang, W., Gao, W.: A low-light image enhancement method for both denoising and contrast enlarging. In: 2015 IEEE International Conference on Image Processing (ICIP), pp. 3730–3734. IEEE (2015)

12. Liang, J., Xu, Y., Quan, Y., Shi, B., Ji, H.: Self-supervised low-light image enhancement using discrepant untrained network priors. IEEE Trans. Circuits Syst. Video Technol. **32**(11), 7332–7345 (2022)

13. Liu, R., Ma, L., Zhang, J., Fan, X., Luo, Z.: Retinex-inspired unrolling with cooperative prior architecture search for low-light image enhancement. In: Proceedings of the IEEE/CVF Conference on Computer Vision and Pattern Recognition, pp. 10561–10570 (2021)

14. Ma, L., Ma, T., Liu, R., Fan, X., Luo, Z.: Toward fast, flexible, and robust low-light image enhancement. In: Proceedings of the IEEE/CVF Conference on Computer Vision and Pattern Recognition, pp. 5637–5646 (2022)

15. Pizer, S.M.: Contrast-limited adaptive histogram equalization: speed and effectiveness stephen m. pizer, r. eugene johnston, james p. ericksen, bonnie c. yankaskas, keith e. muller medical image display research group. In: Proceedings of the First Conference on Visualization in Biomedical Computing, Atlanta, Georgia, vol. 337 (1990)

16. Tu, Z., et al.: Maxim: multi-axis MLP for image processing. In: Proceedings of the IEEE/CVF Conference on Computer Vision and Pattern Recognition, pp. 5769–5780 (2022)

17. Ulyanov, D., Vedaldi, A., Lempitsky, V.: Deep image prior. In: Proceedings of the IEEE Conference on Computer Vision and Pattern Recognition, pp. 9446–9454 (2018)
18. Wang, L., Xiao, L., Liu, H., Wei, Z.: Variational Bayesian method for retinex. IEEE Trans. Image Process. **23**(8), 3381–3396 (2014)
19. Wang, S., Zheng, J., Hu, H.M., Li, B.: Naturalness preserved enhancement algorithm for non-uniform illumination images. IEEE Trans. Image Process. **22**(9), 3538–3548 (2013)
20. Wang, X., Yu, K., Dong, C., Loy, C.C.: Recovering realistic texture in image super-resolution by deep spatial feature transform. In: Proceedings of the IEEE Conference on Computer Vision and Pattern Recognition, pp. 606–615 (2018)
21. Wang, Y., Wang, H., Yin, C., Dai, M.: Biologically inspired image enhancement based on retinex. Neurocomputing **177**, 373–384 (2016)
22. Wang, Y., Wan, R., Yang, W., Li, H., Chau, L.P., Kot, A.: Low-light image enhancement with normalizing flow. In: Proceedings of the AAAI Conference on Artificial Intelligence, pp. 2604–2612 (2022)
23. Zhang, Y., Guo, X., Ma, J., Liu, W., Zhang, J.: Beyond brightening low-light images. Int. J. Comput. Vision **129**(4), 1013–1037 (2021)
24. Zhang, Y., Zhang, J., Guo, X.: Kindling the darkness: a practical low-light image enhancer. In: Proceedings of the 27th ACM International Conference on Multimedia, pp. 1632–1640 (2019)
25. Zhang, Y., Di, X., Zhang, B., Wang, C.: Self-supervised image enhancement network: Training with low light images only. arXiv preprint arXiv:2002.11300 (2020)

Enhancing GAN Compression by Image Probability Distribution Distillation

Lizhou You, Tie Hu, and Fei Chao[✉]

School of Informatics, Xiamen University, Fujian 361005, People's Republic of China
{youlizhou,hutie}@stu.xmu.edu.cn, fchao@xmu.edu.cn

Abstract. This paper presents a novel approach named Image Probability Distribution Distillation (IPDD) for compressing generative adversarial networks (GANs) by distilling knowledge from the global distribution of images. Unlike traditional methods that distill at the pixel level, we propose a holistic approach that captures the overall coherence of images. To achieve this, we introduce a novel teacher discriminator that engages in adversarial training with both the teacher generator and the student generator in asynchronous weighted manner, using variable weights to optimize the Nash equilibrium between them. Our framework explores the uncharted territory of mining global distribution information and federated training of the teacher discriminator, offering potential for enhancing the performance of compressed GANs. Extensive experiments on benchmark datasets demonstrate that our approach significantly reduces GAN complexity while achieving optimal performance.

Keywords: Generative Adversarial Networks · Model Compression · Knowledge Distillation

1 Introduction

Generative Adversarial Networks (GANs) [12] endeavor to achieve an equilibrium between the generator and discriminator components in order to generate outputs that closely resemble real images. The field of computer vision has witnessed remarkable advancements in GANs, particularly in the domains of image synthesis [2, 21, 22, 30, 39], style transfer [9, 10, 37], and image-to-image translation [6, 7, 18, 42], *etc.* However, compared to traditional visual tasks such as classification, image generation presents greater challenges in terms of computational complexity and parameter overhead, making it particularly daunting to deploy GANs on resource-constrained devices. Consequently, both academia and industry have recognized the pressing need for economically viable GAN solutions. GAN compression has emerged as a highly consequential avenue of research, encompassing methodologies such as network pruning [5, 24, 25], neural architecture search [8, 19, 23], weight quantization [34, 35], and knowledge distillation [1, 3, 17, 23, 26, 31, 36], *etc*, among others.

Q. Liu et al. (Eds.): PRCV 2023, LNCS 14435, pp. 76–88, 2024.
https://doi.org/10.1007/978-981-99-8552-4_7

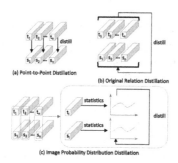

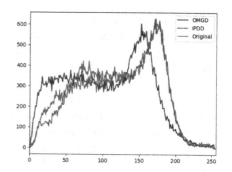

Fig. 1. (a) vanilla knowledge transferring [15] and (b) relational knowledge transferring [29]. (c) Our image-level probability distribution distillation.

Fig. 2. Statistic the pixel distribution of the generated image of CycleGAN [42], OMGD [31] and IPDD (Ours).

When discussing contemporary GAN compression methods, knowledge distillation (KD) has emerged as a particularly effective technique and has almost become a standard complement to other compression approaches. AutoGAN [8] introduces a search space for exploring variations in generator structures and employs a controller to guide the search process. Acceleration is achieved through parameter sharing and dynamic resetting. WaveletKD [40] leverages wavelet transformation to distill generated images, enabling the student generator to acquire high-priority information and enhance the quality of the generated images. However, despite extensive efforts, early methods in GAN compression encounter two limitations that hinder further advancements in the field.

Initially, prevailing approaches [8,15] in the field primarily relied on simplistic distillation techniques. However, these approaches neglect the holistic nature of images, which demands a focus on overall coherence. Motivated by the current prominent trend of Diffusion models [16], we propose a novel distillation methodology that operates at the global distribution level. By conducting distillation at a macroscopic level, our approach enables the student generator to assimilate higher-level semantic information, transcending the limitations of pixel-level similarity. The most crucial aspect lies in the unexplored territory of mining global distribution information and federated training of the teacher discriminator in the context of GAN distillation, which holds promise for enhancing the performance of compressed GANs. We posit that these two issues are intertwined and can be addressed within a unified framework. Therefore, in this paper, we propose the Image Probability Distribution Distillation (IPDD) approach to facilitate the more effective learning of efficient GAN models, as illustrated in Fig. 1(c). Our IPDD framework introduces two key innovations. Firstly, in Sect. 3.2, we model the global distribution between images to capture higher-level information across the entire image. Secondly, in Sect. 3.3, we train the teacher discriminator in an adversarial manner with asynchronously changing weights. The specific architecture of our IPDD framework can be referred to in Fig. 3. Following the generator, we distill the student generator using the probability distribution obtained from statistically analyzing the output images, guided by the probability distribution derived from the teacher generator. We encourage similarity between the distributions of the teacher and student outputs, drawing inspiration from the noise injection and denoising processes in DDPM [16]. Building upon global distribution distillation, we discard a pre-trained and overly powerful discriminator that could disrupt the Nash equilibrium [28] for the student generator. Instead, we train the teacher discriminator from scratch in an online and asynchronous manner, updating the teacher discriminator with variable weight during joint training with the student generator. This approach yields improved Nash equilibrium during adversarial training of the GAN. We conducted extensive experiments on three widely recognized benchmark datasets [38, 42] to demonstrate the effectiveness of our IPDD in enhancing the performance of compressed GANs. The results indicate that our IPDD not only minimizes GAN complexity significantly but also achieves optimal performance. For instance, on the horse2zebra [42], we reduced the MAC of CycleGAN [42] by over $40\times$ and $80\times$ parameters while achieving FID of 49.62, surpassing the current state-of-the-art [31]of 51.92. This work aims to address the challenge of extracting lightweight student generators for the deployment of GANs on resource-limited platforms.

The main contributions of this paper are as follows: (1) A novel image probability distribution distill method tailored for GANs. (2) Online asynchronous weight transformation for improved Nash equilibrium in GANs compression. (3) Significant performance enhancement and model complexity reduction.

The remaining sections of this paper are outlined as follows: We will briefly review relevant studies in Section 0, as discussed in Sect. 2. Then, in Sect. 3, we provide a detailed description of our GAN distillation method, IPDD. In Sect. 4, we discuss the performance comparison of our approach with existing methods such as [8, 19, 20, 23, 24, 31, 33, 34, 40, 42]. Finally, we present the conclusion and future research in Sect. 5.

2   Related Work

Generative Adversarial Networks. GANs [12] have significantly improved the performance of various generative tasks. Among these advancements, Style-GAN [21] stands out as one of the most influential improvements to GAN structures. It introduces disentangled feature vectors as inputs to GANs, replacing pure Gaussian noise. Subsequently, significant efforts have been devoted to improving network loss and network structures [7, 18, 21, 22, 42]. In most image-to-image translation tasks, the aim is to translate images from a source domain to a target domain, making GANs increasingly popular for such applications. Despite the progress made, the improvement in performance has also led to increased computational and memory costs, which are not acceptable for resource-constrained devices. In this paper, we focus on developing an efficient model by cutting down the redundancy in existing GANs.

Knowledge Distillation. KD involves transferring dark knowledge from high-capacity teacher networks to students. This groundbreaking work can be traced back to hiton [15], where the output logic distance between the student and teacher is minimized using ℓ_2 regularization. Similar logit distillation approaches include In addition to output logit [15,41], other forms of knowledge extraction have been explored, such as intermediate feature maps [4,14,32]. In contrast, our IPDD based on statistical analysis of the probability distribution information across the entire image. We transmit higher-dimensional information to the student generator, different from the traditional approach, by exploring the more global aspects of the image, as illustrated in Fig. 1.

GAN Compression. Although GANs excel in synthesizing realistic images, they also come with significant computational costs [24]. Therefore, GAN compression has garnered widespread attention, primarily due to its urgent application on resource-constrained devices. Apart from KD, typical compression methods include weight pruning [5,33], network quantization [34,35], and neural architecture search [8,19,23], among others. SlimGANs [17] proposed slimming pruning, dividing the GAN model into four weight-sharing sub-networks, where larger individuals extract smaller individuals. GAN Compression [23] trained a "once-for-all" student generator using feature distillation and neural architecture search. OMGD [31] performed discriminator-free distillation by introducing a broader and deeper network structure to the teacher and distilling intermediate features and output logits to the student. GCC [25] revisited the effectiveness of discriminators in GAN compression and introduced a generator-discriminator cooperative compression scheme where the discriminator is also pruned to achieve Nash equilibrium. Motivated by GCC, we also consider maintaining Nash equilibrium in GAN training. Instead of performing pruning operations on the discriminator, we propose a federated learning approach where the generator and teacher discriminator collaboratively transform weights in an online asynchronous manner.

3   Methodology

3.1   Background

Generative Adversarial Networks (GANs) have revolutionized the field of generative modeling by introducing a powerful framework for generating realistic synthetic data. GANs consist of two main components: a generator network and a discriminator network. The generator aims to produce synthetic samples that resemble real data, while the discriminator tries to differentiate between real and generated samples. The training objective of a GAN is formulated as a minimax game between the generator and the discriminator:

$$\min_G \max_D V(D, G) = \mathbb{E}_{x \sim p_{\text{data}}(x)}[\log D(x)] + \mathbb{E}_{z \sim p_z(z)}[\log(1 - D(G(z)))] \quad (1)$$

where $p_{\text{data}}(x)$ represents the true data distribution and $p_z(z)$ is the prior noise distribution. The generator aims to minimize this objective by generating samples that deceive the discriminator, while the discriminator aims to maximize it

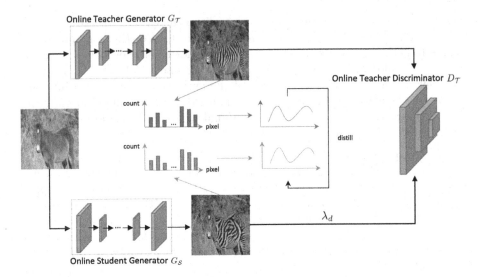

Fig. 3. Framework of our IPDD. We count the distribution of pixels in corresponding image and distill it through divergence calculation between the student and the teacher. Teacher discriminator is online updated when co-trained with the teacher generator and asynchronously updates at λ_d weight when co-trained with the student generator.

by correctly distinguishing real and generated samples. The training process of a GAN involves iteratively updating the generator and discriminator networks. The discriminator is trained to maximize the objective in Eq. 1 by updating its parameters using gradient ascent, while the generator is trained to minimize the objective by updating its parameters using gradient descent. This adversarial training process continues until the generator produces high-quality synthetic samples that are difficult for the discriminator to distinguish from real samples. Achieving this equilibrium [28] is a challenging task and often requires careful hyperparameter tuning and network architecture design.

3.2 Image Probability Distribution Distillation

To delve into the analysis of image probability distributions within the context of GAN compression, we adopt a knowledge distillation approach that centers on the exploitation of the global probability distribution of images. Specifically, we aim to distill the comprehensive knowledge embedded in a well-established teacher GANs into a more compact and streamlined student GANs.

Our methodology hinges upon minimizing the Bregmane divergence, which measures the discrepancy between two probability distributions or vectors based on a convex and differentiable function known as the Bregman generator of generated images produced by the teacher and student networks. By reducing the Bregmane divergence, we encourage the student network to closely approximate the global image probability distribution embodied by the teacher network. This optimization objective can be formulated as:

$$D_B(P\|Q) = g(P) - g(Q) - \langle \nabla g(Q), P - Q \rangle \tag{2}$$

In the formula, $g(\cdot)$ represents the Bregman generator function, $\nabla g(Q)$ represents the gradient of the generator function evaluated at point Q, and $\langle \cdot, \cdot \rangle$ denotes the inner product. Consequently, minimizing the Bregman divergence enables the student to capture and replicate the salient characteristics of the teacher's probability distribution.

Our investigation of image probability distributions and the incorporation of knowledge distillation via the global probability distribution aim to bolster the performance of GAN compression. By leveraging this methodology, we not only achieve efficient model compression but also preserve critical visual attributes and elevate the generative capabilities of the student network, thereby establishing a robust foundation for advancing GAN compression techniques.

3.3 Asynchronous Weighted Discriminator

It is well-known that achieving Nash equilibrium requires the generator and discriminator to possess similar capabilities. Prior methods, such as Chen et al. [3], Lin et al. [27], and Jin et al. [19], deployed a pretrained discriminator. However, when training a generator from scratch, the excessive strength of the discriminator often leads to local optima. Ren et al. [31] addressed this issue by training the teacher generator without starting from scratch and discarding the discriminator during student generator training.

Considering the substantial enhancement in the capacity of the student generator achieved through our global probability distribution distillation, we find that an online teacher discriminator can effectively complement the student generator. Consequently, we employ a teacher discriminator trained from scratch and conduct adversarial training using both the teacher discriminator and student generator. Notably, we utilize an online asynchronous weight update scheme for the discriminator training. Specifically, the discriminator undergoes complete weight updates only during co-training with the teacher generator, while it receives lower weight updates during co-training with the student generator. This enables adaptive weight transformations based on the number of iterations. Thus, our adversarial training incorporating the online teacher discriminator is defined as follows:

$$\mathcal{L}_{ADV} = \mathcal{L}(D_T, G_T) + \lambda_d \mathcal{L}(D_T, G_S), \tag{3}$$

where λ_d adapts the weight transformation of the discriminator.

Through our experimental analysis in Sect. 4.3, we demonstrate superior performance compared to the approach of discarding the discriminator altogether as proposed by Ren et al. [31].

Table 1. Experimental comparison on horse2zebra and summer2winter with Cycle-GAN. The Δ indicates performance increase over the original CycleGAN [42].

Dataset	Method	MACs	#Parameters	FID(Δ)
horse2zebra	Original [42]	56.80G(1.0×)	11.30M(1.0×)	61.53(-)
	Co-Evolution [33]	13.40G(4.2×)	-	96.15(-34.62)
	GAN-Slimming [34]	11.25G(23.6×)	-	86.09(−24.56)
	AutoGAN [8]	6.39G(8.9×)	-	83.60(−22.07)
	Wavelet KD [40]	1.68G(33.8×)	0.72M(15.81×)	77.04(−15.51)
	GANCompression [23]	2.67G(21.3×)	0.34M(33.2×)	64.95(−3.42)
	SCL [20]	2.96G(19.17×)	0.41M(27.5×)	64.64(−3.11)
	DMAD [24]	2.41G(23.6×)	0.28M(40.0×)	62.96(−1.43)
	CAT [19]	2.55G(22.3×)	-	60.18(1.35)
	GCC [25]	2.40G(23.6×)	-	59.31 2.22)
	OMGD [31]	1.41G(40.3×)	0.14M(82.5×)	51.92(9.61)
	IPDD (Ours)	**1.41G(40.3×)**	**0.14M(82.5×)**	**49.62(11.91)**
summer2winter	Original [42]	56.80G(1.0×)	11.30M(1.0×)	79.12(-)
	Co-Evolution [33]	11.10G(5.1×)	-	78.58(0.54)
	AutoGAN [8]	4.34G(13.1×)	-	78.33(0.79)
	DMAD [24]	3.18G(17.9×)	0.30M(37.7×)	78.24(0.88)
	OMGD [31]	1.41G(40.3×)	0.14M(82.5×)	73.79(5.33)
	IPDD(Ours)	**1.41G(40.3×)**	**0.14M(82.5×)**	**73.72(5.40)**

Table 2. Experimental comparison on edges2shoes with Pix2Pix. The Δ indicates the performance increase over the original Pix2Pix [18].

Dataset	Method	MACs	#Parameters	FID(Δf)
edges→shoes	Original [18]	18.60G(1.0×)	54.40M(1.0×)	34.31 (-)
	Wavelet KD [40]	1.56G(11.92×)	13.61M(4.00×)	80.13 (−45.82)
	DMAD [24]	2.99G(6.2×)	2.13M(25.5×)	46.95 (−12.64)
	OMGD [31]	1.219G(15.3×)	3.404M(16.0×)	25.00 (+9.41)
	IPDD (Ours)	**1.219G(15.3×)**	**3.404M(16.0×)**	**24.63 (+9.68)**

4   Experimentation

4.1   Experimental Settings

Models and Datasets. After exploring existing methods [8, 19, 20, 23–25, 31, 33, 40], we further enhance the performance of unpaired image-to-image translation datasets, such as horse2zebra and summer2winter, by employing compressed CycleGAN [42]. Additionally, we apply the compressed Pix2Pix [18] method to the paired image-to-image translation dataset, namely edges2shoes.

Evaluation Metrics. Following previous works for GAN compression [8, 20, 23–25, 31, 33, 34, 40], we consider Fréchet Inception Distance (FID) to measure the similarity between real images and generated images. FID measures the distance between generated and real images based on their abstract feature representations. A lower FID indicates better image quality.

| Original images | Original CycleGAN | GAN-Compression | OMGD | CRD(Ours) |

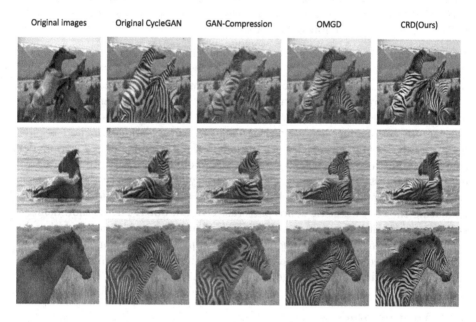

Fig. 4. Visual examples on horse2zebra dataset.

4.2 Result Comparison

CycleGAN. The quantitative experiments of compressed CycleGAN on horse-to-zebra and summer-to-winter are shown in Table 1. Our backbone of Cycle-GAN generator is based on a ResNet [13] style to follow the previous works [8, 11, 19, 23, 31]. From the table, we can see that though reducing network complexity, many methods [8, 20, 23, 33, 34, 40] cause unexpected performance drops on horse2zebra dataset compared with the original CycleGAN. On the contrary, among all methods [19, 24, 25, 31] boosting performance, our IPDD leads to the best FID of 49.62 than 61.53 of the original CycleGAN with 40.3× MACs and 82.5× parameters reduction. Under the same complexity reduction, the existing SoTA OMGD [31] has 51.92 FID score. Similar observations can be found on summer2winter where our IPDD still obtains the best performance increase of 5.51 FID score and results in the most MACs and parameters reduction.

Pix2Pix. In this study, we undertake a comprehensive analysis of the experimental outcomes presented in Table 2, focusing specifically on the compression of Pix2Pix using the edges2shoes dataset. Through this analysis, we make two key observations of significance. Firstly, our proposed IPDD method surpasses existing methodologies in terms of compression rate and performance enhancement. Notably, IPDD achieves a reduction of MAC operations by a factor of 15.3 and a decrease in parameters by a factor of 16.0 when compared to the original Pix2Pix model, while concurrently achieving a substantial FID gain of 9.68. These empirical findings validate the notable advantages offered by our IPDD framework for compressing GANs.

Visualization. Finally, we present the visual image quality results obtained through our compressed model. Figure 4 showcases a curated set of examples from the horse2zebra dataset, encompassing the original CycleGAN [42], the typical GAN Compression [23], the recently proposed state-of-the-art OMGD [31], and our novel IPDD method, which generates both original horse and zebra images. All zebra images are synthesized using the compression models outlined in Table 1. Our approach not only offers the advantage of reduced model complexity but also demonstrates visually compelling outcomes. In contrast to existing compression techniques, which tend to yield a distinctive golden-black style, our IPDD method effortlessly generates discernible white and black stripes, emulating the characteristic zebra patterns. Overall, our IPDD method demonstrates remarkable advantages in generating visually plausible and contextually accurate zebra images while reducing model complexity. The improved pixel distribution achieved by our approach adds an additional level of realism, enhancing the overall visual appeal of the generated results.

4.3 Ablation Study

In this section, we investigate the effectiveness of our IPDD method through a series of ablation experiments using the horse2zebra and CycleGAN architecture as a case study. The primary objective of these experiments is to analyze and evaluate different components of our approach. Figure 2 shows the probability of image distribution obtained after averaging 100 images generated by different models, and it is obvious that the generation similarity of IPDD is closer to the generation distribution of the original CycleGAN [42] than current state-of-the-art [31]. In Table 3, we examine the impact of image probability distribution distillation by employing various divergence measures to distill the output probability distributions of the student model. In Table 4, we introduce an online asynchronous transformation-weighted teacher discriminator method to enhance the training of the student model. It is worth noting that we establish a baseline method in which the student model is trained without employing any form of distillation. Through these ablation studies, we aim to analyze the individual contributions of different components in our IPDD method, assess their impact on the student model's performance, and ultimately demonstrate the effectiveness of our proposed approach.

Table 3. Ablations of different type of divergences.

Method	FID($\downarrow$f)
Baseline (w/o KD)	96.72
Kullback-Leibler	57.63
Jensen-Shannon	53.98
F	51.83
Bregman	49.92

Table 4. Ablations of our aync weighted teacher discriminator.

Method	FID($\downarrow$f)
Pre-trained (freezing)	69.48
Pre-trained (updating)	70.41
Online (w/o disc)	62.75
Online (updating)	58.02
Our online (weighted)	49.92

Divergence. Table 3 presents four different divergences employed in our study, pertaining to our approach. We conducted individual experiments using each divergence measure to assess their effectiveness. It can be derived from the table. Distilling the image distribution using any of the divergence measures yields performance improvements, with Bregman divergence exhibiting the most favorable results. This underscores the significance of distilling distributions.

Asynchronous Weighted Discriminator. In Sect. 3.3, we identified a pretrained teacher discriminator undermines the Nash equilibrium when trained alongside the generator from scratch. To mitigate this, we propose an asynchronous weight update training approach for online teacher discriminator. Experimental results in Table 4 indicate that the pretrained teacher discriminator consistently underperforms compared to the online teacher discriminator. Replacing our asynchronous weight update approach with a simple full update scheme, leads to a performance drop from 49.92 to 58.02. Additionally, following the scenario proposed by OMGD [31] and removing the discriminator during co-training results in a performance decrease to 62.75.

5 Conclusion

In summary, we propose the Image Probability Distribution Distillation (IPDD) method to enhance the performance of compressed GANs. IPDD introduces a novel approach that models the distributional information of the entire image, providing crucial knowledge for GANs. Through global distribution distillation and online asynchronous federated training, our method achieves optimal performance while reducing GAN complexity. Extensive experiments and ablation studies confirm the effectiveness of each component within IPDD.

References

1. Aguinaldo, A., Chiang, P.Y., Gain, A., Patil, A., Pearson, K., Feizi, S.: Compressing GANs using knowledge distillation. arXiv preprint arXiv:1902.00159 (2019)
2. Brock, A., Donahue, J., Simonyan, K.: Large scale GAN training for high fidelity natural image synthesis. arXiv preprint arXiv:1809.11096 (2018)

3. Chen, H., et al.: Distilling portable generative adversarial networks for image translation. In: Proceedings of the AAAI Conference on Artificial Intelligence (AAAI), pp. 3585–3592 (2020)

4. Chen, P., Liu, S., Zhao, H., Jia, J.: Distilling knowledge via knowledge review. In: Proceedings of the IEEE/CVF Conference on Computer Vision and Pattern Recognition (CVPR), pp. 5008–5017 (2021)

5. Chen, X., Zhang, Z., Sui, Y., Chen, T.: GANs can play lottery tickets too. arXiv preprint arXiv:2106.00134 (2021)

6. Chen, Y., Lai, Y.K., Liu, Y.J.: Cartoongan: generative adversarial networks for photo cartoonization. In: Proceedings of the IEEE/CVF Conference on Computer Vision and Pattern Recognition (CVPR), pp. 9465–9474 (2018)

7. Choi, Y., Choi, M., Kim, M., Ha, J.W., Kim, S., Choo, J.: Stargan: unified generative adversarial networks for multi-domain image-to-image translation. In: Proceedings of the IEEE/CVF Conference on Computer Vision and Pattern Recognition (CVPR), pp. 8789–8797 (2018)

8. Fu, Y., Chen, W., Wang, H., Li, H., Lin, Y., Wang, Z.: Autogan-distiller: searching to compress generative adversarial networks. arXiv preprint arXiv:2006.08198 (2020)

9. Gatys, L., Ecker, A.S., Bethge, M.: Texture synthesis using convolutional neural networks. In: Advances in Neural Information Processing Systems (NeurIPS) (2015)

10. Gatys, L.A., Ecker, A.S., Bethge, M.: Image style transfer using convolutional neural networks. In: Proceedings of the IEEE/CVF Conference on Computer Vision and Pattern Recognition (CVPR), pp. 2414–2423 (2016)

11. Gong, X., Shiyu, C., Jiang, Y., Wang, Z.: Autogan: neural architecture search for generative adversarial networks. In: Proceedings of the IEEE/CVF Conference on Computer Vision and Pattern Recognition (CVPR), pp. 3224–3234 (2019)

12. Goodfellow, I., et al.: Generative adversarial nets. In: Advances in Neural Information Processing Systems (NeurIPS) (2014)

13. He, K., Xiangyu, Z., Shaoqing, R., Sun, J.: Deep residual learning for image recognition. In: Proceedings of the IEEE/CVF Conference on Computer Vision and Pattern Recognition (CVPR), pp. 770–778 (2016)

14. Heo, B., Kim, J., Yun, S., Park, H., Kwak, N., Choi, J.Y.: A comprehensive overhaul of feature distillation. In: Proceedings of the IEEE/CVF International Conference on Computer Vision (ICCV), pp. 1921–1930 (2019)

15. Hinton, G., Vinyals, O., Dean, J., et al.: Distilling the knowledge in a neural network. arXiv preprint arXiv:1503.02531 (2015)

16. Ho, J., Ajay, J., Abbeel, P.: Denoising diffusion probabilistic models. In: Advances in Neural Information Processing Systems (NeurIPS) (2020)

17. Hou, L., Yuan, Z., Huang, L., Shen, H., Cheng, X., Wang, C.: Slimmable generative adversarial networks. In: Proceedings of the AAAI Conference on Artificial Intelligence (AAAI), pp. 7746–7753 (2021)

18. Isola, P., Zhu, J.Y., Zhou, T., Efros, A.A.: Image-to-image translation with conditional adversarial networks. In: Proceedings of the IEEE/CVF Conference on Computer Vision and Pattern Recognition (CVPR), pp. 1125–1134 (2017)

19. Jin, Q., et al.: Teachers do more than teach: Compressing image-to-image models. In: Proceedings of the IEEE/CVF Conference on Computer Vision and Pattern Recognition (CVPR), pp. 13600–13611 (2021)

20. Jung, C., Kwon, G., Ye, J.C.: Exploring patch-wise semantic relation for contrastive learning in image-to-image translation tasks. In: Proceedings of the IEEE/CVF Conference on Computer Vision and Pattern Recognition (CVPR), pp. 18260–18269 (2022)
21. Karras, T., Laine, S., Aila, T.: A style-based generator architecture for generative adversarial networks. In: Proceedings of the IEEE/CVF Conference on Computer Vision and Pattern Recognition (CVPR), pp. 4401–4410 (2019)
22. Karras, T., Laine, S., Aittala, M., Hellsten, J., Lehtinen, J., Aila, T.: Analyzing and improving the image quality of stylegan. In: Proceedings of the IEEE/CVF Conference on Computer Vision and Pattern Recognition (CVPR), pp. 8110–8119 (2020)
23. Li, M., Lin, J., Ding, Y., Liu, Z., Zhu, J.Y., Han, S.: GAN compression: efficient architectures for interactive conditional GANs. In: Proceedings of the IEEE/CVF Conference on Computer Vision and Pattern Recognition (CVPR), pp. 5284–5294 (2020)
24. Li, S., Lin, M., Wang, Y., Fei, C., Shao, L., Ji, R.: Learning efficient GANs for image translation via differentiable masks and co-attention distillation. IEEE Trans. Multimedia (TMM) (2022)
25. Li, S., Wu, J., Xiao, X., Chao, F., Mao, X., Ji, R.: Revisiting discriminator in GAN compression: a generator-discriminator cooperative compression scheme. In: Advances in Neural Information Processing Systems (NeurIPS), pp. 28560–28572 (2021)
26. Li, Z., Jiang, R., Aarabi, P.: Semantic relation preserving knowledge distillation for image-to-image translation. In: Vedaldi, A., Bischof, H., Brox, T., Frahm, J.-M. (eds.) ECCV 2020. LNCS, vol. 12371, pp. 648–663. Springer, Cham (2020). https://doi.org/10.1007/978-3-030-58574-7_39
27. Lin, J., Zhang, R., Ganz, F., Han, S., Zhu, J.Y.: Anycost GANs for interactive image synthesis and editing. In: Proceedings of the IEEE/CVF Conference on Computer Vision and Pattern Recognition (CVPR), pp. 14986–14996 (2021)
28. Nash, J.: Non-cooperative games. Ann. Math. 286–295 (1951)
29. Park, W., Kim, D., Lu, Y., Cho, M.: Relational knowledge distillation. In: Proceedings of the IEEE/CVF Conference on Computer Vision and Pattern Recognition (CVPR), pp. 3967–3976 (2019)
30. Radford, A., Metz, L., Chintala, S.: Unsupervised representation learning with deep convolutional generative adversarial networks. arXiv preprint arXiv:1511.06434 (2015)
31. Ren, Y., Wu, J., Xiao, X., Yang, J.: Online multi-granularity distillation for GAN compression. In: Proceedings of the IEEE/CVF International Conference on Computer Vision (ICCV), pp. 6793–6803 (2021)
32. Romero, A., Ballas, N., Kahou, S.E., Chassang, A., Gatta, C., Bengio, Y.: Fitnets: hints for thin deep nets. arXiv preprint arXiv:1412.6550 (2014)
33. Shu, H., et al.: Co-evolutionary compression for unpaired image translation. In: Proceedings of the IEEE/CVF International Conference on Computer Vision (ICCV), pp. 3235–3244 (2019)
34. Wang, H., Gui, S., Yang, H., Liu, J., Wang, Z.: GAN slimming: all-in-one GAN compression by a unified optimization framework. In: Vedaldi, A., Bischof, H., Brox, T., Frahm, J.-M. (eds.) ECCV 2020. LNCS, vol. 12349, pp. 54–73. Springer, Cham (2020). https://doi.org/10.1007/978-3-030-58548-8_4
35. Wang, P., et al.: QGAN: quantized generative adversarial networks. arXiv preprint arXiv:1901.08263 (2019)

36. Wang, X., Zhang, R., Sun, Y., Qi, J.: KDGAN: knowledge distillation with generative adversarial networks. In: Advances in Neural Information Processing Systems (NeurIPS) (2018)
37. Xu, W., Long, C., Wang, R., Wang, G.: DRB-GAN: a dynamic resblock generative adversarial network for artistic style transfer. In: Proceedings of the IEEE/CVF International Conference on Computer Vision (ICCV), pp. 6383–6392 (2021)
38. Yu, A., Kristen, G.: Fine-grained visual comparisons with local learning. In: Proceedings of the IEEE/CVF Conference on Computer Vision and Pattern Recognition (CVPR), pp. 192–199 (2014)
39. Zhang, H., Goodfellow, I., Metaxas, D., Odena, A.: Self-attention generative adversarial networks. In: International Conference on Machine Learning (ICML), pp. 7354–7363 (2019)
40. Zhang, L., Chen, X., Tu, X., Wan, P., Xu, N., Ma, K.: Wavelet knowledge distillation: towards efficient image-to-image translation. In: Proceedings of the IEEE/CVF Conference on Computer Vision and Pattern Recognition (CVPR), pp. 12464–12474 (2022)
41. Zhao, B., Cui, Q., Song, R., Qiu, Y., Liang, J.: Decoupled knowledge distillation. In: Proceedings of the IEEE/CVF Conference on Computer Vision and Pattern Recognition (CVPR), pp. 11953–11962 (2022)
42. Zhu, J.Y., Park, T., Isola, P., Efros, A.A.: Unpaired image-to-image translation using cycle-consistent adversarial networks. In: Proceedings of the IEEE International Conference on Computer Vision (ICCV), pp. 2223–2232 (2017)

HDTR-Net: A Real-Time High-Definition Teeth Restoration Network for Arbitrary Talking Face Generation Methods

Yongyuan Li[1,2], Xiuyuan Qin[3], Chao Liang[4], and Mingqiang Wei[1,2(✉)]

[1] Nanjing University of Aeronautics and Astronautics, Nanjing, China
mqwei@nuaa.edu.cn
[2] Shenzhen Research Institute, Nanjing University of Aeronautics and Astronautics, Shenzhen, China
[3] Soochow University, Suzhou, China
[4] Nanjing University of Science and Technology, Nanjing, China

Abstract. Talking Face Generation (TFG) aims to reconstruct facial movements to achieve high natural lip movements from audio and facial features that are under potential connections. Existing TFG methods have made significant advancements to produce natural and realistic images. However, most work rarely takes visual quality into consideration. It is challenging to ensure lip synchronization while avoiding visual quality degradation in cross-modal generation methods. To address this issue, we propose a universal High-Definition Teeth Restoration Network, dubbed HDTR-Net, for arbitrary TFG methods. HDTR-Net can enhance teeth regions at an extremely fast speed while maintaining synchronization, and temporal consistency. In particular, we propose a Fine-Grained Feature Fusion (FGFF) module to effectively capture fine texture feature information around teeth and surrounding regions, and use these features to fine-grain the feature map to enhance the clarity of teeth. Extensive experiments show that our method can be adapted to arbitrary TFG methods without suffering from lip synchronization and frame coherence. Another advantage of HDTR-Net is its real-time generation ability. Also under the condition of high-definition restoration of talking face video synthesis, its inference speed is 300% faster than the current state-of-the-art face restoration based on super-resolution. Our code and trained models are released at https://github.com/yylgoodlucky/HDTR.

Keywords: High-Definition Teeth Restoration Network · Talking Face Generation · Teeth Restoration · Visual Quality

1 Introduction

Talking Face Generation (TFG) plays an important role in the audio-visual field [26], as it enables the integration of visual and auditory information to enhance the understanding and perception of information for humans.

© The Author(s), under exclusive license to Springer Nature Singapore Pte Ltd. 2024
Q. Liu et al. (Eds.): PRCV 2023, LNCS 14435, pp. 89–103, 2024.
https://doi.org/10.1007/978-981-99-8552-4_8

For TFG, a clear and realistic mouth in the generated image could provide a richer audio-visual experience and thus help the user to understand the semantics better. Traditional TFG methods warp the source image with the help of prior knowledge (e.g., audio), and many of them result in inaccurate output. Thanks to the successful application of deep neural networks, especially Generative Adversarial Networks (GAN) [17] and Convolution Neural Networks (CNN), TFG has made significant progress. Some efforts [22,24,27,29,39,41,44] try to produce natural and realistic talking faces by extracting driving features from speech signals and then integrate them into face animation. Existing TFG methods focus on producing natural realistic and high synchronization mouth shapes. However, it is still challenging to enhance the teeth clarity while ensuring the natural synchronization of mouth shapes.

Analyzing from the data perspective, the low-resolution of images in existing TFG datasets limits the ability of cutting-edge models to generate high-resolution mouth shapes. Analyzing from the network perspective, Obamanet et al. [25] propose a Teeth Proxy to obtain the high-frequency components of the teeth from the candidate frames to improve the clarity of the upper and lower teeth. However, these methods require a rigid selection of the candidate frames and do not optimize the clarity of the surrounding areas of the teeth. Some other efforts [22,42] use additional reference frames to compensate mouth shapes and motion to guide the network for accurate modeling of head posture and mouth synchronization, which produce excellent lip synchronization, but the model still falls short in terms of clarity in predicting teeth and their surrounding regions. We argue that prior knowledge is insufficient to provide and restore fine-grained features about the teeth and their surrounding regions.

Most recent efforts [5,18,31,32] recover high-frequency details from blurred and degraded face images while maintaining the original face features, but it still has gaps in terms of frame coherence and speed. In face restoration, the process is applied to each frame individually, which can result in noticeable pixel discontinuities in the synthesized video. This occurs because face restoration primarily focuses on recovering high-frequency information within each frame, without considering the continuity between frames. As a result, there may be visible inconsistencies or discrepancies in the appearance of consecutive frames, leading to a lack of smoothness or coherence in the synthesized video.

In addition, processing each frame takes a lot of time even on low-resolution images, which limits the practical applicability of face restoration in real-time or time-sensitive scenarios.

We propose a universal real-time High-Definition Teeth Restoration Network (HDTR-Net). HDTR-Net can be applied to arbitrary Talking Face Generation methods for generating quality results (see Fig. 1). Two components are involved in the proposed method: a Fine-Grained Feature Fusion (FGFF) module and a Decoder module. We deliberately design the FGFF module to merge features responsible for extracting the image texture details. The branch below the FGFF module is utilized to leverage the reference image as guidance, enabling the model to effectively restore high-frequency details. In addition, it can be used as a prompt for pixel continuity in the inference stage. The Decoder module

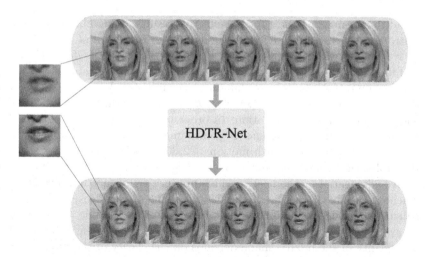

Fig. 1. method effectively enhances the clarity of teeth and surroundings generated by arbitrary talking face videos.

restores the high-frequency details from the extracted fine-grained feature. The main contributions of our work are three-fold:

- We propose a High-Definition Teeth Restoration Network, dubbed HDTR-Net, to enhance the clarity of teeth regions while maintaining texture details. HDTR-Net can be applied to arbitrary talking face generation.
- We propose a Fine-Grained Feature Fusion module, which has the ability to extract fine-grained features effectively.
- HDTR-Net exhibits exceptional speed in terms of repairing capability. For example, our inference speed is 300% faster than the super-resolution based image restoration methods.

2   Related Work

2.1   Talking Face Generation

Talking Face Generation aims to synthesize a sequence of talking face frames according to a sequence of driving audio or text, which is a multi-modal learning method for mapping acoustic features to real facial motions. In recent years, [15,25] have learned implicit mapping functions from audio to corresponding landmarks of the mouth to implement talking face synthesis, but their work is only specific to a particular speaker and audio. It is not sufficient to extend to arbitrary speakers and audio, and the synthesized video has low-quality clarity. [10] proposes a deblurring module, which uses the idea of early super-resolution [12] to transfer the facial feature map from the input image to the generated output using a skip connection to avoid producing a blurred image. Subsequently, [22,27,28,44] train on the large-scale public dataset (e.g., LRS2 [1], LRW [6]) in an end-to-end manner and produce effective results, being able to generalize to arbitrary speakers and audio, but the mouth region is still of low resolution in the synthesized video, resulting in a worse visual experience for the viewer. [4,8,43] firstly allow the speech to predict the face landmarks, and then recover the real face image. Predicting face landmarks couple speech features and mouth motion features obtains better lip synchronization, but the two-stage training method inherently loses information. In order to synthesize more natural and realistic head motions, the reference sequence images are fed into the network to reduce the head pose and mouth motions [22,30,41,42]. Respectively, [42] and [41] decoupled the visual representation space using contrastive learning [20] and generative adversarial methods [17], which produce significant improvements in lip synchronization. To obtain high-fidelity talking face videos, [38] proposes a repair network with an adaptive affine transformation module [37] to achieve clearly synthetic videos with multi-stage training, but the inference speed is slow and the fidelity of the mouth shape depends on the clarity of the input image.

2.2 Face Restoration

Based on the general face hallucination [2,5,9,34,35], most methods utilize a geometric prior and a reference prior together to improve performance. However, geometric priors are estimated from low-quality input images, and such geometric priors cannot provide detailed information about the image. Reference prior [7,18,19] relies on images of the same identity, and its ability to recover high-frequency details is reduced for particular image feature with rich appearance. In particular, [14] estimates face landmarks before restoring a face and estimating facial pose, while for quite minor facial accurate estimation is difficult. [45] proposes a unified framework for multi-resolution and dense correspondence field estimation of faces to recover texture details. Recently, deep learning-based models have advanced significantly in image processing tasks and are at present driving the state-of-the-art in face restoration. [16,23] have better reconstruction performance with a generative approach. [36] uses the discriminative generative network to ultra-resolve images by aligning miniature low-resolution face images. [40] uses convolution to extract features from blurred images to reconstruct high-definition face images, but their restored faces generate unfaithful outcomes. These methods are essentially based on restoration for single images, with the benefit of being able to recover a scaled-up, high-resolution image, however, the temporal coherence present in the video is not taken into account.

3   Method

We propose a novel real-time teeth restoration network called HDTR-Net that can be adaptive to arbitrary TFG methods. The structural details of HDTR-Net are shown in Fig. 2. We first review the structure of the model, which contains two parallel Fine-Grained Feature Fusion (FGFF) modules and a Decoding module. Channel Fusion (CF) and HourGlass are included in FGFF, which focus on extracting fine-grained features.

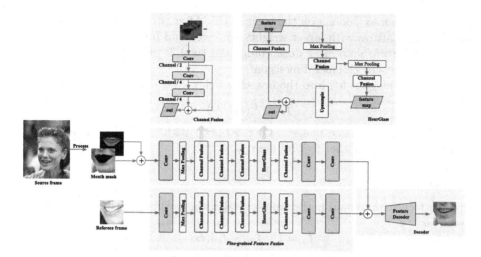

Fig. 2. Pipeline of our HDTR-Net. HDTR-Net consists of two parallel Fine-Grained Feature Fusion (FGFF) modules in the blue rectangles and a Decoder module in the green rectangle. Two important components are Channel Fusion (CF) in the orange rectangular and HourGlass in the grey rectangular included in FGFF. FGFF extracts fine-grained texture features from the processed source image and the reference image, then concatenates the feature map to the Decoder for the restoration of teeth regions. (Color figure online)

3.1 Fine-Grained Feature Fusion

The blue rectangle in Fig. 2 illustrates the structure of Fine-Grained Feature Fusion (FGFF). Given one source image $I_s \in R^{3 \times H \times W}$, after processing the source image for producing the mouth mask image $I_m \in R^{3 \times H \times W}$ and mouth contour image $I_c \in R^{3 \times H \times W}$, I_m and I_c are fed into the Fine-Grained Feature Fusion (FGFF) module for generating the feature map as:

$$f_m = \mathcal{F}((I_m \oplus I_c); \Theta_m) \tag{1}$$

where $\oplus$ is channel concatenation, after processing the source image, I_m and I_c are concatenated into FGFF, which FGFF module $\mathcal{F}(\cdot; \Theta_m)$ with a set of parameters Θ_m transforms I_m and I_c into another feature map f_m.

FGFF contains two important components which are the orange rectangular Channel Fusion (CF) and the grey rectangular HourGlass [21] in Fig. 2. CF takes the input of the feature map and passes through three convolutional layers, with the first convolutional layer output a feature map channel that is one-half of the original channel. The last two convolutional layers output a feature map channel that is one-fourth of the original channel. Finally, the output of each convolutional layer is concatenated at the channel dimension, so that the output of CF remains channel invariant but merges with the feature after multi-layer convolutional. Similarly, HourGlass accepts the input of the feature map, embedding the max pooling on the basis of CF. After two layers of max pooling and CF, the features with larger weights are saved and further merged. Finally, the output feature map after upsampling and the CF output feature map are concatenated at the channel dimension.

Reference image is fed into the branches below FGFF as:

$$f_r = \mathcal{F}(I_r; \Theta_r) \tag{2}$$

where I_r is the reference image, FGFF $\mathcal{F}(\cdot; \Theta_r)$ with a set of parameters Θ_r transforms I_r into another feature map f_r.

3.2 Decoder

Decoder focuses on repairing the mouth mask images utilizing two FGFF module results as:

$$I_o = \mathcal{D}((f_m \oplus f_r); \Theta_d) \tag{3}$$

where $\oplus$ is channel concatenation, decoder $\mathcal{D}(\cdot; \Theta_d)$ with a set of parameters Θ_d transforms f_m and f_r into the output I_o.

3.3 Loss Function

In the training stage, we use three kinds of loss functions to train HDTR-Net, containing reconstruction loss, perception loss [11], and GAN loss [17].

GAN Loss. Frame discriminator predicts the probability whether the generated frame is comparable to ground truth, resulting in the loss as:

$$\mathcal{L}_{GAN} = \mathcal{L}_D + \mathcal{L}_G \tag{4}$$

where

$$\mathcal{L}_D = \frac{1}{2} E \left(D\left(I_g\right) - 1 \right)^2 + \frac{1}{2} E \left(D\left(I_o\right) - 0 \right)^2 \tag{5}$$

where G represents HDTR-Net and D denotes the discriminator, I_g represents the ground truth image, and I_o represents the teeth restoration image.

Reconstruction Loss. To ensure the coherence of image color and mouth shape, we use L1 loss and L2 loss to reconstruct the mouth region as:

$$\mathcal{L}_{rec}(I_g, I_o) = \|I_g - I_o\|_1 + \|I_g - I_o\|_1^2 \tag{6}$$

Perception Loss. In order to make the generated image have a more natural appearance, we use perception loss to capture the high-level feature differences between the generated image and the ground truth. We calculate the perception loss in I_g and I_o by pre-training the VGG network [3], and the perception loss is formulated as:

$$\mathcal{L}_{\text{perc}}(I_g, I_o) = \|\phi(I_g) - \phi(I_o)\|_2^2 \tag{7}$$

where ϕ is a feature extraction network.

The overall loss is formulated as:

$$\mathcal{L} = \lambda_1 \mathcal{L}_{GAN} + \lambda_2 \mathcal{L}_{perc} + \lambda_3 \mathcal{L}_{rec} \tag{8}$$

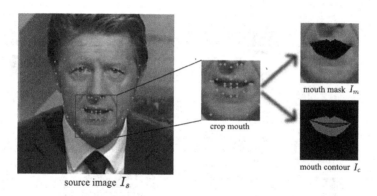

source image I_s

mouth mask I_m

mouth contour I_c

crop mouth

Fig. 3. Details of data processing. Firstly, face landmarks are extracted from source image I_s, then based on the four key points located at the left corner of the mouth, the right corner of the mouth, the tip of the nose, and the jaw, we can determine the boundaries of the mouth region. Using these boundary points, we crop the mouth region from the source image. After processing the cropped mouth region, mouth masks I_m and mouth contour I_c are obtained.

4 Experiment

In this section, we first detail datasets and metrics, comparison methods, and implementation details in our experiment. Then, we show the teeth restoration results of our method. Next, we carry out qualitative and quantitative comparisons with other state-of-the-art works. Finally, we conduct ablation studies.

4.1 Experimental Settings

Dataset and Metrics. We train the proposed method in the train set with high definition processed dataset LRS2, and evaluate with other competitive approaches in the test sets of two prevalent benchmark datasets: LRS2 [1] and LRW [6]. LRS2 contains more than 1000 speakers, with nearly 150,000 instances of words captured and 63,000 different words due to unlimited sentences at the time of capture. LRW selects the 500 most frequent words and clips the speaker's speech, resulting in over 1000 speakers and over 55,000,000 instances of speech.

During the testing phase of our method, there is no ground truth available for direct comparison. To assess the effectiveness of our method, we employ eight image sharpness evaluation metrics: Brenner, Laplacian, SMD, SMD2, Variance, Energy, Vollath, and Entropy. These metrics are chosen to measure the quality of the restored images, aiming to align with human subjective perception. In addition, we measure the time consumption to repair the images.

Comparison Methods. We compare our method with super-resolution based face restoration methods, including ESRGAN [33], GFPGAN [31], and Real-ESRGAN [32]. Real-ESRGAN is an enhanced version of the ESRGAN method.

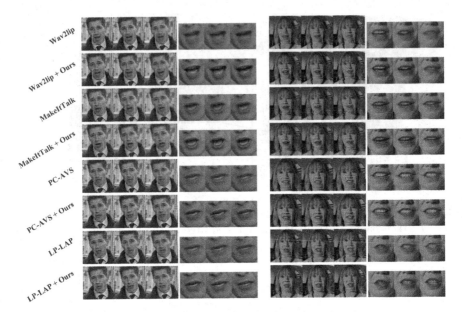

Fig. 4. Our method is adaptable to arbitrary Talking Face Generation methods, and restores the teeth region to obtain high-definition teeth.

These methods input images of different resolutions into the generator to output 1× and multiples of high-resolution enlarged images, but our method's input and output are all 96 × 96 resolution. To ensure a fair comparison, we maintain the same input and output resolution for all the comparison experiments, the same as our method.

Implementation Details. Data processing is shown in Fig. 3. Given source image $I_s \in R^{3 \times H \times W}$, we begin by extracting face landmarks and cropping the mouth region using four key points: the left corner and the right corner of the mouth, the tip of the nose, and the jaw. Then we obtain a rectangular region containing the mouth and surroundings. Using the extracted face landmarks, we construct a mouth mask $I_m \in R^{3 \times H \times W}$ and mouth contour $I_c \in R^{3 \times H \times W}$ using the key points of the lips. To ensure consistency, $I_m \in R^{3 \times H \times W}$ and $I_c \in R^{3 \times H \times W}$ are aligned to a resolution of 96 × 96 pixels, focusing on the mouth region. These aligned frames are then input into HDTR-Net.

During training, HDTR-Net takes three inputs: one masked mouth image $I_m \in R^{3 \times 96 \times 96}$, one mouth contour image $I_c \in R^{3 \times 96 \times 96}$, and one randomly selected image $I_f \in R^{3 \times 96 \times 96}$ from the training dataset as a reference frame. We use Adam optimizer [13] with a default setting to optimize HDTR-Net. The learning rate is set to 0.0001. The batch size is set to 12 on one A100 GPU.

Table 1. Quantitative comparisons with the state-of-the-art methods on image quality.

Metrics	Time ↓	Brenner ↑	Laplacian ↑	SMD ↑	SMD2 ↑	Variance ↑	Energy ↑	Vollath ↑	Entropy ↑
ESRGAN	0.0597	2003566	81.38	**163871**	193160	33401059	725269	31201016	4.56
GFPGAN	1.2332	2302120	140.18	72312	240881	6318184	983560	5108233	4.56
Real-ESRGAN	0.0622	2029371	75.15	66755	199590	6190525	772949	5069267	4.55
Ours	**0.0187**	**3121676**	**102.28**	85607	**331483**	8990130	**1264451**	**7667062**	**4.75**

Table 2. Quantitative results of ablation study.

Metrics	Brenner ↑	Laplacian ↑	SMD ↑	SMD2 ↑	Variance ↑	Energy ↑	Vollath ↑	Entropy ↑
w/o CF	2101920	68.18	68294	253783	6392508	923189	6912671	4.21
w/o ref FGFF	2803462	86.12	74029	248926	7183411	871762	6727348	4.26
w/o percep loss	2003462	70.12	78922	272609	7310412	971504	7027348	4.31
Ours	**3121676**	**102.28**	**85607**	**331483**	**8990130**	**1264451**	**7667062**	**4.75**

4.2   Experimental Results

Restoration Results. The results of teeth restoration are shown in Fig. 4. Our method displays the effectiveness of the restored teeth region in four Talking Face Generation methods, including Wav2lip [22], MakeItTalk [43], PC-AVS [42], and IP-LAP [39]. It can be seen that both sharpness and details are obtained after restoring teeth. Wav2lip and MakeItTalk generate blurred areas of the teeth region. After applying our method, it is clear that our method has significant improvements in terms of sharpness, color accuracy, and level of detail in the resulting images. Although PC-AVS and LP-LAP methods are able to generate satisfactory teeth restoration results, our method further improves the clarity, texture, and color of the tooth region. In addition, our method preserves both frame coherence and lip-synchronization, which is friendly handling of every frame in talking face generation videos.

To further verify the teeth restoration robustness for arbitrary talking face generation methods, we evaluate our method in side face talking face videos and show the restoration results in Fig. 5. We observe that our model produces both sharpness and color teeth details.

Quantitative Results. We compare our model with three recent state-of-the-art methods, including ESRGAN [33], GFPGAN [31], and Real-ESRGAN [32]. Table 1 shows the quantitative results of our method and its competitors. Although ESRGAN obtains the highest variance while our method has a second performance, ours surpassed the state-of-the-art ones among other clarity metrics. With the input and output resolutions kept the same, our method achieves an impressive inference time of only 0.018 s per frame. This makes it more than three times faster than the fastest super-resolution restoration method available.

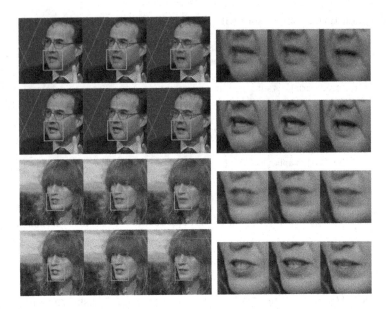

Fig. 5. Visualization results in side-faced speaker. The first and third rows are the original talking face-generated frames, and the second and fourth rows show the results of the dental restoration applied with our method. It can be clearly seen that our model produces both sharpness and color teeth details in a side-faced speaker, demonstrating the effectiveness and robustness of our method.

Qualitative Comparisons. To present the superiority of our method, we provide generated samples compared with ESRGAN, GFPGAN, and Real-ESRGAN. As shown in Fig. 6, face restoration based on super-resolution is not well repaired in frame coherence and tooth texture details, our method has better performance visually in teeth texture details.

4.3   Ablation Study

In order to validate the effect of each component of our method, we conduct ablation study experiments for our HDTR-Net. Specifically, we set 2 conditions: (1) w/o CF: we replace channel fusion with convolution. (2) w/o branches below FGFF module: we remove the branches below FGFF in HDTR-Net. (3) w/o percep loss: we remove perceptual loss in the training stage.

Table 2 illustrates the qualitative results of ablation experiments. In our condition of Ours w/o reference FGFF module, it is clear to see a significant reduction in results without the reference FGFF module. In our condition of Ours w/o perceptual loss, we also observe a decline in all of our evaluation metrics. We argue that adding CF and ref FGFF can obtain rich structural information and generate images with more high-frequency information. In addition, each component operates effectively in our method.

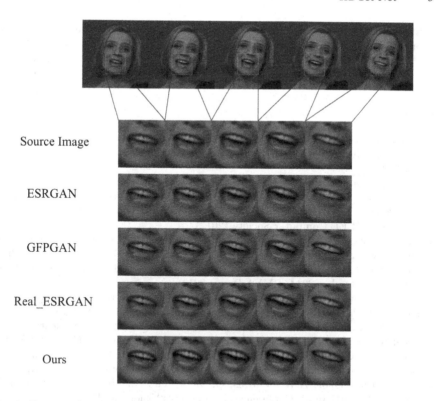

Source Image

ESRGAN

GFPGAN

Real_ESRGAN

Ours

Fig. 6. Compared to face restoration methods based on super-resolution, the visualization results indicate that our method outperforms significantly in teeth texture detail and color parts.

5 Conclusion

In this paper, we propose a real-time High-Definition Teeth Restoration Network (HDTR-Net), including two parallel Fine-Grained Feature Fusion modules and a Decoder module, to realize rich detail and texture in and around the teeth. The Fine-Grained Feature Fusion module is designed to merge low-level edge features and deep-level semantic features in feature dimensions to preserve the image refinement feature. In the inference stage, for two parallel FGFFs, the frame-by-frame guide input of the reference FGFF is capable of repairing teeth while ensuring frame coherence. The Decoder module is designed to merge the extracted feature maps from FGFF and restore the teeth region. With the combination of Fine-Grained Feature Fusion and Decoder, our method preserves more textural details and high-frequency information. Extensive qualitative and quantitative experiments have validated the performance of our method in arbitrary talking face generation methods without suffering lip synchronization and frame coherence. Compared to the super-resolution based face restoration methods, our inference speed is three times faster or even higher.

Acknowledgements. This work was supported by National Natural Science Foundation of China (No. 62172218), Shenzhen Science and Technology Program (No. JCYJ20220818103401003, No. JCYJ20220530172403007), Natural Science Foundation of Guangdong Province (No. 2022A1515010170).

References

1. Afouras, T., Chung, J.S., Senior, A.W., Vinyals, O., Zisserman, A.: Deep audio-visual speech recognition. IEEE Trans. Pattern Anal. Mach. Intell. **44**(12), 8717–8727 (2022)
2. Cao, Q., Lin, L., Shi, Y., Liang, X., Li, G.: Attention-aware face hallucination via deep reinforcement learning. In: 2017 IEEE Conference on Computer Vision and Pattern Recognition, CVPR 2017, Honolulu, HI, USA, 21–26 July 2017, pp. 1656–1664. IEEE Computer Society (2017)
3. Chatfield, K., Simonyan, K., Vedaldi, A., Zisserman, A.: Return of the devil in the details: delving deep into convolutional nets. In: Valstar, M.F., French, A.P., Pridmore, T.P. (eds.) British Machine Vision Conference, BMVC 2014, Nottingham, UK, 1–5 September 2014. BMVA Press (2014)
4. Chen, L., Maddox, R.K., Duan, Z., Xu, C.: Hierarchical cross-modal talking face generation with dynamic pixel-wise loss. CoRR abs/1905.03820 (2019)
5. Chen, Y., Tai, Y., Liu, X., Shen, C., Yang, J.: FSRNet: end-to-end learning face super-resolution with facial priors. In: 2018 IEEE Conference on Computer Vision and Pattern Recognition, CVPR 2018, Salt Lake City, UT, USA, 18–22 June 2018, pp. 2492–2501. Computer Vision Foundation/IEEE Computer Society (2018)
6. Chung, J.S., Senior, A.W., Vinyals, O., Zisserman, A.: Lip reading sentences in the wild. In: 2017 IEEE Conference on Computer Vision and Pattern Recognition, CVPR 2017, Honolulu, HI, USA, 21–26 July 2017, pp. 3444–3453. IEEE Computer Society (2017)
7. Dogan, B., Gu, S., Timofte, R.: Exemplar guided face image super-resolution without facial landmarks. In: IEEE Conference on Computer Vision and Pattern Recognition Workshops, CVPR Workshops 2019, Long Beach, CA, USA, 16–20 June 2019, pp. 1814–1823. Computer Vision Foundation/IEEE (2019)
8. Eskimez, S.E., Maddox, R.K., Xu, C., Duan, Z.: Generating talking face landmarks from speech. In: Deville, Y., Gannot, S., Mason, R., Plumbley, M.D., Ward, D. (eds.) LVA/ICA 2018. LNCS, vol. 10891, pp. 372–381. Springer, Cham (2018). https://doi.org/10.1007/978-3-319-93764-9_35
9. Huang, H., He, R., Sun, Z., Tan, T.: Wavelet-SRNet: a wavelet-based CNN for multi-scale face super resolution. In: IEEE International Conference on Computer Vision, ICCV 2017, Venice, Italy, 22–29 October 2017, pp. 1698–1706. IEEE Computer Society (2017)
10. Jamaludin, A., Chung, J.S., Zisserman, A.: You said that?: synthesising talking faces from audio. Int. J. Comput. Vis. **127**(11–12), 1767–1779 (2019)
11. Johnson, J., Alahi, A., Fei-Fei, L.: Perceptual losses for real-time style transfer and super-resolution. In: Leibe, B., Matas, J., Sebe, N., Welling, M. (eds.) ECCV 2016. LNCS, vol. 9906, pp. 694–711. Springer, Cham (2016). https://doi.org/10.1007/978-3-319-46475-6_43
12. Kim, J., Lee, J.K., Lee, K.M.: Accurate image super-resolution using very deep convolutional networks. In: 2016 IEEE Conference on Computer Vision and Pattern Recognition, CVPR 2016, Las Vegas, NV, USA, 27–30 June 2016, pp. 1646–1654. IEEE Computer Society (2016)

13. Kingma, D.P., Ba, J.: Adam: a method for stochastic optimization. In: Bengio, Y., LeCun, Y. (eds.) 3rd International Conference on Learning Representations, ICLR 2015, San Diego, CA, USA, 7–9 May 2015, Conference Track Proceedings (2015)

14. Kolouri, S., Rohde, G.K.: Transport-based single frame super resolution of very low resolution face images. In: IEEE Conference on Computer Vision and Pattern Recognition, CVPR 2015, Boston, MA, USA, 7–12 June 2015, pp. 4876–4884. IEEE Computer Society (2015)

15. Kumar, R., Sotelo, J., Kumar, K., de Brébisson, A., Bengio, Y.: ObamaNet: photo-realistic lip-sync from text. arXiv preprint arXiv:1801.01442 (2017)

16. Ledig, C., et al.: Photo-realistic single image super-resolution using a generative adversarial network. In: 2017 IEEE Conference on Computer Vision and Pattern Recognition, CVPR 2017, Honolulu, HI, USA, 21–26 July 2017, pp. 105–114. IEEE Computer Society (2017)

17. Lee, C., Cheon, Y., Hwang, W.: Least squares generative adversarial networks-based anomaly detection. IEEE Access **10**, 26920–26930 (2022)

18. Li, X., Li, W., Ren, D., Zhang, H., Wang, M., Zuo, W.: Enhanced blind face restoration with multi-exemplar images and adaptive spatial feature fusion. In: 2020 IEEE/CVF Conference on Computer Vision and Pattern Recognition, CVPR 2020, Seattle, WA, USA, 13–19 June 2020, pp. 2703–2712. Computer Vision Foundation/IEEE (2020)

19. Li, X., Liu, M., Ye, Y., Zuo, W., Lin, L., Yang, R.: Learning warped guidance for blind face restoration. In: Ferrari, V., Hebert, M., Sminchisescu, C., Weiss, Y. (eds.) ECCV 2018. LNCS, vol. 11217, pp. 278–296. Springer, Cham (2018). https://doi.org/10.1007/978-3-030-01261-8_17

20. Nagrani, A., Chung, J.S., Albanie, S., Zisserman, A.: Disentangled speech embeddings using cross-modal self-supervision. In: 2020 IEEE International Conference on Acoustics, Speech and Signal Processing, ICASSP 2020, Barcelona, Spain, 4–8 May 2020, pp. 6829–6833. IEEE (2020)

21. Newell, A., Yang, K., Deng, J.: Stacked hourglass networks for human pose estimation. In: Leibe, B., Matas, J., Sebe, N., Welling, M. (eds.) ECCV 2016. LNCS, vol. 9912, pp. 483–499. Springer, Cham (2016). https://doi.org/10.1007/978-3-319-46484-8_29

22. Prajwal, K.R., Mukhopadhyay, R., Namboodiri, V.P., Jawahar, C.V.: A lip sync expert is all you need for speech to lip generation in the wild. In: Chen, C.W., et al. (eds.) MM 2020: The 28th ACM International Conference on Multimedia, Virtual Event/Seattle, WA, USA, 12–16 October 2020, pp. 484–492. ACM (2020)

23. Sønderby, C.K., Caballero, J., Theis, L., Shi, W., Huszár, F.: Amortised MAP inference for image super-resolution. In: 5th International Conference on Learning Representations, ICLR 2017, Toulon, France, 24–26 April 2017, Conference Track Proceedings. OpenReview.net (2017)

24. Song, Y., Zhu, J., Li, D., Wang, A., Qi, H.: Talking face generation by conditional recurrent adversarial network. In: Kraus, S. (ed.) Proceedings of the Twenty-Eighth International Joint Conference on Artificial Intelligence, IJCAI 2019, Macao, China, 10–16 August 2019, pp. 919–925. ijcai.org (2019)

25. Suwajanakorn, S., Seitz, S.M., Kemelmacher-Shlizerman, I.: Synthesizing obama: learning lip sync from audio. ACM Trans. Graph. **36**(4), 95:1–95:13 (2017)

26. Toshpulatov, M., Lee, W., Lee, S.: Talking human face generation: a survey. Expert Syst. Appl. 119678 (2023)

27. Vougioukas, K., Petridis, S., Pantic, M.: End-to-end speech-driven realistic facial animation with temporal GANs. In: IEEE Conference on Computer Vision and Pattern Recognition Workshops, CVPR Workshops 2019, Long Beach, CA, USA, 16–20 June 2019, pp. 37–40. Computer Vision Foundation/IEEE (2019)

28. Vougioukas, K., Petridis, S., Pantic, M.: Realistic speech-driven facial animation with GANs. Int. J. Comput. Vis. **128**(5), 1398–1413 (2020)

29. Wang, G., Zhang, P., Xie, L., Huang, W., Zha, Y.: Attention-based lip audio-visual synthesis for talking face generation in the wild. CoRR abs/2203.03984 (2022)

30. Wang, J., Qian, X., Zhang, M., Tan, R.T., Li, H.: Seeing what you said: talking face generation guided by a lip reading expert. CoRR abs/2303.17480 (2023)

31. Wang, X., Li, Y., Zhang, H., Shan, Y.: Towards real-world blind face restoration with generative facial prior. In: IEEE Conference on Computer Vision and Pattern Recognition, CVPR 2021, virtual, 19–25 June 2021, pp. 9168–9178. Computer Vision Foundation/IEEE (2021)

32. Wang, X., Xie, L., Dong, C., Shan, Y.: Real-ESRGAN: training real-world blind super-resolution with pure synthetic data. In: IEEE/CVF International Conference on Computer Vision Workshops, ICCVW 2021, Montreal, BC, Canada, 11–17 October 2021, pp. 1905–1914. IEEE (2021)

33. Wang, X., et al.: ESRGAN: enhanced super-resolution generative adversarial networks. In: Leal-Taixé, L., Roth, S. (eds.) ECCV 2018. LNCS, vol. 11133, pp. 63–79. Springer, Cham (2019). https://doi.org/10.1007/978-3-030-11021-5_5

34. Xu, X., Sun, D., Pan, J., Zhang, Y., Pfister, H., Yang, M.: Learning to super-resolve blurry face and text images. In: IEEE International Conference on Computer Vision, ICCV 2017, Venice, Italy, 22–29 October 2017, pp. 251–260. IEEE Computer Society (2017)

35. Yu, X., Fernando, B., Hartley, R., Porikli, F.: Super-resolving very low-resolution face images with supplementary attributes. In: 2018 IEEE Conference on Computer Vision and Pattern Recognition, CVPR 2018, Salt Lake City, UT, USA, 18–22 June 2018, pp. 908–917. Computer Vision Foundation/IEEE Computer Society (2018)

36. Yu, X., Porikli, F.: Ultra-resolving face images by discriminative generative networks. In: Leibe, B., Matas, J., Sebe, N., Welling, M. (eds.) ECCV 2016. LNCS, vol. 9909, pp. 318–333. Springer, Cham (2016). https://doi.org/10.1007/978-3-319-46454-1_20

37. Zhang, Z., Ding, Y.: Adaptive affine transformation: a simple and effective operation for spatial misaligned image generation. In: Magalhães, J., Bimbo, A.D., Satoh, S., Sebe, N., Alameda-Pineda, X., Jin, Q., Oria, V., Toni, L. (eds.) MM 2022: The 30th ACM International Conference on Multimedia, Lisboa, Portugal, 10–14 October 2022, pp. 1167–1176. ACM (2022)

38. Zhang, Z., Hu, Z., Deng, W., Fan, C., Lv, T., Ding, Y.: DINet: deformation inpainting network for realistic face visually dubbing on high resolution video. CoRR abs/2303.03988 (2023)

39. Zhong, W., et al.: Identity-preserving talking face generation with landmark and appearance priors. CoRR abs/2305.08293 (2023)

40. Zhou, E., Fan, H., Cao, Z., Jiang, Y., Yin, Q.: Learning face hallucination in the wild. In: Bonet, B., Koenig, S. (eds.) Proceedings of the Twenty-Ninth AAAI Conference on Artificial Intelligence, 25–30 January 2015, Austin, Texas, USA, pp. 3871–3877. AAAI Press (2015)

41. Zhou, H., Liu, Y., Liu, Z., Luo, P., Wang, X.: Talking face generation by adversarially disentangled audio-visual representation. In: The Thirty-Third AAAI Conference on Artificial Intelligence, AAAI 2019, The Thirty-First Innovative Applications of Artificial Intelligence Conference, IAAI 2019, The Ninth AAAI Symposium on Educational Advances in Artificial Intelligence, EAAI 2019, Honolulu, Hawaii, USA, 27 January–1 February 2019, pp. 9299–9306. AAAI Press (2019)
42. Zhou, H., Sun, Y., Wu, W., Loy, C.C., Wang, X., Liu, Z.: Pose-controllable talking face generation by implicitly modularized audio-visual representation. In: IEEE Conference on Computer Vision and Pattern Recognition, CVPR 2021, virtual, 19–25 June 2021, pp. 4176–4186. Computer Vision Foundation/IEEE (2021)
43. Zhou, Y., Li, D., Han, X., Kalogerakis, E., Shechtman, E., Echevarria, J.: Makeittalk: speaker-aware talking head animation. CoRR abs/2004.12992 (2020)
44. Zhu, H., Huang, H., Li, Y., Zheng, A., He, R.: Arbitrary talking face generation via attentional audio-visual coherence learning. In: Bessiere, C. (ed.) Proceedings of the Twenty-Ninth International Joint Conference on Artificial Intelligence, IJCAI 2020, pp. 2362–2368. ijcai.org (2020)
45. Zhu, S., Liu, S., Loy, C.C., Tang, X.: Deep cascaded bi-network for face hallucination. In: Leibe, B., Matas, J., Sebe, N., Welling, M. (eds.) ECCV 2016. LNCS, vol. 9909, pp. 614–630. Springer, Cham (2016). https://doi.org/10.1007/978-3-319-46454-1_37

Multi-stream-Based Low-Latency Viewport Switching Scheme for Panoramic Videos

Yong Wang, Hengyu Man, Xingtao Wang[✉], and Xiaopeng Fan

Harbin Institute of Technology, Harbin, China
{22S103197,20B903061}@stu.hit.edu.cn, {xtwang,fxp}@hit.edu.cn

Abstract. Panoramic videos have emerged as a widely-used media format as they provide immersive and interactive experience, but their large resolution and high frame rate pose great challenges for video compression efficiency. The challenges are relieved by tile-based viewport adaptive streaming, in which the viewport switching latency is critical to users' quality of experience. Existing methods for reducing the switching latency are limited to single high-quality stream. In this paper, we propose a smooth viewport switching strategy, where a panoramic video is encoded into a low-quality stream and multiple high-quality streams. The low-quality stream provides basic quality for the entire video, while the high-quality streams delivery real-time high-quality rendering for the current viewport. The multiple high-quality streams have different starting times during encoding, resulting in interleaved keyframes. These interleaved keyframes serve as random access points when the viewport switches. Smooth switching is achieved through selecting the stream corresponding to the keyframe closest to the switching point. The experimental results show that our method is significantly superior to existing methods.

Keywords: panoramic video · smooth viewport switching · low latency

1 Introduction

With the development of multimedia technologies and virtual reality display devices, panoramic videos have emerged as a popular and widely-used media format [10]. Compared to traditional flat videos, panoramic videos provide users with a wide Field of View (FOV) and the ability to change viewing direction, resulting in a more immersive and interactive experience that surpasses the realism and engagement of flat videos [8,13]. However, to ensure the immersive viewing experience, high-quality panoramic videos require both very large resolutions (up to 8K) and very high frame rates (up to 60 fps) [2]. Therefore, the demand of network bandwidth for high-quality panoramic video is very large, which poses great challenges for current network [7,12,15].

This work was supported in part by the National Key R&D Program of China (2021YFF0900500).

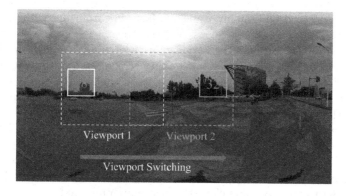

Fig. 1. The example of a viewport switch.

In order to alleviate this problem and improve users' quality of experience (QoE), one of the most popular approaches is tile-based viewport adaptive streaming [8,10,13]. Viewport is the region watched by the user. In tile-based viewport adaptive streaming, a panoramic video is divided into small parts called tiles in the spatial domain, and is encoded into high and low-quality streams. In the visual scene within the viewport, high-quality stream is delivered and decoded, while low-quality stream is applied in the visual scene outside the viewport. Thus, the bandwidth is reduced without compromising the user's QoE. However, when the viewport switches, the high-quality video stream file corresponding to the new viewport needs to be promptly requested, decoded, and rendered. Figure 1 shows an example of viewport switching, where the user's viewport consists of multiple tiles. Tile 28 and tile 32 represent parts of the old viewport and the new viewport, respectively. The time from the user triggering the switching operation to presenting the high-quality video content on the screen is called the switching latency. During the latency period, the high-quality video content in the new viewport is unavailable, which severely influences user's QoE [4,15]. Therefore, reducing the viewport switching latency can make the switching process smoother, real-time, and seam-less, which is critical to users' QoE.

In tile-based viewport adaptive streaming, the viewport switching latency is reduced based on frequent random access points (RAPs) [5]. Concretely, if each segment has only one random access point, when the viewport switches, the client cannot immediately display the corresponding segment in the new viewport [10], but keeps playing low-quality content until the current segment ends. One common approach for this issue is to use smaller Group of Pictures (GOP) during encoding, which increases the number of intra-coded frames (I-frames) that can serve as RAPs. However, using smaller GOP will reduce encoding efficiency and increase the required transmission bandwidth. Deploying multiple decoders for high-quality tiles allows the use of an additional decoder to decode the video content of the new viewport region. However, if the density of RAPs remains low, there will still be a problem of high switching latency, as the client may need to decode many frames before synchronizing with the current playback time.

In this paper, we propose a fast viewport switching strategy, in which the panoramic video is encoded into a low-quality stream and multiple high-quality streams. The low-quality stream provides basic quality for the entire video, while the high-quality streams are responsible for delivering high-quality rendering for the current viewport. The panoramic video is spatially divided into tiles and temporally segmented into fixed time periods (segments). The multiple high-quality streams have different starting times during encoding, resulting in interleaved keyframes. These interleaved keyframes serve as RAPs when the viewport switches. Fast switching can be achieved through selecting the stream corresponding to the keyframe closest to the switching point.

2   Related Works

This section first reviews related works on reducing the switching latency, and then introduces the Omnidirectional Media Application Format (OMAF) [6] as well as Dynamic Adaptive Streaming over HTTP (DASH) [11] standards, which are closely related to the proposed scheme.

2.1   Tile-Based Viewport Adaptive Streaming

There have been numerous methods proposed for tile-based viewport adaptive streaming [3,5,10,14]. In a panoramic video system that adopts the viewport area quality priority transmission scheme, low-quality video content is exposed in the user's viewport when the user switches the viewport, resulting in a sharp decline in the user's video experience quality. The reason is that the predictive-coded frame (P-frame) cannot be encoded independently but relies on the I-frame in the same group of picture (GOP). In order to solve this problem, Song *et al.* [10] propose a low-latency panoramic video transmission strategy called FFS-360DASH. They encode an additional I-frame backup for each P-frame in the GOP as a RAP of the code stream during encoding. In theory, this method can reduce the switching latency to the duration of one frame. However, the data size of I-frames is much larger than that of P-frames, so using I-frames will increase transmission latency. Overcoming this problem, Yang *et al.* [14] develop a new low-latency panoramic video transmission scheme, which additionally encodes a random access inter-prediction frame (RAPF) for each P-frame, and uses RAPF as the user random entry point when switching viewports. Compared with the I-frame, the coding efficiency of the RAPF frame is higher, and the transmission latency caused by it is smaller. Ghaznavi *et al.* [5] propose a panoramic video streaming scheme based on Shared Coded Pictures (SCP). SCP is used as a switching point instead of I-frames to enable viewport switching between different quality versions of the content. Duanmu *et al.* [3] propose a two-tier dynamic 360-degree video streaming framework, which encodes a 360-degree video into two tiers and adaptively streams the two tiers of the video to cope with the variations in the network bandwidth and user view direction changes.

These methods for reducing the switching latency have yielded encouraging results, but are limited to single high-quality stream. In this paper, we propose a smooth viewport switching strategy, where the panoramic video is encoded into a low-quality stream and multiple high-quality streams.

2.2   MPEG-DASH and OMAF

This subsection reviews MPEG-DASH and OMAF briefly, as we focus on improvements to the DASH-based implementation of OMAF system.

MPEG-DASH (or DASH) [11] is an adaptive bitrate streaming media technology, which can transmit streaming media through the HTTP web server, and is known as fast startup and low latency. DASH divides the video content into a sequence of small HTTP-based file segments, and each segment can be encoded at a different bitrate to meet the network requirements of different clients. For example, the DASH client can automatically select the corresponding most matching bitrate segment file to download and play back according to the current network conditions without causing pauses or rebuffering. At present, DASH is widely used for video streaming on the Internet.

OMAF is a standard format for panoramic video and virtual reality (VR) content. It is a set of specifications defined by the Moving Picture Experts Group (MPEG) with the aim of achieving interoperability of panoramic video and VR content across various devices and platforms. It defines the representation of panoramic video, viewport switching, adaptive streaming, and multi-channel audio functionalities to achieve consistent playback and interactive experiences across different devices and application scenarios.

2.3   MCTS Coding Scheme

Based on the idea of blocking and viewport switching, a technology called motion-constrained tile set (MCTS) coding is used in the OMAF codec scheme. It implements tile-based independent codecs by adding some restrictions. Specifically, when performing intra prediction of HEVC tile, it is forbidden to perform intra prediction across the tile boundary. At the same time, when performing inter prediction, it is also forbidden to refer to tile at other positions in the encoded frame. Through the restrictions set in MCTS encoding, the coupling and dependency of encoding and decoding between different tiles can be eliminated. In other words, independent encoding and decoding can be performed on video areas of all tile partitions in parallel.

After MCTS encoding, we can separately extract the data stream of one or more tiles for decoding. Therefore, after obtaining the user's viewport information, you can select a combination of tiles related to the field of view, and different combinations can cover different viewports. In this way, while enhancing the parallel processing capability, the coupling between blocks is also eliminated, and decoding distortion is reduced.

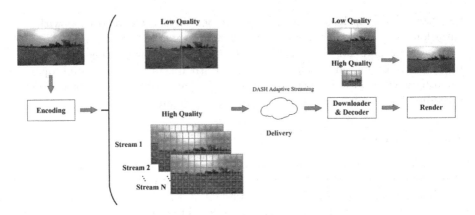

Fig. 2. The framework of the multi-stream panoramic video transmission system.

3 Methodology

In this paper, we propose a multi-stream-based low-latency viewport switching scheme, where the panoramic video is encoded into a low-quality stream and multiple high-quality streams. The multiple high-quality streams have different starting times during encoding, resulting in interleaved keyframes. These interleaved keyframes serve as RAPs when the viewport switches. Smooth switching is achieved by selecting the stream corresponding to the keyframe closest to the switching point. The framework of the proposed method mainly contains a server and a client as shown in Fig. 2. The server conducts the tile-based encoding by the SVT-HEVC encoding library, supporting MCTS. The encoded bit stream includes projection, rotation, RWPK, and SEI information. Then, the bit stream is packaged by OMAF-compliant to produce OMAF-compliant DASH content.

After the encoding process, the panoramic video is divided into tiles that do not overlap each other in the spatial domain. Each tile is divided into video segments of fixed duration in the time domain. The first frame of each segment is guaranteed to be the key frame by setting the encoding parameters. The client mainly accesses and parses the multimedia content from the server and reconstructs the video stream according to the requirements, and then plays it. Details of each process will be described in this section.

3.1 Tile-Based Panoramic Video Encoding

At the server, the panoramic video is encoded into multiple high-quality streams and one low-quality stream. The client requests the low-quality video stream and the high-quality video stream of the current viewport area to perform rendering. The high-quality stream ensures that high-quality video content is presented within the user's viewport. Low-quality video stream serves the purpose of preventing a drastic decrease in the user's video experience quality when switching viewports to a previously invisible area by avoiding the absence of any video

Fig. 3. The illustration of tile division. Left: Low quality stream tile division. Right: High quality stream tile division.

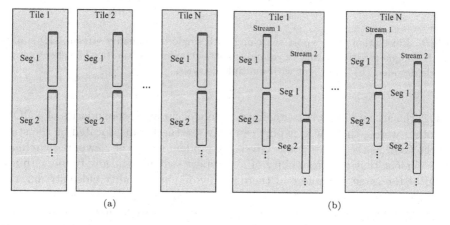

Fig. 4. Segments for each tile. In each segment, the red frames represent I-frames, while the remaining frames are P-frames. (a) Segment partition for the single high-quality stream. (b) Segment partition for dual high-quality streams. (Color figure online)

content in the new viewport. Both high-quality and low-quality video streams can adopt spatial partitioning (using tiles) and temporal segmentation strategies during encoding.

Figure 3 shows an example of spatial tile partitioning for a 4K panoramic video. When encoding the low-quality video stream, the original panoramic video is first downsampled, and then encoded using a lower bitrate. When encoding the high-quality video stream, it is necessary to maintain the original resolution and use a higher bitrate to ensure user Quality of Experience (QoE).

3.2   Multiple High Quality Streams

Figure 4(a) shows the division of each tile in the video into fixed time segments. Each segment consists of a group of pictures (GOP), and the first frame of each segment is an I-frame. When the viewport switches, the client will request high-quality video streams from the server that correspond to the tiles covering the new viewport. However, due to the dependency of P-frames on I-frames in a

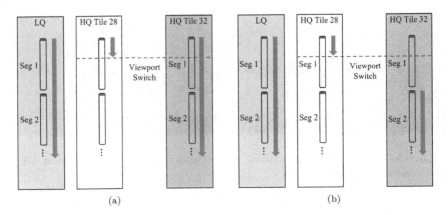

Fig. 5. Illustration of viewport switching with single high-quality stream. HQ stands for High Quality stream, and LQ stands for Low Quality stream. (a) The client with a single decoder. (b) The client with multiple decoders.

GOP, the actual viewport switching can only occur at the boundaries of a GOP. In other words, the new viewport can only use the I-frame located at the start of the GOP as the access point. Given the random nature of viewport switching, it is evident that the probability of a viewport switching at any frame within a GOP is the same. Therefore, if there is only one high-quality video stream, the average latency of viewport switching is half of decoding the content of a GOP.

Overcoming this, we encode multiple streams of high-quality video for the video content. These video streams have the same encoding parameters and quality, but their keyframes are interleaved with each other. This arrangement of interleaved keyframes forms RAPs within the streams. When a user switches the viewport, the system selects the video stream associated with the keyframe closest to the switching point for presentation. As shown in Fig. 4(b), we encode two high-quality video streams where the keyframes of these two streams have a time difference of half a GOP.

In the scenario where only one high-quality stream is encoded, there are two potential situations that need to be considered. As shown in Fig. 5(a), if only one decoder is available for decoding the high-quality stream, the client will not be able to immediately play the content of the new viewport area while the viewport changes. This is because the decoding progress between the new and old viewports within the current segment is inconsistent. Figure 5(b) illustrates the viewport switching in the situation where multiple decoders are allocated for the high-quality video stream. In detail, the additional decoders are responsible for decoding the video content of the new viewport. Therefore, the video content of the new viewport can be decoded in the current segment without waiting for the next segment.

Figure 6 shows the illustration of viewport switching with dual high-quality streams denoted as Stream1 and Stream2. These two streams have similar quality, but their keyframes are staggered by half of a segment length. When the client

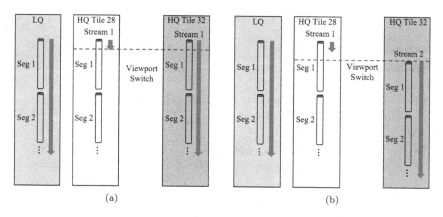

Fig. 6. Illustration of viewport switching with dual high-quality stream. HQ stands for High Quality stream, and LQ stands for Low Quality stream. The blue arrows indicate the frames being decoded. (a) The stream does not change after switching viewport. (b) The stream changes after switching viewport. (Color figure online)

detects a viewport switching, the video stream that is closest to the switching point is adaptively selected based on its relative position within a GOP. Figures 6(a) and (b) show the cases of switching to Stream1 and Stream2 respectively after viewport switching occurs at different switching points. It is evident that when the switching point is located in the middle of a segment, Stream2 has an I-frame closer to the switching point. By switching to Stream2, the frames to be decoded can be reduced, thereby reducing the switching delay.

4 Experimental Results and Discussion

In this section, we provide the experimental results and discussion of the proposed scheme. We perform experiments according to the description in Sect. 3 and evaluate the performance of panoramic video switching with viewport switching latency.

4.1 Experiment Setup

The testing system consists of a server and a client component. The server and client machine configurations are listed in Table 1. The server is responsible for encoding the video and packaging it according to DASH, and it uses Nginx [9] to serve the video files to the client. The client utilizes the FFmpeg tool to decode the video files received from the server, and uses the OpenGL library for video rendering and playback.

To conduct the experiments, several test sequences [1] from Joint Video Exploration Team (JVET) for 360-video are selected. All video sequences have a resolution of 8K (8192 × 4096) and a frame rate of 30fps, and a duration of 10 s. In addition, these videos are mapped using ERP projection and stored in YUV

Table 1. Machine configurations.

Component	Server	Client
Operating System	CentOS Linux release 7.9.2009 (Core)	Ubuntu 18.04.6 LTS
CPU	Intel(R) Xeon(R) Platinum 8180 CPU @ 2.50GHz	Intel(R) Xeon(R) CPU E5-2678 v3 @ 2.50GHz
RAM	128 GB	32 GB
GPU	ASPEED Technology, Inc. ASPEED Graphics Family (rev 30)	ASPEED Technology, Inc. ASPEED Graphics Family (rev 30)

4:2:0 format. These videos will be processed into high-quality and low-quality segments. For encoding the high-quality video segments, we downscale the 8K video sequences to 7680 × 3840 resolution and divided them into 12 × 6 tiles. For encoding the low-quality video segments, the video sequences are downscaled to a smaller resolution (2560 × 1280) and divided into fewer tiles (2 × 2).

As shown in Fig. 3, we divided the panoramic video into 12 × 6 tiles (360° × 180°), and the user's viewport is 4 × 4 tiles (120° × 120°). For a 8K video (7680 × 3840), the size of the user's viewport is 2560 × 2560. To test the switching performance of the system, we shift the viewport to the left by the distance of one tile (30°). Additionally, since switching at different positions within a segment can result in different switching latency, we chose the segment corresponding to the third second (frames 60–89) and performed 10 switches at each frame to calculate the average switching latency and enhance the reliability of the experimental results.

To eliminate the impact of client device performance, we specify single-threaded decoding when using FFmpeg. Meanwhile, since the focus of this paper is only on the decoding time, we disable the rendering functionality during the experiments. We use viewport switching latency to evaluate the performance of the proposed method. Specifically, we record the time and number of frames spent transitioning from low-quality to high-quality after the viewport is switched. The shorter the transition time, the better the performance of the method. In the experiments, we compare the performance of single-stream, dual-stream, triple-stream, and six-stream methods to validate the effectiveness of our approach.

4.2   Analysis of the Results

The main objective of this paper is to reduce the latency of user viewport switching. Through experiments, our proposed method achieves this goal. Additionally, the multiple high-quality video streams are encoded with the same encoding parameters, so there is no variation in video quality. Furthermore, by adaptively switching to the nearest video stream to the switching point, some video frames

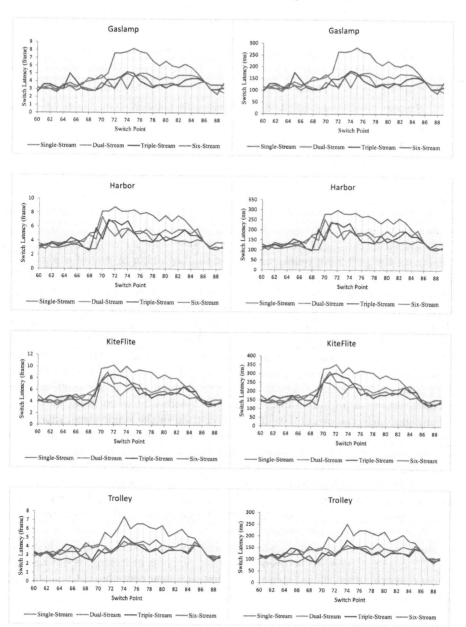

Fig. 7. The results of four videos. Left: Switching latency represented in frames. Right: Switching latency represented in time.

Table 2. Average switching latency.

Video	Single-Stream		Dual-Stream		Triple-Stream		Six-Stream	
	Frame	Time	Frame	Time	Frame	Time	Frame	Time
Gaslamp	5.11	178.09	3.99	141.07	3.70	131.53	3.45	123.61
Harbor	5.82	200.81	4.58	159.87	4.37	153.49	4.02	141.72
KiteFlite	6.59	226.77	5.31	185.53	5.39	187.53	5.23	182.72
Trolley	4.67	163.18	3.83	136.07	3.50	125.09	3.37	120.71

before the switching point do not need to be transmitted or decoded, thereby reducing the overall video bitrate. However, encoding multiple high-quality video streams requires more storage space.

As shown in the Fig. 7, there is typically lower latency when there is an I-frame near the switching point, such as at the beginning or end of a segment. Conversely, higher latency is observed when there is no I-frame nearby. This phenomenon occurs because the client needs to obtain the frame corresponding to the current playback position to restore high quality, and P-frames rely on I-frames for decoding. When the switching point is located in the middle of a segment, more frames need to be decoded to reach the playback position, resulting in higher latency. The multi-stream approach indirectly increases the density of I-frames, ensuring that regardless of the switching point's location, there is always an I-frame closer to the switching point than in the single-stream method, serving as a random insertion point. Specifically, when using the multi-stream approach, the overall latency is lower than in the single-stream method, and as the density of I-frames increases, the latency decreases. For example, the triple-stream method performs better than the dual-stream method.

Table 2 presents the average switching latency at any point within a segment. Compared to the single-stream method, the dual-stream method already reduces the time delay by approximately 20%, corresponding to around 30 ms, which is equivalent to one frame duration. Furthermore, the multi-stream methods exhibit even lower average latency.

Based on the experimental results, our method successfully reduces the number of frames that need to be decoded during viewport switching, resulting in a decrease in overall switching latency. This improvement contributes to enhancing the user experience.

5   Conclusion

In this paper, we propose a tile-based multi-stream smooth switching solution for panoramic video services. To quickly respond to user's viewport switching requests, we encode multiple streams of high-quality video with interleaved keyframes at the server, where the interleaved keyframes serve as random access points for switching. According to the experiments conducted, we have effectively reduced the viewport switching latency when the switching point is located in the middle of a segment. It is found that the performance of the method improves as the density of random access points increases. As future work, we will optimize the system implementation and explore better rendering methods to further reduce the system's switching latency.

References

1. Boyce, J., Alshina, E., Abbas, A., Ye, Y.: JVET-G1030: JVET common test conditions and evaluation procedures for 360° video (2018)
2. Corbillon, X., Devlic, A., Simon, G., Chakareski, J.: Viewport-adaptive navigable 360-degree video delivery. In: IEEE International Conference on Communications, vol. abs/1609.08042 (2017)
3. Duanmu, F., Kurdoglu, E., Liu, Y., Wang, Y.: View direction and bandwidth adaptive 360 degree video streaming using a two-tier system. In: 2017 IEEE International Symposium on Circuits and Systems (ISCAS), pp. 1–4. IEEE (2017)
4. de la Fuente, Y.S., Bhullar, G.S., Skupin, R., Hellge, C., Schierl, T.: Delay impact on MPEG OMAF's tile-based viewport-dependent 360 video streaming. IEEE J. Emerg. Sel. Top. Circuits Syst. 9(1), 18–28 (2019)
5. Ghaznavi-Youvalari, R., Zare, A., Aminlou, A., Hannuksela, M.M., Gabbouj, M.: Shared coded picture technique for tile-based viewport-adaptive streaming of omnidirectional video. IEEE Trans. Circuits Syst. Video Technol. 29(10), 3106–3120 (2018)
6. Hannuksela, M.M., Wang, Y.K.: An overview of omnidirectional media format (OMAF). Proc. IEEE 109(9), 1590–1606 (2021)
7. Koch, C., Rak, A.T., Zink, M., Steinmetz, R., Rizk, A.: Transitions of viewport quality adaptation mechanisms in 360 degree video streaming. In: ACM Workshop on Network and Operating Systems Support for Digital Audio and Video, pp. 14–19 (2019)
8. Nguyen, D.V., Tran, H.T., Pham, A.T., Thang, T.C.: An optimal tile-based approach for viewport-adaptive 360-degree video streaming. IEEE J. Emerg. Sel. Top. Circuits Syst. 9(1), 29–42 (2019)
9. Reese, W.: Nginx: the high-performance web server and reverse proxy. Linux J. 2008(173), 2 (2008)
10. Song, J., Yang, F., Zhang, W., Zou, W., Fan, Y., Di, P.: A fast FoV-switching DASH system based on tiling mechanism for practical omnidirectional video services. IEEE Trans. Multimedia 22(9), 2366–2381 (2019)
11. Stockhammer, T.: Dynamic adaptive streaming over HTTP-standards and design principles. In: ACM Conference on Multimedia Systems, pp. 133–144 (2011)
12. Wang, S., Tan, X., Li, S., Xu, X., Yang, J., Zheng, Q.: A QoE-based 360 video adaptive bitrate delivery and caching scheme for C-RAN. In: International Conference on Mobility, Sensing and Networking, pp. 49–56. IEEE (2020)
13. Xie, L., Xu, Z., Ban, Y., Zhang, X., Guo, Z.: 360ProbDASH: improving QoE of 360 video streaming using tile-based http adaptive streaming. In: ACM International Conference on Multimedia, pp. 315–323 (2017)

14. Yang, M., Liang, H., Yang, F.: Real-time adaptive switching mechanism towards viewport-adaptive omnidirectional video streaming. In: IEEE International Conference on Multimedia & Expo Workshops, pp. 1–6. IEEE (2021)

15. Yaqoob, A., Bi, T., Muntean, G.M.: A survey on adaptive 360 video streaming: solutions, challenges and opportunities. IEEE Commun. Surv. Tutor. **22**(4), 2801–2838 (2020)

Large Kernel Convolutional Attention Based U-Net Network for Inpainting Oracle Bone Inscription

Hong Yang[1], Xiang Chang[2], Zhihua Guo[1], Fei Chao[1,2(✉)], Changjing Shang[2], and Qiang Shen[2]

[1] School of Informatics, Xiamen University, Fujian 361005, People's Republic of China
yanghong@stu.xmu.edu.cn, {guozhihua,fchao}@xmu.edu.cn
[2] Department of Computer Science, Aberystwyth University, Wales SY23 3DB, UK
{xic9,CnsCns,qqsCns}@aber.ac.uk

Abstract. Oracle Bone Inscription (OBI) image inpainting is important for inheriting the long history of Chinese culture. The complexity of OBI characters leads to many unique writing characteristics, which increase the difficulty of restoring irregular structures and multiple types of strokes. In this work, we formulate a large kernel convolutional attention based U-Net framework to restore OBI images. The framework consists of two modified U-Nets arranged in a series, which perform an edge inpainting function and an overall image inpainting function, respectively. The implementations of the two U-Nets are identical and each U-Net is composed of an encoder that downsamples input images, followed by eight large kernel convolutional attention blocks, and a decoder that upsamples the image back to its original size. In addition, an adversarial learning algorithm with local and global discriminative networks is used to train the proposed framework to obtain OBI inpainting results. Compared with state-of-the-art image inpainting methods, our experimental results show that the proposed method can achieve the best inpainting results in OBI images; in particular, the method is also suitable to handle the large-area mask tasks.

Keywords: Oracle bone inscription · image inpainting · irregular structures · large kernel convolution

1 Introduction

Oracle bone inscription (OBI) involves many of the oldest characters in the world, which are engraved on animal bones or tortoise shells. OBI Characters have a profound impact on the formation and development of Chinese characters; in addition, OBI characters also recorded important historical and cultural information [1]. However, many OBI characters have been damaged over the centuries, so that the characters lose their original shapes and become unrecognizable. Therefore, OBI image inpainting is crucial for discovering the origin of

© The Author(s), under exclusive license to Springer Nature Singapore Pte Ltd. 2024
Q. Liu et al. (Eds.): PRCV 2023, LNCS 14435, pp. 117–129, 2024.
https://doi.org/10.1007/978-981-99-8552-4_10

Chinese characters and even ancient Chinese history. Currently, only a few image inpainting studies on modern Chinese characters have appeared; unfortunately, OBI-oriented inpainting methods have been rarely proposed.

Direct applying methods of image and Chinese characters inpainting have achieved good results; however, two difficulties hinder the applications of these methods in OBI inpainting. First, due to the complex shapes and irregular structures of OBI characters, obtaining good inpainting results by regular Chinese inpainting methods is ambitious. Second, the semantic information of many OBI characters has not been fully interpreted; thus, it is very difficult to further complete OBI inpainting tasks by introducing semantic information. However, we notice that edge information of OBI strokes contains the connectivity of strokes, which is useful to restore damaged strokes. In addition, within an OBI character image, a damaged area's surrounding pixel information may also supply useful inpainting information.

To solve the above difficulties, we introduce a large kernel convolutional attention based U-Net network framework, which involves a large kernel convolution attention with a larger receptive field to repair damaged parts of OBI characters. The framework consists of two modified U-Net networks, which perform an edge inpainting function and an overall image inpainting function, respectively. In addition, we adopt a global discriminator and a local discriminator to conduct adversarial learning for both the entire and damaged parts of an OBI character, respectively. Moreover, a perceptual loss generated by a pre-trained OBI recognition VGG19 network is developed to strengthen semantic information. The training process of the proposed method is inspired by Nazeri et al.'s "EdgeConnect" method [5]: first, the proposed method learns the edge inpainting network of OBI images; second, the overall inpainting network with the edge graph of real OBI images is pre-trained; finally, the trained edge inpainting network and overall inpainting network are jointly trained to obtain the best restoring effects.

Our work's main innovations are summarized as follows: (1) It is the first time to use a large kernel convolutional attention based U-Net network framework for OBI inpainting. (2) By using guidances of the perceptual loss through both the global and local discriminative networks, OBI inpainting with irregular structures is enhanced.

2   Method

2.1   Overview

We propose a large kernel convolutional attention based U-Net network framework, as shown in Fig. 1, for OBI character image inpainting. The framework consists of two modified U-Nets placed arranged in a series, which perform an edge inpainting function and an overall image inpainting function, respectively. The implementations of the two U-Nets are identical to each other and each U-Net is composed of an encoder that downsamples the input image, followed by eight large kernel convolutional attention blocks and a decoder that upsamples

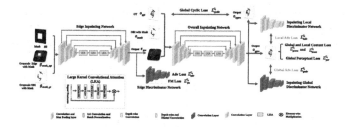

Fig. 1. Architecture of proposed framework.

the image back to its original size. Also, a global and a local discriminative networks are created to conduct adversarial learning processes on global and local missing parts of OBI images.

Since the edge information of OBI strokes has a huge impact on image inpainting processes, we adopt a training method from Nazeri et al's "Edge-Connect", which divides an entire inpainting training process into three steps: **First**, a generative network is applied as an edge inpainting network and a feature map discriminator is used to build adversarial learning to train the edge inpainting network. **Second**, the generative network is combined with the global and local discriminative networks to act as an OBI overall inpainting network, which is trained by using edge maps of the ground truths and masked OBI images. **Third**, the trained edge inpainting network and OBI overall inpainting network are connected together for a joint training. In addition, the edge map in the OBI overall inpainting network is the result generated from the edge inpainting network.

2.2 Large Kernel Attention Block

We use Large Kernel Attention (LKA) blocks with different kernel sizes and attentions to obtain larger receptive fields and establish correlations among pixels for missing parts of OBI images. LKA adopts large kernel convolutions to obtain larger receptive fields and longer-range pixel dependencies. In addition, we take advantage of the large receptive fields in the large kernel convolutions and the small calculation amounts of depth-wise convolutions, dilated convolutions, and 1×1 convolutions to faster and better obtain restored OBI images. Therefore, each LKA block contains a 1×1 convolutional Batch-Norm layer (Conv-BN), a depth-wise convolution (DW-Conv), a depth-wise dilated convolution (DW-D-Conv), 1×1 convolution, and 1×1 convolution (Conv-BN).

Meanwhile, with the assistance of the attention mechanism, LKA uses feature vectors obtained by the large-kernel convolution layer as the importance values (*AttenScore*) of the surrounding pixels. Then, the importance matrix is multiplied by the feature vector of a complete image that has not been processed by

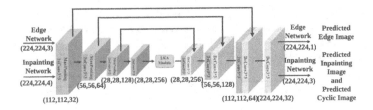

Fig. 2. Architecture of U-Net.

LKA to obtain the pixel attention value ($PixelScore$) of the image. Therefore, $AttenScore$ and $PixelScore$ are respectively defined as:

$$AttenScore = Conv_{1 \times 1}(Conv_{1 \times 1}$$
$$(DW - D - Conv(DW - Conv(Conv_{1 \times 1}(x)))) \tag{1}$$

$$PixelScore = AttenScore * x \tag{2}$$

where $Conv_{1 \times 1}$ denotes the 1×1 convolution operation, $DW - Conv$ denotes the depth-wise convolution, $DW - D - Conv$ denotes the depth-wise dilated convolution, and x denotes the feature vector of the entire image that is not processed by the LKA.

In addition, we analyze the importance of the large kernel size in LKA to obtain pixel dependencies at longer distances in missing parts of OBI images. The OBI image's large receptive field information is then obtained by using different large kernel convolution sizes. The specific results are shown in the ablation experiments.

2.3   U-Net Inpainting Generative Network

We introduce a U-Net based inpainting network embedded with LKA blocks as a generative network, which is shown in Fig. 2. Note that two U-Net based inpainting networks are used for both the edge inpainting and overall inpainting tasks. Therefore, the missing parts of OBI images are selectively filled in the surrounding pixel values through the LKA blocks. In Fig. 1, to perform image feature pre-processing before and after OBI inpainting, each U-Net structure is composed of an encoder that downsamples input images, followed by eight LKA blocks and a decoder that upsamples the images back to their original size. In our training process, the generative adversarial learning method is used to train both the edge inpainting and overall image inpainting networks. The input of the edge inpainting network is a damaged OBI grayscale map g_{mask_gt}, the missing OBI edge grayscale map g_{mask_egt}, and the mask m. The input of the overall inpainting network in the pre-training stage is the edge map of real OBI images and damaged OBI images. Note that: In the joint training, the input of the overall inpainting network is a damaged OBI image and an edge map obtained from the edge inpainting network.

2.4 Global and Local Discriminative Networks

The proposed OBI overall inpainting network integrates both global and local information of OBI image feature maps. Therefore, we build a global discriminant network and a local discriminant network, which are respectively used for both the overall restoring of entire images and local restoring of missing parts. As shown in Fig. 3, the network structures of both the global discriminator and local discriminator are identical to each other. The global discriminator performs end-to-end discriminations on entire restoring results, while the local discriminator performs adversarial learning on restoring results of missing parts. This end-to-end discriminate method more accurately determines image details and is beneficial to restoring complex OBI characters.

2.5 Loss Functions

The training process consists of three steps: 1) training of the edge inpainting network, 2) training of the overall inpainting network, and 3) joint training of both the edge-inpainting network and overall inpainting network.

Training of Edge Inpainting Network. In particular, two loss functions used in the first two training processes are different to each other. To train the edge inpainting network, the loss function must include a generative adversarial network's adversarial loss and discriminator-based feature matching loss Thus, the output of the edge inpainting network, e_{pre}, is defined as:

$$e_{pre} = G_e(g_{mask\_gt}, g_{mask\_egt}, m) \tag{3}$$

where G_e denotes the edge inpainting network.

The adversarial loss in edge discriminator network, $L_{adv}^{D_e}$, is then obtained through the adversarial learning of the edge map of a ground truth e_{gt} and the predicted edge map e_{pre} combined with the grayscale image of the ground truth g_{gt}. Thus, $L_{adv}^{D_e}$ is defined as:

$$L_{adv}^{D_e} = E[\log D_e(e_{gt}, g_{gt})] + E[\log(1 - D_e(e_{pre}, g_{gt})] \tag{4}$$

where D_e denotes the output of the edge discriminator network. Furthermore, the L_1 distance between the feature predicted by the edge discriminative network and the edge map of the ground truth serves as the perceptual loss. Thus, with the assistance of $L_{adv}^{D_e}$, the adversarial loss of the edge inpainting network, $L_{adv}^{G_e}$ is defined as:

$$L_{adv}^{G_e} = -E[\log D_e(e_{pre}, g_{gt})]. \tag{5}$$

The feature matching loss, $L_{fm}^{G_e}$, is defined as:

$$L_{fm}^{G_e} = E[\sum_{i=1}^{\mathfrak{N}} ||D_e^i(e_{gt}) - D_e^i(e_{pre})||] \tag{6}$$

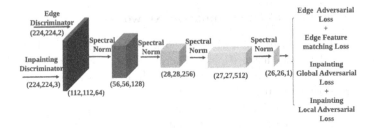

Fig. 3. Architecture of discriminator network.

where $\mathfrak{N}$ denotes the amount of the convolution layers of the discriminator and D_e^i is the activation in the i-th layer of the discriminator.

The final loss function of the edge inpainting network, L^{G_e}, is then defined as:

$$L^{G_e} = \lambda_{eadv} L_{adv}^{G_e} + \lambda_{fm} L_{fm}^{G_e} \tag{7}$$

where λ_{eadv} and λ_{fm} are the $L_{adv}^{G_e}$ and $L_{fm}^{G_e}$ hyperparameters, respectively, and all these hyperparameters are set to 0.1.

Training of Overall Inpainting Network. In the OBI overall inpainting training stage, the overall inpainting Network contains four loss functions: 1) adversarial loss consisted of global loss $L_{gadv}^{G_i}$ and local loss $L_{ladv}^{G_i}$, 2) content loss consisted of global loss $L_{gcont}^{G_i}$ and local loss $L_{lcont}^{G_i}$, 3) perceptual loss $L_{per}^{G_i}$, and 4) cyclic loss $L_{cyclic}^{G_i}$ [8]. In the image inpainting stage, the global discriminators deal with complex OBI inpainting. Therefore, the adversarial loss and content loss are composed of two types of errors: 1) global restoring and 2) local restoring. The definitions of these loss functions as introduced as follows:

The output of the overall inpainting network is:

$$o_{gpre}, o_{cypre} = G_i((cat(o_{mask}, e_{gt})), (cat(o_{ggt}, e_{gt}))) \tag{8}$$

where o_{gpre} denotes the overall inpainting network's predicted image, o_{mask} denotes a masked OBI ground truth image, o_{ggt} denotes an OBI ground truth image without masks, and o_{cypre} denotes ground truth predicted image generated by the U-Net in overall inpainting network. Thus, the global discriminator adversarial loss $L_{adv}^{D_i^g}$ and local discriminator adversarial loss $L_{adv}^{D_i^l}$ are respectively defined as:

$$L_{adv}^{D_i^g} = E[\log(D_i^g(o_{ggt})] + E[1 - \log(D_i^g(o_{gpre}))] \tag{9}$$

$$L_{adv}^{D_i^l} = E[\log(D_i^l(o_{lgt})] + E[1 - \log(D_i^l(o_{lpre}))] \tag{10}$$

where o_{lgt} denotes the missing part of the ground truth image and o_{lpre} denotes the predicted image part generated by the overall inpainting network for the missing part. Therefore, $L_{adv}^{D_i}$ is obtained by:

$$L_{adv}^{D_i} = L_{adv}^{D_i^g} + L_{adv}^{D_i^l} \tag{11}$$

where D_i^g and D_i^l are the outputs of the global and local discriminators, respectively.

Therefore, the global loss $L_{gadv}^{G_i}$ and local loss $L_{ladv}^{G_i}$ are then respectively defined as:

$$L_{gadv}^{G_i} = -E[\log D_i^g(o_{gpre}, o_{ggt})] \tag{12}$$

$$L_{ladv}^{G_i} = -E[\log D_i^l(o_{lpre}, o_{lgt})]. \tag{13}$$

By combining Eqs. (12) and (13), then we have the adversarial loss:

$$L_{adv}^{G_i} = L_{gadv}^{G_i} + L_{ladv}^{G_i}. \tag{14}$$

In addition, the global loss $L_{gcont}^{G_i}$ and local loss $L_{lcont}^{G_i}$ are respectively defined as:

$$L_{gcont}^{G_i} = E[||o_{gpre} - o_{ggt}||_2] \tag{15}$$

$$L_{lcont}^{G_i} = E[||o_{lpre} - o_{lgt}||_2]. \tag{16}$$

Thus, $L_{cont}^{G_i}$ is defined by:

$$L_{cont}^{G_i} = L_{gcont}^{G_i} + L_{lcont}^{G_i}. \tag{17}$$

The perceptual loss defined in Pihlgren et al's work [7] is used to calculate the square root values between the generated predicted OBI image's VGG19 output and its corresponding ground truth's VGG19 output. The perceptual loss $L_{per}^{G_i}$ is defined as:

$$L_{per}^{G_i} = E[||\varphi(o_{gpre}) - \varphi(o_{ggt})||_2] \tag{18}$$

where φ denotes the pre-trained VGG19 network.

To enable the U-Net of the overall inpainting network to better encode and restore images, we use the cyclic loss [2]:

$$L_{cyclic}^{G_i} = E[||o_{cypre} - o_{ggt}||_2]. \tag{19}$$

The total loss L^{G_i} of the generative network is:

$$L^{G_i} = \lambda_{iadv} L_{adv}^{G_i} + \lambda_{cont} L_{cont}^{G_i} + \lambda_{per} L_{per}^{G_i} + \lambda_{cyclic} L_{cyclic}^{G_i} \tag{20}$$

where λ_{iadv}, λ_{cont}, λ_{per} and λ_{cyclic} respectively denote the $L_{adv}^{G_i}$, $L_{cont}^{G_i}$, $L_{per}^{G_i}$ and $L_{cyclic}^{G_i}$ hyperparameters, which are set to 0.1, 1.0, 10.0, and 10.0, respectively.

Joint Training. In the joint training process of both the edge inpainting network and overall inpainting network, the predicted edge map first is obtained by the edge inpainting network through $g_{mask\_gt}, g_{mask\_egt}$ and m in Eq. (3), then the overall inpainting network uses the predicted edge map to gain the o_{gpre} and o_{cyper} through Eq. (8); after that, Eqs. (5), (6), (7), (12), (13), (15), (16), (18), (19) and (20) are executed to update the edge inpainting network and overall inpainting network, respectively.

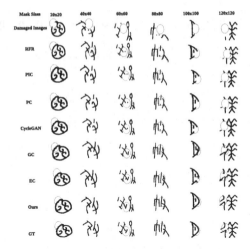

Fig. 4. Comparison results with the other six algorithms and ours.

Fig. 5. Ablation comparison results of various large kernel convolution sizes on OBI inpainting and various network structures.

3   Experimentation

3.1   Experimental Datasets and Settings

In our experiment, we used a public OBI dataset provided by the "Yin Qi Wen Yuan"[1], which involves 83,245 OBI sample images within 3,881 categories. We used 10% images in the dataset as our test set, and we randomly generated masks with different six sizes to build the test set. In addition, we used OBI images generated by using masks of different sizes to train our model. The image size is set to 224×224 in training, and the mask size is set to 20×20, 40×40, 60×60, 80×80, 100×100, and 120×120, respectively. The experimentation contained two parts: 1) comparisons with state-of-the-art image inpainting methods in OBI inpainting tasks; and 2) ablation studies on our proposed method with different setups.

3.2   Evaluation Metrics

We applied five evaluation metrics that are widely used in the image inpainting area: 1) mean absolute errors (MAE), 2) peak signal-to-noise ratio (PSNR) 3) structural similarity index (SSIM) 4) perceived distance (PD) that measures the

---

[1] http://jgw.aynu.edu.cn/ajaxpage/home2.0/index.html.

Table 1. Quantitative results of ours and other six methods on handwritten oracles dataset. (↓) indicates lower is better and (↑) indicates higher is better.

	Mask Size	20	40	60	80	100	120
MAE(↓)	RFR	0.0098	0.0116	0.0143	0.0184	0.0232	0.0288
	CycleGAN	0.0023	0.0043	0.0078	0.0128	0.0183	0.0238
	PC	0.0067	0.0087	0.0117	0.0162	0.0216	0.0278
	PIC	0.0042	0.0063	0.0098	0.0144	0.0197	0.0249
	GC	0.0003	0.0014	0.0035	0.007	0.0115	0.0164
	EC	0.0002	0.0011	0.003	0.0063	0.0106	0.0155
	Ours	**0.0001**	**0.0008**	**0.0023**	**0.0052**	**0.0094**	**0.0146**
PSNR(↑)	RFR	21.6583	20.699	19.5666	18.2994	17.1472	16.0848
	CycleGAN	31.1622	26.6305	22.9779	20.1416	18.3147	16.9921
	PC	26.2415	23.9645	21.7821	19.7313	18.0702	16.6643
	PIC	27.4512	24.4455	21.7466	19.5591	17.9666	16.8291
	GC	39.2881	31.6044	26.6023	23.0709	20.626	18.9049
	EC	46.211	34.8323	28.2365	24.0484	21.3014	19.3859
	Ours	**46.4164**	**36.7611**	**30.6461**	**26.1582**	**23.141**	**20.983**
SSIM(↑)	RFR	0.9688	0.9621	0.9517	0.9368	0.9183	0.8961
	CycleGAN	0.9954	0.9889	0.9779	0.9625	0.9436	0.9173
	PC	0.9838	0.9767	0.9653	0.9491	0.9289	0.9051
	PIC	0.9916	0.9843	0.9722	0.9553	0.9349	0.9135
	GC	0.9992	0.9964	0.9903	0.9794	0.9639	0.945
	EC	0.9996	0.9972	0.9918	0.9819	0.9677	0.9499
	Ours	**0.9997**	**0.9981**	**0.9943**	**0.9865**	**0.9741**	**0.9569**
PD(↓)	RFR	1.8361	2.1816	2.3585	2.7035	3.1182	3.3914
	CycleGAN	0.0261	0.0864	0.2202	0.3113	0.5785	1.8745
	PC	0.3219	0.3468	0.4999	0.8372	1.3975	2.1719
	PIC	0.0894	0.1586	0.3672	0.7796	1.4245	2.2872
	GC	0.0021	0.0116	0.0355	0.1023	0.2713	0.6131
	EC	0.0009	0.0111	0.0359	0.0868	0.201	0.4365
	Ours	**0.0005**	**0.0054**	**0.0207**	**0.0641**	**0.1799**	**0.4284**
LPIPS(↓)	RFR	0.3479	0.3695	0.3777	0.3879	0.3934	0.4222
	CycleGAN	0.0058	0.0135	0.0258	0.0406	0.0613	0.1023
	PC	0.0265	0.0346	0.0474	0.0661	0.0903	0.1196
	PIC	0.0145	0.0231	0.0365	0.0559	0.0804	0.1094
	GC	0.0012	0.0047	0.0114	0.0231	0.0409	0.0649
	EC	0.0006	0.0036	0.0092	0.0191	0.0341	0.0555
	Ours	**0.0004**	**0.0026**	**0.0067**	**0.0152**	**0.0299**	**0.0518**

distance in high-dimensional space by the perceptual loss defined in Eq. (18), and 5) average values of learned perceptual image patch similarity (LPIPS) that evaluate the quality of generated images, to evaluate each method's output quality.

3.3   Experimental Results and Quantitative Evaluations

Since this proposed work is the first study on OBI image inpainting, we used Chang's CycleGAN [2], Li et al.'s RFR [3], Liu et al.'s PC [4], Zheng et al.'s PIC [10], Yu et al.'s GC [9], and Nazeri et al.'s EdgeConnect (EC) [5] in the field of image inpainting for comparisons. Note that other image inpainting methods that must involve text inputs were not included in this comparison. We used six different sizes of masks to evaluate our method against these compared methods after training with different mask sizes. The comparison results are shown in Fig. 4: GC, EC, and our methods can achieve better inpainting results than the rest four methods (CycleGAN, PIC, PC, and RFR). One potential reason is that the rest four methods belonged to the one-stage restoring, which is difficult to restore the complicated strokes and structures in OBI.

We evaluated our method and the six compared methods for OBI image inpainting by the five metrics defined in Sect. 3.2 with six different mask sizes. From Table 1, we have achieved significant advantages in all the five indicators of MAE, PSNR, SSIM, PD, and LPIPS. If the mask was relatively small, the advantage of our method was significant compared with the other six methods. If the mask gradually increased, our method still exhibited better performance than the other methods. Due to the various degrees of attention to surrounding pixels in our inpainting network, our final results were remarkable.

3.4   Ablation Study

To explore the effectiveness of proposed method, we conducted ablation studies on the effects of 1) various large kernel convolution sizes on OBI inpainting, 2) various network structures in the proposed framework, and 3) local discriminator. Because a LKA block consists of a depth-wise large-kernel convolution and a depth-wise and dilated large-kernel convolution, we set the size of the large-kernel convolution into various levels: The depth-wise large kernel convolution size was set to 5, 7, 9, 11; the corresponding padding sizes were set to 2, 3, 4, 5; the depth-wise and dilated large kernel convolution sizes were set to 7, 9, 11, 13; and the corresponding padding sizes were set 9, 12, 15, 18. The corresponding dilated size was set to 3. In addition, to reflect the advantages of our method, we used the U-Net with residual convolution (UCNN) blocks as a comparison. Also, we replaced the restoring module with the UConformer [6] consisting of transformer and CNN as another to form our ablation experiments. Furthermore, we evaluated the network performance with or without the local discriminator.

These methods' results shown in Table 2 and Fig. 5 indicate that when the depth-wise large kernel convolution size and depth-wise and dilated large kernel convolution size were set to 9 and 11 (LKA(9-11)), respectively, the proposed method with such a size combination owned significant advantages over the other large kernel size combination in all the five metrics. In addition, the optimal size combination-based method can achieve better performance than those of both UCNN and UConformer. Therefore, these ablation experimental results proved the feasibility and efficiency of the proposed method.

The ablation results of the local discriminator are shown in Table 3 and Fig. 6. Because the PD and LPIPS indicators can better represent the evaluation of the restored image; therefore, these two indicators were used to compare the ablation experiments of the local discriminator. From the comparison results, we can know that the inpainting network without the local discriminator network (LKA(9-11)g) misjudged the local missing part. Thus, the local discriminator network is essential to the proposed work.

Table 2. Ablation results of various large kernel convolution sizes and various network structures.

	Mask Size	20	40	60	80	100	120
MAE(↓)	LKA(5-7)	0.0002	0.0009	0.0024	0.0055	0.0101	0.0152
	LKA(7-9)	0.0002	0.0008	0.0024	0.0055	0.0098	0.0152
	LKA(9-11)	**0.0001**	**0.0008**	**0.0023**	**0.0052**	**0.0094**	**0.0146**
	LKA(11-13)	0.0002	0.0009	0.0026	0.0057	0.0099	0.0151
	UCNN	0.0002	0.0009	0.0026	0.0058	0.0102	0.0153
	UConformer	0.0003	0.0013	0.0036	0.0073	0.0123	0.0184
PSNR(↑)	LKA(5-7)	45.7867	36.257	30.5288	25.8551	22.6593	20.5815
	LKA(7-9)	46.015	36.4067	30.3383	25.9179	22.9737	20.8643
	LKA(9-11)	**46.4164**	**36.7611**	**30.6461**	**26.1582**	**23.141**	**20.983**
	LKA(11-13)	45.3785	35.8864	29.9551	25.7128	22.8364	20.8043
	UCNN	45.7682	36.2299	30.1239	25.685	22.7552	20.6859
	UConformer	42.8436	33.0098	27.4474	23.739	21.2522	19.4459
SSIM(↑)	LKA(5-7)	0.9997	0.998	0.994	0.9858	0.972	0.9545
	LKA(7-9)	0.9997	0.998	0.9941	0.986	0.9734	0.9559
	LKA(9-11)	**0.9997**	**0.9981**	**0.9943**	**0.9865**	**0.9741**	**0.9569**
	LKA(11-13)	0.9996	0.9978	0.9937	0.9855	0.9729	0.9558
	UCNN	0.9997	0.9979	0.9936	0.985	0.972	0.9549
	UConformer	0.9993	0.9965	0.9905	0.98	0.9643	0.9431
PD(↓)	LKA(5-7)	0.0008	0.0084	0.0296	0.0736	0.2058	0.4532
	LKA(7-9)	0.0006	0.0059	0.0206	0.0673	0.1779	0.4386
	LKA(9-11)	**0.0005**	**0.0054**	**0.0207**	**0.0641**	**0.1799**	**0.4284**
	LKA(11-13)	0.0008	0.0067	0.02386	0.07537	0.1963	0.4494
	UCNN	0.0007	0.0073	0.0273	0.0834	0.2061	0.4534
	UConformer	0.0018	0.0125	0.0379	0.1054	0.2676	0.6153
LPIPS(↓)	LKA(5-7)	0.0005	0.0027	0.0073	0.0162	0.0321	0.0536
	LKA(7-9)	0.0005	0.0026	0.0068	0.0157	0.0309	0.0535
	LKA(9-11)	**0.0004**	**0.0026**	**0.0067**	**0.0152**	**0.0299**	**0.0518**
	LKA(11-13)	0.0006	0.0029	0.0073	0.0164	0.0314	0.0535
	UCNN	0.0005	0.0027	0.0076	0.0177	0.0336	0.0556
	UConformer	0.0013	0.005	0.0111	0.0223	0.0403	0.0673

Table 3. Ablation results of local and global discriminators.

	Mask Size	20	40	60	80	100	120
PD(↓)	LKA(9-11)g	0.0009	0.0093	0.0408	0.1276	0.3114	0.638
	LKA(9-11)	**0.0005**	**0.0054**	**0.0207**	**0.0641**	**0.1799**	**0.4284**
LPIPS(↓)	LKA(9-11)g	0.0005	0.0029	0.0086	0.0193	0.0353	0.0576
	LKA(9-11)	**0.0004**	**0.0026**	**0.0067**	**0.0152**	**0.0299**	**0.0518**

Damaged Image   LKA(9-11)g   LKA(9-11)   GT

Fig. 6. Ablation comparison results of local discriminator. The LKA(9-11)g means LKA(9-11) without local discriminator network.

4   Conclusion

This paper proposed a U-Net network-based framework with large kernel convolutional attention kernels for oracle bone inscription (OBI) inpainting, which conducted adversarial learning on missing parts and the whole image through global and local discriminators. Experiments showed that our method had advantages in OBI inpainting compared with the other image inpainting methods.

References

1. Assael, Y., et al.: Restoring and attributing ancient texts using deep neural networks. Nature **603**(7900), 280–283 (2022)
2. Chang, B., Zhang, Q., Pan, S., Meng, L.: Generating handwritten Chinese characters using CycleGAN. In: 2018 IEEE Winter Conference on Applications of Computer Vision, pp. 199–207. IEEE (2018)
3. Li, J., Wang, N., Zhang, L., Du, B., Tao, D.: Recurrent feature reasoning for image inpainting. In: Proceedings of the IEEE/CVF Conference on Computer Vision and Pattern Recognition, pp. 7760–7768 (2020)
4. Liu, G., Reda, F.A., Shih, K.J., Wang, T.C., Tao, A., Catanzaro, B.: Image inpainting for irregular holes using partial convolutions. In: Proceedings of the European Conference on Computer Vision, pp. 85–100 (2018)
5. Nazeri, K., Ng, E., Joseph, T., Qureshi, F.Z., Ebrahimi, M.: Edgeconnect: generative image inpainting with adversarial edge learning. CoRR (2019). https://doi.org/10.48550/ARXIV.1901.00212. https://arxiv.org/abs/1901.00212
6. Peng, Z., et al.: Conformer: local features coupling global representations for visual recognition. In: Proceedings of the IEEE/CVF International Conference on Computer Vision, pp. 367–376 (2021)
7. Pihlgren, G.G., Sandin, F., Liwicki, M.: Improving image autoencoder embeddings with perceptual loss. In: 2020 International Joint Conference on Neural Networks, pp. 1–7. IEEE (2020)

8. Wang, H., Lin, G., Hoi, S.C., Miao, C.: Cycle-consistent inverse GAN for text-to-image synthesis. In: Proceedings of the 29th ACM International Conference on Multimedia, pp. 630–638 (2021)

9. Yu, J., Lin, Z., Yang, J., Shen, X., Lu, X., Huang, T.S.: Free-form image inpainting with gated convolution. In: Proceedings of the IEEE/CVF International Conference on Computer Vision, pp. 4471–4480 (2019)

10. Zheng, C., Cham, T.J., Cai, J.: Pluralistic image completion. In: Proceedings of the IEEE/CVF Conference on Computer Vision and Pattern Recognition, pp. 1438–1447 (2019)

L²DM: A Diffusion Model for Low-Light Image Enhancement

Xingguo Lv[1], Xingbo Dong[2(✉)], Zhe Jin[2], Hui Zhang[1], Siyi Song[1], and Xuejun Li[1]

[1] Anhui Provincial International Joint Research Center for Advanced Technology in Medical Imaging, School of Computer Science and Technology, Anhui University, Hefei 230093, China
[2] Anhui Provincial Key Laboratory of Secure Artificial Intelligence, School of Artificial Intelligence, Anhui University, Hefei 230093, China
xingbo.dong@ahu.edu.cn

Abstract. Low-light image enhancement is a challenging yet beneficial task in computer vision that aims to improve the quality of images captured under poor illumination conditions. It involves addressing difficulties such as color distortions and noise, which often degrade the visual fidelity of low-light images. Although tremendous CNN-based and ViT-based approaches have been proposed, the potential of diffusion models in this domain remains unexplored. This paper presents L²DM, a novel framework for low-light image enhancement using diffusion models. Since L²DM falls into the category of latent diffusion models, it can reduce computational requirements through denoising and the diffusion process in latent space. Conditioning inputs are essential for guiding the enhancement process, therefore, a new ViT-based network called ViTCondNet is introduced to efficiently incorporate conditioning low-light inputs into the image generation pipeline. Extensive experiments on benchmark LOL datasets demonstrate L²DM's state-of-the-art performance compared to diffusion-based counterparts. The L²DM source code is available on GitHub for reproducibility and further research.

Keywords: Computational photography · Low-light image enhancement · Diffusion models · Latent diffusion models

1 Introduction

Computational photography is a field that merges digital image processing and computer vision methods to enhance and manipulate photographs. One crucial aspect of computational photography is low-light image enhancement. When capturing digital images in conditions of poor illumination, such as indoors, at night, or with improper camera exposure settings, issues like color distortions and noise can arise, diminishing the overall quality of the image. To enhance low-light images, tremendous Low-light Image Enhancement (LLIE) solutions

Q. Liu et al. (Eds.): PRCV 2023, LNCS 14435, pp. 130–145, 2024.
https://doi.org/10.1007/978-981-99-8552-4_11

have been proposed, which can be generally classified into traditional approaches, convolutional neural network (CNN) approaches and vision transformer (ViT) approaches.

As a traditional approach, histogram equalization (HE) techniques [1,23,49] aim to rearrange pixel values in order to achieve a uniform distribution. Methods based on dehaze models [33] adjust pixel values to conform to a natural distribution. Retinex-theory-based approaches [10,43,47,65] utilize models that assume an image can be decomposed into illumination and reflectance components. These methods are generally regarded as traditional approaches.

In the last decade, Convolutional Neural Networks (CNNs) have achieved impressive progress in low-level image processing applications [11,20,50,68]. For instance, Chen et al. [3] employed a U-Net [35] to process RAW sensor data and generate high-perceptual-quality RGB images. Xu et al. [50] proposed a pipeline for low-light image enhancement that incorporates a frequency-based decomposition and enhancement model. Similarly, Dong et al. [7] introduced a CNN-based network based on the U-Net architecture, aiming to virtually eliminate the color filter and achieve improved image processing performance.

In recent years, Transformers have demonstrated significant advantages compared to CNN-based approaches in the field of low-level vision tasks. This is primarily due to their ability to leverage spatial and channel-wise attention mechanisms, as originally introduced in [41]. One example is the framework proposed by Xu et al. in [53], which exploits the Signal-to-Noise Ratio (SNR) as prior information to guide the process of feature fusion. By incorporating a novel self-attention model, this SNR-aware transformer effectively avoids incorporating tokens from image regions with significant noise or very low SNR values. Consequently, it dynamically enhances pixels using spatial-varying operations, resulting in state-of-the-art performance in low-level vision tasks.

Recently, diffusion models [37] have gained significant attention as a type of probabilistic generative model capable of generating high-resolution images with a wide range of diversity [6]. These models have shown great promise in image reconstruction and offer easier training and improved sample quality compared to traditional generative models and variational autoencoders (VAEs). However, due to the nature of probabilistic generative models, which aim to learn a representative distribution rather than a deterministic solution like the combination of U-Net and L2 loss, they often exhibit inferior performance on traditional metrics such as PSNR/SSIM for image enhancement tasks. Consequently, the use of diffusion models for low-light image enhancement remains relatively unexplored.

Motivated by the preceding discussion, we present a diffusion model for low-light image enhancement, namely L^{2}DM (Low-Light Diffusion Model). L^{2}DM draws inspiration from latent diffusion models (LDMs) [34], which leverage denoising and diffusion models in the latent space of autoencoders to enhance image synthesis and substantially decrease computational demands. To address the task of low-light image enhancement and mitigate potential degradation in peak signal-to-noise ratio (PSNR) performance, we further introduce a novel

ViT-based network, namely ViTCondNet, to handle low-light images conditioning inputs. The contributions of our work can be summarized as follows:

1. We proposed a Diffusion Model enabled framework (i.e. L^2DM) for low-light image enhancement. The proposed framework attempts latent diffusion in low-light image enhancement task, which results in the reduction of the computational overhead, particularly beneficial to the higher dimensional data. Simultaneously offering more faithful and detailed reconstructions.
2. We introduced a ViT-based network, namely ViTCondNet to effectively handles conditioning inputs. ViTCondNet manages to achieve a decent low-light image enhancement performance and improves poor PSNR performance while allowing for the synthesis of normal light images.
3. We conducted the extensive experiments on publicly available benchmark datasets and the results demonstrate that L^2DM exhibits state-of-the-art performance compared to diffusion-based counterparts. The PSNR on the LOL dataset reaches 24.54, while on the LOLV2-real and synthetic subsets, it achieves 24.80 and 23.96, respectively. The source code of our work is available at github.com/Yore0/L2DM.

2   Related Work

In this section, we primarily concentrate on data-driven methods that employ convolutional neural networks (CNN) and vision transformers (ViT) to enhance low-light images.

With the advancement of deep learning technology, CNN-based approaches have been at the forefront, where researchers have leveraged the hierarchical structure of convolutional layers to effectively extract pertinent features from low-light images. In the Zero-DCE method [11], a lightweight CNN network is proposed to estimate pixel-wise and high-order curves for image enhancement. Zero-DCE achieves efficient and effective enhancement across various lighting conditions without the need for reference images. It outperforms state-of-the-art methods and shows potential applications in dark face detection. However, compared to supervised approaches, Zero-DCE exhibits a performance gap. In [30], DeepLPF is introduced as a deep neural network approach for automatic image enhancement. It uses spatially local filters (Elliptical, Graduated, Polynomial) and achieves superior results on benchmark datasets with fewer parameters. In [55], DRBN is proposed for low-light image enhancement, using paired low/normal-light images and adversarial learning. DSLR [25] utilizes Laplacian pyramid for global and local enhancement. EnlighenGAN [17] enhances low-light images using unsupervised GANs without paired training data.

In a recent work by Xu et al. [54], a novel framework is introduced for enhancing low-light images by simultaneously addressing their appearance and structure. This approach incorporates edge detection to effectively model the image structure and employs a structure-guided enhancement module to enhance the overall image quality. In another study by Wang et al. [45], a normalizing flow model called LLFlow is proposed to tackle the challenging task of enhancing low-light images. By modeling the one-to-many relationship between low-light and normally exposed images, LLFlow achieves notable improvements in terms of brightness enhancement, noise reduction, artifact removal, and color enhancement.

The CNN-based models generally produce visually satisfactory enhancement results in most scenarios. However, when it comes to global modeling, the ViT architecture outperforms CNNs by focusing on the entire image rather than just local regions. In a study by Xu et al. [53], the authors exploit the Signal-to-Noise Ratio (SNR) as prior information to guide the feature fusion process. As a result, it dynamically enhances pixels using spatial-varying operations, leading to state-of-the-art performance in low-level vision tasks. In another work by Yuan et al. [59], an end-to-end low-light image enhancement network combining transformers and CNNs is proposed. This network accurately captures both global and local features to restore normal light images, resulting in improved brightness, reduced noise, and preserved texture and color information. Similarly, in a study by Jiang et al. [16], a Stage-Transformer-Guided Network (STGNet) is introduced to effectively enhance low-light images by addressing region-specific distortions. It employs a multi-stage approach with efficient transformers that capture degradation distributions at different scales and orientations. Learnable degradation queries are used for adaptive feature selection. Additionally, a histogram loss and other loss functions are utilized to exploit global contrast and local details, further enhancing the quality of the enhanced images.

Recently, denoising diffusion probabilistic models (DDPM) [37] have garnered considerable attention as a type of probabilistic generative model capable of generating high-resolution images. These models have found applications in various domains, including image generation [34], inpainting [36], colorization [36], and image segmentation [34]. Reader is suggested to refer to [4] for a comprehensive survey.

While DDPMs have achieved impressive results, they typically require a high number of iterations, leading to slow performance. In contrast, Latent diffusion models (LDMs) [34] address this issue by training the diffusion process in the latent space of pre-trained autoencoders. Furthermore, LDMs incorporate cross-attention layers to achieve near-optimal complexity reduction and preserve fine details. This enhancement in visual fidelity not only improves the performance but also mitigates the slow performance associated with traditional DDPMs.

In a nutshell, DDPMs have shown promise in image reconstruction, offering easier training and improved sample quality compared to traditional generative models and VAEs. However, as probabilistic generative models learn representative distributions rather than deterministic solutions like U-Net with L2 loss, they often underperform on traditional metrics like PSNR/SSIM for image enhancement. Consequently, the exploration of diffusion models for low-light image enhancement remains limited. Given their impressive capabilities in image generation, in this work we investigate their potential in enhancing low-light images.

3   Proposed Method

3.1   Preliminaries

Diffusion models draw inspiration from non-equilibrium thermodynamics principles. They utilize a Markov chain of diffusion processes to introduce random noise to data and subsequently learn a reverse process that reconstructs the desired data samples from the noise. Numerous generative models based on diffusion have been proposed, such as diffusion probabilistic models [37], noise-conditioned score network [39], and denoising diffusion probabilistic models (DDPM) [14].

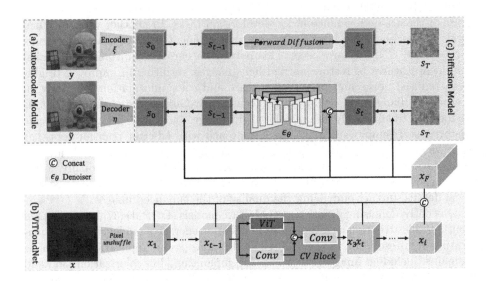

Fig. 1. Overview of our proposed model, L^2DM. (a) Autoencoder module is a trainable component used to transform images from the pixel space to the latent space, significantly reducing computational requirements during model training. (b) Condition module is a newly proposed trainable framework that achieves excellent feature extraction capabilities while significantly reducing the number of parameters. (c) Extracted condition feature map is concatenated with the s_t and fed into the time-conditional U-Net.

In the forward diffusion process, an image data sampled from a real data distribution undergoes degradation over T timesteps. At each timestep, Gaussian noise is added incrementally, gradually reducing the quality of the image. This process can be formulated as:

$$q(y_t|y_{t-1}) = \mathcal{N}(y_t; \sqrt{\beta}y_{t-1}, (1 - \beta_t)I), \tag{1}$$

$$q(y_t|y_0) = \mathcal{N}(y_t; \sqrt{\bar{\alpha}}y_0, (1 - \bar{\alpha}_t)I), \tag{2}$$

where $\{\beta_1, \ldots, \beta_T\}$ is a prescribed variance schedule defined over timestep T, $\alpha_t = 1 - \beta_t$ and $\bar{\alpha}_t = \prod_{k=1}^{t} \alpha_k$. By iteratively introducing noise steps with small variances, the image data progressively converges to a distribution that closely resembles a standard Gaussian in form of:

$$q(y_T|y_0) = \mathcal{N}(y_t; 0, I). \tag{3}$$

In the reverse diffusion process, we start with a completely random standard normal distribution, denoted as y_t, and gradually remove the Gaussian noise to recover y_0 using t intermediate steps. Each reverse step utilizes a distribution p modeled by a neural network with parameters θ, which can be expressed as follows:

$$p_\theta(y_{t-1}|y_t) = \mathcal{N}(y_{t-1}; \mu_\theta(y_t, t), \widetilde{\beta}_t I), \tag{4}$$

where $\widetilde{\beta}_t = \frac{1-\bar{\alpha}_{t-1}}{\bar{\alpha}_t}\beta_t$. The mean μ_θ can be written as:

$$\mu_\theta(y_t, t) = \frac{1}{\sqrt{\alpha_t}}(y_t - \frac{\beta_t}{\sqrt{1-\bar{\alpha}_t}}\epsilon_\theta(y_t, t)), \tag{5}$$

where ϵ_θ represents the estimated residual noise. To determine the parameters of the distribution $p_\theta(\cdot)$, we optimize the variational lower bound of the log-likelihood of $p_\theta(y_0)$. This can be expressed as follows:

$$L_{DM} = \mathbb{E}_{\epsilon \sim \mathcal{N}(0,I), t \sim [1,T]}[\|\epsilon - \epsilon_\theta(y_t, t)\|_2^2], \tag{6}$$

The network p_θ is trained to model the Gaussian distribution $\mathcal{N}(0, I)$ at each timestep t.

3.2   Autoencoder Module

Training a full-resolution diffusion model is impractical due to its extremely high computational demands, requiring days of training on multiple GPUs. Taking inspiration from [34], instead of processing the original image input in the pixel space, which is only suitable for small-resolution image processing, we employ an autoencoder that compresses the pixel-space features into a latent feature space. The autoencoder is trained using a combination of perceptual loss [63], patch-based [15] adversarial objectives [8,58].

More precisely, given an input image $y \in \mathbb{R}^{H \times W \times 3}$ in RGB space, the encoder ξ encodes x into a latent representation $s = \xi(y)$, where $s \in \mathbb{R}^{h \times w \times 3}$. We set the encoder downsampling factors $f = H/h = W/w = 4$. The decoder η reconstructs the representation from the latent $\widetilde{y} = \eta(s)$, and note that $\widetilde{y}$ has the same shape as y. Within the decoder, vector quantization (VQ) regularization, which uses a vector quantization layer [40] is imposed in order to avoid arbitrarily high-variance latent spaces. The autoencoder model is trained in an adversarial manner following [8]. A patch-based discriminator D_ω is optimized to differentiate original images y from reconstructions $\widetilde{y} = \eta(\xi(y))$. To avoid arbitrarily scaled latent spaces, we regularize the latent s to be zero centered and obtain small variance by introducing an regularizing loss term L_{reg}. We regularize the latent space with a vector quantization layer by learning a high codebook dimensionality $|\mathcal{S}|$ [40]. The full objective L_{ae} to train the autoencoder model can be written as:

$$L_{ae} = \min_{\eta,\xi} \max_{\omega} \left(L_{rec}(y, \eta(\xi(y))) - L_{adv}(\eta(\xi(y))) + \log D_{\omega}(y) + L_{reg}(y; \eta, \xi)\right), \quad (7)$$

where $L_{rec} = ||y - \widetilde{y}||_2^2$ is a reconstruction loss.

3.3 ViTCondNet

Similar to other types of generative models [29,38], diffusion models have the inherent capability to model conditional distributions in the form of $p(y|x)$. This can be accomplished by introducing a conditional variable x to control the synthesis process using condition inputs, such as text [32], semantic maps [15,31], or images [15]. For the specific task of low-light image enhancement, we utilize the captured low-light image as the condition input.

In this work, a new conditional feature extraction framework is designed to extract the representations of conditional images and fuse them with Diffusion Model. As shown in Fig. 1(b), layered structure of our condition module is based on Convolution-ViT (CV) block [52] and a concatenation that concatenates all the hierarchical features processed by CV blocks. Given an input of condition image $x \in \mathbb{R}^{H \times W \times 3}$, where H and W denote the spatial height and width of the condition image. A Pixel-unshuffle Layer unshuffles x into 48 channels by setting the downscale factor to 4, to construct $x_1 \in \mathbb{R}^{\frac{H}{4} \times \frac{W}{4} \times 48}$. Then the same operation is performed on x_1 with CV block to obtain x_2, and so on to obtain x_i. Finally, $x_1, x_2, \cdots, x_i$ are concatenated in the channel dimension by concat operation, and a feature map $x_F \in \mathbb{R}^{\frac{H}{4} \times \frac{W}{4} \times (48*i)}$ is obtained. Through a series of ablation experiments, we have demonstrated that this feature extraction framework can achieve excellent feature extraction results even with minimal parameters, making it highly effective and easy to train.

3.4 Main Architecture

The backbone of $\epsilon_\theta(x, y_t, t)$ in Eq. 6 is implemented by a time-conditional U-Net [35]. With a well-trained autoencoder model composed of ξ and η, we have gained access to an efficient and low-dimensional latent space, in which high-frequency and imperceptible details are abstracted. As shown in Fig. 1(a)(c), the ground truth image is compressed into s_0 in the latent space by the encoder module ξ. During the forward diffusion process, Gaussian noise is gradually added to s_0 for T iterations, resulting in a final image, s_T, which is completely Gaussian-distributed. s_T is concatenated with the features x_F, extracted by our ViTCondNet in the channel dimension. The concatenated data is then fed into the U-Net, and the desired data distribution is gradually restored through loss calculation with respect to the standard normal distribution. The loss function of our network is:

$$L = \mathbb{E}_{\epsilon \sim \mathcal{N}(0,I), t \sim [1,T]}[||\epsilon - \epsilon_\theta(x, y_t, t)||_1], \quad (8)$$

where x represents the conditional features obtained after passing through the ViTCondNet. We abandon the widely used L2 loss in diffusion models and instead employ the L1 loss. Considering the image enhancement task and the significant amount of noise pixels introduced during the forward diffusion process, we anticipate the presence of more outlier pixels. L1 loss, being sensitive to outliers, is capable of penalizing these exceptional pixels effectively. Therefore, by employing L1 loss, we can precisely control the extent of penalization for outlier pixels, leading to improved enhancement results. This facilitates better modeling and handling of outlier pixels, ultimately contributing to superior image enhancement outcomes. The experimental results also validate the effectiveness of our approach.

4   Experiments

4.1   Setup

Schedules. First, we set timestep $T = 1000$ and the base learning rate of the model to 1×10^{-5}. We follow a linear increase for the variance schedule β_t, ranging from 0.0015 to 0.0155. To assess the system's performance, we evaluate it based on two metrics: peak signal-to-noise ratio (PSNR) and structural similarity (SSIM) [46]. For the number of concatenation features x_i in ViTCondNet, we empirically set i to 4.

Datasets. We conduct experiments on LOL [47] and LOLV2 [57] datasets. The LOL dataset contains 485 training and 15 testing paired images, with each pair comprising a low-light and a normal-light image. The LOLV2 dataset contains two parts, Real-captured and Synthetic. Real part has noise distributions not present in the LOL dataset, and Synthetic part has illumination distributions not present in the LOL dataset.

Training. During the training phase, we randomly cropped pairs of low-light and normal-light images to a size of 256×256. For LOL and LOLV2-real datasets, we set the batch size to 16. For the LOLV2-sync dataset, we use a larger batch size of 32. All training processes are completed on NVIDIA GeForce RTX 3090.

Evaluation. Combined with DPM-solver [27], sampling process can be completed within 20 steps. We must note that due to the diffusion model denoising from a completely random Gaussian distribution, each generated sample will not be exactly the same. This leads to fluctuations in the assessment of the results within a range. During the validation process, we define the batch size as $b = 10$, which represents the number of repetitions of the same image. We select one high quality sample from each batch and calculate the average PSNR and SSIM over 5 batches. The images in the LOL and LOL-real datasets have a resolution of 400×600, which doesn't match the size required by our network. Therefore, during the validation phase, we uniformly resize the images in the test set to a size of 384×576 while maintaining the aspect ratio.

Other Details. We pre-trained the complete architecture of L^2DM on the COCO dataset. We excluded images from the COCO training set that had inappropriate sizes, resulting in a set of 117,188 images, which we used as the ground truth. We then applied darkening transformations to these images and added random-sized noise to each pixel. These processed images were used as the conditional inputs. After training the model weights on the COCO dataset, we only need to fine-tune the model on the target dataset to achieve excellent results.

Table 1. Quantitative results on the **LOL dataset** in terms of PSNR and SSIM. ↑ denotes that larger values lead to better quality. Models marked with an asterisk ∗ indicate that the model is implemented based on the diffusion model. The best results are shown in **bold**, and the second-best results are <u>underlined</u>.

Method	Zero-DCE [11]	LIME [13]	RetinexNet [47]	EG [17]	RUAS [26]	DRBN [55]
PSNR↑	14.86	16.76	16.77	17.48	18.23	20.13
SSIM↑	0.54	0.56	0.56	0.65	0.72	<u>0.83</u>
Method	KinD [65]	KinD++ [64]	NE [18]	Bread [12]	RCTNet [21]	IAT [5]
PSNR↑	20.87	21.30	21.52	22.96	22.67	23.38
SSIM↑	0.80	0.82	0.76	0.82	0.79	0.81
Method	HWMNet [9]	LLFlow [45]	StarDiffusion* [60]	DDRM* [19]	LDM* [34]	L²DM*(Ours)
PSNR↑	24.24	**24.99**	20.77	16.41	21.41	<u>24.54</u>
SSIM↑	<u>0.83</u>	**0.92**	0.80	0.65	0.75	<u>0.83</u>

Table 2. Quantitative comparison on the **LOLV2-real dataset**. ↑ denotes that larger values lead to better quality. Models marked with an asterisk ∗ indicate that the model is implemented based on the diffusion model. The best results are shown in **bold**, and the second-best results are <u>underlined</u>.

Methods	DeepUPE [43]	RF [22]	DeepLPF [30]	KIND [65]	FIDE [50]	LPNet [24]
PSNR↑	13.27	14.05	14.10	14.74	16.85	17.80
SSIM↑	0.452	0.458	0.480	0.641	0.678	0.792
Methods	3DLUT [62]	MIR-Net [61]	UNIE [18]	LCDR [42]	LLFlow [45]	A3DLUT [44]
PSNR↑	17.59	20.02	20.85	18.57	19.36	18.19
SSIM↑	0.721	0.820	0.724	0.641	0.705	0.745
Methods	Band [56]	EG [17]	Retinex [26]	Sparse [57]	DSN [66]	RCTNet [21]
PSNR↑	20.29	18.23	18.37	20.06	19.23	20.51
SSIM↑	0.831	0.617	0.723	0.815	0.736	0.831
Methods	UTVNet [67]	SCI [28]	Uretinex [48]	SNR [53]	SMG [54]	L²DM*(Ours)
PSNR↑	20.37	20.28	21.16	21.48	<u>24.62</u>	**24.80**
SSIM↑	0.834	0.752	0.840	<u>0.849</u>	**0.867**	0.817

Table 3. Quantitative comparison on the **LOLV2-synthetic dataset.** ↑ denotes that larger values lead to better quality. Models marked with an asterisk ∗ indicate that the model is implemented based on the diffusion model. The best results are shown in **bold**, and the second-best results are <u>underlined</u>.

Methods	KIND [65]	SID [3]	DeepUPE [43]	FIDE [51]	RF [22]	DeepLPF [30]
PSNR↑	13.29	15.04	15.08	15.20	15.97	16.02
SSIM↑	0.578	0.610	0.623	0.612	0.632	0.587
Methods	Retinex [26]	3DLUT [62]	HWMNet [9]	LCDR [42]	A3DLUT [44]	Bread [12]
PSNR↑	16.55	18.04	18.79	18.91	18.92	19.28
SSIM↑	0.652	0.800	0.817	0.825	0.838	0.831
Methods	LPNet [24]	LLFlow [45]	DSN [66]	UTVNet [67]	MIR-Net [61]	UNIE [18]
PSNR↑	19.51	19.69	21.22	21.62	21.94	21.84
SSIM↑	0.846	0.871	0.827	0.904	0.876	0.884
Methods	Sparse [57]	SCI [28]	RCTNet [21]	Uretinex [48]	Band [56]	L²DM*(Ours)
PSNR↑	22.05	22.20	22.44	22.89	<u>23.22</u>	**23.96**
SSIM↑	<u>0.905</u>	0.887	0.891	0.895	**0.927**	0.786

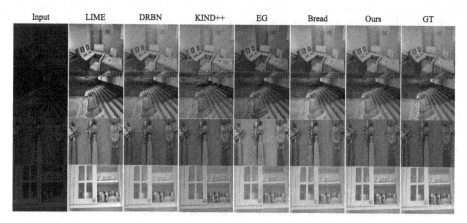

Fig. 2. Qualitative comparison with state-of-the-art methods on the **LOL dataset.**

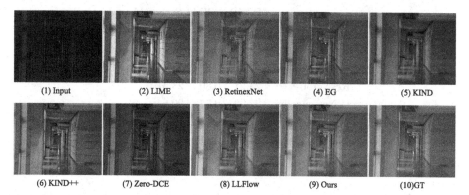

Fig. 3. Qualitative comparison with state-of-the-art methods on the **LOLV2-real dataset.**

4.2 Comparsion with SOTA Methods

LOL Dataset. We first compare L^2DM with SOTA methods on the LOL dataset. As mentioned in the previous subsection, the generated samples exhibit uncertainty. Therefore, our validation results are averaged over 5 tests. As shown in Table 1, we differentiate between methods based on conventional U-Net architecture and those based on the diffusion model. Our proposed model achieves state-of-the-art performance among all diffusion-based methods, surpassing the second-best method by a significant margin of 3.13 in terms of PSNR. Figure 2 shows qualitative comparisons with other methods, the differences between our generated images and the ground truth images are barely discernible.

LOLV2 Dataset. For the LOLV2 dataset, since there are limited results available for diffusion-based methods, we compare our method with state-of-the-art approaches based on CNN and Transformer models. The quantitative results of LOLV2 real part and synthetic are shown in Table 2 and Table 3, respectively. Our proposed model achieves the top position in terms of PSNR on the LOLV2 dataset. However, there is still a gap compared to mainstream methods in terms of SSIM. This is an area that we aim to improve in our future work. Figure 3 presents the qualitative comparison results on the LOLV2-real dataset, where our method showcases excellent enhancement performance.

4.3 Ablation Studies

In this subsection, we conduct ablation studies on the main components of L^2DM to better demonstrate the effectiveness of each module of our system. The experiments were conducted on the LOL dataset.

Table 4. Ablation study on the L^2DM using LOL dataset. *Param* stands for the parameter size of the condition module, while "–" means the hyperparameter variation does not affect the number of parameters in the conditional module.

	Metrics		
	PSNR	SSIM	*Param*(M)
Baseline	24.54	0.83	0.07
KL-reg	19.83	0.76	–
SpatialRescaler	17.51	0.65	None
ResNet50	23.21	0.80	23.51
U-Net-encoder	22.02	0.77	20.64
L2	23.89	0.81	–
Charbonnier	24.10	0.81	–

Regularization. In the autoencoder module, we replaced VQ regularization with Kullback-Leibler (KL) regularization. KL regularization applies a slight

KL penalty to the learned latent towards a standard normal distribution. From Table 4, we can observe that using VQ regularization achieves better generation results.

Condition Module. We compared our proposed condition module ViTCond-Net ($i = 4$) with some conditional modules, such as SpatialRescaler in LDM [34], ResNet and U-Net. We utilize the encoder component of a U-Net architecture that incorporates an Attention mechanism to extract features. The comparison results are shown in Table 4. Our condition module outperforms others in terms of both model parameter size and image generation quality.

Loss Function. We replaced the L1 loss with the L2 loss and the Charbonnier loss [2]. However, our experiments revealed that the L1 loss outperforms both the L2 and Charbonnier losses.

5 Conclusion

In this paper, we present L^{2}DM, an end-to-end trainable low-light image enhancement model based on the diffusion model, showcasing exceptional image generation performance. L^{2}DM incorporates an autoencoder module that significantly reduces the computational demands of the diffusion model, along with an innovative lightweight conditional module known as ViTCondNet. This combination empowers the model to process low-light images as input and generate high-quality enhanced images. Furthermore, the underwhelming performance of diffusion models in terms of PSNR on low-light enhancement datasets has resulted in their limited adoption in the field. Our approach addresses this issue by demonstrating that diffusion-based methods can achieve remarkable outcomes across diverse datasets. Through extensive experiments and comparisons on publicly available datasets, we provide further validation of our approach's performance. It is our intention for L^{2}DM to serve as a baseline for low-light image enhancement and to encourage the application of diffusion models in low-level visual tasks.

Acknowledgments. This work was supported by the National Natural Science Foundation of China (Grant Nos. 62376003, 62306003, 62372004, 62302005).

References

1. Arici, T., Dikbas, S., Altunbasak, Y.: A histogram modification framework and its application for image contrast enhancement. IEEE Trans. Image Process. **18**(9), 1921–1935 (2009)
2. Charbonnier, P., Blanc-Feraud, L., Aubert, G., Barlaud, M.: Two deterministic half-quadratic regularization algorithms for computed imaging. In: Proceedings of IEEE International Conference on Image Processing, vol. 2, pp. 168–172. IEEE (1994)

3. Chen, C., Chen, Q., Xu, J., Koltun, V.: Learning to see in the dark. In: Proceedings of IEEE/CVF Conference on Computer Vision and Pattern Recognition, pp. 3291–3300 (2018)
4. Croitoru, F.A., Hondru, V., Ionescu, R.T., Shah, M.: Diffusion models in vision: a survey. IEEE Trans. Pattern Anal. Mach. Intelli. (2023)
5. Cui, Z., et al.: Illumination adaptive transformer. arXiv preprint arXiv:2205.14871 (2022)
6. Dhariwal, P., Nichol, A.: Diffusion models beat GANs on image synthesis. Adv. Neural. Inf. Process. Syst. **34**, 8780–8794 (2021)
7. Dong, X., et al.: Abandoning the Bayer-filter to see in the dark. In: Proceedings of IEEE/CVF Conference on Computer Vision and Pattern Recognition, pp. 17431–17440 (2022)
8. Esser, P., Rombach, R., Ommer, B.: Taming transformers for high-resolution image synthesis. In: Proceedings of the IEEE/CVF Conference on Computer Vision and Pattern Recognition, pp. 12873–12883 (2021)
9. Fan, C.M., Liu, T.J., Liu, K.H.: Half wavelet attention on M-Net+ for low-light image enhancement. In: 2022 IEEE International Conference on Image Processing (ICIP), pp. 3878–3882. IEEE (2022)
10. Fan, M., Wang, W., Yang, W., Liu, J.: Integrating semantic segmentation and retinex model for low-light image enhancement. In: Proceedings of the 28th ACM International Conference on Multimedia (ACMMM). pp. 2317–2325 (2020)
11. Guo, C., et al.: Zero-reference deep curve estimation for low-light image enhancement. In: Proceedings of IEEE/CVF Conference on Computer Vision and Pattern Recognition, pp. 1780–1789 (2020)
12. Guo, X., Hu, Q.: Low-light image enhancement via breaking down the darkness. Int. J. Comput. Vision **131**(1), 48–66 (2023)
13. Guo, X., Li, Y., Ling, H.: LIME: low-light image enhancement via illumination map estimation. IEEE Trans. Image Process. **26**(2), 982–993 (2016)
14. Ho, J., Jain, A., Abbeel, P.: Denoising diffusion probabilistic models. Adv. Neural. Inf. Process. Syst. **33**, 6840–6851 (2020)
15. Isola, P., Zhu, J.Y., Zhou, T., Efros, A.A.: Image-to-image translation with conditional adversarial networks. In: Proceedings of the IEEE Conference on Computer Vision and Pattern Recognition, pp. 1125–1134 (2017)
16. Jiang, N., Lin, J., Zhang, T., Zheng, H., Zhao, T.: Low-light image enhancement via stage-transformer-guided network. IEEE Trans. Circuits Syst. Video Technol. (2023)
17. Jiang, Y., et al.: EnlightenGAN: deep light enhancement without paired supervision. IEEE Trans. Image Process. **30**, 2340–2349 (2021)
18. Jin, Y., Yang, W., Tan, R.T.: Unsupervised night image enhancement: when layer decomposition meets light-effects suppression. In: Avidan, S., Brostow, G., Cissé, M., Farinella, G.M., Hassner, T. (eds.) ECCV 2022, Part XXXVII. LNCS, vol. 13697, pp. 404–421. Springer, Cham (2022). https://doi.org/10.1007/978-3-031-19836-6_23
19. Kawar, B., Elad, M., Ermon, S., Song, J.: Denoising diffusion restoration models. arXiv preprint arXiv:2201.11793 (2022)
20. Kim, G., Kwon, D., Kwon, J.: Low-lightgan: low-light enhancement via advanced generative adversarial network with task-driven training. In: 2019 IEEE International Conference on Image Processing (ICIP), pp. 2811–2815. IEEE (2019)
21. Kim, H., Choi, S.M., Kim, C.S., Koh, Y.J.: Representative color transform for image enhancement. In: Proceedings of the IEEE/CVF International Conference on Computer Vision, pp. 4459–4468 (2021)

22. Kosugi, S., Yamasaki, T.: Unpaired image enhancement featuring reinforcement-learning-controlled image editing software. In: Proceedings of the AAAI Conference on Artificial Intelligence, vol. 34, pp. 11296–11303 (2020)

23. Lee, C., Lee, C., Kim, C.S.: Contrast enhancement based on layered difference representation of 2D histograms. IEEE Trans. Image Process. **22**(12), 5372–5384 (2013)

24. Li, J., Li, J., Fang, F., Li, F., Zhang, G.: Luminance-aware pyramid network for low-light image enhancement. IEEE Trans. Multimedia **23**, 3153–3165 (2020)

25. Lim, S., Kim, W.: DSLR: deep stacked laplacian restorer for low-light image enhancement. IEEE Trans. Multimedia **23**, 4272–4284 (2020)

26. Liu, R., Ma, L., Zhang, J., Fan, X., Luo, Z.: Retinex-inspired unrolling with cooperative prior architecture search for low-light image enhancement. In: Proceedings of the IEEE/CVF Conference on Computer Vision and Pattern Recognition, pp. 10561–10570 (2021)

27. Lu, C., Zhou, Y., Bao, F., Chen, J., Li, C., Zhu, J.: DPM-solver: a fast ode solver for diffusion probabilistic model sampling in around 10 steps. arXiv preprint arXiv:2206.00927 (2022)

28. Ma, L., Ma, T., Liu, R., Fan, X., Luo, Z.: Toward fast, flexible, and robust low-light image enhancement. In: Proceedings of the IEEE/CVF Conference on Computer Vision and Pattern Recognition, pp. 5637–5646 (2022)

29. Mirza, M., Osindero, S.: Conditional generative adversarial nets. arXiv preprint arXiv:1411.1784 (2014)

30. Moran, S., Marza, P., McDonagh, S., Parisot, S., Slabaugh, G.: DeepLPF: deep local parametric filters for image enhancement. In: Proceedings of the IEEE/CVF Conference on Computer Vision and Pattern Recognition, pp. 12826–12835 (2020)

31. Park, T., Liu, M.Y., Wang, T.C., Zhu, J.Y.: Semantic image synthesis with spatially-adaptive normalization. In: Proceedings of the IEEE/CVF Conference on Computer Vision and Pattern Recognition, pp. 2337–2346 (2019)

32. Reed, S., Akata, Z., Yan, X., Logeswaran, L., Schiele, B., Lee, H.: Generative adversarial text to image synthesis. In: International Conference on Machine Learning, pp. 1060–1069. PMLR (2016)

33. Ren, W., et al.: Deep video dehazing with semantic segmentation. IEEE Trans. Image Process. **28**(4), 1895–1908 (2018)

34. Rombach, R., Blattmann, A., Lorenz, D., Esser, P., Ommer, B.: High-resolution image synthesis with latent diffusion models. In: Proceedings of the IEEE/CVF Conference on Computer Vision and Pattern Recognition, pp. 10684–10695 (2022)

35. Ronneberger, O., Fischer, P., Brox, T.: U-Net: convolutional networks for biomedical image segmentation. In: Navab, N., Hornegger, J., Wells, W.M., Frangi, A.F. (eds.) MICCAI 2015. LNCS, vol. 9351, pp. 234–241. Springer, Cham (2015). https://doi.org/10.1007/978-3-319-24574-4_28

36. Saharia, C., et al.: Palette: image-to-image diffusion models. In: ACM SIGGRAPH 2022 Conference Proceedings, pp. 1–10 (2022)

37. Sohl-Dickstein, J., Weiss, E., Maheswaranathan, N., Ganguli, S.: Deep unsupervised learning using nonequilibrium thermodynamics. In: International Conference on Machine Learning, pp. 2256–2265. PMLR (2015)

38. Sohn, K., Lee, H., Yan, X.: Learning structured output representation using deep conditional generative models. In: Advances in Neural Information Processing Systems, vol. 28 (2015)

39. Song, Y., Ermon, S.: Generative modeling by estimating gradients of the data distribution. In: Advances in Neural Information Processing Systems, vol. 32 (2019)

40. Van Den Oord, A., Vinyals, O., et al.: Neural discrete representation learning. In: Advances in Neural Information Processing Systems, vol. 30 (2017)
41. Vaswani, A., et al.: Attention is all you need. In: Advances in Neural Information Processing Systems, vol. 30 (2017)
42. Wang, H., Xu, K., Lau, R.W.: Local color distributions prior for image enhancement. In: Avidan, S., Brostow, G., Cissé, M., Farinella, G.M., Hassner, T. (eds.) ECCV 2022, Part XVIII. LNCS, vol. 13678, pp. 343–359. Springer, Cham (2022). https://doi.org/10.1007/978-3-031-19797-0_20
43. Wang, R., Zhang, Q., Fu, C.W., Shen, X., Zheng, W.S., Jia, J.: Underexposed photo enhancement using deep illumination estimation. In: Proceedings of the IEEE/CVF Conference on Computer Vision and Pattern Recognition (CVPR), pp. 6849–6857 (2019)
44. Wang, T., Li, Y., Peng, J., Ma, Y., Wang, X., Song, F., Yan, Y.: Real-time image enhancer via learnable spatial-aware 3D lookup tables. In: Proceedings of the IEEE/CVF International Conference on Computer Vision, pp. 2471–2480 (2021)
45. Wang, Y., Wan, R., Yang, W., Li, H., Chau, L.P., Kot, A.: Low-light image enhancement with normalizing flow. In: Proceedings of the AAAI Conference on Artificial Intelligence, vol. 36, pp. 2604–2612 (2022)
46. Wang, Z., Bovik, A.C., Sheikh, H.R., Simoncelli, E.P.: Image quality assessment: from error visibility to structural similarity. IEEE Trans. Image Process. **13**(4), 600–612 (2004)
47. Wei, C., Wang, W., Yang, W., Liu, J.: Deep retinex decomposition for low-light enhancement. arXiv preprint arXiv:1808.04560 (2018)
48. Wu, W., Weng, J., Zhang, P., Wang, X., Yang, W., Jiang, J.: URetinex-Net: Retinex-based deep unfolding network for low-light image enhancement. In: Proceedings of the IEEE/CVF Conference on Computer Vision and Pattern Recognition, pp. 5901–5910 (2022)
49. Wu, X., Liu, X., Hiramatsu, K., Kashino, K.: Contrast-accumulated histogram equalization for image enhancement. In: 2017 IEEE International Conference on Image Processing (ICIP), pp. 3190–3194. IEEE (2017)
50. Xu, K., Yang, X., Yin, B., Lau, R.W.: Learning to restore low-light images via decomposition-and-enhancement. In: Proceedings of the IEEE/CVF Conference on Computer Vision and Pattern Recognition, pp. 2281–2290 (2020)
51. Xu, K., Yang, X., Yin, B., Lau, R.W.: Learning to restore low-light images via decomposition-and-enhancement. In: Proceedings of the IEEE/CVF Conference on Computer Vision and Pattern Recognition (CVPR), pp. 2281–2290 (2020)
52. Xu, W., Dong, X., Ma, L., Teoh, A.B.J., Lin, Z.: Rawformer: an efficient vision transformer for low-light raw image enhancement. IEEE Signal Process. Lett. **29**, 2677–2681 (2022)
53. Xu, X., Wang, R., Fu, C.W., Jia, J.: SNR-aware low-light image enhancement. In: Proceedings of the IEEE/CVF Conference on Computer Vision and Pattern Recognition, pp. 17714–17724 (2022)
54. Xu, X., Wang, R., Lu, J.: Low-light image enhancement via structure modeling and guidance. In: Proceedings of the IEEE/CVF Conference on Computer Vision and Pattern Recognition, pp. 9893–9903 (2023)
55. Yang, W., Wang, S., Fang, Y., Wang, Y., Liu, J.: From fidelity to perceptual quality: a semi-supervised approach for low-light image enhancement. In: Proceedings of the IEEE/CVF Conference on Computer Vision and Pattern Recognition, pp. 3063–3072 (2020)

56. Yang, W., Wang, S., Fang, Y., Wang, Y., Liu, J.: Band representation-based semi-supervised low-light image enhancement: bridging the gap between signal fidelity and perceptual quality. IEEE Trans. Image Process. **30**, 3461–3473 (2021)
57. Yang, W., Wang, W., Huang, H., Wang, S., Liu, J.: Sparse gradient regularized deep retinex network for robust low-light image enhancement. IEEE Trans. Image Process. **30**, 2072–2086 (2021)
58. Yu, J., et al.: Vector-quantized image modeling with improved VQGAN. arXiv preprint arXiv:2110.04627 (2021)
59. Yuan, N., et al.: Low-light image enhancement by combining transformer and convolutional neural network. Mathematics **11**(7), 1657 (2023)
60. Yuan, Y., et al.: Learning to kindle the starlight. arXiv preprint arXiv:2211.09206 (2022)
61. Zamir, S.W., et al.: Learning enriched features for real image restoration and enhancement. In: Vedaldi, A., Bischof, H., Brox, T., Frahm, J.-M. (eds.) ECCV 2020. LNCS, vol. 12370, pp. 492–511. Springer, Cham (2020). https://doi.org/10.1007/978-3-030-58595-2_30
62. Zeng, H., Cai, J., Li, L., Cao, Z., Zhang, L.: Learning image-adaptive 3D lookup tables for high performance photo enhancement in real-time. IEEE Trans. Pattern Anal. Mach. Intell. **44**(4), 2058–2073 (2020)
63. Zhang, R., Isola, P., Efros, A.A., Shechtman, E., Wang, O.: The unreasonable effectiveness of deep features as a perceptual metric. In: Proceedings of the IEEE Conference on Computer Vision and Pattern Recognition, pp. 586–595 (2018)
64. Zhang, Y., Guo, X., Ma, J., Liu, W., Zhang, J.: Beyond brightening low-light images. Int. J. Comput. Vision **129**, 1013–1037 (2021)
65. Zhang, Y., Zhang, J., Guo, X.: Kindling the darkness: a practical low-light image enhancer. In: Proceedings of the 27th ACM International Conference on Multimedia (ACMMM), pp. 1632–1640 (2019)
66. Zhao, L., Lu, S.P., Chen, T., Yang, Z., Shamir, A.: Deep symmetric network for underexposed image enhancement with recurrent attentional learning. In: Proceedings of the IEEE/CVF International Conference on Computer Vision, pp. 12075–12084 (2021)
67. Zheng, C., Shi, D., Shi, W.: Adaptive unfolding total variation network for low-light image enhancement. In: Proceedings of the IEEE/CVF International Conference on Computer Vision, pp. 4439–4448 (2021)
68. Zhu, M., Pan, P., Chen, W., Yang, Y.: EEMEFN: low-light image enhancement via edge-enhanced multi-exposure fusion network. In: Proceedings of the AAAI Conference on Artificial Intelligence, vol. 34, pp. 13106–13113 (2020)

Multi-domain Information Fusion for Key-Points Guided GAN Inversion

Ruize Xu[1], Xiaowen Qiu[2], Boan He[2], Weifeng Ge[2], and Wenqiang Zhang[1,2(✉)]

[1] Academy for Engineering and Technology, Fudan University, Shanghai 200433, China
rzxu21@m.fudan.edu.cn
[2] School of Computer Science, Fudan University, Shanghai 200433, China
{19307130022,20210240067,wfge,wqzhang}@fudan.edu.cn

Abstract. In recent years, GAN inversion has emerged as a powerful technique for bridging the gap between real and fake image domains, and it has become increasingly important for enabling pre-trained GAN models for real image editing applications. However, current GAN inversion methods are limited by network parameters and model structures, and there is still room for improvement in accurate reconstruction and latent editing tasks. In this paper, we propose a two-stage model that fine-tunes a pre-trained Masked Autoencoder in the first stage and utilizes multi-layers information fusion to obtain an initial global latent code. We then use this latent code as global queries for the subsequent cross-attention-based fusion of local key patch, key point feature, and residual image information in the second stage, guided by facial landmarks. This allows our model to better embed images in the $W+$ space and perform related attribute editing, achieving better results than current state-of-the-art methods. We conduct extensive experiments to demonstrate the capabilities of our model, as well as the roles of relevant modules, and study the effects of different domain information on inversion.

Keywords: GAN Inversion · Image Editing · Facial Key-points

1 Introduction

Generative adversarial networks (GANs) have seen rapid development in recent years, particularly with the introduction of technologies such as BigGAN [4], PGGAN [14], and StyleGAN [15–17]. This has opened up exciting possibilities for image editing, allowing for the modification of desired attributes while preserving other details. Although significant improvements have been achieved in the generation of results using diffusion models primarily driven by stable diffusion, combined with parameter tuning methods such as LoRA and controlnet network for respective control, GAN inversion still maintains an advantage in terms of generation speed and controllability of editing (Fig. 1).

The first step in achieving accurate attribute editing is to obtain a high-quality and highly editable latent code. There are various methods such as optimization-based,

Supplementary Information The online version contains supplementary material available at https://doi.org/10.1007/978-981-99-8552-4_12.

Fig. 1. Results of our method in image inversion, face attribute editing, style mixing, and faces morphing task. The first row displays the original image and our inversion & editing result, where we use InterfaceGAN [29], GANSpace [9], and StyleCLIP [26] to edit facial attributes. The second row shows the style mixing [16] result, where we take the style B of source B from the coarse $(4^2 - 8^2)$, middle $(16^2 - 32^2)$, and fine $(64^2 - 256^2)$ resolutions and the rest from source A. The third row shows the faces morphing result, where we obtain the latent code of image A and image B through inversion and achieve the transition between the two faces by interpolation.

learning-based, and hybrid methods, which focus on the image itself and extract its features to obtain a latent code through hierarchical mapping or direct gradient back propagation. Traditional StyleGAN [16] maps the Z space of PGGAN [14] to W space for editing, where all 18 layers of vectors in W space are the same and contain less information. pSp [27] proposes to map images to the $W+$ space, which contains richer information and allows different latent codes for each layer. This leads to improve reconstruction quality but reduces editability [31]. From this perspective, GAN inversion can be viewed as a task involving restricted information compression. One of the most effective methods involves mapping real images to a $W+$ space, which exhibits subtle differences between layers but is able to effectively decouple attributes controlled by latent codes between layers. This approach has the advantage of containing a high amount of information that enables effective image reconstruction and editability. However, currently, it remains a challenging task to improve the editability while simultaneously reducing reconstruction distortion.

We propose a unified mapping approach to map an image to each layer of latent codes. Besides, We employ a pre-trained Masked Autoencoder (MAE) [10] as our feature extractor, given its superior ability to extract information and its pre-training on ImageNet [6]. The latent code is then mapped through its class token, which fuses global information. Despite the class token's ability to roughly reconstruct the image, it lacks accuracy in local details such as the corners of the eyes and mouth, and the skin color of the face. To encode rich local information into the latent code, we propose using key point information to obtain local patch tokens of facial key regions encoded by the MAE. Residual information has been shown to improve the quality of reconstructed images in multiple reconstruction attempts, as demonstrated by Restyle [3] and HFGI [35]. Additionally, contour information obtained from key points can improve the reconstruction of facial expressions and help us better combine information from the local and residual domains to improve the reconstruction effect. To integrate information from these three domains, we propose using an attention method [25, 32], which enables the latent codes to selectively attend to important parts in each domain.

In this paper, we propose a two-stage Encoder-based GAN inversion method. In the first stage, an MAE encoder is utilized to obtain the initial latent code. This code is then inputted into a fixed StyleGAN [17] to generate an initial inversion image. Residual image information is obtained by subtracting the initial inversion image from the original image. Local detailed information is expanded using key-point information, which we refer to as key patch tokens, while facial contour information is extracted from the key-point image features. In the second stage, we enhance information exchange through self-attention and cross-attention mechanisms. We use the initial latent code as queries and combine it with the three types of information we obtained. This allows the initial latent code to absorb more details from different domains, resulting in the accurate reconstruction of the original image with high fidelity.

This paper makes the following contributions:

- We use facial landmark information [18] to guide image reconstruction, propose key-point patch loss and achieve high-quality reconstruction results by combining it with a pre-trained MAE [10] model.
- We propose a GAN inversion method that integrates information from local key point patch tokens, facial contour information, and residual image, and we investigate the impact of each domain on the quality of reconstruction.
- We apply our method to various off-the-shelves techniques, including Interface-GAN [29], GANSpace [9], StyleCLIP [26], style mixing [16], and faces morphing. We quantitatively compare our method with current state-of-the-art techniques and demonstrate that it achieves robust reconstruction results with sufficient decoupling between the layers of the latent space, resulting in high editability.

2   Related Works

2.1   GAN Inversion

There are three main types of GAN inversion methods: optimization-based [1, 2, 16, 20, 23, 42], learning-based [3, 12, 19, 31, 37, 41], and hybrid [8, 28, 42]. Optimization-based methods minimize the distance between the generator output and the target output via iterative gradient descent, such as Adam [1, 13] used the Covariance Matrix Adaptation (CMA) algorithm for gradient-free optimization. However it may take minutes to process a single image. Learning-based methods [19, 31, 37, 41] design an encoder to learn the mapping between images and latent codes, which are fast to compute but may result in lower quality outputs. IDGI [41] proposes an in-domain method for inversion, and is applicable in both Z and W spaces. Methods like pSp [27] and GH-Feat [37] capture multi-scale information from images and map it hierarchically to the latent code, serving as the backbone for subsequent work. Restyle [3] and other methods based on pSp's encoder iteratively optimize the latent code, and style transformer [12] utilizes DETR's [5] transformer block to interactively combine multi-scale information with a learnable latent vector via cross-attention. Hybrid methods [8, 28, 42] combine the advantages of optimization-based and learning-based approaches, using the encoder method to obtain an initial latent code and then optimizing it further via optimization-based methods for faster and better results. In terms of the latent space, there are several types, including Z space [14], W space [16], W^+ space [17], S space [21,36], P space, etc. e4e [31] studied the trade-off between reconstruction and editing in different latent spaces, adding regularization and adversarial losses to constrain the generation space of the latent code. HFGI [35] and StyleMapGAN [19] expand the spatial dimension of the latent code to provide more information for subsequent generation. But it is necessary to adjust the generator simultaneously. Our method strictly relies on the $W+$ space and combines multiple types of information to optimize the output of the initial encoder, achieving accurate image reconstruction.

2.2 Latent Space Manipulation

There are two main types of methods for searching for editable directions and corresponding semantic attributes in the latent space. The first type is supervised methods [26,29,34,36,38], which rely on well-trained attribute classifiers or image labels to guide the search for the corresponding editing direction. The second type is unsupervised methods [9,30,33,39], which do not require additional models for attribute evaluation. More information in Supplement Material (Fig. 2).

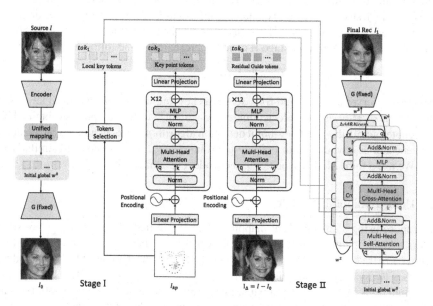

Fig. 2. Our Multi-domain Information Fusion Inversion network is a two-stage model that generates the initial latent code w_0 and the initial reconstruction image I_0 in stage I, where patch tokens with rich local information guided by key points are selected using a tokens selector. In stage II, the residual image I_Δ is obtained, and the key point image, which encodes facial contour information, is combined with local key point patch tokens from stage I and sent to the multi-domain information fusion module as key-value pairs. These pairs are fused with the query of the initial latent code to obtain the final latent code w_3, which preserves the original facial structure in the inversion result.

3   Methodology

Our pipeline consists of two stage. In stage I, we utilize a fine-tuned pre-trained Masked Autoencoder and the StyleGAN decoder to generate a coarse initial latent code w^0 and its corresponding inversion result with obvious discrepancy compared to the source image. Then in the stage II, we fuse information obtained from three domains into the initial latent code, generating a refined latent code w^3 whose corresponding inversion result that shows less discrepancy compared to the source image.

3.1   Overall Architecture

In this study, we note that z represents the encoded tokens acquired through MAE Encoder E_1, with z^{cls} denoting the class token and z^i representing the patch token where i signifies the total number of patches. During stage I, we feed the source image $I \in \mathbb{R}^{H \times W \times 3}$ into MAE encoder E_1 to retrieve the class token z^{cls} and map it to the initial latent code $w^0 = F(E_1(I))$, $F(\cdot)$ denotes Unified Mapping operation. Subsequently, we utilize a StyleGAN generator G to generate an initial inversion outcome I_0. In stage II, we refine w^0 utilizing three transformer blocks to produce the final latent code w^3. In each transformer block, w^{n-1} initially interacts with itself through multi-head self-attention, followed by interaction with tokens tok_n (where n represents the corresponding transformer block) containing information from one of the three domains via multi-head cross-attention. The standard multi-head attention operation is represented by $Attn(\cdot)$. Therefore, stage II can be represented as the following formula:

$$w^n = Attn(w^{n-1}, tok_n, tok_n), n = 1, 2, 3 \tag{1}$$

where w^{n-1} serves as query and tok_n serves as key and value. tok_1 are local key point patch tokens of $z^i = E_1(I)$ selected under the guidance of 68 key-points (obtained by $S(\cdot)$ denotes an off-the-shelf key-point detection algorithm). tok_2 generated by a second MAE encoder E_2, whose input is a key-point image I_{kp}. tok_3 is acquired similarly to tok_2. We use a third MAE encoder E_3 to encode a residual image I_Δ into tok_3.

3.2   Unified Mapping Module

We propose using a shared latent mapping module, Unified Mapping ($F(\cdot)$), for style vectors and key point patch tokens, which differs from the approach in pSp [27] that employs separate modules for each StyleGAN layer. The Unified Mapping module comprises two groups of one-dimensional convolutional layers with PReLU [11] activation functions that capture contextual information and enhance representation capabilities. Applying the Unified Mapping module to the class token yields w^0, while expanding the information of selected key point patch tokens to tok_1 increases the input information's dimension by three times. Concatenating six class tokens from every two layers of the Encoder, from shallow to deep, along the zero dimension after applying Unified Mapping results in an 18×512 dimensional latent code. This operation is also applied to the selected key patch tokens.

$$w^0 = concat(F(z_i^{cls})), tok_1 = concat(F(S(z^j))) \tag{2}$$

For detailed network implementation of E_1, please refer to the Supplement material. The following notation is used, where i denotes the layer of the MAE encoder, j denotes the patch from all local patches, and $S(\cdot)$ denotes the key patch selector.

3.3 Multi Domain Information Fusion

To provide stronger guidance, we aid our inversion process with information from three domains, including key-point patches tok_1, key-point image I_{kp} and residual image I_Δ.

Key-Point Patches. Previous works, such as IntereStyle [24], have highlighted the importance of the face region by utilizing an interest region mask. We take a step further in this direction incorporating 68 facial key points [18] information. We select patches where these key points fall in as the key-point patches, and map them to tok_1 through the Unified Mapping operation, which is the same as mapping the class token z^{cls} to w^0. This process increases the dimension of tok_1 by 18 times compared to before. tok_1 is then used to refine the latent code through the cross-attention module in the first transformer block of the stage II network.

Key-Point Image. To enhance the fidelity of the inverted image with respect to the original facial structure, we introduce a key-point image I_{kp} that captures the 68 facial landmarks. This key-point image includes the 68 key points detected in the face region, categorized by color for the contour, eyebrows, nose, eyes, and mouth, against a gray background. I_{kp} is encoded using E_2, a pre-trained MAE encoder incorporated into the pipeline via cross-attention modules in the second transformer block of stage II.

Residual Image. We use a residual image to improve the performance of our GAN inversion network. As shown in previous works [35], the residual image, calculated as the difference between the source image and the inversion result of stage I, can help to reconstruct high-frequency details. To obtain the residual image I_Δ, we subtract the inversion result I_0 from the source image I. Then, we encode I_Δ using another pre-trained MAE encoder E_3, which is also fine-tuned in stage II.

$$tok_2 = E_2(I_{kp}), tok_3 = E_3(I_\Delta) \tag{3}$$

3.4 Key-Point Patch Loss

The key-point patch loss (KPL) is a novel loss function designed for training GAN inversion networks. It prioritizes the reconstruction of the perceptually important areas in a face image by calculating the L1 distance between the key-point patches of a given source image and its corresponding inversion result. Here is the formula for KPL:

$$\mathcal{L}_{kp} = \frac{1}{N} \sum_{i=1}^{N} \left\| I_{mask} \cdot I_1^{kp_i} - I_{mask} \cdot I^{kp_i} \right\|_1 \tag{4}$$

where N is the number of key-points, I_{mask} is a binary mask generated online based on the indices of the key-point patches, I is the source image, I_1 is the final reconstructed image, kp_i represents the i-th key-point patch, and $|| \cdot ||_1$ denotes the L1 distance.

To calculate the key-point patch loss, we first generate a binary mask I_{mask} online based on the indices of the key-point patches. Then, we multiply the source image I and the final reconstructed image I_1 by the binary mask I_{mask}, respectively. By using the key-point patch loss, we can provide an additional gradient for the perceptually important areas in a face image, which helps to prioritize the reconstruction of these areas during GAN inversion network training.

3.5 Training Approaches for Inversion

Our training strategy and optimizer are similar to the pSp [27] framework. During training, the parameters of the decoder are fixed and do not participate in the training process. Our training strategy involves a two-stage process. To fully reconstruct the input image I, we first calculate the L2 loss between the reconstructed image I_0, and the input image I, $\mathcal{L}_2 = \|I - I_0\|_2$. We also use the LPIPS [40] loss, which uses an inception network $F(\cdot)$ (in this case, AlexNet) to extract multi-level feature maps from the source image I and predicted image, and calculate the L2 distance between these feature maps, $\mathcal{L}_{LPIPS} = \|F(I) - F(I_0)\|_2$. To ensure image identity, we also use the pre-trained ArcFace network [7] $R(\cdot)$ to calculate the cosine similarity of the key areas, $\mathcal{L}_{id} = 1 - \langle R(I), R(I_0) \rangle$. In the stage I, we train the modified MAE encoder. Since it is built on top of a pre-trained model, it can quickly generate initial reconstructed images.

$$\mathcal{L}_{stage1} = \lambda_1 \mathcal{L}_2 + \lambda_2 \mathcal{L}_{lpips} + \lambda_3 \mathcal{L}_{id} \tag{5}$$

After training in stage I, we fix the parameters of the MAE encoder and proceed to train the multi-domain information fusion module in the stage II by adding key-point patch loss $\mathcal{L}_{kp}$, which stage II loss can be represented as:

$$\mathcal{L}_{stage2} = \lambda_1 \mathcal{L}_2 + \lambda_2 \mathcal{L}_{lpips} + \lambda_3 \mathcal{L}_{id} + \lambda_4 \mathcal{L}_{kp} \tag{6}$$

Table 1. Quantitative comparison for inversion quality and editing quality.

Method	Inversion					Edit (FID)		
	MSE↓	LPIPS↓	FID↓	ID Similarity↑	Times (s)	Age	Smile	Mohawk
pSp [27]	0.0378	0.1842	29.53	0.7808	0.0459 ± 0.0028	40.8	37.6	42.1
e4e [31]	0.0522	0.2300	37.52	**0.8185**	0.0478 ± 0.0037	36.3	34.5	40.1
StyleT [12]	0.0366	0.1686	25.06	0.3541	0.0375 ± 0.0024	34.2	31.1	35.7
stage1	0.0495	0.1862	27.80	0.7012	**0.0369 ± 0.0019**	35.8	32.4	37.3
stage2	**0.0357**	**0.1653**	**24.62**	0.7230	0.0780 ± 0.0090	**32.3**	**30.2**	**34.4**

4 Experiments

4.1 Implementation Details

Datasets. We adopt the FFHQ [16] as our training set, which comprises of 70,000 high-quality face images, and the CelebA-HQ [14] dataset as our validation set, which consists of 30,000 high-definition facial images. More details in Supplement Material.

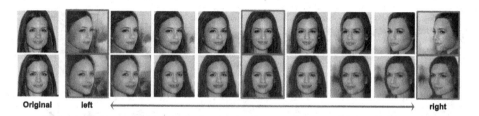

Original   left ←——————————————————————————————————————→ right

Fig. 3. Compared with pSp [27] in terms of edit distance. First row is ours, second row is pSp [27]. InterfaceGAN [29] is used to edit the pose attribute of a face and control the factor α from -10 to 10, $w_{edit} = w + \alpha N_{edit}$.

Training. More details in Supplement Material.

4.2 Comparison with Inversion Method

Quantitative Comparison. We present the quantitative comparisons of the inversion performance in Table 1. Our method achieves state-of-the-art performance in terms of Mean Squared Error (MSE) [40], Learned Perceptual Image Patch Similarity (LPIPS) [40], and Fréchet Inception Distance (FID) [22]. However, due to our two-stage network, our inference time is slower compared to other methods. In comparison to the style transformer [12], which also utilizes the Transformer architecture, we have achieved more than a 50% improvement in ID similarity.

Quality Comparison. In Fig. 4, we show a comparison between our method and others. Our images are more similar to the original images in terms of facial features, facial details, overall color, and skin texture. We observed that style transformer [12] produces the least similar results to the original images, with noticeable differences in facial features. This is also reflected in their low values of ID similarity. On the contrary, our method accurately restores the details corresponding to the face, as shown in Fig. 4. Compared with other methods, our method accurately reconstructs the eye contour and facial texture. Additionally, other methods exhibit slight blurring during reconstruction and differences in the mouth corners, eye corners, and gaze direction compared to the original.

Editing Comparison. In Fig. 4, we show our editing results compared to other methods in columns 6–9. Our method demonstrates stronger editability, as the pSp method shows minimal changes when editing the hairstyle using StyleCLIP [26], while our method generates reasonable results by incorporating the original hairstyle, while maintaining facial structure consistency. In Figure 3, we further compare the long-range editing result of our method and pSp [27]. Our method generates reasonable results from left to right, achieving more drastic turns in the far left and far right than the pSp [27], without significant changes in facial features. In contrast, pSp exhibits distortions and artifacts in long-range editing.

User Study. More information in the supplementary material.

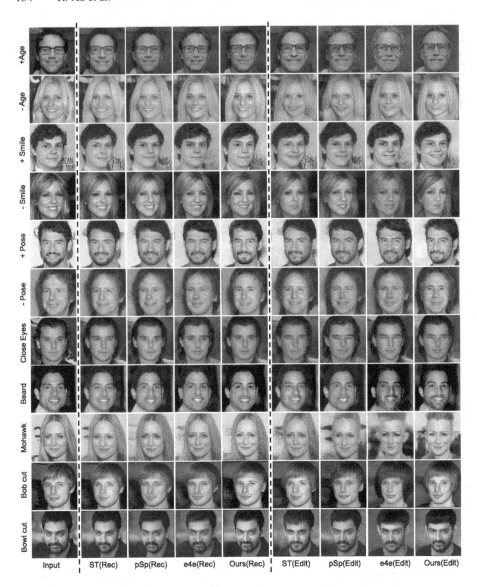

Fig. 4. Visual comparison with state-of-the-art methods: style transformer [12], pSp [27], and e4e [31]. The first column displays the original input image, while columns 2 to 5 show the inversion results. Columns 6 to 9 demonstrate the edited images. Rows 1 to 6 show editing results using InterfaceGAN [29], rows 7 to 8 show results with GANSpace [9], and rows 9 to 11 show results with StyleCLIP [26]. It should be noted that we directly used the pre-trained editing method without any modification.

4.3 Ablation Study and Analyse

Due to space limitations, we conduct detailed ablation experiments in the supplementary material. The ablation results indicate a decrease in MSE, LPIPS, and FID without KPL. The visual results demonstrate that, without KPL, the reconstructed faces exhibit inconsistencies in important facial regions such as the corners of the mouth and eyes when compared to the original images. Furthermore, the overall color and temperature of the reconstructed images are darker and less accurate compared to the original images, as observed in a substantial number of reconstructed images. For MDIF module, the local and residual information have the most significant impact on the results, with the removal of these resulting in a considerable drop in all indicators. And removing the residual domain and contour domain at this point results in worse image inversion, with the residual domain being responsible for reconstructing subtle differences in the image and guiding semantic information. The facial contour information, on the other hand, is used to combine information from the first two domains to assist in their fusion. If the MDIF module is completely removed, the stage I image becomes blurrier, losing some crucial details.

5 Conclusion

In conclusion, there is scope for enhancing GAN inversion tasks by incorporating fine-grained features and utilizing residual and contour information as guidance. Our proposed MDIF method effectively integrates detailed information into the latent code, which is initially initialized with global information. The key patch loss is crucial in supervising the reconstruction of critical image regions. Our high-quality inversion results are compatible with other editing methods and demonstrate robustness and editability. We hope that our method will stimulate extensive discussions and promote future research and practical applications in this field.

References

1. Abdal, R., Qin, Y., Wonka, P.: Image2StyleGAN: how to embed images into the stylegan latent space? In: Proceedings of the IEEE/CVF International Conference on Computer Vision, pp. 4432–4441 (2019)
2. Abdal, R., Qin, Y., Wonka, P.: Image2StyleGAN++: how to edit the embedded images? In: Proceedings of the IEEE/CVF Conference on Computer Vision and Pattern Recognition, pp. 8296–8305 (2020)
3. Alaluf, Y., Patashnik, O., Cohen-Or, D.: Restyle: a residual-based StyleGAN encoder via iterative refinement. In: Proceedings of the IEEE/CVF International Conference on Computer Vision, pp. 6711–6720 (2021)
4. Brock, A., Donahue, J., Simonyan, K.: Large scale GAN training for high fidelity natural image synthesis. arXiv preprint arXiv:1809.11096 (2018)
5. Carion, N., Massa, F., Synnaeve, G., Usunier, N., Kirillov, A., Zagoruyko, S.: End-to-end object detection with transformers. In: Vedaldi, A., Bischof, H., Brox, T., Frahm, J.-M. (eds.) ECCV 2020. LNCS, vol. 12346, pp. 213–229. Springer, Cham (2020). https://doi.org/10. 1007/978-3-030-58452-8_13

6. Deng, J., Dong, W., Socher, R., Li, L.J., Li, K., Fei-Fei, L.: Imagenet: a large-scale hierarchical image database. In: 2009 IEEE Conference on Computer Vision and Pattern Recognition, pp. 248–255. IEEE (2009)

7. Deng, J., Guo, J., Xue, N., Zafeiriou, S.: Arcface: additive angular margin loss for deep face recognition. In: Proceedings of the IEEE/CVF Conference on Computer Vision and Pattern Recognition, pp. 4690–4699 (2019)

8. Guan, S., Tai, Y., Ni, B., Zhu, F., Huang, F., Yang, X.: Collaborative learning for faster stylegan embedding. arXiv preprint arXiv:2007.01758 (2020)

9. Härkönen, E., Hertzmann, A., Lehtinen, J., Paris, S.: Ganspace: discovering interpretable GAN controls. Adv. Neural. Inf. Process. Syst. **33**, 9841–9850 (2020)

10. He, K., Chen, X., Xie, S., Li, Y., Dollár, P., Girshick, R.: Masked autoencoders are scalable vision learners. In: Proceedings of the IEEE/CVF Conference on Computer Vision and Pattern Recognition, pp. 16000–16009 (2022)

11. He, K., Zhang, X., Ren, S., Sun, J.: Delving deep into rectifiers: surpassing human-level performance on imagenet classification. In: Proceedings of the IEEE International Conference on Computer Vision, pp. 1026–1034 (2015)

12. Hu, X., Huang, Q., Shi, Z., Li, S., Gao, C., Sun, L., Li, Q.: Style transformer for image inversion and editing. In: Proceedings of the IEEE/CVF Conference on Computer Vision and Pattern Recognition, pp. 11337–11346 (2022)

13. Huh, M., Zhang, R., Zhu, J.-Y., Paris, S., Hertzmann, A.: Transforming and projecting images into class-conditional generative networks. In: Vedaldi, A., Bischof, H., Brox, T., Frahm, J.-M. (eds.) ECCV 2020. LNCS, vol. 12347, pp. 17–34. Springer, Cham (2020). https://doi.org/10.1007/978-3-030-58536-5_2

14. Karras, T., Aila, T., Laine, S., Lehtinen, J.: Progressive growing of GANs for improved quality, stability, and variation. In: International Conference on Learning Representations (2018)

15. Karras, T., et al.: Alias-free generative adversarial networks. Adv. Neural. Inf. Process. Syst. **34**, 852–863 (2021)

16. Karras, T., Laine, S., Aila, T.: A style-based generator architecture for generative adversarial networks. In: Proceedings of the IEEE/CVF Conference on Computer Vision and Pattern Recognition, pp. 4401–4410 (2019)

17. Karras, T., Laine, S., Aittala, M., Hellsten, J., Lehtinen, J., Aila, T.: Analyzing and improving the image quality of stylegan. In: Proceedings of the IEEE/CVF Conference on Computer Vision and Pattern Recognition, pp. 8110–8119 (2020)

18. Kazemi, V., Sullivan, J.: One millisecond face alignment with an ensemble of regression trees. In: Proceedings of the IEEE Conference on Computer Vision and Pattern Recognition, pp. 1867–1874 (2014)

19. Kim, H., Choi, Y., Kim, J., Yoo, S., Uh, Y.: Exploiting spatial dimensions of latent in GAN for real-time image editing. In: Proceedings of the IEEE/CVF Conference on Computer Vision and Pattern Recognition, pp. 852–861 (2021)

20. Lipton, Z.C., Tripathi, S.: Precise recovery of latent vectors from generative adversarial networks. arXiv preprint arXiv:1702.04782 (2017)

21. Liu, Y., Li, Q., Sun, Z., Tan, T.: Style intervention: how to achieve spatial disentanglement with style-based generators? arXiv preprint arXiv:2011.09699 (2020)

22. Lucic, M., Kurach, K., Michalski, M., Gelly, S., Bousquet, O.: Are GANs created equal? A large-scale study. In: Advances in Neural Information Processing Systems, vol. 31 (2018)

23. Ma, F., Ayaz, U., Karaman, S.: Invertibility of convolutional generative networks from partial measurements. In: Neural Information Processing Systems (2018)

24. Moon, S.J., Park, G.M.: IntereStyle: encoding an interest region for robust StyleGAN inversion. In: Avidan, S., Brostow, G., Cissé, M., Farinella, G.M., Hassner, T. (eds.) ECCV 2022, Part XV. LNCS, vol. 13675, pp. 460–476. Springer, Cham (2022). https://doi.org/10.1007/978-3-031-19784-0_27

25. Niu, Z., Zhong, G., Yu, H.: A review on the attention mechanism of deep learning. Neuro-computing **452**, 48–62 (2021)
26. Patashnik, O., Wu, Z., Shechtman, E., Cohen-Or, D., Lischinski, D.: Styleclip: text-driven manipulation of stylegan imagery. In: Proceedings of the IEEE/CVF International Conference on Computer Vision, pp. 2085–2094 (2021)
27. Richardson, E., et al.: Encoding in style: a stylegan encoder for image-to-image translation. In: Proceedings of the IEEE/CVF Conference on Computer Vision and Pattern Recognition, pp. 2287–2296 (2021)
28. Roich, D., Mokady, R., Bermano, A.H., Cohen-Or, D.: Pivotal tuning for latent-based editing of real images. ACM Trans. Graph. (TOG) **42**(1), 1–13 (2022)
29. Shen, Y., Gu, J., Tang, X., Zhou, B.: Interpreting the latent space of GANs for semantic face editing. In: Proceedings of the IEEE/CVF Conference on Computer Vision and Pattern Recognition, pp. 9243–9252 (2020)
30. Shen, Y., Zhou, B.: Closed-form factorization of latent semantics in GANs. In: Proceedings of the IEEE/CVF Conference on Computer Vision and Pattern Recognition, pp. 1532–1540 (2021)
31. Tov, O., Alaluf, Y., Nitzan, Y., Patashnik, O., Cohen-Or, D.: Designing an encoder for style-gan image manipulation. ACM Trans. Graph. (TOG) **40**(4), 1–14 (2021)
32. Vaswani, A., et al.: Attention is all you need. In: Advances in Neural Information Processing Systems, vol. 30 (2017)
33. Voynov, A., Babenko, A.: Unsupervised discovery of interpretable directions in the GAN latent space. In: International Conference on Machine Learning, pp. 9786–9796. PMLR (2020)
34. Wang, R., et al.: Attribute-specific control units in stylegan for fine-grained image manipulation. In: Proceedings of the 29th ACM International Conference on Multimedia, pp. 926–934 (2021)
35. Wang, T., Zhang, Y., Fan, Y., Wang, J., Chen, Q.: High-fidelity GAN inversion for image attribute editing. In: Proceedings of the IEEE/CVF Conference on Computer Vision and Pattern Recognition, pp. 11379–11388 (2022)
36. Wu, Z., Lischinski, D., Shechtman, E.: Stylespace analysis: disentangled controls for style-gan image generation. In: Proceedings of the IEEE/CVF Conference on Computer Vision and Pattern Recognition, pp. 12863–12872 (2021)
37. Xu, Y., Shen, Y., Zhu, J., Yang, C., Zhou, B.: Generative hierarchical features from synthesizing images. In: Proceedings of the IEEE/CVF Conference on Computer Vision and Pattern Recognition, pp. 4432–4442 (2021)
38. Yao, X., Newson, A., Gousseau, Y., Hellier, P.: A latent transformer for disentangled face editing in images and videos. In: Proceedings of the IEEE/CVF International Conference on Computer Vision, pp. 13789–13798 (2021)
39. Yüksel, O.K., Simsar, E., Er, E.G., Yanardag, P.: LatentCLR: a contrastive learning approach for unsupervised discovery of interpretable directions. In: Proceedings of the IEEE/CVF International Conference on Computer Vision, pp. 14263–14272 (2021)
40. Zhang, R., Isola, P., Efros, A.A., Shechtman, E., Wang, O.: The unreasonable effectiveness of deep features as a perceptual metric. In: Proceedings of the IEEE Conference on Computer Vision and Pattern Recognition, pp. 586–595 (2018)
41. Zhu, J., Shen, Y., Zhao, D., Zhou, B.: In-domain GAN inversion for real image editing. In: Vedaldi, A., Bischof, H., Brox, T., Frahm, J.-M. (eds.) ECCV 2020. LNCS, vol. 12362, pp. 592–608. Springer, Cham (2020). https://doi.org/10.1007/978-3-030-58520-4_35
42. Zhu, J.-Y., Krähenbühl, P., Shechtman, E., Efros, A.A.: Generative visual manipulation on the natural image manifold. In: Leibe, B., Matas, J., Sebe, N., Welling, M. (eds.) ECCV 2016. LNCS, vol. 9909, pp. 597–613. Springer, Cham (2016). https://doi.org/10.1007/978-3-319-46454-1_36

Adaptive Low-Light Image Enhancement Optimization Framework with Algorithm Unrolling

Qichang He[1,2], Lingyu Liang[1,2,3(✉)], Wocheng Xiao[1,4], and Mingju Liang[4]

[1] South China University of Technology, Guangzhou, China
lianglysky@gmail.com
[2] Guangdong Artificial Intelligence and Digital Economy Laboratory (Pazhou Lab Guangzhou), Guangzhou, China
[3] Ministry of Education Key Laboratory of Computer Network and Information Integration, Southeast University, Nanjing, China
[4] Guangdong Foshan Lianchuang Graduate School of Engineering, Foshan, China

Abstract. Images captured in a dark environment may cause low visibility and lose significant details leading to poor performance of vision-based recognition systems. Recently, deep learning-based methods have been proposed for low-light image enhancement (LIE) with different priors or training schemes. However, even those LIE methods may introduce visual artifacts into the enhanced images. This paper proposes an adaptive LIE optimization framework that allows to re-optimize different deep learning-based LIE methods based on an adaptive quality evaluation (QE). Specifically, we design an interpretable and learnable LIE-QE module for LIE. To find the optimal structure of the LIE-QE module, we propose an algorithm unrolling method to design the LIE-QE module, where the each layer of the decomposition component of the LIE-QE module can be interpreted as WLS edge-aware smoothing. Both qualitative and quantitative experiments were conducted, and the evaluation verified the effectiveness of the proposed learnable deep unrolled LIE-QE module for LIE. The results shows that the proposed LIE framework can effectively improve different deep learning-based LIE methods indicating the potential of the optimization framework with LIE-QE module to re-optimize existing DNN-based LIE methods.

Keywords: Low-light Image Enhancement · Algorithm Unrolling

1 Introduction

In our daily life, low-light images are fairly common due to insufficient illumination or limited exposure time. Low-light images not only have poor visibility, but also degrade the performance of vision-based recognition systems that designed for high-quality images.

Low-light image enhancement (LIE) has been an active research topic for decades, and many classic methods are based on intensity transform [18], filtering

© The Author(s), under exclusive license to Springer Nature Singapore Pte Ltd. 2024
Q. Liu et al. (Eds.): PRCV 2023, LNCS 14435, pp. 158–170, 2024.
https://doi.org/10.1007/978-981-99-8552-4_13

Fig. 1. Visual artifacts introduced using some recent low-light enhancement methods, where (a) is the ground-truth image, (b) is the low-light image, and (c)–(f) are the results of RetinexNet [22], KinD [26], DALE [9] and Zero-DCE++ [12], respectively.

operations or Retinex theory that utilize intrinsic properties of scene or objects to obtain better LIE [4,7,10,11,14].

Recently, many learning-based methods have been proposed for LIE with deep neural network (DNN). One line of works modify existing DNN with different priors or training schemes for LIE, such as Zero-Reference [6], Zero-DCE++ [12], DALE [9], EnlightenGAN [8], LLFlow [20] and SNR-Aware transformer [24]. Another line of works takes intrinsic image decomposition or Retinex theory [10,11] as a main illumination constraints to design DNN structure for LIE, and integrates additional schemes to further optimize the networks, such as RetinexNet [22], LightenNet [13], KinD [26], KinD++ [25], URetinex-Net [23] and LIE with semantic segmentation [2].

However, experiments indicate that even though the state-of-the-art LIE methods may inevitably introduce some visual artifacts into the enhanced output, like underexposure, color shift or loss of details, as shown in Fig. 1. It indicates that LIE is still an unsolved problem. It would be significant to develop a framework to further improve the performance of existing DNN-based methods without completely re-designing the trained DNN structure.

Therefore, this paper proposes an adaptive LIE optimization framework that enables to re-optimize different deep learning-based LIE methods based on an adaptive quality evaluation (QE) of the enhanced low-light images. Specifically, we design the LIE optimization framework with an interpretable and learnable LIE-QE module. We employ the algorithm unrolling [5,15,17] method to determine the optimal structure of the LIE-QE module, where the each layer of the decomposition component of the LIE-QE module can be interpreted as WLS edge-aware filtering [3].

Both qualitative and quantitative experiments were conducted to evaluate the effectiveness of the proposed learnable deep unrolling LIE-QE module for LIE. The experimental results demonstrate that the proposed LIE framework can significantly further improve the performance of different deep learning-based LIE methods, such as RetinexNet [22], KinD [26], DALE [9] and Zero-DCE++ [12]. Furthermore, the results also indicate the potential of the optimization framework with LIE-QE module to re-optimize a variety of existing DNN-based LIE methods.

The main contributions are summarized as follows:

1. An adaptive LIE optimization framework is constructed that can re-optimize different deep learning-based LIE methods.

Fig. 2. The overall structure of our unrolling adaptive LIE optimization framework.

2. An interpretable and learnable LIE-QE module is proposed to evaluate the quality of LIE methods, and the DNN structure of the LIE-QE module can be interpreted as WLS edge-aware filtering from an algorithm unrolling perspective.
3. Qualitative and quantitative experiments indicates that our LIE optimization framework with the LIE-QE module can effectively improve deep learning-based LIE methods, including RetinexNet [22], KinD [26], DALE [9] and Zero-DCE++ [12].

2   LIE Optimization Framework with Algorithm Unrolling

We design an adaptive LIE optimization framework to optimize the LIE methods and guide them generate higher quality results that are more visually pleasing to humans. The framework consists of the low-light image enhancement method and unrolling LIE-QE module, as shown in Fig. 2. The quality evaluation serves as the quality constraint to optimize and improve the LIE module, leading to higher quality and better performance which is more preferable to the human visual system. The LIE optimization framework can be applied to a variety of deep-learning LIE methods, such as KinD [26], ZeroDce [6] and the detailed evaluations were shown in Sec. 3.

2.1   Unrolling LIE-QE Module

As shown in Fig. 2, our Unrolling LIE-QE model mainly contains the Unrolling Decomposition Module (UDM) that is the key component of Unrolling LIE-QE and the Feature Extraction Module.

Unrolling Decomposition Module. Our Unrolling Decomposition Module (UDM) is a novel decomposition module which unrolls the decomposition iterations into a neural network by algorithm unrolling. UDM contains n Unrolling blocks, with each block representing an iterative step of the solution, as shown in Fig. 3. UDM can be interpreted as WLS edge-aware smoothing [3], combining the prior knowledge of WLS [3] with the advantages of a data-driven neural network. This leads to better performance and interpretability of the module.

We designed an objective function as the WLS filtering [3] and unroll the structure of iterative WLS, whose the constraint term is a related term of the

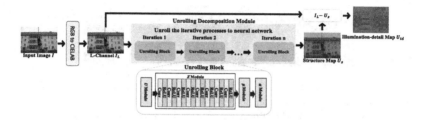

Fig. 3. The overall structure of Unrolling Decomposition Module(UDM).

output U:

$$\begin{cases} min.\{(U-I)^2 + \lambda w(Z)\}, \\ \quad s.t.\, U = Z. \end{cases} \tag{1}$$

where, I is the input image, U is the output, $w(Z)$ is a constraint term of the output U, λ is the weight of $w(Z)$ and Z is an auxiliary variable to repalce U for making the problem easy to solve.

Alternating direction method of multipliers (ADMM) [1] is used to solve Eq. (1). Then, we obtain the following equations:

$$\begin{cases} U^{k+1} = \dfrac{2I + \mu^k Z^k - \alpha^k}{2 + \mu^k}, & \text{(2a)} \\[2mm] Z^{k+1} = h(U^{k+1} + \alpha^k; W_Z), & \text{(2b)} \\[2mm] \mu^{k+1} = 2\mu^k, & \text{(2c)} \\[2mm] \alpha^{k+1} = \alpha^k + \mu^{k+1}(U^{k+1} - Z^{k+1}). & \text{(2d)} \end{cases}$$

Here, Z^k, μ^k and α^k denote the auxiliary variable Z and the constraint coefficients μ and α in the kth iteration, respectively. U^{k+1}, Z^{k+1}, μ^{k+1} and α^{k+1} denote U, Z, and α in the $k+1$th iteration. $h()$ is a nonlinear shrink function designed as fully convolutional neural network.

UDM unrolls the iterative update solution of Eq. (2a), (2b), (2c) and (2d) into a neural network using the algorithm unrolling, as shown in Fig. 3. The iteration is converted into a data-driven training process using neural network training techniques. U^k, Z^k, μ^k and α^k obtained in the kth iteration, are inputted for u^{k+1}, Z^{k+1}, μ^{k+1} and α^{k+1} respectively. That iterations are repeated until the final decomposition result u is got. The U module, Z module, μ module, and α module of the Unrolling block in Fig. 3 correspond to Eq. (2a), (2b), (2c) and (2d), respectively.

Algorithm unrolling allows each hidden layer in the network to have a certain meaning, which can be interpreted as a certain step in the iteration. Each Unrolling block in UDM acts as a WLS filter [3], showing a similar effect in decomposition as the WLS filter [3].

As shown in Fig. 4, UDM is capable of decomposing the illumination-independent structure map and the illumination detail map from the low-light image. The illumination detail map contains the illumination information of the

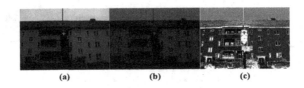

Fig. 4. Outputs of Unrolling Decomposition Module. (a) is the low-light image, (b) is the structure map and (c) is the illumination-detail map from UDM.

image, showing the regions impaired by the low-light environment. This illumination detail map can be used to evaluate the quality of low-light enhanced images and optimize LIE methods more effectively.

Moreover, the processing time of UDM in decomposition is 0.00789 s per image, which overcomes the disadvantage of the high-time-complexity iterations.

The input image I is converted from RGB color space to CIELAB color space to obtain the L-channel I_L containing illumination information. UDM decomposes the L channels of the input image to obtain the illumination detail map u_{id} and the structure map u_s:

$$U_s = UDM(I_L), \tag{3}$$

where I_L is the L-channel image of input image I, U_s is the structure map of I.

After obtaining structure map U_s using UDM, it can be subtracted from the original low-light image I_L to obtain the illumination-detail map u_{id} containing information about the illumination in the image:

$$U_{id} = I_L - U_s, \tag{4}$$

where U_{id} is the illumination-detail map of I.

Feature Extract. The input enhanced image I_{en} and the reference I_{ref}, as well as their respective structure map U_{s_en}, U_{s_ref} and illumination detail map U_{id_en}, U_{id_ref} are fed into the Feature Extraction Module. The module extracts perceptual features, structure features and illumination features of I_{en} and I_{ref} using pretrained VGGNet [19]. It benefits the quality evaluation for low-light image enhancement. We choose the outputs of the five latent layers of the VGGNet [19] as our features.

Quality Evaluation. The final image quality evaluation is calculated by taking the weighted sum of the feature similarities, which are obtained by $L2$ distance between the perceptual features, structure features and illumination features of I_{en} and I_{ref}. This evaluation can be used to optimize LIE methods more effectively.

2.2   Loss of the LIE Optimization Framework

The loss of our framework is given by:

$$\mathcal{L} = \mathcal{L}_{en}(I_{dark}, I_{en}) + \beta * \mathcal{L}_q(I_{en}, I_{ref}), \tag{5}$$

where $\mathcal{L}_{en}$ is the enhancement loss of the framework from LIE module, $\mathcal{L}_q$ is the quality constraint. β weights the influence of $\mathcal{L}_q$ which is set as 0.2 in our experiments. I_{dark} is low-light image inputted to the LIE method, I_{en} is enhancement image of the LIE method and I_{ref} is the reference image.

$\mathcal{L}_{en}$ depends on the LIE methods used in our framework, calculated by the low-light image I_{dark} and enhanced image I_{en}. It is typically as same as the loss function of the selected LIE method in general. To ensure the versatility of our framework and preserve the original characteristics and advantages of the selected LIE method, it is recommended to keep $\mathcal{L}_{en}$ consistent with the original loss function of the chosen method.

$\mathcal{L}_q$ is the result of Unrolling LIE-QE calculated by enhanced image I_{en} which is the output of the LIE method and reference image I_{ref}. $\mathcal{L}_q$ is the quality constraint, which is used to optimize the LIE method for better and higher quality enhanced results. It guides the LIE method to produce results that are more visually pleasing to humans.

3   Experiment

We evaluated our method using LOL dataset [22] which is a real-world dataset comprising low-light and normal-light image pairs, as well as Large Scale Low-light Synthetic(LSLS) dataset [16]. We selected low/normal-light image pairs in LOL dataset [22] and synthesized dark images and their high-contrast images from LSLS [16] for our experiments.

3.1   Evaluation of the Proposed Framework

To evaluate the effectiveness of our framework, we implemented LIE-QE of the framework using WLS [3] instead of UDM. The framework implemented with WLS [3] is tested to optimize KinD [26] and Zero-DCE++ [12] in comparison with the original KinD [26] and Zero-DCE++ [12].

The experiments were conducted on LOL dataset [22] and LSLS dataset [16]. We used PSNR and SSIM [21] as the image quality metrics in the experiment. The results are presented in Table 1, where "WLS-Optimized" indicates that the method is optimized using the framework with the LIE-QE module implemented using WLS [3]. PSNR and SSIM [21] of the WLS-Optimized images are higher than the original ones, indicating that our framework can lead to better enhancement and higher quality. It demonstrates that our framework has the ability to optimize LIE methods.

Table 1. Experiment results of the framework implemented with WLS [3].

Method	LOL Dataset [22]		LSLS Dataset [16]	
	PSNR(↑)	SSIM [21](↑)	PSNR(↑)	SSIM [21](↑)
KinD [26]	19.650	0.821	17.385	0.765
WLS-Optimized KinD	**19.734**	**0.821**	**17.389**	**0.765**
Zero-DCE++ [12]	14.861	0.559	15.132	0.648
WLS-Optimized Zero-DCE++	**16.051**	**0.569**	**17.032**	**0.683**

3.2 Evaluation of Unrolling Decomposition Module

Decomposition Results. We tested the effectiveness in decomposition of UDM compaired with WLS [3]. As shown in Fig. 5, our UDM is effective in decomposing the illumination-independent structure map and the illumination-detail map from the low-light image. The illumination map from UDM contains the illumination information of the image, showing the regions of image affected by the low-light environment. It can guide the quality evaluation of low-light enhanced images and optimization for low-light enhancement algorithms more effectively. Moreover, the decomposition results indicate that UDM is similar to WLS [3]. This suggests that UDM can be interpreted as WLS [3], increasing the interpretability of the model.

Fig. 5. Decomposition results of UDM. (a) is the low-light image, (b) is the reference image, (c) is the structure map from UDM, (d) is the illumination-detail map from UDM and (e) is the illumination-detail map from WLS [3].

Ablation Studies. We conducted ablation studies to evaluate the important role of UDM playing in our unrolling LIE optimization framework. We compared the performance of UDM and WLS in our framework by implementing them in the LIE-QE module to optimize the LIE methods.

The results in Fig. 6, Fig. 7 and Table 2, demonstrate that our framework with UDM outperforms the framework with WLS [3] in terms of optimization for LIE methods. "UDM-optimized" means that the method is optimized by Unrolling LIE optimization framework whose LIE-QE implemented with UDM

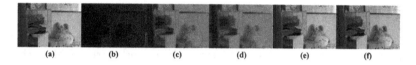

Fig. 6. Experiment results of ablation studies in LOL dataset [22]. (a) are the reference images, (b) are the low-light images, (c) are the results of WLS-Optimized RetinexNet, (d) are the results of UDM-Optimized RetinexNet, (e) are the results of WLS-Optimized SNR and (f) are the results of UDM-Optimized SNR.

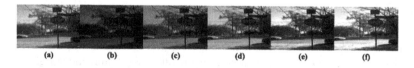

Fig. 7. Experiment results of ablation studies in LSLS dataset [16]. (a) are the reference images, (b) are the low-light images, (c) are the results of WLS-Optimized ZeroDCE++, (d) are the results of UDM-Optimized ZeroDCE++, (e) are the results of WLS-Optimized SNR and (f) are the results of UDM-Optimized SNR.

Table 2. Ablation studies results of UDM and WLS [3].

Method	LOL Dataset [22]		LSLS Dataset [16]	
	PSNR($\uparrow$)	SSIM [21]($\uparrow$)	PSNR($\uparrow$)	SSIM [21]($\uparrow$)
WLS-Optimized RetinexNet	16.747	0.682	8.059	0.626
UDM-Optimized RetinexNet	**17.111**	**0.694**	**8.673**	**0.632**
WLS-Optimized KinD	19.734	0.821	17.389	0.765
UDM-Optimized KinD	**19.756**	**0.822**	**17.394**	**0.766**
WLS-Optimized Zero-DCE++	16.051	0.569	17.032	0.683
UDM-Optimized Zero-DCE++	**16.854**	**0.571**	**17.510**	**0.693**
WLS-Optimized SNR	23.285	0.825	16.683	0.637
UDM-Optimized SNR	**25.485**	**0.857**	**17.061**	**0.667**

and "WLS-optimized" means the method is optimized by the framework with LIE-QE implemented with WLS [3]. The experiment results indicate that UDM can extract illumination information from image better and more effectively than WLS [3]. UDM was found to be more effective at extracting illumination information from the image, resulting in better image quality and more natural light and details in the enhanced images, which are more preferable for the human visual system. The ablation studies confirmed that UDM is beneficial to LIE and plays an important role in the effectiveness of our framework in improving the performance of LIE methods.

3.3   Comparison with Related Methods

Since UDM has better decomposition and optimization performance than WLS [3], we selected Unrolling LIE Optimization Framework with UDM as the LIE optimization framework. We applied our framework to optimize four deeplearning LIE methods, namely RetinexNet [22], KinD [26], Zero-DCE++ [12], and SNR [24], on both the LOL [22] and LSLS [16] datasets. We used PSNR and SSIM [21] as the image quality metrics. In the results, "UDM-optimized" means that the method is optimized by the framework whose LIE-QE is implemented with UDM.

Qualitative Evaluation. As demonstrated in Fig. 8 and Fig. 9, the UDM-Optimized enhancement results have more natural light and more light details compared with the original ones. They are more preferable for human visual system than before.

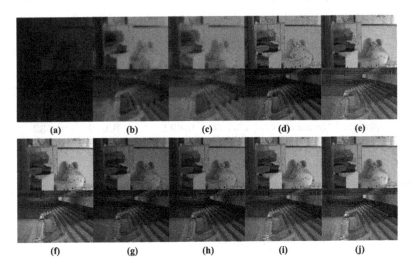

Fig. 8. Optimization experiment for existing low-light image enhancement methods in LOL dataset [22]. (a) are the low-light images, (b) are the results of RetinexNet [22], (c) are the results of optimized RetinexNet, (d) are the results of KinD [26], (e) are the results of optimized KinD, (f) are the reference images, (g) are the results of Zero-DCE++ [12], (h) are the results of optimized Zere-DCE++, (i) are the results of SNR [24] and (j) are the results of optimized SNR.

Quantitative Evaluation. In Table 3, PSNR and SSIM [21] of the UDM-Optimized images optimized by Unrolling LIE Optimization Framework are higher than the originals, indicating better quality. The enhanced image quality of the four LIE methods is improved after optimization with our framework.

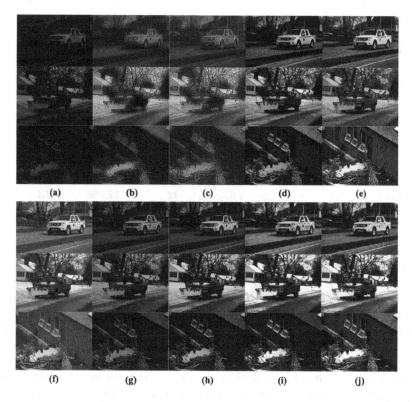

Fig. 9. Optimization experiment for existing low-light image enhancement methods in LSLS dataset [16]. (a) are the low-light images, (b) are the results of RetinexNet [22], (c) are the results of optimized RetinexNet, (d) are the results of KinD [26], (e) are the results of optimized KinD, (f) are the reference images, (g) are the results of Zero-DCE++ [12], (h) are the results of optimized Zere-DCE++, (i) are the results of SNR [24] and (j) are the results of optimized SNR.

The results show that LIE methods optimized by our framework have better enhancement for the low-light images and higher quality results.

Moreover, both quantitative and qualitative experiments of the framework demonstrated that our framework is applicable to a variety of deep-learning LIE methods, which indicating the potential of our framework to improve the performance of various LIE methods.

Table 3. Optimization results of our Unrolling LIE Optimization Framework.

Method	LOL Dataset [22]		LSLS Dataset [16]	
	PSNR(↑)	SSIM [21](↑)	PSNR(↑)	SSIM [21](↑)
RetinexNet [22]	16.774	0.559	8.447	0.611
UDM-Optimized RetinexNet	**17.111**	**0.694**	**8.673**	**0.632**
KinD [26]	19.650	0.821	17.385	0.765
UDM-Optimized KinD	**19.756**	**0.822**	**17.394**	**0.766**
Zero DCE++ [12]	14.861	0.559	15.132	0.648
UDM-Optimized Zero DCE++	**16.854**	**0.571**	**17.510**	**0.693**
SNR [24]	24.610	0.842	17.006	0.664
UDM-Optimized SNR	**25.485**	**0.857**	**17.061**	**0.667**

4   Conclusion

We propose a novel Unrolling-based adaptive LIE optimization framework, to address some visual artifacts like underexposure and color shift of state-of-the-art LIE methods, which can result in higher quality enhanced images. Our framework incorporates the output of the Unrolling LIE-QE model into the loss function as a quality constraint during optimization. Our UDM implemented in LIE-QE takes advantage of algorithm unrolling, which unrolls the iteration into a neural network, to decompose structure map and the illumination-detail map that contains the illumination information result in better enhancement from the input image. The experimental results demonstrate that our method can effectively improve and re-optimize the LIE methods to produce higher quality and visually-pleasing results. The results indicate the potential of framework with LIE-QE module to re-optimize various existing DNN-based LIE methods.

Acknowledgements. This research was supported by the Fundamental Research Funds for the Central Universities, the Open Fund of Ministry of Education Key Laboratory of Computer Network and Information Integration (Southeast University) (K93-9-2021-01), and the Science and Technology Program of Pazhou Lab.

References

1. Boyd, S., Parikh, N., Chu, E., Peleato, B., Eckstein, J., et al.: Distributed optimization and statistical learning via the alternating direction method of multipliers. Found. Trends® Mach. Learn. **3**(1), 1–122 (2011)
2. Fan, M., Wang, W., Yang, W., Liu, J.: Integrating semantic segmentation and retinex model for low-light image enhancement. In: Proceedings of the ACM MM, pp. 2317–2325 (2020)

3. Farbman, Z., Fattal, R., Lischinski, D., Szeliski, R.: Edge-preserving decompositions for multi-scale tone and detail manipulation. ACM Trans. Graph. **27**(3), 1–10 (2008)
4. Fu, X., Zeng, D., Huang, Y., Zhang, X.P., Ding, X.: A weighted variational model for simultaneous reflectance and illumination estimation. In: Proceedings of CVPR, pp. 2782–2790 (2016)
5. Gregor, K., LeCun, Y.: Learning fast approximations of sparse coding. In: Proceedings of the ICML, pp. 399–406 (2010)
6. Guo, C., et al.: Zero-reference deep curve estimation for low-light image enhancement. In: Proceedings of CVPR, pp. 1780–1789 (2020)
7. Guo, X., Li, Y., Ling, H.: Lime: low-light image enhancement via illumination map estimation. IEEE Trans. Image Process. **26**(2), 982–993 (2016)
8. Jiang, Y., et al.: EnlightenGAN: deep light enhancement without paired supervision. IEEE Trans. Image Process. **30**, 2340–2349 (2021)
9. Kwon, D., Kim, G., Kwon, J.: Dale: dark region-aware low-light image enhancement. arXiv preprint arXiv:2008.12493 (2020)
10. Land, E.H.: The retinex theory of color vision. Sci. Am. **237**(6), 108–129 (1977)
11. Land, E.H., McCann, J.J.: Lightness and retinex theory. JOSA **61**(1), 1–11 (1971)
12. Li, C., Guo, C., Chen, C.: Learning to enhance low-light image via zero-reference deep curve estimation. IEEE Trans. Pattern Anal. Mach. Intell. **44**, 4225–4238 (2021)
13. Li, C., Guo, J., Porikli, F., Pang, Y.: Lightennet: a convolutional neural network for weakly illuminated image enhancement. Pattern Recogn. Lett. **104**, 15–22 (2018)
14. Li, M., Liu, J., Yang, W., Sun, X., Guo, Z.: Structure-revealing low-light image enhancement via robust retinex model. IEEE Trans. Image Process. **27**(6), 2828–2841 (2018)
15. Liu, R., Ma, L., Zhang, J., Fan, X., Luo, Z.: Retinex-inspired unrolling with cooperative prior architecture search for low-light image enhancement. In: Proceedings of CVPR, pp. 10561–10570 (2021)
16. Lv, F., Li, Y., Lu, F.: Attention guided low-light image enhancement with a large scale low-light simulation dataset. IJCV **129**(7), 2175–2193 (2021)
17. Monga, V., Li, Y., Eldar, Y.C.: Algorithm unrolling: interpretable, efficient deep learning for signal and image processing. IEEE Signal Process. Mag. **38**(2), 18–44 (2021)
18. Pizer, S.M., et al.: Adaptive histogram equalization and its variations. Comput. Vis. Graphics Image Process. **39**(3), 355–368 (1987)
19. Simonyan, K., Zisserman, A.: Very deep convolutional networks for large-scale image recognition. arXiv preprint arXiv:1409.1556 (2014)
20. Wang, Y., Wan, R., Yang, W., Li, H., Chau, L.P., Kot, A.: Low-light image enhancement with normalizing flow. In: Proceedings of AAAI on Artificial Intelligence, vol. 36, pp. 2604–2612 (2022)
21. Wang, Z., Bovik, A.C., Sheikh, H.R., Simoncelli, E.P.: Image quality assessment: from error visibility to structural similarity. IEEE Trans. Image Process. **13**(4), 600–612 (2004)
22. Wei, C., Wang, W., Yang, W., Liu, J.: Deep retinex decomposition for low-light enhancement. In: Proceedings of the BMVC, pp. 127–136 (2018)
23. Wu, W., Weng, J., Zhang, P., Wang, X., Yang, W., Jiang, J.: Uretinex-net: Retinex-based deep unfolding network for low-light image enhancement. In: Proceedings of CVPR, pp. 5901–5910 (2022)

24. Xu, X., Wang, R., Fu, C.W., Jia, J.: Snr-aware low-light image enhancement. In: Proceedings of CVPR, pp. 17714–17724 (2022)
25. Zhang, Y., Guo, X., Ma, J., Liu, W., Zhang, J.: Beyond brightening low-light images. IJCV **129**, 1013–1037 (2021)
26. Zhang, Y., Zhang, J., Guo, X.: Kindling the darkness: a practical low-light image enhancer. In: Proceedings of the ACM MM, pp. 1632–1640 (2019)

Feature Matching in the Changed Environments for Visual Localization

Qian Hu, Xuelun Shen, Zijun Li, Weiquan Liu[✉], and Cheng Wang[✉]

Fujian Key Laboratory of Sensing and Computing for Smart Cities, School of
Informatics, Xiamen University, Xiamen, China
{qianhu,xuelun,lizijun,wqliu,cwang}@stu.xmu.edu.cn
{wqliu,cwang}@xmu.edu.cn

Abstract. Robust feature matching is a fundamental capability for
visual SLAM. It remains, however, a challenging task, particularly for
changed environments. Some researchers use semantic segmentation to
remove potentially dynamic objects which cause changes in the indoor
environment. However, removing these objects may reduce the quantity
and quality of feature matching. We observed that objects are moved
but the room layout does not change. Inspired by this, we proposed to
leverage the room layout information for feature matching. Considering
current image matching datasets do not have obvious changes caused
by dynamic objects and lack layout information, we created a dataset
named Changed Indoor 10k (CR10k) to evaluate if we can utilize layout information for image matching in the changed environment. Our
dataset contains apparent movements of large objects, and room layout
can be extracted from it. We evaluate the performance of existing image
matching methods on our dataset and ScanNet dataset. In addition, we
propose Layout Constraint Matching (LCM) which is robust to changed
environments and the LCM outperforms conventional approaches on the
task of pose estimation.

Keywords: Feature Matching · Pose Estimation · Room Layout
Estimation · Layout Constraint

1 Introduction

Current feature matching methods deal directly with changing environments
have limitations. Because most of these methods assume that the objects in the
environment are stationary. However, there are a large number of moving objects
in a real-world environment, such as tables and objects moved by people. These
objects will cause changes in the indoor environment and bring false observations
(see Fig. 1), thus the robustness of the system will be reduced. Although the
RANSAC [5] algorithm can filter out outliers, it can still seriously affect the
performance of pose estimation in the case of moving objects occupying most of
the image area. Current image matching in changed environments uses semantic
segmentation to mask potential moving objects (e.g., walking people, doors and

© The Author(s), under exclusive license to Springer Nature Singapore Pte Ltd. 2024
Q. Liu et al. (Eds.): PRCV 2023, LNCS 14435, pp. 171–183, 2024.
https://doi.org/10.1007/978-981-99-8552-4_14

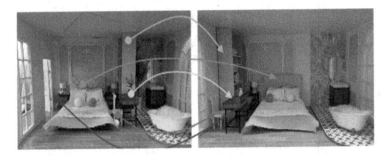

Fig. 1. The size of the miniature room is $50 \times 27 \times 31$ cm. The furniture in the house, such as bed, desk, chair, etc. can be moved. We want to robustly estimate the camera relative pose in the changed environment.

windows that are repeatedly opened and closed). Then, approaches based on static environment assumptions match feature points. Objects may be stationary for a while, removing them may cause static areas in images to fail to provide enough features. The quantity and quality of feature matching may be reduced.

We observe that the indoor objects are changing with the interaction of people and the environment, but the room layout does not change over time, which provides a reference for matching between two images. Given two images, if the same object is in different positions in the room layout, the object must be moved, and the matches that fall on it are mismatches. Therefore, we propose using layout information to constraint matching. Extracting layout means getting semantic planar (e.g., ceiling, floor, and walls). A match is incorrect if corresponding points fall on different semantic planar, which we called layout constraint.

Motivated by the above observations, we propose LCM, a local feature matching approach which is robust to changed environments. Inspired by the seminal work SuperGlue [13], we utilize self-attention and cross-attention layers to transform features. Putative matches are first extracted from transformed features via the matching layer. Matches with high confidence are selected from these putative matches. We extract room layouts from images and utilize putative matches to conduct layout correspondence between image pairs. We use layout constraint to update matches after obtaining layout correspondence.

Current public datasets focus on static environments, only a few datasets contain dynamic objects (e.g. TUM dataset [18]). These datasets not contain the obvious movement of large objects. Hence, we created a dataset named Changed Indoor 10k (CR10k). The proposed dataset contains obvious movements of large objects and the room layout can be extracted from it. We experimentally evaluated some image matching methods and LCM on our dataset. The contributions of this paper are as follows:

1) To improve the robustness of feature matching in indoor environments, we propose LCM which is robust to match in changing indoor scenery. Firstly, LCM leverages multilayer perceptron to map layout features, keypoint coordi-

nates, and feature descriptors to a high-dimensional space. Secondly, feature aggregation is performed using self-attention and cross-attention. Finally, the layout information is used to guide the network for feature matching.

2) We propose a new dataset that can be used for image matching and pose estimation in changed indoor scenery. The proposed dataset contains obvious movements of large objects, which poses certain challenges for feature matching.

2   Related Work

2.1   Room Layout Estimation

Room layout estimation aims to extract the semantic boundaries (among walls, floor, and ceiling) and classify each pixel. Early room layout estimation [2,10] usually extracted handcrafted features to estimate vanishing points and use a structured regressor to generate layout hypotheses. Recently, researchers use Convolutional Neural Networks (CNNs) to estimate room layout. Mallya et al. [8] used the Fully Convolutional Networks (FCN) to predict the location of the semantic boundaries among walls, floor, and ceiling and generate a layout by sampling the vanishing line in the area near the predicted boundary. [11] utilize FCN to obtain a layout that is close to ground truth, and then obtain a finer layout estimation through layout contour straightness, surface smoothness, etc. Lin et al. [7] proposed adding adaptive edge penalty and smoothing during the training process, which improves smoothness without post-processing. In this paper, the room layout is extracted by this method.

2.2   Local Feature Matching

The classic pipeline of feature matching contains keypoint detection, feature description, feature matching and outliers filtering. Early methods trained feature detector and feature descriptor separately [14,22]. Recently, joint learning of keypoints detection and feature description in one forward pass has gradually become the mainstream [1,9]. Because joint learning of keypoints detection and feature description not only facilitates the learning of globally optimal parameter models, but also improves the performance of both keypoint detector and descriptor. The feature matching stage aims to establish the correspondence between the two feature point sets. Traditional methods such as mutual nearest neighbor (MNN) search have been widely used. Recently, graph neural network methods have attracted the attention of researchers [12,17]. Sarlin et al. [13] proposed to use a graph neural network to predict the cost of keypoints matching. They constructed an attentional graph neural network based on self-attention and cross-attention and improve the quality of match. Inspired by Sarlin et al., we use self-attention and cross-attention for feature aggregation. Additionally, room layout constraints are utilized to learn feature matching.

As the correspondences obtained in the feature matching stage often contain outliers, researchers usually use RANSAC [5] to solve the outlier rejection problem. However, RANSAC produces very little effect in challenging changed environments. Therefore, some deep learning methods focus on outlier rejection, such as PointCN [20], OANet [21]. In this paper, we compare LCM against MNN, OANet and PointCN.

2.3 Datasets for Matching

There are many challenging images matching datasets [3,6,18,19], including motion blur, viewpoint change greatly, and illumination changes. However, few researchers have paid attention to image matching in changed environments. Although there are some datasets containing dynamic objects, they only contain the movements of small objects. Existing methods are robust to these datasets since static areas in these datasets can provide enough reliable features. Considering these, we create a dataset named CR10 for image matching in changed environments. CR10 contains the movements of large objects, which will reduce the quality of matching. The proposed dataset can be used to evaluate the robustness of image matching methods.

We conducted a summary comparison of the available datasets with CR10k (see Table 1). The only dataset that contains a large number of object position changes is RIO-10 [19], but it does not provide the poses of the image pairs. Therefore, RIO-10 cannot be used for camera pose estimation tasks. CR10k not only contains a large number of object position changes but also provides the ground truth of the camera pose, which is suitable for evaluating the performance of feature matching methods in changing environments.

Table 1. Datasets for Image Matching

Dataset	Indoor or Outdoor	Characteristics	Ground Truth
ScanNet [3]	Indoor	static environment	with
MegaDepth [6]	Outdoor	few dynamic objects	with
RIO-10 [19]	Indoor	object deformations, position change	without
TUM [18]	Indoor	a few dynamic objects	with
CR10k	Indoor	oblivious movement of objects	with

3   Image Matching Dataset for Changed Indoor Environments

3.1   Design of the Dataset

We built a dataset named CR10k, which contains 591 indoor images, and 10215 image pairs. Since moving large objects in real life is difficult, we constructed a

Fig. 2. Example image pairs. Bed and wardrobe are moved by people in (a). In (b), the nightstand is moved by people.

miniature room to simulate a room (see Fig. 1). We move objects in the model house to simulate changes in the position of the objects caused by human activities. Figure 2 shows example image pairs. Each pair of images contains both static and semi-static objects. Static objects are objects that never move, such as walls, and windows. Semi-static objects are objects that keep stationary over a while and move for a while. If we mask semi-static objects, static areas maybe not provide enough features to match. In Fig. 2 (a), the bed should be masked because it is moved. If we mask all beds in the dataset, Fig. 2 (b) does not have enough features to match, and image matching will fail in these cases. In Fig. 2, large objects (e.g. bed, wardrobe) moved but their appearance doesn't change. Feature matching methods tend to be confused by them. We design the dataset primarily to evaluate image matching methods in this challenging environment. The proposed dataset contains obvious movements of large/small and rigid objects and the room layout can be extracted from it.

3.2 Detailed Specifications

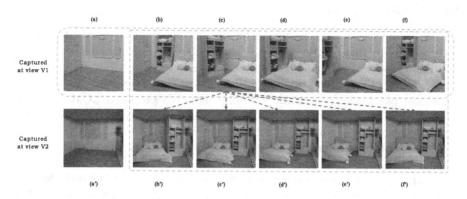

Fig. 3. The first row of images is captured in the same view. There are only static objects in (a), (b) ∼ (f) has both static and dynamic objects. An image from (b) ∼ (f) can form a image pair with any of the images from (b') ∼ (f'). There are 25 image pairs in this figure.

There are of 1000 × 1000 pixels and in a JPG format of 24 bits per pixel. We used the Canon Camera EOS M100 to capture all the images. Figure 3 shows the process of forming image pairs. We first take an image of only static objects in the model house, then put furniture in the model house. Finally, we take another image after changing the position of the objects. Therefore, we know which objects are moved by comparing (a) and (b)~ (f). Images captured at the same view have the same camera intrinsic and extrinsic. We conduct keypoints matching between two images captured at different views, and further estimate the camera relative pose from matches.

3.3   Obtaining Ground Truth Camera Pose

To evaluate image matching and camera pose estimation methods, we need to know the camera intrinsic and extrinsic. We perform 3D reconstruction from images containing only static objects captured at different views to obtain camera intrinsic and extrinsic. We use COLMAP which is a state-of-the-art Structure from Motion (SFM) system [15] (for reconstructing camera poses and sparse point clouds) and Multi View Stereo (MVS) system [16] (for generating dense depth maps) to get camera parameters. Hence, we can get the ground truth of relative camera pose.

4   Method

This section presents our end-to-end deep learning algorithm for image matching for pose estimation. In contrast to the previous approaches, we use layout information as prior knowledge to guide feature matching (see Fig. 4).

4.1   Network Architecture

We use LayoutSeg [7] to extract room layout. LayoutSeg modifies the ResNet101 to extract the features of the input images and classify pixels. LayoutSeg outputs semantic planar, such as ceiling, floor, and walls.

SuperGlue [13] uses an attention mechanism to match keypoints and achieves impressive performance in static environments. We proposed LCM based on SuperGlue. LCM adds layout feature encoding and room layout constraint modules to SuperGLue's network structure (see Fig. 4). Given a pair of images, keypoints position and description are extracted first by keypoint detector and descriptor. Then, we use self-attention and cross-attention layer to transform features. The matching layer outputs a partial assignment which denotes the confidence of each match. Finally, we construct layout correspondence and penalize putative matches that violate the room layout constraint.

In layout feature encoding modules, we use a Multi-layer Perceptron(MLP) to map keypoint positions p_i, descriptiond_i and layout feature l_i onto a high-dimensional feature space. Layout feature l_i is the feature of the semantic plane

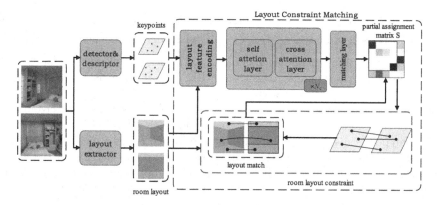

Fig. 4. Overview of the proposed method. Detector, descriptor and layout extractor are used to extract keypoints, descriptors, and room layout. We use Superpoint [4] to extract keypoints positions and descriptions. LayoutSeg [7] are utilized to extract room layout. We use N_C self-attention and cross-attention layers to transform features and transformed features are matched by matching layer. LCM ends up with a partial assignment matrix. Matches in the partial assignment matrix are selected according to the confidence threshold. We leverage matches with high confidence to conduct layout correspondence between image pairs. Finally, we update partial assignment through the result of the layout match, which can guide the network learning matches that meet the layout constraints.

Fig. 5. Layout correspondence score

to which the keypoint belongs. In this way, the semantic, appearance and position information of the keypoints can be aggregated. The mathematical expression is as follows:

$$x_i = d_i + MLP(p_i, l_i) \tag{1}$$

We use x_i denotes the feature of keypoint from image A, x_j denotes the feature of keypoint from image B. Using following expression, We can obtain a matrix $\bar{S}$. $\bar{S}$ describes the degree of matching between keypoint A and keypoint B.

$$\bar{S} = x_i \cdot x_j \tag{2}$$

In room layout constraint modules, we design a layout correspondence matrix LCS(see Fig. 5), where (m, n) denotes the m-th region in image A against the n-th region in image B. For the set of semantic planar of the image A, $L_A = \{L_{l1}, L_{l2}, L_{l3}, L_{l4}, L_{l5}\}$ and the set of image B, $L_B = \{L_{r1}, L_{r2}, L_{r3}, L_{r4}, L_{r5}\}$,

$L_m \in L_A$, $L_n \in L_B$, $LCS(m,n)$ represents the score of L_m semantic consistent with L_n. We select matches $M_{Pre}\{i',j'\}$ with high confidence to find layout correspondence. If i' in planar $L_m(L_m \in L_A)$ and j' in planar $L_n(L_n \in L_B)$, then planar L_m is semantic consistent with L_{rn}. If there is K matches support L_m semantic consistent with L_n, $LCS_{(lm,rn)} = K$. $F(LCS)$ represents the maximum indexes of matrix LCS by row, which provides the robust layout correspondences. The accuracy of layout correspondence will affect the performance of the keypoints matching. We use layout confidence L_C as a metric for evaluating layout correspondence. If the number of all keypoint matches in the image pair is N_A, and the number of keypoint matches supporting the current layout correspondence is N_S, the layout match confidence L_C is as follows:

$$L_C = \frac{N_S}{N_A} \tag{3}$$

We update the assignment matrix using L_C layout correspondence so that the matches satisfy layout constraints.

4.2   Loss Function

The final loss $\mathcal{L}_A$ consists of matching loss and layout constraint loss. We trained LCM in a supervised manner from ground truth matches $M_{gt} = (i,j)$. The matching loss function is as follows:

$$\mathcal{L}_M = -\sum_{(i,j)\in M_{gt}} \log S_{(i,j)} - \sum_{(i\in I)} \log S_{(i,N+1)} - \sum_{(j\in J)} \log S_{(M+1,j)} \tag{4}$$

I is the keypoints set of the image A, J is the keypoints set in image B. When the keypoint in image A does not have a corresponding point in image B, forcing a match with other keypoints can produce a significant cost. Therefore, we expand $\bar{S}$ to S. S is partial assignment matrix of $(M+1) \times (N+1)$. The $M+1$ row and the $N + 1$ column represent the absence of a corresponding point. M means image A have M keypoints, N means image B have N keypoints. $M + 1$ and $N + 1$ row are dustbins which used in SuperGlue [13] to assign unmatched keypoints. To make the neural network robust to dynamic environments, a layout constraint loss is designed to penalize mismatches. $L_{i'}$ represents the semantic planar that the ith keypoint of the image A belongs. $L_{j'}$ represents the semantic planar that the jth keypoint of the image A belongs. We construct the loss function $\mathcal{L}_C$ is

$$\mathcal{L}_C = \begin{cases} 0 & L_{i'} = L_{j'} \\ \sum_{(i',j')\in M_{Pre}} L_C \times e^{S_{(i',j')}} & L_{i'} \neq L_{j'} \end{cases} \tag{5}$$

If we still use room layout constraint when the layout confidence level is low, the neural network hard to output high-quality matches. So, we use room layout constraints only if the layout confidence is above a threshold TL_C. The final loss is as follows:

$$\mathcal{L}_\mathcal{A} = \begin{cases} \mathcal{L}_\mathcal{M} & L_C < TL_C \\ \mathcal{L}_\mathcal{M} + L_C & L_C >= TL_C \end{cases} \tag{6}$$

5 Experiment

5.1 Metrics and Datasets

As in the previous work [13], we use AUC (the area under the cumulative error curve) of the pose error at the thresholds (5°, 10°, 20°) as metrics to measure the matching method. The AUC of the pose error in percentage is reported. The pose error here refers to the maximum angular error in rotation and translation. After using RANSAC to filter out outliers, matches are used to estimate the essence matrix to obtain the relative pose of the camera. We also use precision as the metric. A match is considered correct when its epipolar distance is less than the threshold. We evaluated LCM on CR10k and ScanNet. ScanNet is a large scale indoor image dataset consisting of RGB images with the ground truth of the camera pose and depth images. ScanNet is explicitly defined as training set, validation set and test set. The dataset is captured from a real static environment, while CR10k is captured from a model house. The evaluation of the LCM on ScanNet will allow us to determine whether the LCM can be applied to real scenes. Therefore, we evaluate LCM on ScanNet too.

5.2 Results

Table 2. Indoor Pose Estimation

Dataset	Method	AUC5° ↑	AUC10° ↑	AUC20° ↑	Precision↑
CR10k	MNN	6.07	9.17	13.71	25.12
	MNN + OANet [21]	3.51	10.44	21.38	64.85
	MNN + PointCN [20]	3.62	10.82	22.86	66.14
	SuperGlue [13]	7.53	18.23	31.66	82.67
	Ours(LCM)	**10.78**	**25.13**	**42.23**	**86.13**
ScanNet	MNN	9.06	20.99	35.62	61.14
	MNN + OANet [21]	4.79	11.23	20.55	65.37
	MNN + PointCN [20]	5.63	14.18	26.46	77.46
	SuperGlue [13]	**16.16**	33.81	51.84	84.40
	Ours(LCM)	15.83	**34.28**	**53.63**	**86.60**

Our results are shown in Table 2. The AUC of the pose error in percentage is reported. Compared to MNN, OANet, and SuperGlue, LCM enables significantly higher pose estimation and precision on CR10k. Compared with SuperGlue, LCM improved by 3.25%, 6.9%, 19.23% and 3.46% in respectively. According to the visualization results, LCM also produces more correct matches and fewer mismatches (see Fig. 6), because it utilizes layout constraint to filter mismatches and guide neural network focus on keypoints that fall on static objects. Compared with image matching methods that only based on appearance features, LCM makes it possible to estimate correct matches and accurate poses in challenging changed environments. This demonstrates that LCM is robust to changed environments after using layout information to guide matching.

Table 2 shows LCM enables slight improvement against SuperGlue on Scan-Net. On the one hand, On the one hand, matches that violate the room layout constraint are not easily produced in the ScanNet. Therefore, the room layout contstraint has little effect on feature matching. On the other hand, layout feature encoding can bring more information for feature matching. According to the result on ScanNet, it can be determined that LCM is suitable for feature matching in static environments.

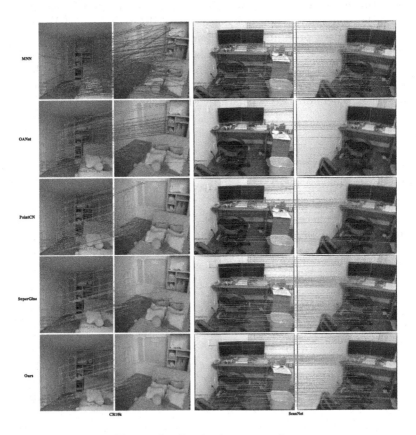

Fig. 6. Qualitative image matches.

5.3 Implementation Details

Because SuperPoint is capable of producing repeatable and sparse keypoints, which is very efficient for matching, we use it to extract feature point pixel coordinates and feature descriptors. To ensure the number of feature points, 1024 feature points are extracted from each image for matching. We set match threshold to 0.2 and the epipolar distance threshold to $5e-4$. Because dataset CR10k is small, training directly on CR10k would be overfitting. The self-attention layer, cross-attention layer and matching layer are loaded with pre-training parameters. Then, we train LCM on the dataset CR10k until convergence. Training LCM requires ground truth matching pairs. We establish initial matching pairs on CR10k using a pre-trained SuperGlue model. However, these pairs may contain incorrect matches. If the same point on a closet is still matched after changing its position, these matching pairs will not satisfy the epipolar constraint. We can use the epipolar constraint to remove these incorrect matching pairs and obtain correct ones for training. To ensure a fair comparison, we have fine-tuned the baselines on the CR10k dataset.

6 Conclusion

This paper presents a dataset that can be used to evaluate the performance of image matching methods in challenging changed environments. The proposed dataset provides camera pose ground truth to evaluate image matching in terms of the accuracy of relative camera pose. Our method uses layout information to guide the network to learn the matches, which is robust to changed environments. Our experiments show that LCM achieves a significant improvement compared to baselines, enabling highly accurate relative pose estimation in an indoor changed environment. What's more, LCM can be used for feature matching in static environments without judging whether moving objects are in the environments.

Acknowledgment. This work is supported by the China Postdoctoral Science Foundation (No.2021M690094) and the FuXiaQuan National Independent Innovation Demonstration Zone Collaborative Innovation Platform (No. 3502ZCQXT2021003).

References

1. Bai, X., Luo, Z., Zhou, L., Fu, H., Quan, L., Tai, C.L.: D3feat: joint learning of dense detection and description of 3D local features. In: 2020 IEEE Conference on Computer Vision and Pattern Recognition (2020)
2. Coughlan, J.M., Yuille, A.L.: Manhattan world: compass direction from a single image by Bayesian inference. In: Proceedings of the Seventh IEEE International Conference on Computer Vision, vol. 2 (1999)
3. Dai, A., Chang, A.X., Savva, M., Halber, M., Funkhouser, T.A., Nießner, M.: Scannet: richly-annotated 3D reconstructions of indoor scenes. In: 2017 IEEE Conference on Computer Vision and Pattern Recognition (2017)

4. DeTone, D., Malisiewicz, T., Rabinovich, A.: Superpoint: self-supervised interest point detection and description. In: 2018 IEEE Conference on Computer Vision and Pattern Recognition Workshops (2018)

5. Fischler, M.A., Bolles, R.C.: Random sample consensus: a paradigm for model fitting with applications to image analysis and automated cartography. Commun. ACM **24**, 381–395 (1981)

6. Li, Z., Snavely, N.: Megadepth: learning single-view depth prediction from internet photos. In: 2018 IEEE Conference on Computer Vision and Pattern Recognition (2018)

7. Lin, H.J., Huang, S.W., Lai, S.H., Chiang, C.K.: Indoor scene layout estimation from a single image. In: 2018 24th International Conference on Pattern Recognition (2018)

8. Mallya, A., Lazebnik, S.: Learning informative edge maps for indoor scene layout prediction. In: 2015 IEEE International Conference on Computer Vision (2015)

9. Ono, Y., Trulls, E., Fua, P.V., Yi, K.M.: Lf-net: Learning local features from images. In: Neural Information Processing Systems (2018)

10. Pero, L.D., Bowdish, J., Fried, D., Kermgard, B., Hartley, E., Barnard, K.: Bayesian geometric modeling of indoor scenes. In: 2012 IEEE Conference on Computer Vision and Pattern Recognition (2012)

11. Ren, Y., Li, S., Chen, C., Kuo, C.C.J.: A coarse-to-fine indoor layout estimation (cfile) method. In: 2016 Asian Conference on Computer Vision (2016)

12. Revaud, J., Leroy, V., Weinzaepfel, P., Chidlovskii, B.: Pump: pyramidal and uniqueness matching priors for unsupervised learning of local descriptors. In: 2022 IEEE/CVF Conference on Computer Vision and Pattern Recognition, pp. 3916–3926 (2022)

13. Sarlin, P.E., DeTone, D., Malisiewicz, T., Rabinovich, A.: Superglue: learning feature matching with graph neural networks. In: 2020 IEEE Conference on Computer Vision and Pattern Recognition (2020)

14. Savinov, N., Seki, A., Ladicky, L., Sattler, T., Pollefeys, M.: Quad-networks: unsupervised learning to rank for interest point detection. In: 2017 IEEE Conference on Computer Vision and Pattern Recognition (2017)

15. Schönberger, J.L., Frahm, J.M.: Structure-from-motion revisited. In: 2016 IEEE Conference on Computer Vision and Pattern Recognition (2016)

16. Schönberger, J.L., Zheng, E., Frahm, J.-M., Pollefeys, M.: Pixelwise view selection for unstructured multi-view stereo. In: Leibe, B., Matas, J., Sebe, N., Welling, M. (eds.) ECCV 2016. LNCS, vol. 9907, pp. 501–518. Springer, Cham (2016). https://doi.org/10.1007/978-3-319-46487-9_31

17. Shen, Z., Sun, J., Wang, Y., He, X.H., Bao, H., Zhou, X.: Semi-dense feature matching with transformers and its applications in multiple-view geometry. IEEE Trans. Pattern Anal. Mach. Intell. **45**, 7726–7738 (2022)

18. Sturm, J., Engelhard, N., Endres, F., Burgard, W., Cremers, D.: A benchmark for the evaluation of RGB-D slam systems. In: 2012 IEEE International Conference on Intelligent Robots and Systems (2012)

19. Wald, J., Sattler, T., Golodetz, S., Cavallari, T., Tombari, F.: Beyond controlled environments: 3D camera re-localization in changing indoor scenes. In: Vedaldi, A., Bischof, H., Brox, T., Frahm, J.-M. (eds.) ECCV 2020. LNCS, vol. 12352, pp. 467–487. Springer, Cham (2020). https://doi.org/10.1007/978-3-030-58571-6_28

20. Yi, K.M., Trulls, E., Ono, Y., Lepetit, V., Salzmann, M., Fua, P.V.: Learning to find good correspondences. In: 2018 IEEE Conference on Computer Vision and Pattern Recognition (2018)
21. Zhang, J., et al.: Learning two-view correspondences and geometry using order-aware network. In: 2019 IEEE International Conference on Computer Vision (2019)
22. Zhang, L., Rusinkiewicz, S.M.: Learning to detect features in texture images. In: 2018 IEEE Conference on Computer Vision and Pattern Recognition (2018)

To Be Critical: Self-calibrated Weakly Supervised Learning for Salient Object Detection

Jian Wang[1], Tingwei Liu[1], Miao Zhang[1,2], and Yongri Piao[1(✉)]

[1] Dalian University of Technology, Dalian, China
{dlyimi,tingwei}@mail.dlut.edu.cn, {miaozhang,yrpiao}@dlut.edu.cn
[2] Key Lab for Ubiquitous Network and Service Software of Liaoning Province, Dalian, China

Abstract. Weakly-supervised salient object detection (WSOD) aims to develop saliency models using image-level annotations. Despite of the success of previous works, explorations on an effective training strategy for the saliency network and accurate matches between image-level annotations and salient objects are still inadequate. In this work, **1)** we propose a self-calibrated training strategy by explicitly establishing a mutual calibration loop between pseudo labels and network predictions, liberating the saliency network from error-prone propagation caused by pseudo labels. **2)** we prove that even a much smaller dataset (merely 1.8% of ImageNet) with well-matched annotations can facilitate models to achieve better performance as well as generalizability. This sheds new light on the development of WSOD and encourages more contributions to the community. Comprehensive experiments demonstrate that our method outperforms all the existing WSOD methods by adopting the self-calibrated strategy only. Steady improvements are further achieved by training on the proposed dataset. Additionally, our method achieves 94.7% of the performance of fully-supervised methods on average. And what is more, the fully supervised models adopting our predicted results as "ground truths" achieve successful results (95.6% for BASNet and 97.3% for ITSD on F-measure), while costing only 0.32% of labeling time for pixel-level annotation. The code and dataset are available at https://github.com/DUTyimmy/SCW.

Keywords: Salient object detection · Weakly supervised learning · Deep learning

1 Introduction

Salient object detection (SOD) aims to segment objects in an image that visually attract human attention most. It plays an important role in many computer

This work was supported by the National Natural Science Foundation of China under Grant 62172070 and Grant 61976035.

Supplementary Information The online version contains supplementary material available at https://doi.org/10.1007/978-981-99-8552-4_15.

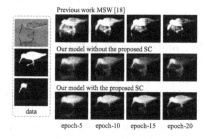

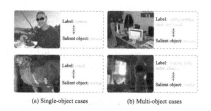

Fig. 1. The visual saliency predictions during the training process of different models. **SC** represents proposed self-calibrated training strategy. Column **data** represents image, ground truth and pseudo label.

Fig. 2. Cross-domain inconsistency between ImageNet dataset and salient object detection. (a) and (b) represent single-object and multi-object cases, respectively.

visions, such as image segmentation [20] and visual tracking [12]. Recently, deep learning based methods [4,5] have proved its superiority and achieved remarkable progress. Success of those methods, however, heavily relies a large number of highly accurate pixel-level annotations, which are time-consuming and labor-intensive to collect. A trade-off between testing accuracy and training annotation cost has long existed in the SOD task.

To alleviate this predicament, several attempts have been made to explore different weakly supervised formats, such as noisy label [22,27], scribble [34] and image-level annotation (*i.e.*, classification label). Image-level annotation based WSOD methods usually adopt a two-stage scheme, which leverages a classification network to generate pseudo labels and then trains a saliency network on these labels. In this paper, we focus on this most challenging problem of developing WSOD by only using image-level annotation.

Some pioneering works [29,35] pursue accurate pseudo labels to train a saliency network and achieve good performance. However, given the fact that pseudo labels are still a far cry from the ground truths, the error remaining unaddressed in the pseudo labels can propagate to the generated predictions. This is consistent with the fact that as the number of epochs increases, the parameters of the model are updated and the prediction curve goes from underfitting to optimal to overfitting. Interestingly, we observe that the relatively good results containing global representations of saliency can be predicted at the early training process (*e.g.*, epoch-5), while the predictions are more prone to error at the latter training process (*e.g.*, epoch-20), as shown in the first two rows in Fig. 1. This inspires us to go one step further exploring how this global representation can be evolved as the model is properly trained.

Moreover, previous works adopt existing large-scale datasets, *e.g.*, ImageNet [6] and COCO [21], to perform WSOD. However, an observable fact should not be ignored that there is an inherent inconsistency between image classification and SOD task. For example, many classification labels do not match the salient objects in both single-object and multi-object cases in ImageNet, as illustrated in Fig. 2. Such cross-domain inconsistency caused by those mismatched samples impairs the generalizability of models and prevents WSOD methods from achieving optimal results.

In this work, our core insight is that we can design a self-calibrated training strategy and establish a saliency-based classification dataset to address the aforementioned challenges, respectively. To be specific, we **1)** aim to calibrate our network with progressively updated labels to curb the spread of errors in low-quality pseudo labels during the training process. **2)** collect a new dataset on existing saliency dataset DUTS [29], ensure the reliable matches for which image-level annotations are correctly corresponding to salient objects. The source code will be released upon publication. Concretely, our contributions are as follows:

- We propose a self-calibrated training strategy to prevent the network from propagating the negative influence of error-prone pseudo labels. A mutual calibration loop is established between pseudo labels and network predictions to promote each other.
- We open up a fresh perspective on that even a much smaller dataset (merely 1.8% of ImageNet) with well-matched image-level annotations allows WSOD to achieve better performance. This encourages more existing data to be correctly annotated and further paves the way for the booming future of WSOD.
- Our method outperforms existing WSOD methods on all metrics over five benchmark datasets, and meanwhile achieves averagely 94.7% performance of state-of-the-art fully supervised methods. We also demonstrate that our method retains its competitive edge on most metrics even without our proposed dataset.
- We extend the proposed method to other fully supervised SOD methods. Our offered pseudo labels enable these methods to achieve comparatively high accuracy (95.6% for BASNet [25] and 97.3% for ITSD [39] on F-measure) while being free of pixel-level annotations, costing only 0.32% of labeling time for pixel-level annotation.

2   Related Work

2.1   Salient Object Detection

Early SOD methods mainly focus on detecting salient objects by utilizing hand-craft features and setting various priors, such as center prior [15], boundary prior [33]. Recently, deep learning based methods demonstrate its advantages and achieve remarkable improvements. Plenty of promising works [13,28,31] are proposed and present various effective architectures. Among them, Hou *et al.* [13] present short connections to integrate the low-level and high-level features, and predict more detailed saliency maps. Wu *et al.* [31] propose a novel cascaded partial decoder framework and utilize generated relatively precise attention map to refine high-level features. In [28], researchers propose to explore boundary of the salient objects to predict a more detailed prediction. Although appealing performance these methods have achieved, vast high-quality pixel-level annotations are needed for training their models, which are time-consuming and laborious.

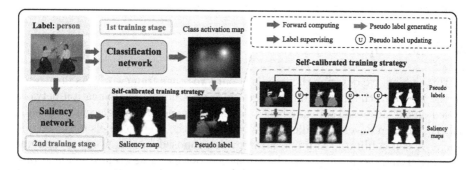

Fig. 3. Overall framework of our proposed method. In the first stage, classification labels are used to supervise classification network to generate CAMs and further produce pseudo labels. In the second stage, we train a saliency network with the above pseudo labels and propose a self-calibrated strategy to correct labels and predictions progressively.

2.2 Weakly Supervised Salient Object Detection

For achieving a trade-off between labeling efficiency and model performance, researchers aim to perform salient object detect with low-cost annotations. To this end, WSOD is presented and achieves an appealing performance with image-level annotations only.

Wang *et al.* [29] design a foreground inference network (FIN) to predict saliency maps from image-level annotations, and introduce a global smooth pooling (GSP) to combine the advantages of global average pooling (GAP) and global max pooling (GMP), which explicitly computes the activation of salient objects. In [18], Li *et al.* also perform WSOD based on image-level annotations. And they adopt a recurrent self-training strategy and propose a conditional random field based graphical model to cleanse the noisy pixel-wise annotations by enhancing the spatial coherence as well as salient object localization. Based on a traditional method MB+ [36], more accurate saliency maps are generated in less than one second per image. Zeng *et al.* [35] intelligently utilize multiple annotations (*i.e.*,, classification and caption annotations) and design a multi-source weak supervision framework to integrate information from various annotations. Benefited from multiple annotations and an interactive training strategy, a really simple saliency network can also achieve appealing performance. All the above methods target to train a classification network (on existing large-scale multiple objects dataset, *i.e.*, ImageNet [6] or Microsoft COCO [21]) to generate class activation maps (CAMs) [38], then perform different refinement methods to generate pseudo labels. Supervised by these pseudo labels directly, a saliency network is trained and predicts the final saliency maps.

Different from the aforementioned works, we argue that: **1)** Developing an effective training strategy encourages more accurate predictions even under the supervision of inaccurate pseudo labels which would mislead the networks. **2)** Establishing accurate matches between classification labels and salient objects could facilitate the further development of WSOD.

3   The Proposed Method

In this section, we describe the details of our two-stage framework. As illustrated in the Fig. 3, in the first training stage, we train a normal classification network based on the proposed saliency-based dataset, to generate more accurate pseudo labels. We then develop a saliency network using the pseudo labels in the second stage. A self-calibrated training strategy is proposed in this stage to immune network from inaccurate pseudo labels and encourage more accurate predictions.

3.1   From Image-Level to Pixel-Level

Class activation maps (CAMs) [38] localize the most discriminative regions in an image using only a normal classification network and build a preliminary bridge from image-level annotations to pixel-level segmentation tasks. In this paper, we adopt CAMs following the same setting of [2], to generate pixel-level pseudo labels in the first training stage. To better understand our proposed approach, we will describe the generation of CAMs in a brief way.

For a classification network, we discard all the fully connected layers and apply an extra global average pooling (GAP) layer as well as a convolution layer as previous works do. In the training phase, we take images in classification dataset as input, and compute its classification scores Cls as follows:

$$Cls = w_s{}^T * GAP(F^5) + b_s, \tag{1}$$

where F^5 represents the features from the last convolution block, $GAP(\cdot)$ denotes the global average pooling operation and w_s^T as well as b_s are the learnable parameters of the convolution layer. In the inference phase, we compute the CAMs of images in DUTS-Train dataset as follows:

$$C_{AM} = \sum_{k=1}^{N} Cls_k * \frac{Relu(w_s^T * F^5 + b_s)_k}{m} \tag{2}$$

where $Relu(\cdot)$ and denote the relu activation function, m is the maximum value of the input tensor. w_s^T and b_s are the shared parameters learnd in the training phase, Cls_k represents the classification scores for category k and N represents the total number of categories. In this phase, multi-scale inference strategy is adopted, which rescales the original image into four sizes and computes the average CAMs as the final output. As Ahn *et al.* [2] have pointed out, CAMs mainly concentrate on the most discriminative regions and are too coarse to serve as pseudo labels. Various refinements have been conducted to generate pseudo labels. Different from [29,35] using a clustering algorithm, a plug-and-play module PAMR [3] is adopted in our method.

It performs refinement using the low-level color information of RGB images, which can be inserted into our framework flexibly and efficiently. Following the settings of [29,35], we also adopt CRF [17] for a further refinement. Note that it is only used to generate pseudo labels in our method.

3.2 Self-calibrated Training Strategy

In the second training stage, a saliency network is trained with the pseudo labels generated in the first training stage. As is mentioned above, the relatively good results containing global representations of saliency is gradually degraded as the training process continues. A straightforward method to tackle this dilemma is setting a validation set to pick the best result during the training process. However, we argue that it may lead to sub-optimal results because **1)** despite good saliency representations are learned at the early training stage, the predictions are coarse and lack detail as the loss function is still converging (as shown in the 2nd row in Fig. 1). **2)** the capability of networks to learn saliency representation is not fully excavated. **3)** we believe that WSOD should not use any pixel-level ground truth in the training process, even as a validation set. Following this main idea, we propose to establish a mutual calibration loop during the training process in which error-prone pseudo labels are recursively updated and calibrate the network for better predictions in turn. **Insight:** As is discussed in the Sect. 1, under the supervision of noisy pseudo labels, the saliency network goes from relatively good to overfitting. On the one hand, in our weakly supervised settings, this "overfitting" manifests itself as the network being affected by the noisy pseudo labels and learning the inaccurate noise information in them, which heavily restricts the performance of WSOD. It is also worth to mention that this is fundamentally different from the "overfitting" in supervised learning, the latter means that the network learns the biased information in a less comprehensive training set. On the other hand, we conclude reasons of the relatively good point before overfitting as: 1) Although many pseudo labels are noisy and inaccurate, the whole pseudo labels still describe general saliency cues. It can provide a roughly correct guidance for the saliency network. 2) Before the loss converges, the saliency network is prone to learn the regular and generalized saliency cues rather than the irregular and noisy information in pseudo labels. Such kind of robustness is also discussed in [11]. Motivated by the above analyses, we propose a self-calibrated training strategy to effectively utilize the robustness and tackle the negative overfitting.

To be specific, supervised by inaccurate pseudo labels Y, we take the predictions P of the saliency network as saliency seeds. As is illustrated in Fig. 3, coarse but more accurate seeds are predicted during the first few epochs regardless of the inaccurate supervision of error-prone pseudo labels. We take these seeds as correction terms to calibrate and update the original pseudo labels Y, while performing refinement again with PAMR. Detailed procedure is present in the supplementary material Algorithm 1, here we set a threshold μ to 0.4 for the binarization operation on refined predictions P'. We conduct self-calibrated strategy throughout the training process, that is, it is performed on each training batch. The loss function for this training stage can be described as:

$$L(P, Y) = -\sum_{i=1} ((1 - \lambda)y_i + \lambda p_i') * \log p_i - ((1-\lambda)(1-y_i)+\lambda(1-p_i')) * \log(1-p_i), \quad (3)$$

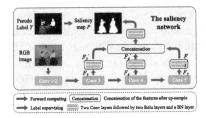

Fig. 4. Detailed structure of our saliency network. We adopt a simple encoder-decoder architecture and take prediction P as our final result.

Fig. 5. DUTS-Cls is a saliency-based dataset with image-level annotations, containing 44 categories and 5959 images.

where λ is the weighting factor that is illustrated in the supplementary material Algorithm 1. The intuition is that as the training process goes on, the saliency prediction is more accurate and larger weight should be given. y_i, p_i and p_i' represent the elements of Y, P and refined predictions P', respectively.

As is illustrated in the Fig. 3, equipped with our proposed self-calibrated training strategy, inaccurate pseudo labels are progressively updated, and in turn supervise the network. This mutual calibration loop finally encourages both accurate pseudo labels and predictions.

3.3   Saliency Network

As for the saliency network, we adopt a simple encoder-decoder architecture without any auxiliary modules, which is usually served as baseline for fully-supervised SOD methods [13,31]. As illustrated in Fig. 4, for an image from DUTS-Train dataset, we take features F_3, F_4 and F_5 from the encoder, to generate F_3', F_4' and F_5' through two convolution layers, and then adopt a bottom-up strategy to perform feature fusion, which can be denoted as:

$$P = \sigma(Conv(Cat(Up(F_5'), Up(F_4'), F_3'))), \quad (4)$$

where $\sigma(\cdot)$ represents the sigmoid function, $Conv(\cdot)$ and $Cat(\cdot)$ denote the convolution and concatenation operation, respectively. $Up(\cdot)$ represents upsampling feature maps to the same size.

In the decoder, the number of output channels of all the middle convolution layers are set to 64 for acceleration. Note that our final prediction P is predicted in an end-to-end manner in the test phase without any post-processing.

4   Dataset Construction

To explore the advantages of accurate matches between image-level annotations and corresponding salient objects, we establish a saliency-based classification dataset, which ensures all the classification labels correspond to the salient objects. Following this main idea, we relabel an existing widely-adopted saliency training set DUTS-Train [29] with well-matched image-level annotations, namely DUTS-Cls dataset. It fits with WSOD better than existing large-scale classification datasets due to the accurate matches, and facilitates the further improvements for WSOD.

Table 1. Quantitative comparisons of E-measure (E_s), S-measure (S_α), F-measure (F_β) and MAE (M) metrics over five benchmark datasets. - means unavailable results, Ours- and Ours represent our method trained on ImageNet and proposed DUTS-Cls dataset, respectively. The best two results are marked in **boldface** and magenta.

Methods	ECSSD				DUTS-Test				HKU-IS				DUT-OMRON				PASCAL-S			
	S_α	E_s	F_β	M	S_α	E_s	F_β	M	S_α	E_s	F_β	M	S_α	E_s	F_β	M	S_α	E_s	F_β	M
WSS [29]	.811	.869	.823	.104	.748	.795	.654	.100	.822	.896	.821	.079	.725	.768	.603	.109	.744	.791	.715	.139
ASMO [18]	.802	.853	.797	.110	.697	.772	.614	.116	-	-	-	-	.752	.776	.622	.101	.717	.772	.693	.149
MSW [35]	.827	.884	.840	.096	.759	.814	.684	.091	.818	.895	.814	.084	.756	.763	.609	.109	.768	.790	.713	.133
Ours-	.836	.887	.838	.083	.770	**.830**	**.689**	.079	.836	.907	.822	.064	.743	.807	.643	.085	.778	.818	.742	.111
Ours	**.858**	**.901**	**.853**	**.071**	**.776**	.829	.688	**.077**	**.850**	**.918**	**.835**	**.058**	**.766**	**.817**	**.667**	**.078**	**.781**	**.824**	**.749**	**.108**

To be specific, we select and label images in DUTS-Train with image-level annotations, while discarding rare categories because only several images are contained. The proposed DUTS-Cls dataset contains 44 categories and 5959 images. As is illustrated in Fig. 5, it reaches a relative equilibrium in terms of image numbers of each category and covers most common categories. It is worth mentioning that labeling image-level annotations is quite fast, which only takes less than 1 s per image. Compared to about 3 min [23] for labeling a pixel-level ground truth, it takes less than 0.56% of the time and labor cost for a sample. Annotating DUTS-Cls dataset (5959 samples) only costs 0.32% of labeling time than annotating the whole DUTS-Train dataset (10553 samples) with pixel-level ground truth. This indicates that exploring WSOD with image-level annotation is quite efficient. Moreover, the DUTS-Cls dataset with well-matched image-level annotations offers a better choice for WSOD than ImageNet, and we genuinely hope it could contribute to the community and encourage more existing data to be correctly annotated at image level.

Table 2. Quantitative results of ablation studies. **Dataset** represents different training sets used in the first training stage. **Strategy** denotes training strategy used in the second stage, - indicates the baseline model without any training strategy and +SC represents adopting our proposed self-calibrated strategy. The best results are marked in **boldface**.

Dataset		Strategy		ECSSD		DUTS-Test		HKU-IS		DUT-OMRON		PASCAL-S	
ImageNet	DUTS-Cls	-	+ SC	$F_\beta\uparrow$	$M\downarrow$	$F_\beta\uparrow$	$M\downarrow$	$F_\beta\uparrow$	$M\downarrow$	$F_\beta\uparrow$	$M\downarrow$	$F_\beta\uparrow$	$M\downarrow$
✓			✓	0.776	0.121	0.642	0.094	0.773	0.090	0.568	0.111	0.694	0.140
	✓		✓	0.836	0.096	0.675	0.085	0.822	0.075	0.648	0.083	0.735	0.126
✓			✓	0.838	0.083	**0.689**	0.079	0.822	0.064	0.643	0.085	0.742	0.111
	✓		✓	**0.853**	**0.071**	0.688	**0.077**	**0.835**	**0.058**	**0.667**	**0.078**	**0.749**	**0.108**

5 Experiments

5.1 Implementation Details

We implement our method on the Pytorch toolbox with a single RTX 2080Ti GPU. The backbone adopted in our method is DenseNet-169 [14], which is same as the latest work [35]. During the first training stage, we train a classification network on our proposed DUTS-Cls dataset. In this stage, we adopt the Adam optimization algorithm [16], the learning rate is set to 1e−4 and maximum epoch is set to 20. In the second training stage, we only take the RGB images from DUTS-Train as our training set. In this stage, we use Adam optimization algorithm with the learning rate 3e−6 and maximum epoch 25. The batch size of both training stages is set to 20 and all the training and testing images are resized to 256 × 256.

5.2 Datasets and Evaluation Metrics

For a fair comparison, we train our model on ImageNet and our proposed DUTS-Cls dataset respectively, denoted as Ours- and Ours in Table 1. We conduct comparisons on five following widely-adopted test datasets. ECSSD [32]: contains 1000 images which cover various scenes. DUT-OMRON [33]: includes 5168 challenging images consisting of single or multiple salient objects with complex contours and backgrounds. PASCAL-S [20]: is collected from the validation set of the PASCAL VOC semantic segmentation dataset [8], and contains 850 challenging images. HKU-IS [19]: includes 4447 images, many of which contain multiple disconnected salient objects. DUTS [29]: is the largest salient object detection benchmark, which contains 10553 training samples (DUTS-Train) and 5019 testing samples (DUTS-Test). Most images in DUTS-Test are challenging with various locations and scales.

To evaluate our method in a comprehensive and reliable way, we adopt four well-accepted metrics, including S-measure [9], E-measure [10], F-measure [1] as well as Mean Absolute Error (MAE).

5.3 Comparison with State-of-the-Arts

We compare our method with all the existing image-level annotation based WSOD methods: WSS [29], ASMO [18] and MSW [35]. To further demonstrate the effectiveness of our weakly supervised methods, we also compare the proposed method with nine state-of-the-art fully supervised methods including DSS [13], R^3Net [7], DGRL [30], BASNet [25], PFA [37], CPD [31], ITSD [39] and MINet [24], all of which are trained on pixel-level ground truth and based on DNNs. For a fair comparison, we use the saliency maps provided by authors and perform the same evaluation code for all methods.

Quantitative Evaluation. Table 1 shows the quantitative comparison on four evaluation metrics over five datasets. It can be seen that our method outperforms all the weakly supervised methods on all metrics. Especially, 31.0% improvement on HKU-IS and 28.4% on DUT-OMRON are achieved on MAE metric. Our method also improves the performance on two challenging datasets DUT-ORMON and PASCAL-S by a large margin, which indicates that our method can explore more accurate saliency cues even in complex scenes. **Additionally**, the proposed saliency-based dataset with well-matched image-level annotations enables our method to achieve better performance, while far less training samples (less than 1.45% of the latest work MSW [35]) are required. To prove the effect of our method in a more objective manner, we also train our method on ImageNet dataset following the previous works. The results of "Ours-" shown in Table 1 demonstrate that our method can outperform existing methods on most metrics even without the proposed dataset thanks to the effective strategy. **Moreover**, we also compare our method with nine state-of-the-art fully supervised methods. It can be seen in supplementary materials Fig. 1 that our method, even with the image-level annotations only and a simple baseline network without any auxiliary modules, can also achieve 94.7% accuracy of fully supervised methods on average. **Finally**, in Table 3, we provide more comparisons with existing methods on Supervision type, Backbone, Image resolution, Flops and Parameter numbers. Despite the performance superiority, it can be seen that our method saves more computational costs than the other existing methods. **Qualitative evaluation.** In supplementary materials Fig. 1, we show the qualitative comparisons of our method with existing three WSOD methods as well as six state-of-the-art fully supervised methods. It can be seen that our method could discriminate salient objects from various challenging scenes (such as small objects case in the 1st row and complex background cases in the 2nd and 3rd rows) and achieve more complete and accurate predictions. **Moreover**, compared with the fully supervised methods, our method also predicts comparable and even better results in some cases, such as the complete house and log in the 5th and 6th rows. But we would like to point out that our results also need to be improved in term of the boundary of the salient objects.

Table 3. Comprisons with existing works on supervision types (Sup.), backbone, image resolution (Reso.), Flops as well as parameter numbers (Param.). The supervision type I indicates using image-level category annotations only, and I&C represents developing WSOD on both image-level category annotations and caption annotations simultaneously.

Methods	Sup.	Backbone	Reso.	FLOPs	Param.
WSS [29]	I	VGG16	256	20.12G	14.73M
ASMO [18]	I	ResNet101	256	44.47G	42.50M
MSW [35]	I&C	DNet169	256	14.21G	12.54M
Ours	I	DNet169	256	6.99G	8.51M

Table 4. The effectiveness of our proposed self-calibrated strategy on ECSSD dataset. **+ SC** indicates simply applying our self-calibrated strategy during the training process. The best results are marked in **boldface**.

Method	Strategy	$S_\alpha\uparrow$	$E_s\uparrow$	$F_\beta\uparrow$	MAE $\downarrow$
BSCA [26]	-	0.846	0.884	0.814	0.084
	+ SC	**+0.007**	**+0.009**	**+0.018**	**-0.008**
MR [33]	-	0.839	0.884	0.823	0.085
	+ SC	**+0.014**	**+0.010**	**+0.016**	**-0.009**
MSW [35]	-	0.827	0.884	0.840	0.096
	+ SC	**+0.017**	**+0.012**	**+0.014**	**-0.014**

5.4 Ablation Studies

Effect of the Self-calibrated Strategy. We conduct experiments on both ImageNet (1st and 3rd rows) and DUTS-Cls (2nd and 4th rows) settings in Table 2. It can be seen that the proposed self-calibrated strategy can not only enhance the performance of our method on ImageNet setting greatly, but also achieve great improvements even on the DUTS-Cls setting, especially on MAE metrics. **Besides**, the effectiveness of the proposed self-calibrated strategy can also be demonstrated by the visual results in Fig. 6. It can be seen that the proposed strategy can keep and enhance the globally good representations during the training process, and predict accurate saliency maps even supervised by error-prone pseudo labels. **Moreover**, for a comprehensive evaluation, 1) We change the pseudo label by using two traditional SOD methods BSCA [26] and MR [33], and then train our model with and without the proposed strategy respectively, the results are shown in the first four rows in Table 4. 2) We further apply our method on the lasted work MSW [35] by just adding our proposed strategy in the last two rows in Table 4. These results strongly prove that the self-calibrated strategy can not only work well on our method, but also effective for other pseudo labels and other works. **Effect of the DUTS-Cls dataset.** We introduce a saliency-based dataset with well-matched image-level annotations to offer a better choice for WSOD. The first two rows in Table 2 demonstrate that DUTS-Cls dataset encourages the baseline model to achieve remarkable improvements, compared to ImageNet dataset. And as is illustrated in the last two rows in Table 2, it also proves its superiority by a steady improvement on most metrics even if good enough performance is already achieved by adopting the self-calibrated strategy. This is consistent with our argument that the cross-domain inconsistency does impede the performance of WSOD, and a saliency-based dataset can settle this matter better. **Additionally**, we visualize the CAMs trained on ImageNet (named CAM_I) and DUTS-Cls (named CAM_D) in Fig. 7, it can be seen that CAM_D have higher activation level within the salient

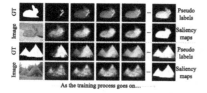

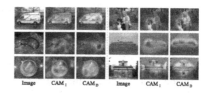

Fig. 6. Visual analysis of the effectiveness of our proposed self-calibrated strategy during the training process, noting that the ground truth is just for exhibition and not used in our framework.

Fig. 7. Visual analysis of effect of DUTS-Cls datset. CAM_I and CAM_D represent the CAMs generated by training on ImageNet and our DUTS-Cls dataset, respectively. Heatmap is adopted for better visualization.

objects trained on well-matched DUTS-Cls dataset. **Last but not least**, to further prove the effectiveness of the proposed DUTS-Cls dataset objectively, we also train the latest work MSW [35] on the DUTS-Cls dataset. As is shown in supplementary materials Fig. 4, by simply replacing ImageNet with DUTS-Cls, considerable improvements are achieved in less training iterations. It is worth to mention that the DUTS-Cls dataset reaches less than 1.8% percent of ImageNet in terms of sample size. This strongly demonstrates the effectiveness and generalizability of the well-matched DUTS-Cls dataset for WSOD.

6   Conclusion

In this paper, we propose a novel self-calibrated training strategy and introduce a saliency-based dataset with well-matched image-level annotations for WSOD. The proposed strategy establishes a mutual calibration loop between pseudo labels and network predictions, which effectively prevents the network from propagating the negative influence of error-prone pseudo labels. We also argue that cross-domain inconsistency exists between SOD and existing large-scale classification datasets, and impedes the development of WSOD. To offer a better choice for WSOD and encourage more contributions to the community, we introduce a saliency-based classification dataset DUTS-Cls to settle this matter well. In addition, our method can serve as an alternative to provide pixel-level labels for fully supervised SOD methods while maintaining comparatively high performance, costing only 0.32% of labeling time for pixel-level annotation.

References

1. Achanta, R., Hemami, S., Estrada, F., Susstrunk, S.: Frequency-tuned salient region detection. In: 2009 IEEE Conference on Computer Vision and Pattern Recognition, pp. 1597–1604 (2009)
2. Ahn, J., Cho, S., Kwak, S.: Weakly supervised learning of instance segmentation with inter-pixel relations. In: Proceedings of the IEEE Conference on Computer Vision and Pattern Recognition, pp. 2209–2218 (2019)

3. Araslanov, N., Roth, S.: Single-stage semantic segmentation from image labels. In: Proceedings of the IEEE/CVF Conference on Computer Vision and Pattern Recognition, pp. 4253–4262 (2020)
4. Cong, R., Lei, J., Fu, H., Huang, Q., Cao, X., Hou, C.: Co-saliency detection for RGBD images based on multi-constraint feature matching and cross label propagation. IEEE Trans. Image Process. 27(2), 568–579 (2018)
5. Cong, R., Lei, J., Fu, H., Porikli, F., Huang, Q., Hou, C.: Video saliency detection via sparsity-based reconstruction and propagation. IEEE Trans. Image Process. 28(10), 4819–4931 (2019)
6. Deng, J., Dong, W., Socher, R., Li, L.J., Li, K., Fei-Fei, L.: Imagenet: a large-scale hierarchical image database. In: 2009 IEEE Conference on Computer Vision and Pattern Recognition, pp. 248–255. IEEE (2009)
7. Deng, Z., et al.: R3net: recurrent residual refinement network for saliency detection. In: Proceedings of the 27th International Joint Conference on Artificial Intelligence, pp. 684–690 (2018)
8. Everingham, M., Van Gool, L., Williams, C.K., Winn, J., Zisserman, A.: The pascal visual object classes (VOC) challenge. Int. J. Comput. Vision 88(2), 303–338 (2010)
9. Fan, D.P., Cheng, M.M., Liu, Y., Li, T., Borji, A.: Structure-measure: a new way to evaluate foreground maps. In: Proceedings of the IEEE International Conference on Computer Vision, pp. 4548–4557 (2017)
10. Fan, D.P., Gong, C., Cao, Y., Ren, B., Cheng, M.M., Borji, A.: Enhanced-alignment measure for binary foreground map evaluation. arXiv preprint arXiv:1805.10421 (2018)
11. Fan, J., Zhang, Z., Tan, T.: Employing multi-estimations for weakly-supervised semantic segmentation. In: 2020 IEEE/CVF European Conference on Computer Vision, ECCV 2020. vol. 12362, pp. 332–348 (2020)
12. Hong, S., You, T., Kwak, S., Han, B.: Online tracking by learning discriminative saliency map with convolutional neural network. In: International Conference on Machine Learning, pp. 597–606 (2015)
13. Hou, Q., Cheng, M.M., Hu, X., Borji, A., Tu, Z., Torr, P.H.: Deeply supervised salient object detection with short connections. In: Proceedings of the IEEE Conference on Computer Vision and Pattern Recognition, pp. 3203–3212 (2017)
14. Huang, G., Liu, Z., Weinberger, K.Q.: Densely connected convolutional networks. In: 2017 IEEE Conference on Computer Vision and Pattern Recognition (CVPR), pp. 2261–2269 (2017)
15. Jiang, Z., Davis, L.S.: Submodular salient region detection. In: Proceedings of the IEEE Conference on Computer Vision and Pattern Recognition, pp. 2043–2050 (2013)
16. Kingma, D.P., Ba, J.: Adam: a method for stochastic optimization. arXiv preprint arXiv:1412.6980 (2014)
17. Krähenbühl, P., Koltun, V.: Efficient inference in fully connected CRFs with gaussian edge potentials. In: Advances in Neural Information Processing Systems, pp. 109–117 (2011)
18. Li, G., Xie, Y., Lin, L.: Weakly supervised salient object detection using image labels. arXiv preprint arXiv:1803.06503 (2018)
19. Li, G., Yu, Y.: Visual saliency based on multiscale deep features. In: Proceedings of the IEEE Conference on Computer Vision and Pattern Recognition, pp. 5455–5463 (2015)
20. Li, Y., Hou, X., Koch, C., Rehg, J.M., Yuille, A.L.: The secrets of salient object segmentation. In: Proceedings of the IEEE Conference on Computer Vision and Pattern Recognition, pp. 280–287 (2014)

21. Lin, T.-Y., et al.: Microsoft COCO: common objects in context. In: Fleet, D., Pajdla, T., Schiele, B., Tuytelaars, T. (eds.) ECCV 2014. LNCS, vol. 8693, pp. 740–755. Springer, Cham (2014). https://doi.org/10.1007/978-3-319-10602-1_48
22. Nguyen, T., et al.: DeepUSPS: deep robust unsupervised saliency prediction via self-supervision. In: Advances in Neural Information Processing Systems, vol. 32, pp. 204–214 (2019)
23. Niu, Y., Geng, Y., Li, X., Liu, F.: Leveraging stereopsis for saliency analysis. In: 2012 IEEE Conference on Computer Vision and Pattern Recognition, pp. 454–461 (2012)
24. Pang, Y., Zhao, X., Zhang, L., Lu, H.: Multi-scale interactive network for salient object detection. In: Proceedings of the IEEE/CVF Conference on Computer Vision and Pattern Recognition, pp. 9413–9422 (2020)
25. Qin, X., Zhang, Z., Huang, C., Gao, C., Dehghan, M., Jagersand, M.: BASNet: Boundary-aware salient object detection. In: Proceedings of the IEEE Conference on Computer Vision and Pattern Recognition, pp. 7479–7489 (2019)
26. Qin, Y., Lu, H., Xu, Y., Wang, H.: Saliency detection via cellular automata. In: Proceedings of the IEEE Conference on Computer Vision and Pattern Recognition, pp. 110–119 (2015)
27. Siva, P., Russell, C., Xiang, T., Agapito, L.: Looking beyond the image: unsupervised learning for object saliency and detection. In: Proceedings of the IEEE Conference on Computer Vision and Pattern Recognition, pp. 3238–3245 (2013)
28. Su, J., Li, J., Zhang, Y., Xia, C., Tian, Y.: Selectivity or invariance: boundary-aware salient object detection. In: 2019 IEEE/CVF International Conference on Computer Vision, ICCV 2019, pp. 3798–3807. IEEE (2019)
29. Wang, L., et al.: Learning to detect salient objects with image-level supervision. In: Proceedings of the IEEE Conference on Computer Vision and Pattern Recognition, pp. 136–145 (2017)
30. Wang, T., et al.: Detect globally, refine locally: a novel approach to saliency detection. In: Proceedings of the IEEE Conference on Computer Vision and Pattern Recognition, pp. 3127–3135 (2018)
31. Wu, Z., Su, L., Huang, Q.: Cascaded partial decoder for fast and accurate salient object detection. In: Proceedings of the IEEE Conference on Computer Vision and Pattern Recognition, pp. 3907–3916 (2019)
32. Yan, Q., Xu, L., Shi, J., Jia, J.: Hierarchical saliency detection. In: Proceedings of the IEEE Conference on Computer Vision and Pattern Recognition, pp. 1155–1162 (2013)
33. Yang, C., Zhang, L., Lu, H., Ruan, X., Yang, M.H.: Saliency detection via graph-based manifold ranking. In: Proceedings of the IEEE Conference on Computer Vision and Pattern Recognition, pp. 3166–3173 (2013)
34. yue Yu, S., Zhang, B., Xiao, J., Lim, E.: Structure-consistent weakly supervised salient object detection with local saliency coherence. arXiv:abs/2012.04404 (2020)
35. Zeng, Y., Zhuge, Y., Lu, H., Zhang, L., Qian, M., Yu, Y.: Multi-source weak supervision for saliency detection. In: Proceedings of the IEEE Conference on Computer Vision and Pattern Recognition, pp. 6074–6083 (2019)
36. Zhang, J., Sclaroff, S., Lin, Z., Shen, X., Price, B., Mech, R.: Minimum barrier salient object detection at 80 fps. In: Proceedings of the IEEE International Conference on Computer Vision, pp. 1404–1412 (2015)
37. Zhao, T., Wu, X.: Pyramid feature attention network for saliency detection. In: Proceedings of the IEEE Conference on Computer Vision and Pattern Recognition, pp. 3085–3094 (2019)

38. Zhou, B., Khosla, A., Lapedriza, A., Oliva, A., Torralba, A.: Learning deep features for discriminative localization. In: Proceedings of the IEEE Conference on Computer Vision and Pattern Recognition, pp. 2921–2929 (2016)
39. Zhou, H., Xie, X., Lai, J.H., Chen, Z., Yang, L.: Interactive two-stream decoder for accurate and fast saliency detection. In: Proceedings of the IEEE/CVF Conference on Computer Vision and Pattern Recognition, pp. 9141–9150 (2020)

Image Visual Complexity Evaluation Based on Deep Ordinal Regression

Xiaoying Guo[1,2(✉)], Lu Wang[3], Tao Yan[2,3], and Yanfeng Wei[4]

[1] School of Automation and Software Engineering, Shanxi University, Taiyuan, Shanxi, China
guoxiaoying@sxu.edu.cn
[2] Institute of Big Data Science and Industry, Shanxi University, Taiyuan, Shanxi, China
[3] School of Compute and Information Technology, Shanxi University, Taiyuan, Shanxi, China
[4] Institute of Management and Decision, School of Economics and Management,
Shanxi University, Taiyuan, Shanxi, China

Abstract. Image complexity is an important indicator in computer vision that helps people to more accurately evaluate and understand visual image information. Compared with traditional regression algorithms, ordinal regression methods are better suited to handling relationships and structures among ordinal data, providing more accurate reference for image complexity assessment. Currently, IC9600 dataset has made significant progress for providing the largest image complexity dataset, and ICNet provided a baseline model to evaluate the complexity score of images, but it neglects the ordinal property of complexity scores. This paper focuses on exploring a method to evaluate image complexity based on deep ordinal regression. We propose an evaluation model (ICCORN) that combines convolutional neural network ICNet and ordinal regression approach CORN. The model firstly extracts global semantic information and local detail information, and then considers ordinal relationship between complexity scores in the prediction process. The model demonstrates a high degree of correlation with human perception, as indicated by an increased Pearson correlation coefficient of 0.955. Furthermore, other evaluation metrics have also yielded favorable results.

Keywords: Image complexity · Complexity evaluation · Ordinal regression · Visual perception

1 Introduction

"Image complexity" refers to the description of the inherent complexity level of images. In recent years, researchers have proposed the concept of computable image complexity to simulate human visual perception by computers, quantifying the perception of image

This work is partially supported by the Youth Program of the National Natural Science Foundation of China (61603228, 62006146), Natural Science Research Pro-gram of Shanxi Prov-ince Basic Research Program (202203021221029), Youth Project of Applied Basic Research Program of Shanxi Province (20210302123030), and Funds for central-government-guided local science and technology development (YDZJSX20231C001).

Q. Liu et al. (Eds.): PRCV 2023, LNCS 14435, pp. 199–210, 2024.
https://doi.org/10.1007/978-981-99-8552-4_16

complexity research [1], which is a part of affective computing. Visual complexity, as a basic aspect of visual perception, is extremely important for human being to understand and perceive the visual stimuli. The connotation of image complexity can be concluded as three aspects [2]: Firstly, as an evaluation index, it reflects the difficulty of image watermarking technology, image compression, and image segmentation; secondly, as an image feature, it is used for image classification and retrieval; thirdly, as an image visual perception emotion, it is used for research on image cognition, aesthetics, and the correlation of image memory.

Ordinal Regression (OR) is a machine learning paradigm that deals with the ordinal relationships between label categories. It aims to partition data based on ordinal labels, which are often encountered in real-world scenarios and have both discrete and ordinal characteristics. OR has been widely applied in various fields, such as age estimation [3], image aesthetic quality assessment [4], dating estimation [5], image retrieval [6] and medical diagnosis [7]. The goal of OR is to automatically discriminate the level of samples with fixed discrete grades, where higher scores correspond to better quality. Unlike typical regression problems, the output of OR is a finite set of discrete values, and more importantly, for multi-class classification problems, OR aims to achieve high classification accuracy while minimizing the difference between predicted and true labels. Therefore, considering the ordinal information in OR can result in smaller absolute mean errors compared to simple classification methods.

Traditional methods for evaluating image complexity often rely on simple categorization (e.g., high complexity (HC), medium complexity (MC), low complexity (LC)) or regression (e.g., 1–5, 1–100), without considering the ordered relationship between different levels (see Fig. 1). To address this issue, this study focuses on exploring the problem of using convolutional neural networks for ordered evaluation of visual complexity in images. In assessing the complexity scores of images, we introduce the concept of ordinal regression by incorporating ordered regression (dividing the score data into multiple levels: (1–5) and aim to learn the ordered information ($1 < 2 < 3 < 4 < 5$). This enables the model to better capture the complexity scores of images.

The main contributions of this paper are as follows:

(1) By adding deeper layers to improve the basic network model of ICNet [8], the model is better able to capture complex patterns and semantic information within the input data, ultimately resulting in more accurate and informative representations.
(2) Proposing a new deep ordinal regression model (ICCORN), which incorporates ordinal information into the prediction of image complexity. Experimental results on three benchmark data sets demonstrate the effectiveness and the stability of our ICCORN in image complexity.

The rest of this paper is organized as follows: Sect. 2 reviews related work of image complexity and ordinal regression. Section 3, introduces the proposed method (ICCORN). Section 4 analyzes and discusses the experimental results. Section 5 provides a summary of our work.

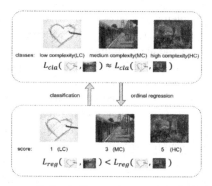

Fig. 1. Comparison between Ordinal Regression and Multi-Classification. For multi-classification, the difference between LC and MC in the feature space is similar to the difference between LC and HC in the feature space, while OR makes the difference between LC and MC much smaller than the difference between LC and HC in the feature space.

2 Related Work

2.1 Image Complexity Evaluation

Image complexity has always been a hot and difficult problem in the field of image engineering. In the computer science field, researchers are mainly dedicated to making image complexity computable, and have proposed image complexity calculation methods based on information theory [9], image compression [10], traditional machine learning-based image feature combinations [11], and deep learning-based image complexity calculations.

Recent research has focused on using deep neural networks to perceive image complexity. Nagle and Lavie [12] found that deep convolutional networks are more correlated with complexity ratings than traditional feature combinations and mid-level features. Other studies, such as Saraee et al. [13] used deep neural networks to predict complex regions in images and evaluate image complexity using both low-level and high-level features. Kyle-Davidson's [14] work also introduced two datasets containing scene scores and corresponding complexity annotations collected from human observers. Kyle-Davidson et al. [15] developed ComplexityNet, a deep learning model that integrates metrics for scene clutter and symmetry to generate a single score complexity rating and a two-dimensional per-pixel complexity map for images. Feng et al. [8] constructed an image complexity evaluation dataset containing 9600 well-annotated images and provided a baseline model, ICNet, to predict image complexity scores and estimate complexity heatmaps in a weakly supervised manner.

2.2 Ordinal Regression

With the gradual enhancement of computer vision, there is an increasing demand for image complexity evaluation models that consider the ordinal relationship between levels in predictions and classifications. Niu et al. [3] were the first to use deep convolutional neural networks to solve ordinal regression problems by transforming OR into a series of binary classification sub-problems. Che et al. [16] proposed a framework named Rank-CNN based on convolutional neural network (CNN), which aims to classify facial images into age groups. By doing so, the age ranking estimation problem with a range from 1 to K is decomposed into a series of binary classification subproblems (K-1 classifiers). Cao et al. [17] addressed the issue of inconsistent classifiers in ordinal regression CNNs by proposing a new framework called Consistent Ranking Logits (CORAL), while Shi et al. [18] introduced a Rank Consistency Ordinal Regression Network framework, CORN (Conditional Ordinal Regression for Neural Networks) that achieves ranking consistency through a novel training scheme.

3   The Proposed Method

In this section, we propose a new deep ordinal regression model (ICCORN) to estimate image complexity. The ICCORN model consists of a global branch and a detail branch as inputs. The global branch uses ResNet152 to extract overall background and semantic information from low-resolution images, while the detail branch uses ResNet18 to capture local detailed representations from high-resolution images, the ICCORN model produces an estimated score and a heat map for image complexity. The score prediction component consists of an ordinal regression layer and a regression layer, where the former provides sequence information to assist the latter in determining the final image complexity score. Meanwhile, in the heat map prediction component, a pixel-level regression is employed to generate the heat map (as shown in Fig. 2). Ordinal regression is integrated throughout the entire ICCORN model and provides sequence information. The global branch module and the ordinal regression module will be discussed in detail in the following sections.

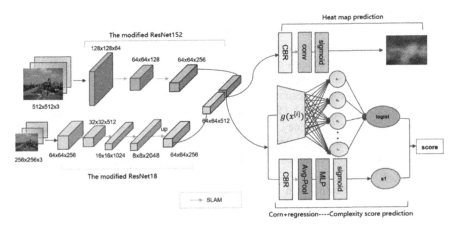

Fig. 2. ICCORN model comprises not only the feature inputs from the detail and global branches, but also the label inputs from ordinal labels and regression labels, ultimately generating image complexity scores and activation maps.

3.1 Improved the ICNet Model

Global image properties and semantic features are crucial for complexity evaluation, with humans relying on both low-level and high-level content [14]. In deep learning, deeper networks can extract more refined semantic information than shallower ones. The ICNet [8] model has two branches modified from ResNet18, and includes the spatial layout attention module (SLAM [8]). We retained the remaining components of the original model and upgraded the network framework by modifying the global branch to ResNet152, as illustrated in Fig. 3. Deep neural networks provide semantic information to shallow neural networks, while shallow neural networks extract detailed local information.

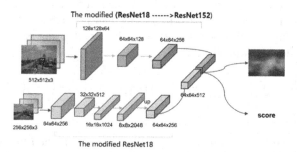

Fig. 3. The global branch of ICNet has been modified from ResNet18 to ResNet152 in order to extract more comprehensive semantic information.

3.2 Ordinal Regression Model

Preliminaries. The ordinal regression problem consists of an input vector x and an output prediction label y, where $x^{[i]} \in \mathcal{X} \subseteq \mathbb{R}^k$ denote the input of the i-th training sample and $y^{[i]}$ its corresponding class label, where $y^{[i]} \in \mathcal{Y} = \{r_1, r_2, ...r_K\}$ with a natural label ordering $r_1 \prec r_2 \prec, \cdots \prec r_K$. Given a dataset with N training examples, $D = \{x^{[i]}, y^{[i]}\}_{i=1}^N$, The goal is to find a classification rule or function $h : \mathcal{X} \to \mathcal{Y}$ that minimizes a loss function $L(h)$.

Rank-Consistent Ordinal Regression based on Conditional Probabilities. Given a training set $D = \{x^{[i]}, y^{[i]}\}_{i=1}^N$, CORN extends labeling to ordinal labels $y^{[i]}$, binary label $y_k^{[i]} \in \{0, 1\}$ denote whether $y^{[i]}$ exceeds rank r_K. $K - 1$ learning tasks associated with ranks $r_1, r_2, ...r_K$ are also utilized by CORN in the output layer, which is demonstrated in Fig. 4.

CORN estimates a sequence of conditional probabilities by using conditional training subsets. In this way, the k-th binary task $f_k(x^{[i]})$ output presents the conditional probability:

$$f_k\left(x^{[i]}\right) = \hat{P}\left(y^{[i]} > r_k | y^{[i]} > r_{k-1}\right) \tag{1}$$

where the events are nested: $\{y^{[i]} > r_k\} \subseteq \{y^{[i]} > r_{k-1}\}$. Especially, when $k = 1$, $f_k(x^{[i]})$ represents the initial unconditional probability $f_1(x^{[i]})=\hat{P}(y^{[i]} > r_1)$.

Applying the chain rule for probabilities to the model outputs allows the computation of transformed, unconditional probabilities:

$$\hat{P}\left(y^{[i]} > r_K\right) = \prod_{j=1}^{k} f_j\left(x^{[i]}\right) \tag{2}$$

Since $\forall j, 0 \leq f_j(X^{[i]}) \leq 1$, we have
$\hat{P}(y^{[i]} > r_1) \geq \hat{P}(y^{[i]} > r_2) \geq ... \geq \hat{P}(y^{[i]} > r_{K-1})$,
which guarantees rank consistency among the $K - 1$ binary tasks.

CORN Loss Function. The predicted value of the j-th node in the output layer of the network (shown in Fig. 4) is denoted by $f_j(x^{[i]})$, and $|S_j|$ denotes the size of the j-th conditional training set. To train a CORN neural network using backpropagation, we minimize the following loss function:

$$L(X,y) = -\frac{1}{\sum_{j=1}^{K-1}|S_j|} \sum_{j=1}^{K-1} \sum_{i=1}^{|S_j|} \left[log\left(f_j\left(x^{[i]}\right)\right) \cdot \triangleleft\left\{y^{[i]} > r_j\right\} + log\left(1 - f_j\left(x^{[i]}\right)\right) \cdot \triangleleft\left\{y^{[i]} \leq r_j\right\}\right] \tag{3}$$

Please note that in the expression of $f_j(x^{[i]})$, $x^{[i]}$ refers to the i-th training example in S_j. In order to simplify the notation, we omit an additional index j, which would otherwise distinguish between $x^{[i]}$ in different conditional training sets.

In order to enhance the numerical stability of the loss gradients in the process of training, we have implemented an alternative formulation of the loss that involves the net inputs of the last layer (also known as logits), denoted by Z as shown in Fig. 4, and $log\left(\sigma(z^{[i]})\right) = log\left(f_j(x^{[i]})\right)$:

$$L(Z,y) = -\frac{1}{\sum_{j=1}^{K-1}|S_j|} \sum_{j=1}^{K-1} \sum_{i=1}^{|S_j|} \left[log\left(\sigma\left(z^{[i]}\right)\right) \cdot \triangleleft\left\{y^{[i]} > r_j\right\} + \left(log\left(\sigma\left(z^{[i]}\right)\right) - z^{[i]}\right) \cdot \triangleleft\left\{y^{[i]} \leq r_j\right\}\right] \tag{4}$$

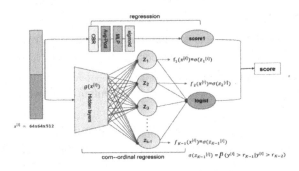

Fig. 4. Prediction of Image Complexity Scores.

The Ordinal Regression of ICCORN. In the network, the ordinal regression (CORN) is combined with regression to calculate the final image complexity score, as shown in Fig. 4. The input part is the feature map ($64 \times 64 \times 512$) concatenated from the detail branch and the global branch. In the ordinal regression part, the input data are firstly modified by dividing the original regression data into categories (1–5), and the five classifications data and regression data are jointly used as the ground-truth for model training. Ordinal regression loss is added to the total loss function, and finally, the two are combined to obtain the score of image complexity.

3.3 Total Loss Function

During the training process, mean squared error (MSE) is used to calculate the loss for the regression part.

$$\mathcal{L}_1 = \frac{1}{N} \sum_{j=1}^{N} (S_1 - S_{gt})^2 \tag{5}$$

For the heatmap prediction part, weakly supervised learning is used to calculate the distance between the averaged scalar value of the predicted complexity heatmaps and the ground-truth values using mean squared error (MSE) to compute the loss.

$$\mathcal{L}_2 = \frac{1}{N} \sum_{j=1}^{N} (S_2 - S_{gt})^2 \tag{6}$$

where N refers to the total number of samples in a batch, S_1 represents the global predicted score, S_{gt} represents the ground-truth score, and S_2 represents the scalar value obtained after averaging the complexity generation heatmap.

The loss function for the ordinal regression part is represented by Eq. 4 in the previous section, and is denoted as $\mathcal{L}_3$ which indicates the ordinal regression loss.

The total loss is represented by the combination of $\mathcal{L}_1$, $\mathcal{L}_2$, and $\mathcal{L}_3$.

$$\mathcal{L} = a\mathcal{L}_1 + b\mathcal{L}_2 + c\mathcal{L}_3 \tag{7}$$

Furthermore, let a, b, and c be 0.8, 0.1, and 0.1, respectively.

4   Experiment and Results

4.1 Dataset

IC9600 Dataset. The IC9600 [8] dataset is widely used for evaluating image complexity and is currently the largest well-annotated dataset available. It contains 9,600 images covering eight semantic categories, each with five levels of complexity. The scores are obtained through subjective evaluation by multiple people to represent the human perception of image complexity.

VISC-C Dataset. The VISC-C [14] dataset contains 800 images with a resolution of 700x700 pixels, divided into eight categories corresponding to common scene types. Each image is assigned a complexity score between 0 and 100, reflecting its visual complexity, and also has 10 two-dimensional complexity annotations (5 complex and 5 simple).

PASCAL VOC_4000 Dataset. The PASCAL VOC_4000 [12] dataset is a subset of the PASCAL VOC dataset [19], consisting of 4,000 randomly selected images observed by 53 observers to obtain 75,020 forced-choice pairs. Using the TrueSkill algorithm, the dataset generates visual complexity ratings for the 4,000 images (Fig. 5).

Fig. 5. Three datasets and the display of ordinal labels for different datasets.

4.2 Experimental Setup

Implementation Details. This article is based on the PyTorch framework and uses a pre-trained model on the ImageNet dataset to initialize the parameters of each stage of the dual-branch extraction. During the training process, SGD is used as the optimizer, with a batch size of 64 and a learning rate of 0.0001. The loss function combines three types of loss functions (regression, ordinal regression, heatmap).

Evaluation Metrics. The evaluation metrics used in this article include Pearson correlation coefficient (PCC), Spearman correlation coefficient (SRCC), root mean square error (RMSE), and root mean absolute error (RMAE).

PCC can be represented as follows:

$$\rho(X, Y) = \frac{E[(X - \mu_X)(Y - \mu_Y)]}{\sqrt{\sum_{i=1}^{N}(X_i - \mu_X)^2}\sqrt{\sum_{i=1}^{N}(Y_i - \mu_Y)^2}} \tag{8}$$

where x and y represent predicted score and true score, μ_X and μ_Y are their means, N is the total number of images. SRCC is calculated as follows:

$$\rho' = 1 - 6\frac{\sum_{i=1}^{N}(r_i - r_i')^2}{N^3 - N} \tag{9}$$

Here, r_i and r_i' represents the rank difference between the predicted and true scores when they are sorted in descending order for the i-th image. Additionally, the computation of RMSE is as follows:

$$r(X, Y) = \sqrt{\frac{1}{N}\sum_{i=1}^{N}(X_i - Y_i)^2} \tag{10}$$

RMSE can be represented as follows:

$$m(X, Y) = \sqrt{\frac{1}{N} \sum_{i=1}^{N} |X_i - Y_i|} \tag{11}$$

4.3 Experimental Results Analysis

In this section, experimental results on the IC9600, VISC-C, and PASCAL VOC_4000 dataset are presented to validate the effectiveness of the proposed model.

Comparison with Other Methods. As shown in Table 1, on the large-scale IC9600 dataset, ICNet18_152, obtained by only upgrading the global branch to ResNet152, outperforms both ICNet152, which upgrades both branches to deeper networks, and ICNet18_swin, which upgrades only the global branch to Swin Transformer, which has shown excellent performance in classification and image segmentation but not necessarily in regression of specific image complexity. On the small-sized VISC-C dataset (Table 2) and medium-sized PASCAL VOC_4000 dataset (Table 3), the proposed method achieved better results.

Table 1. Comparisons of Different Methods on IC9600 Dataset.

model	PCC↑	SRCC↑	RMSE↓	RMAE↓
ICNet [8]	0.949	0.94	0.053	0.205
ICNet152	0.9529	0.949	0.0515	0.1974
ICNet152 + CORN	0.9537	0.950	0.0486	0.1923
ICNet + swin	<u>0.8851</u>	0.8748	0.0780	0.2436
ICNet18_152	**0.9550**	0.951	0.0502	0.1926
ICCORN(ICNet18_152 + CORN)	0.9535	**0.951**	**0.0477**	**0.1910**

Table 2. Comparisons of Different Methods on VISC-C dataset.

model	PCC↑
CNet(ResNet-152) [15]	0.693
CNet-C (ResNet-152) [15]	0.729
CNet-CS (ResNet-152) [15]	0.716
ICNet18_152	0.767
ICCORN(ICNet18_152 + CORN)	**0.784**

Table 3. Comparisons of Different Methods on PASCAL VOC_4000 dataset.

model	PCC↑
VGG-16(pre-trained) [12]	0.65
Inception V3(pre-trained) [12]	0.83
Inception V3(random weights) [12]	0.73
ICNet18_152	0.84
ICCORN(ICNet18_152 + CORN)	**0.86**

Ablation Experiments. For the ordinal regression network, CORAL [17], CORN [18], and the cumulative link model (odd) are compared, and CORN is selected as the ordinal regression part combined with the network, as shown in Table 4. Table 5 shows the comparison of results on different datasets with and without ordinal regression. The results indicate that adding ordinal regression improves performance in all aspects, demonstrating its effectiveness for estimating image orderliness.

Table 4. Comparisons of the network with ordinal regression models add.

model	PCC↑	SRCC↑	RMSE↓	RMAE↓
ICNet [8]	0.949	0.94	0.053	0.205
ICNet + CORN	0.9486	**0.946**	**0.0515**	**0.1974**
ICNet + CORAL	**0.9504**	**0.946**	0.0524	0.2018
ICNet + odd	0.9483	0.9446	0.0528	0.2083

Table 5. Comparisons on different datasets after incorporating ordinal regression.

dataset	model	PCC↑	SRCC↑	RMSE↓	RMAE↓
IC9600 [8]	ICNet18_152	**0.955**	**0.951**	0.0502	0.1926
	ICNet18_152 + CORN	0.9535	**0.951**	**0.0477**	**0.1910**
VISC-C [14]	ICNet18_152	0.7666	0.7405	0.3369	0.1388
	ICNet18_152 + CORN	**0.7838**	**0.7657**	**0.3323**	**0.1346**
Fintan et al. [12]	ICNet18_152	0.8350	0.8372	0.0912	0.2636
	ICNet18_152 + CORN	**0.8553**	**0.8736**	**0.0798**	**0.2462**

Additionally, it was found in experiments that incorporating ordinal regression on the IC9600 dataset can reduce overfitting, allowing the metrics to remain stable as the number of training epochs increases. As shown in Fig. 6, after adding ordinal regression, the Pearson correlation coefficient (PCC) and Spearman correlation coefficient (SPCC) exhibit a relatively stable trend in the later stages of training, whereas the data without ordinal regression gradually decreases.

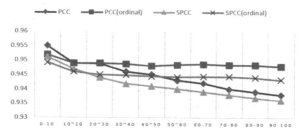

Fig. 6. Comparisons of PCC and SRCC with and without ordinal regression.

This paper shows some visualization results predicted by our model in Fig. 7. The dashed box in the figure highlights the cases of failure. The ordinal regression model performs poorly in visualizing complexity when the complexity is high. Additionally, both models tend to overestimate the scores when the complexity ratings are low.

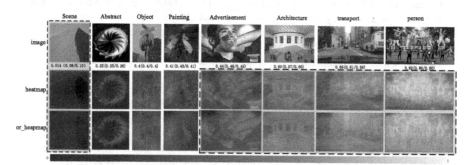

Fig. 7. Visualization results. In each pair of images, the upper (lower) image represents the input image (predicted complexity heatmap). The number in parentheses to the left indicates the result without incorporating ordinal regression (OR), while the number to the right indicates the result with OR. The number outside the parentheses is the true score (normalized to 0–1).

5   Conclusions

The ICCORN model proposes a new method for predicting image complexity, using an improved ICNet and ordinal regression module CORN. The model includes a global branch and detail branch, which extract overall background and local detail representations from low and high-resolution images, respectively. The model predicts the score and heat map of image complexity. Our proposed model outperforms state-of-the-art models in evaluating image complexity scores. In the future work, we will explore the impact of image complexity regions on different areas and use eye-tracking tools to improve computer image complexity evaluation.

References

1. Nadal, M., Munar, E., Marty, G., Cela-Conde, C.J.: Visual complexity and beauty appreciation: explaining the divergence of results. Empir. Stud. Arts **28**(2), 173–191 (2010)
2. Guo, X., Li, W., Qian, Y., Bai, R., Jia, C.: A review of computational methods for image complexity assessment. Acta Electron. Sin. **48**(4), 819–826 (2020)
3. Niu, Z., Zhou, M., Wang, L., Gao, X.: Ordinal regression with multiple output CNN for age estimation. In: Proceedings of the IEEE Conference on Computer Vision and Pattern Recognition (CVPR), pp. 4920–4928 (2016). https://doi.org/10.1109/CVPR.2016.532
4. Díaz, R., Marathe, A.: Soft labels for ordinal regression. In: 2019 IEEE/CVF Conference on Computer Vision and Pattern Recognition (CVPR), pp. 4733–4742 (2019)
5. Lee, Y.J., Efros, A.A., Hebert, M.: Style-aware mid-level representation for discovering visual connections in space and time. In: Proceedings of the 2013 IEEE International Conference on Computer Vision (ICCV), pp. 1857–1864 (2013)
6. Xiao, Y., Liu, B., Hao, Z.: Multiple-instance ordinal regression. IEEE Trans. Neural Netw. Learn. Syst. **29**(9), 4398–4413 (2018)
7. Vargas, V.M., Gutierrez, P.A., Hervas-Martinez, C.: Cumulative link models for deep ordinal classification. Neurocomputing **401**, 48–58 (2020)
8. Feng, T., et al.: IC9600: a benchmark dataset for automatic image complexity assessment. IEEE Trans. Pattern Anal. Mach. Intell. **45**(7), 8577–8593 (2023)
9. Cardaci, M., Di Gesù, V., Petrou, M., Tabacchi, M.E.: A fuzzy approach to the evaluation of image complexity. Fuzzy Sets Syst. **160**, 1474–1484 (2009)
10. Mayer, S., Landwehr, J.: When complexity is symmetric: the interplay of two core determinants of visual aesthetics. Adv. Cogn. Psychol. **10**, 71–80 (2014)
11. Guo, X., Qian, Y., Li, L., Asano, A.: Assessment model for perceived visual complexity of painting images. Knowl.-Based Syst. **159**, 110–119 (2018)
12. Nagle, F., Lavie, N.: Predicting human complexity perception of real-world scienes. R. Soc. Open Sci. **7**(191487), 1–14 (2020). https://doi.org/10.1098/rsos.191487
13. Saraee, E., Jalal, M., Betke, M.: Visual complexity analysis using deep intermediate-layer features. Comput. Vis. Image Understand. **195**, 1–13 (2020)
14. Kyle-Davidson, C., Zhou, E.Y., Walther, D.B., Bors, A.G., Evans, K.K.: Characterising and dissecting human perception of scene complexity. Cognition **231**, 105319 (2023)
15. Kyle-Davidson, C., Bors, A.G., Evans, K.K.: Predicting human perception of scene complexity. In: IEEE International Conference on Image Processing (ICIP), pp. 1281–1285 (2022)
16. Chen, S., Zhang, C., Dong, M., Le, J., Rao, M.: Using ranking-CNN for age estimation. In: Computer Vision and Pattern Recognition (CVPR), pp. 5183–5192 (2017)
17. Cao, W., Mirjalili, V., Raschka S.: Consistent rank logits for ordinal regression with convolutional neural networks. arXiv preprint arXiv:1901.078846 (2019)
18. Shi, X., Cao, W.: Deep neural networks for rank-consistent ordinal regression based on conditional probabilities. arXiv preprintarXiv:2111.08851 (2021)
19. Everingham, M., Van Gool, L., Williams, C.K., Winn, J., Zisserman, A.: The pascal visual object classes (voc) challenge. Int. J. Comput. Vision **88**, 303–338 (2010)

Low-Light Image Enhancement Based on Mutual Guidance Between Enhancing Strength and Image Appearance

Linlin Hu[1,2], Shijie Hao[1,2(✉)], Yanrong Guo[1,2], Richang Hong[1,2], and Meng Wang[1,2]

[1] Key Laboratory of Knowledge Engineering with Big Data, Ministry of Education, Hefei University of Technology, Hefei, China
`2021171244@mail.hfut.edu.cn, yrguo@hfut.edu.cn`
[2] School of Computer Science and Information Engineering, Hefei University of Technology, Hefei, China
`hfut.hsj@gmail.com`

Abstract. The existing low light image enhancement (LLIE) methods primarily aim at adjusting the overall brightness of the image, which are prone to produce the over-enhancement issue, such as over-exposure and edge halo. Therefore, it is desirable to improve the visibility of originally dark regions of an image, while preserving the naturalness of the originally bright regions. Based on this motivation, we propose a simple but effective mutual guidance module, which builds a mutual guidance process between a pixel-wise enhancing strength map and an edge-aware lightness map. Based on this module, the image appearance information such as illumination and structure can be effectively propagated onto the enhancing strength map. By integrating this module into the ZeroDCE++ model, the over-enhancement issue like over-exposure and edge halo can be greatly alleviated. We have conducted extensive experiments to validate the effectiveness and the superiority of our model. Compared with many state-of-the-art unsupervised and supervised LLIE methods, our model achieves a much better visual effect as it consistently keeps the naturalness during the enhancement process. Our model also has better or comparable performance than its counterparts in quantitative comparison with various image quality assessment metrics.

Keywords: Low-light image enhancement · Mutual guidance module · Unsupervised model · Over-enhancement

1 Introduction

Images taken under unsatisfying illumination conditions, such as the low-light or back-light environment, often have low visual quality. Therefore, the task of low-light image enhancement (LLIE) has attracted many concerns from both research and industrial communities. Recently, great progress has been made on the LLIE task based on the deep learning paradigm, and many novel LLIE models have been built [6, 10].

This work was supported by the National Natural Science Foundation of China under Grant No. 62172137.

Q. Liu et al. (Eds.): PRCV 2023, LNCS 14435, pp. 211–223, 2024.
https://doi.org/10.1007/978-981-99-8552-4_17

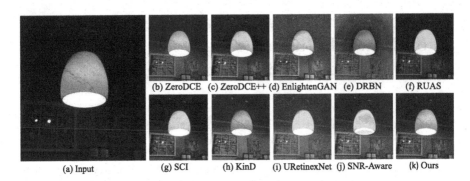

Fig. 1. An example for showing the necessity of preserving the visual naturalness during the LLIE process. Better to have a zoomed-in view of the figures in this paper.

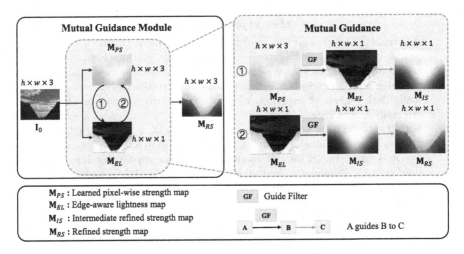

Fig. 2. An illustration on the mutual guidance module's structure.

Despite of the success, in a general sense, current LLIE models based on fully supervised learning [4,24], still have some common limitations that hider the development of this field. The first one is that the pairwise training images are not easy to obtain. It is difficult to capture sufficient bright/dark image pairs with strict pixel-wise correspondence, especially for the outdoor environment. The second limitation is that computational efficiency still needs improvement. Many LLIE models aims to learn an end-to-end pixel-wise mapping, in which an encoder-decoder backbone network structure is adopted. To ensure the visual quality of enhancement, the model size and computational complexity are usually relatively high, which poses challenges for the implementation on edge devices.

To address the above limitation, several novel LLIE models based on the unsupervised learning paradigm have been proposed. For example, the EnlightenGAN model [9] greatly facilitates the training process in real-world applications, as it does not require the collection of bright/dark image pairs. The ZeroDCE model [5] provides a new technical roadmap to solve the above issues. On one hand, it needs no normal light images as the reference during the training process. Instead, the model fully utilizes the prior information about the illumination distribution of normal-light images via several subtly designed loss function terms. On the other hand, ZeroDCE directly learns the parameters for an explicit enhancing function, which is different from most of the other LLIE models trying to learn an end-to-end pixel mapping. Therefore, the model size and complexity can be fundamentally reduced.

However, current LLIE models mainly concentrate on improving the global visibility of low-light images, while spending fewer efforts on ensuring the visual naturalness of enhanced results, leading to the over-enhancement issue. For example, overexposure widely exist in the originally bright regions (e.g. Fig. 1 (f), (g), (i) and (j)). As for the reason, the illumination distribution of an original image is not fully utilized in controlling the enhancing strength. Therefore, enhancement models usually fail to impose smaller or even no enhancing strength on them according the illumination distribution. In addition, artifacts like edge halo are usually encountered in LLIE results (e.g. Fig. 1(b), (d), (e), (h), (i) and (j)). This can be attributed to the fact that most LLIE models are less aware of the image structures at a finer scale, leading to the inconsistency between the enhancing strength and the edge structures of an image.

To address the above issue, we propose a simple but effective module that enables our LLIE model to be fully aware of image appearance information that are critical for a success enhancement, such illumination and structure. For a low-light image, the enhancing strength can be gradually refined by the proposed module through a mutual guidance process between the pixel-wise strength map ($\mathbf{M}_{PS}$) and the edge-aware lightness map ($\mathbf{M}_{EL}$), as exemplified in Fig. 2. The feasibility comes from the fact that these two maps are highly correlated with each other. On one hand, $\mathbf{M}_{PS}$ indicates the general strength distribution imposed on the lightness map at the macro level. On the other hand, at the micro level, $\mathbf{M}_{EL}$ expects that the strength distribution should be consistent with the edge structures as much as possible. Based on this observation, we establish the mutual guidance process described as Fig. 2. At first, $\mathbf{M}_{PS}$ guides the refinement of $\mathbf{M}_{EL}$, producing an intermediate refined strength map ($\mathbf{M}_{IS}$). Then, $\mathbf{M}_{EL}$ guides the further refinement $\mathbf{M}_{IS}$, obtaining the desired refinement strength map ($\mathbf{M}_{RS}$), which becomes aware of the illumination and structure. The above process is realized through the guided filtering, making this module computationally efficient and free of training. Based on the mutual guidance module, we improve the ZeroDCE++ model [11] by incorporating the module into each enhancing iteration. We find that the over-enhancement and edge halo issues can be effectively solved by the new model. In addition, the enhancement process converges quickly and tends to be more stable.

The technical contributions are summarized as follows. First, we build the mutual guidance module by fully utilizing the highly correlated enhancing strength map and the edge-ware lightness map. Second, the mutual guidance module is corporated into the ZeroDCE++ model. Our improved model keeps all the merits of the original ZeroDCE++, such as pariwise-data-free and lightweight. More importantly, it effectively avoids the over-enhancement issue, making the enhanced result more natural.

The rest of the paper is organized as follows. Section 2 briefly introduces the related LLIE works. In Sect. 3, we describe the details of our LLIE model and the proposed mutual guidance module. In Sect. 4, we report qualitative and quantitative experiment results to validate the effectiveness of our model. Section 5 finally concludes the paper.

2   Related Works

The data-driven LLIE models are built upon deep neural networks, and they are optimized by screening sufficient training images under different kinds of supervision paradigms. The supervised LLIE models [2–4,13,24] have paved the way for the era of low-light image enhancement with deep learning models. However, the supervised LLIE models usually requires sufficient dark-bright image pairs with strict pixel-wise correspondence. Actually, in real world applications, it is difficult to obtain such image pairs especially in outdoor environments, which limits the generalization ability of the supervised LLIE methods. To solve this issue, researchers resort to relieving the requirement of collecting pairwise training images during model construction. EnlightenGAN [9] enables the learning of an end-to-end enhancement network under the generative adversarial framework, which does not require the paired images for training have pixel-wise correspondence. Yang et al. [22] proposed a recursive band network (DRBN) as a semi-supervised learning method for low light image enhancement, in which only a small number of pairwise training images are required to extract coarse-to-fine band representations. As a step further, Guo et al. [5] proposed a novel unsupervised LLIE model ZeroDCE, which only requires dark images for training and greatly facilitates its practicability. ZeroDCE++ [11] further extends ZeroDCE via simplifying its model structure. Liu et al. [12] proposed an unsupervised Retinex-based unfolding framework RUAS, and achieves good performance on model efficiency. Ma et al. [14] proposed a self-calibrating illumination learning module (SCI) which is characterized as faster, more flexible and robust. In general, the recently proposed unsupervised LLIE methods successfully solve data restriction issue, and obtain very promising performance on model efficiency. However, most of the data-driven LLIE methods pay less attention to preserving the naturalness during the enhancing process, and usually introduce the over-enhancement issue. The model proposed in this paper belongs to the unsupervised learning LLIE group. Different from the existing ones, our model mainly targets at solving the over-enhancement issue via building the mutual guidance module. The module fully exploits the off-the-shelf image appearance information, such as the illumination distribution and the edge structure of an image for processing itself, which is often neglected in other data-driven methods.

3 Method

3.1 The Overall Framework of Our Model

The overall framework of our LLIE model is presented in Fig. 3. Based on the input image $\mathbf{I}_0$, the DCE-Net proposed in [11] learns an initial pixel-wise strength map $\mathbf{M}_{PS}$ (or named as $\mathbf{M}_{RS_0}$). In the meantime, an edge-aware illumination estimation $\mathbf{M}_{EL_0}$ is extracted from $\mathbf{I}_0$. Then, the enhancement process is implemented in an iterative way. For example, in the first iteration, $\mathbf{M}_{RS_0}$ and $\mathbf{M}_{EL_0}$ are taken as the inputs of the first mutual guidance module, producing refined strength map $\mathbf{M}_{RS_1}$. The intermediate enhanced image $\mathbf{I}_1$ can be obtained by a deterministic function $\mathbf{I}_1 = LE(\mathbf{I}_0, \mathbf{M}_{RS_1})$. Typically, the enhancement function LE at the k-th iteration can be described as:

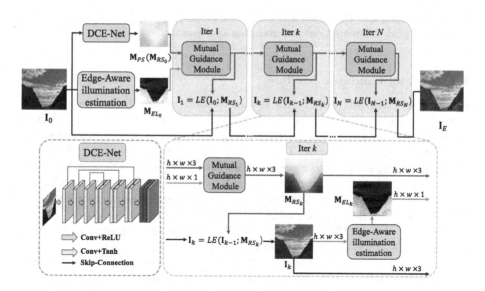

Fig. 3. The overall framework of our model.

$$\mathbf{I}_k(p) = \mathbf{I}_{k-1}(p) + \mathbf{M}_{RS_k}(p)(1 - \mathbf{I}_{k-1}(p)) \tag{1}$$

where p is a pixel position, k equals to $1, 2, ..., N$, and $\mathbf{M}_{RS_k}$ is the output of the k-th mutual guidance module, jointly determined by $\mathbf{M}_{RS_{k-1}}$ and $\mathbf{M}_{EL_{k-1}}$. The enhanced image $\mathbf{I}_N$ produced by the N-th iteration is taken as the final enhancement result $\mathbf{I}_E$.

From the above description, we can see that our model can be seen as an improved version of ZeroDCE++ by equipping it with the mutual guidance module. During the iterative enhancement process, the mutual guidance module keeps refining the enhancing strength at the pixel level, forcing $\mathbf{M}_{RS_k}$ to be always aware of the image appearance information, such as the illumination distribution and structure. To train this model, we follow ZeroDCE++ by using the same training dataset (Part 1 of the SICE dataset [2]), the same loss function terms (spatial consistency loss, exposure control loss, illumination smoothness loss, and color constancy loss) and their weights. The efficiency of the our model can be kept, as ZeroDCE++ is very efficient, and the mutual guidance also has a low computational cost. In addition, the introduction of the mutual guidance module effectively improve the convergence of the iterative enhancement process, which is empirically validated in the experiments.

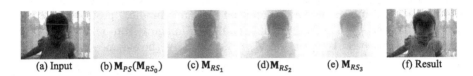

(a) Input   (b) $\mathbf{M}_{PS}(\mathbf{M}_{RS_0})$ (c) $\mathbf{M}_{RS_1}$ (d) $\mathbf{M}_{RS_2}$ (e) $\mathbf{M}_{RS_3}$ (f) Result

Fig. 4. An example of the initial and the refined enhancing strength maps obtained by the mutual guidance module in each iteration of our model.

3.2   Mutual Guidance Module

In this subsection, we describe the details of constructing the mutual guidance module, which is presented in Fig. 2. This module aims to refine the learned enhancing strength map $\mathbf{M}_{PS}$, making it aware of the lightness distribution and image structure at a fine scale. This can be a difficult task if we choose deep neural networks via learning to reconstruct so many fine-grained details. However, these details are off-the-shelf in the image $\mathbf{I}_0$ itself. To fully utilize this information, we establish a mutual guidance process to propagate the image appearance information, i.e., the edge-aware illumination distribution, onto $\mathbf{M}_{PS}$.

First, we choose $\mathbf{M}_{PS}$ as the guidance to refine the edge-aware illumination map $\mathbf{M}_{EL}$, and obtain an intermediate refined strength map $\mathbf{M}_{IS}$. Then, we choose $\mathbf{M}_{EL}$ as the guidance to refine $\mathbf{M}_{IS}$, and obtain the desired refined enhancing strength map $\mathbf{M}_{RS}$. The refinement of the above two steps is implemented with the guided filter [8], which can be computationally very efficient. By taking a closer observation of Fig. 2, the first-round guidance forces $\mathbf{M}_{EL}$ to be similar with the learned strength $\mathbf{M}_{PS}$ at a macro scale, while the second-round guidance forces the intermediate strength $\mathbf{M}_{IS}$ to be similar with $\mathbf{M}_{EL}$ at a fine scale. In this way, the obtained strength map $\mathbf{M}_{RS}$ finally becomes fully aware of the image illumination and structure. This process only utilizes the supervision information from the image for processing itself, and requires no training aforehand. Of note, the edge-aware illumination map $\mathbf{M}_{EL}$ can be directly extracted from $\mathbf{I}_0$, which is described in the next subsection. When incorporating the mutual guidance module into the LLIE model, $\mathbf{M}_{EL}$ is kept updated based on the intermediate enhanced image $\mathbf{I}_k$ during the iterative process.

In Fig. 4, we provide an example to demonstrate the effectiveness of using the mutual guidance module. On one hand, by comparing the original images and their enhanced results, we can see that the visibility of originally dark regions has been well improved, while the image contrast and details of the originally bright regions are well preserved. On the other hand, by observing the enhancing strength maps, we can see that $\mathbf{M}_{RS_k}$ keeps updated along with the iterative enhancing process. Compared with $\mathbf{M}_{RS_0}$, $\mathbf{M}_{RS_1}$ successfully indicates originally bright regions, helping the model avoid the over-enhancement in the initial stage. In the next two iterations, $\mathbf{M}_{RS_2}$ and $\mathbf{M}_{RS_3}$ keep refining their enhancing strength at a much finer scale, and avoid the unnecessary enhancement on the already lightened regions. This observation empirically validate the usefulness of updating $\mathbf{M}_{EL}$ during the iterations.

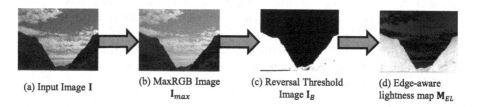

(a) Input Image $\mathbf{I}$ (b) MaxRGB Image $\mathbf{I}_{max}$ (c) Reversal Threshold Image $\mathbf{I}_B$ (d) Edge-aware lightness map $\mathbf{M}_{EL}$

Fig. 5. An example of extracting the edge-aware lightness map M_{EL} from I.

3.3 Estimation of the Edge-Aware Lightness Map

From the construction of the mutual guidance module, we can see that the extraction of the edge-aware illumination map $\mathbf{M}_{EL}$ plays an important role. We present an example of extracting $\mathbf{M}_{EL}$ from an image in Fig 5. Given an image $\mathbf{I}$ ($\mathbf{I}$ can be the input image $\mathbf{I}_0$ or an intermediate enhanced image $\mathbf{I}_k$), we first extract its maxRGB image $\mathbf{I}_{max}$, which is considered as useful in estimating its illumination distribution:

$$\mathbf{I}_{max}(p) = \max_{c \in \{R,G,B\}} \mathbf{I}^c(p) \tag{2}$$

where p stands for a pixel position, and c stands for color channel. Then, a reversal binary image $\mathbf{I}_B$ is obtained through a simple binaryzation imposed on $\mathbf{I}_{max}$. The pixel value in $\mathbf{I}_B$ is set as zero if the pixel value in $\mathbf{I}_{max}$ of the corresponding position is not smaller than a threshold T. Otherwise, this pixel value in $\mathbf{I}_B$ is set as one. However, $\mathbf{I}_B$ only has a very rough estimation on $\mathbf{I}$'s illumination distribution, and it is not structure-aware. In the following, we simply use the guided filter [8] again to obtain $\mathbf{M}_{EL}$ and make it fully aware of the illumination and structure, in which $\mathbf{I}_{max}$ acts as the guidance role [7]. The illumination distribution and the fine-grained image structure in $\mathbf{I}$ are transferred onto $\mathbf{I}_B$ via the guided filtering process. We find $T = 0.2$ works well in all our experiments.

4   Experiments

4.1   Experimental Settings

Datasets. To train our model, we followed ZeroDCE [5,11] by using 2422 images with multiple exposures in Part 1 of the SICE dataset. Of note, the hyperparameters, e.g. the loss term weights, were kept the same as the off-the-shelf ones in ZeroDCE++. To test the model performance, we chose seven unpaired datasets(7UD) (DICM, Fusion, HQimage, LIME, VV, MEF, NPE, 321 images in total) and one paired dataset (MIT Adobe FiveK test dataset, 500 images) that are widely used in the LLIE task [10].

Evaluation Metrics. Since the experimental data include pairwise and unpaired datasets, several commonly used full reference quality metrics and non-reference quality evaluation metrics were chosen to evaluate the quantitative performance. The full reference quality metrics include Peak Signal-to-Noise Ratio (PSNR, dB), Structural Similarity (SSIM) [18], MS-SSIM [19], MSE and LPIPS [23]. The non-reference quality evaluation metrics include EME [1], LOE [17], NIQE [16] and BRISQUE [15].

Compared Methods. In the quantitative comparison, we compared our method with other six popular methods, including five unsupervised methods: ZeroDCE [5], ZeroDCE++ [11], EnlightenGAN [9], RUAS [12], and SCI [14]; one semi-supervised method: DRBN [22]. In the visual comparison, in addition to the above six methods, we also introduced three supervised methods, i.e., KinD [24], URetinexNet [20] and SNR-Aware method [21].

Implementation Details. Our method was implemented with Pytorch version 1.10 and Nvidia RTX 2080 Ti with 24G memory. The training epoch number was set to 100. We used the Adam optimizer, and the learning rate and decay coefficient were both empirically set to 0.0001. The iteration number N was set as 3 in all the experiments for the visual and the quantitative comparison.

4.2   Experimental Results

1) Quantitative comparison. To fully exploring the generalization ability, we choose the MIT Adobe FiveK test dataset for evaluation, which is unseen during the training of our model. Table 1 demonstrates the quantitative performance of all seven methods. We have the following observations from the table. First, in general, our model achieves satisfying quantitative performance under various evaluation metrics. Specifically, on one hand, our model obtains the best performance under all five full-reference metrics among all seven models. The reason can be attributed to the mutual guidance module used in our method, which effectively preserves the originally bright regions of the input image. On the other hand, our model also achieves satisfying performance under the three

Table 1. The full reference and non-reference quality evaluation metric values on the MIT Adobe FiveK testing dataset. Their sizes are uniformly resized to 512 × 512. The best and the second best results are highlighted in red and blue, respectively.

		ZeroDCE++[11]	ZeroDCE[5]	EnlightenGAN[9]	DRBN[22]	RUAS[12]	SCI[14]	Ours
Full-reference metrics	PSNR↑	11.9416	10.5777	11.4701	12.7158	5.3046	7.1390	16.7238
	SSIM↑	0.5223	0.4393	0.5323	0.5660	0.3529	0.4105	0.5933
	MS-SSIM↑	0.9988	0.9982	0.9987	0.9990	0.9946	0.9963	0.9996
	MSE↓	4.698	6.913	5.280	3.942	20.295	13.755	1.684
	LPIPS↓	0.1869	0.2462	0.2262	0.2521	0.5549	0.3877	0.1509
No-reference metrics	EME↑	5.7241	4.4413	5.7117	5.4109	2.34	3.9045	7.6424
	LOE↓	210.8541	258.502	404.7867	202.9693	707.3587	187.2514	141.5003
	NIQE↓	3.6448	3.9383	3.4922	3.4591	3.8254	4.1032	3.5746

Table 2. The scores of the BRISQUE↓ metric on 7UD. The best and the second best results are highlighted in red and blue, respectively.

	ZeroDCE++ [11]	ZeroDCE [5]	EnlightenGAN [9]	DRBN [22]	RUAS [12]	SCI [14]	Ours
BRISQUE↓	27.1301	29.5830	27.1762	33.4999	34.9013	30.5955	27.0252

no-reference metrics. Specifically, it obtains the best EME and LOE values, and ranks the third place under the NIQE metric. Second, as the mutual guidance module is integrated with the zero-reference-based models, we further take a closer observation on our results as well as ZeroDCE and ZeroDCE++ results. From the table, we can see that our model is better than these two counterparts under almost all the evaluation metrics. In addition, thanks to the regularization on the enhancing strength map made by the mutual guidance module, our model only needs three iterations (both ZeroDCE and ZeroDCE++ need eight iterations) during the image enhancement process to obtain these results.

To further demonstrate the effectiveness of our model, we use another no-reference metric BRISQUE to evaluate the performance on 7UD. From the results shown in Table 2, first, our model obtains best performance by averaging the scores across 7UD. Second, by comparing ZeroDCE and ZeroDCE++ with our model, we can see that our BRISQUE values better than the two counterparts, showing the effectiveness of introducing the mutual guidance module.

2) Visual comparison In Fig. 6, we provide the enhanced results produced by different LLIE methods of an image from the MIT Adobe FiveK dataset. In addition to the unsupevised LLIE methods, we present the results of several supervised LLIE methods, including KinD [24], URetinexNet [20] and SNR-Aware method [21]. Since the input image has ground truth image, we also report the PSNR and SSIM values in the figure. On one hand, we can see that the over-enhancement widely exist in the results of the nine methods for comparison. For example, the details in the upper part become very unclear in the results produced by RAUS and SCI due to the over-exposure effect. The artifacts in SNR-Aware result largely lower the visual quality. In contrary, our result has a more natural visual effect that the originally dark regions are lightened while the originally bright regions are well preserved. The PSNR and SSIM scores of our result also verify this point, showing that our result mostly resembles the ground truth image produced by human retouching.

4.3   Analysis of Our Method

1) The iteration number. As we adopt an iterative enhancement process, it is important to investigate the details in this process. As shown in Fig. 7, we set the iteration number N as 8, and provide all the intermediate results produced from each iteration. We can see that the originally dark regions become more visible at the first four iterations. After that, the enhancement of these regions becomes unobvious in the next four iterations. On the other hand, the originally bright regions in this image are kept almost the same during all eight iterations. This example empirically validates the properties of the fast convergence and the naturalness preservation of our method. In addition, we also provide some quantitative results to study the influence of the iteration number. In Table 3, we report the BRISQUE scores under each iteration, which shows that our model achieves acceptable BRISQUE scores (smaller than 28) during all the iterations. In real world applications, we can empirically set the iteration number as 3.

Table 3. The scores of the BRISQUE↓ metric on 7UD under original images and each iteration. The best and the second best results are highlighted in red and blue, respectively.

	Inputs	Iter 1	Iter 2	Iter 3	Iter 4	Iter 5	Iter 6	Iter 7	Iter 8
BRISQUE↓	28.4526	27.6695	27.2356	27.0252	27.2369	27.5454	27.7864	27.9151	27.9777

| (a) Input (PSNR/SSIM) | (b) ZeroDCE (12.86/0.6850) | (c) EnlightenGAN (14.05/0.7079) | (d) DRBN (13.35/0.6316) | (e) RUAS (4.77/0.2418) | (f) SCI (11.83/0.5883) |

| (g) KinD (13.22/0.5997) | (h) ZeroDCE++ (12.06/0.6794) | (i) URetinexNet (11.69/0.6398) | (j) SNR-Aware (7.28/0.4610) | (k) Ours (20.06/0.8151) | (l) Ground truth |

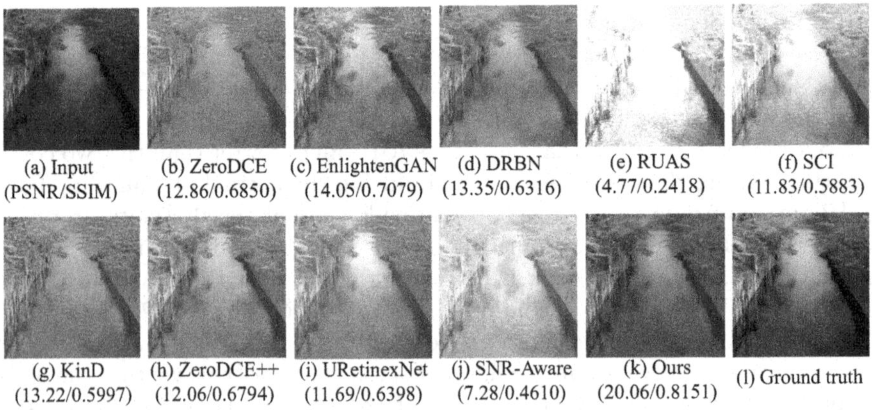

Fig. 6. An example of the visual comparison. The experimental image is from the MIT Adobe FiveK dataset.

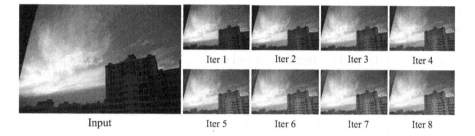

Fig. 7. An example of studying the influence of the iteration number on the enhancement process.

Table 4. The PSNR, SSIM, MSE($\times 10^3$), LPIPS values of the three models tested on the MIT Adobe FiveK dataset (500 images). The best and the second best results are highlighted in red and blue, respectively.

	PSNR↑	SSIM↑	MS-SSIM↑	MSE↓	LPIPS↓
Model-A	14.6782	0.3687	0.9995	2.480	0.2047
Model-B	15.8979	0.4033	0.9996	1.980	0.1819
Ours	16.7238	0.5933	0.9996	1.684	0.1509

2) The mutual guidance process. To fully investigate the effectiveness of the mutual guidance process, we construct two versions of onesided guidance module. Specifically, in a typical k-th iteration, the first version only uses the first-round guidance step, i.e., the $\mathbf{M}_{IS_k}$ map is directly taken as the updated enhancing strength map $\mathbf{M}_{RS_k}$. The second one only uses the second-round guidance step. As there is no intermediate map $\mathbf{M}_{IS}$, the $\mathbf{M}_{EL_{k-1}}$ extracted from $\mathbf{I}_{k-1}$ guides the refinement of $\mathbf{M}_{RS_{k-1}}$, producing the refined enhancing strength map $\mathbf{M}_{RS_k}$. We term these two incomplete models as Model-A and Model-B. Combining the completed version of our model, we evaluate their quantitative performance on the MIT Adobe FiveK dataset, of which the results are shown in Table 4. We can see that our model is consistently better than the ones based on onesided guidance in terms of all the five evaluation metrics which further demonstrates the effectiveness of the mutual guidance process.

5   Conclusion

In this paper, we target at solving the over-enhancement issue widely existed in current LLIE methods. We construct the mutual guidance module, and incorporate it into the ZeroDCE++ model. Via fully exploiting the off-the-shelf image appearance information such as illumination and structure, the module is able to refine the enhancing strength at the pixel level through establishing mutual guidance between the enhancing strength map and the edge-aware illumination map. Via using this module, the originally dark regions can be effectively lightened, and the originally bright regions can be well preserved. Therefore, the over-enhancement issue can be effectively avoided. In experiments, we evaluate our LLIE model by comparing it with many state-of-the-art unsupervised and supervised LLIE models. Both the visual comparison and the quantitative comparison demonstrate the effectiveness and superiority of our model. In the future research, we plan to further improve the generalization ability of the mutual guidance process, and encode it into other LLIE models in an implicit way.

References

1. Agaian, S.S., Silver, B., Panetta, K.A.: Transform coefficient histogram-based image enhancement algorithms using contrast entropy. IEEE Trans. Image Process. **16**(3), 741–758 (2007)
2. Cai, J., Gu, S., Zhang, L.: Learning a deep single image contrast enhancer from multi-exposure images. IEEE Trans. Image Process. **27**(4), 2049–2062 (2018)
3. Chen, C., Chen, Q., Xu, J., Koltun, V.: Learning to see in the dark. In: Proceedings of Computer Vision and Pattern Recognition, pp. 3291–3300 (2018)
4. Chen, W., Wenjing Wang, W.Y.: Deep retinex decomposition for low-light enhancement. In: Proceedings of British Machine Vision Conference (2018)
5. Guo, C., et al.: Zero-reference deep curve estimation for low-light image enhancement. In: Proceedings of Computer Vision and Pattern Recognition, pp. 1780–1789 (2020)
6. Guo, X., Hu, Q.: Low-light image enhancement via breaking down the darkness. Int. J. Comput. Vision **131**(1), 48–66 (2023)
7. Hao, S., Guo, Y., Wei, Z.: Lightness-aware contrast enhancement for images with different illumination conditions. Multimedia Tools Appl. **78**, 3817–3830 (2019)
8. He, K., Sun, J.: Fast guided filter. arXiv preprint arXiv:1505.00996 (2015)
9. Jiang, Y., et al.: EnlightenGAN: deep light enhancement without paired supervision. IEEE Trans. Image Process. **30**, 2340–2349 (2021)
10. Li, C., et al.: Low-light image and video enhancement using deep learning: a survey. IEEE Trans. Pattern Anal. Mach. Intell. **44**(12), 9396–9416 (2022)
11. Li, C., Guo, C., Loy, C.C.: Learning to enhance low-light image via zero-reference deep curve estimation. IEEE Trans. Pattern Anal. Mach. Intell. **44**(8), 4225–4238 (2022)
12. Liu, R., Ma, L., Zhang, J., Fan, X., Luo, Z.: Retinex-inspired unrolling with cooperative prior architecture search for low-light image enhancement. In: Proceedings of Computer Vision and Pattern Recognition, pp. 10561–10570 (2021)
13. Lore, K.G., Akintayo, A., Sarkar, S.: LLNet: a deep autoencoder approach to natural low-light image enhancement. Pattern Recogn. **61**, 650–662 (2017)
14. Ma, L., Ma, T., Liu, R., Fan, X., Luo, Z.: Toward fast, flexible, and robust low-light image enhancement. In: Proceedings of Computer Vision and Pattern Recognition, pp. 5637–5646 (2022)
15. Mittal, A., Moorthy, A.K., Bovik, A.C.: No-reference image quality assessment in the spatial domain. IEEE Trans. Image Process. **21**(12), 4695–4708 (2012)
16. Mittal, A., Soundararajan, R., Bovik, A.C.: Making a "completely blind" image quality analyzer. IEEE Signal Process. Lett. **20**(3), 209–212 (2012)
17. Wang, S., Zheng, J., Hu, H.M., Li, B.: Naturalness preserved enhancement algorithm for non-uniform illumination images. IEEE Trans. Image Process. **22**(9), 3538–3548 (2013)

18. Wang, Z., Bovik, A.C., Sheikh, H.R., Simoncelli, E.P.: Image quality assessment: from error visibility to structural similarity. IEEE Trans. Image Process. **13**(4), 600–612 (2004)
19. Wang, Z., Simoncelli, E.P., Bovik, A.C.: Multiscale structural similarity for image quality assessment. In: The Thrity-Seventh Asilomar Conference on Signals, Systems & Computers, 2003, vol. 2, pp. 1398–1402. IEEE (2003)
20. Wu, W., Weng, J., Zhang, P., Wang, X., Yang, W., Jiang, J.: URetinex-Net: Retinex-based deep unfolding network for low-light image enhancement. In: Proceedings of Computer Vision and Pattern Recognition, pp. 5901–5910 (2022)
21. Xu, X., Wang, R., Fu, C.W., Jia, J.: SNR-aware low-light image enhancement. In: Proceedings of Computer Vision and Pattern Recognition, pp. 17714–17724 (2022)
22. Yang, W., Wang, S., Fang, Y., Wang, Y., Liu, J.: From fidelity to perceptual quality: a semi-supervised approach for low-light image enhancement. In: Proceedings of Computer Vision and Pattern Recognition, pp. 3063–3072 (2020)
23. Zhang, R., Isola, P., Efros, A.A., Shechtman, E., Wang, O.: The unreasonable effectiveness of deep features as a perceptual metric. In: Proceedings of Computer Vision and Pattern Recognition, pp. 586–595 (2018)
24. Zhang, Y., Zhang, J., Guo, X.: Kindling the darkness: a practical low-light image enhancer. In: Proceedings of ACM Multimedia, pp. 1632–1640 (2019)

Semantic-Guided Completion Network for Video Inpainting in Complex Urban Scene

Jianan Wang[1], Hanyu Xuan[2], and Zhiliang Wu[1(✉)]

[1] Nanjing University of Science and Technology, Nanjing 210094, China
wu_zhiliang@njust.edu.cn
[2] Anhui University, Anhui 230039, China

Abstract. Video inpainting aims to fill damaged areas in video frames with appropriate content. Complex scene contain cluttered or ambiguous semantics and objects, making video inpainting in such scenarios a challenging yet meaningful task. Current methods are limited by the lack of sufficient video information, resulting in blurred results and temporal artifacts. In this paper, we design a novel semantic-guided completion network that uses the semantic information of the videos to complete missing regions in the complex urban scene. Specifically, we first leverage the semantic information to model the structure and content of the video and improve U-Net network to complete the broken semantic image. Then, we propose a module based on spatial-adaptive normalization to guide the generation of the damaged part of the video pixels by combining semantic information. Our model's ability to generate reasonable and accurate content is demonstrated through both quantitative and qualitative results on two publicly available urban scene datasets.

Keywords: Semantic-Guided Completion Network · Video Inpainting · Complex Urban Scene

1 Introduction

Video inpainting, as introduced by Bertalmio et al. [1], fills damaged regions in video frames with plausible content. At present, video inpainting is used in many fields, such as repairing damaged videos [8,24], removing objects [32], video denoising [5] and video retargeting [13], etc. However, despite the significant benefits of this technology, video inpainting in complex scene still faces great challenges. Videos captured in complex environments often have many occluded or cluttered semantics and objects, making it difficult to complete such videos using current technologies.

Early work [8,18,21] use patch-based optimization strategies. For example, some methods search for similar patch blocks from appropriate areas of the video to synthesize damaged content. Nonetheless, there are many problems with these methods, such as high computational complexity [3] and lacking high-level understanding of videos [31]. To address these common problems, recent work

Q. Liu et al. (Eds.): PRCV 2023, LNCS 14435, pp. 224–236, 2024.
https://doi.org/10.1007/978-981-99-8552-4_18

Fig. 1. Feature visualization comparison between real images and semantically segmented images.

usually adopt deep learning approaches for video inpainting [14,29,30]. These methods generally extract features with high similarity to the current frame from adjacent or related video frames, and then fill the damaged regions with these features using various types of context alignment methods. However, when the video texture is complex or the missing area is large, the inpainting results are often unsatisfactory.

In this article, we mainly address the problem of completing videos captured in complex scene, which typically contain multiple objects and a large amount of cluttered semantic information. Features extracted from real images contain a large amount of noise and complex textures (see from Fig. 1), which can greatly interfere with the quality of video inpainting. In contrast, the corresponding semantic graph has smooth regions and simple textures, which are conducive to generating high-quality video restoration results. To address the challenges presented by complex scene, we propose a novel framework. Unlike previous works, we incorporate information from the semantic segmentation maps and enforce their consistency by learning to accurately generate both the semantic video frames and the actual pixels in the damaged video frames. Furthermore, we propose a module based on spatial-adaptive normalization(SPADE [20]), which utilizes several video frames as a reference to transform the predicted semantic segmentation information into a video with the same style as ground truth. To further evaluate our network, we conduct experiments on two publicly urban scene datasets, namely Cityscapes [6] and Camvid [2]. These datasets are captured under diverse and unstructured environmental conditions and contain much semantic information, making them ideal benchmarks for completing videos captured in complex scene. The quantitative and qualitative results demonstrate that our proposed framework can effectively repair videos in complex scene and generate more realistic and reasonable results.

Overall, our work makes the following contributions:

- We design a novel network for the restoration of videos captured in complex scene that leverages the semantic segmentation information of videos to generate more realistic and reasonable restoration results.
- We propose a new module based on spatially-adaptive normalization that utilizes several video frames as a reference to transform the predicted semantic

segmentation information into the video frames with the same style as ground truth.

- We use two publicly datasets, Cityscapes and Camvid, to verify the video repair results of our framework in complex scene. After comparing the experimental results both quantitatively and qualitatively, we find that our method significantly outperforms the existing models in all experiments.

2   Related Work

To develop more realistic and reasonable video inpainting techniques, researchers have made significant efforts to fill damaged areas with spatio-temporal consistency content in videos [14,18]. In this paper, we will discuss two representative approaches: patch-based models and deep convolutional models.

Patch-Based Models. Early research on video inpainting mainly treat the whole process as a patch-based optimization problem. Some methods [18] typically use global optimization to sample similar or identical spatial or spatio-temporal patches from appropriate regions to synthesize the contents of the damage. Some methods improve performance by providing clips of the foreground and background of the video or by jointly estimating the optical flow and appearance of the object during the video inpainting [8,21]. Patch-based optimization algorithms can achieve good results in some scene, but when the motion of the damaged area is complex, the repair effect of these algorithms is unsatisfactory.

Deep Convolutional Models. With the advancement of hardware, deep learning models [22,23] are increasingly being applied in the field of video inpainting. Xu et al. [33] propose the first video inpainting method guided by optical flow. They use a newly designed optical flow estimator to generate an optical flow field with spatio-temporal consistency. This optical flow field is then used to guide the pixels of adjacent frames to propagate and fill in the missing regions of the current video frame. Wang et al. [25] design a data-driven video inpainting network that can infer reasonable spatial structures and recover corresponding temporal details. Liu et al. [16] propose an end-to-end deep learning network consisting of a two-stage feature fusion module which can collect useful spatial information from adjacent frames and generate temporally consistent content for the damaged region of the current frame. Zeng et al. [34] propose a novel transformer module for video inpainting which can extract spatio-temporal information from all frames in a video sequence through multi-scale attention and fill the damaged areas of these frames simultaneously. Additionally, they propose a spatio-temporal adversarial loss function to further optimize the model during training and generate high-quality inpainting results. However, these methods usually present fuzzy and cluttered inpainting results when the scene is complex or the missing regions are large.

3   Methods

3.1   Problem Formulation

Let $X_1^T := \{x_1, x_2, \cdots, x_T\}$ denote a damaged video sequence with width W, height H and T frames. Let $S_1^T := \{s_1, s_2, \cdots, s_T\}$ be the corrupted semantic segmentation images and $M_1^T := \{m_1, m_2, \cdots, m_T\}$ be the related masks. For the each mask m_i, a value of zero indicates undamaged areas, while a value of one indicates missing pixels. Video inpainting can be framed as a self-supervised task that randomly creates $\left(X_1^T, S_1^T, M_1^T\right)$ pairs as input and reconstruct ground truth $Y_1^T := \{y_1, y_2, \cdots, y_T\}$ and the related semantic video frames $Z_1^T := \{z_1, z_2, \cdots, z_T\}$. Through continuous learning and training of the deep network, we make the output results of the network $\widehat{X}_1^T := \{\widehat{x}_1, \widehat{x}_2, \cdots, \widehat{x}_T\}$ and $\widehat{S}_1^T := \{\widehat{s}_1, \widehat{s}_2, \cdots, \widehat{s}_T\}$ can be approximated by the actual value. The specific network design shows in Fig. 2(a). To facilitate the description of our method, we

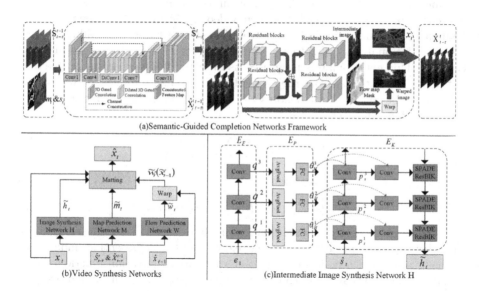

Fig. 2. (a) Our model consists of two parts: semantic video completion network(SVCNet) and video synthesis network(VSNet). The left part, SVCNet, is a U-Net-like network structure composed of eleven convolutional layers with a gated temporal shift module, and it completes the missing semantic in the damaged video frames. The right part, VSNet, synthesizes the missing regions in the damaged video using the completed semantic video frames as a guide. (b) VSNet is mainly composed of three subnetworks: the soft map prediction module M, the optical flow prediction module W, and the intermediate image generator H. (c) The intermediate image generator H consists of three parts: a reference feature extractor E_F, a multi-layer perceptron E_P and SPADE generator E_K. H can use several real video frames as reference to transform the predicted semantic segmentation information into an intermediate image with the same style as ground truth.

divide our model into two parts: semantic video completion network(SVCNet) and video synthesis network(VSNet).

3.2   Semantic Video Completion Network

As shown in Fig. 1, the texture of semantic videos is simpler compared to real videos, and even for complex scene, there are only dozens of different semantic categories. Completing semantic videos can reduce wastage of time and resources. Therefore, we propose a semantic video completion network consisting of eleven convolutional layers with a gated temporal shift module. The whole network has a U-Net-like structure, including up-sampling, dilation, and down-sampling layers. However, unlike U-Net, since the down-sampling layer has many masked regions, there are no skip connections in that layer. For up-sampling and down-sampling layers, we apply bilinear interpolation before convolutions. Additionally, we use 3D gated convolution, which is an extension of the method proposed in paper [3], to replace the ordinary convolution in the U-Net architecture. This approach can effectively handle free-form video inpainting tasks where the mask's boundaries are uncertain. In each convolution layer, we apply an additional gating convolutional filter W_g to the input features F_t to obtain a gating map $Gating_t$. This gating map is then used as an attention mechanism on the output features $Features_t$ from the original convolutional filter W_f, based on the validity of the mask. The t denotes the video frame number. Mathematically, this operation can be expressed as:

$$Gating_t = \sum\sum W_g \cdot F_t, \tag{1}$$

$$Features_t = \sum\sum W_f \cdot F_t, \tag{2}$$

$$Output_t = \sigma\left(Gating_t\right)\phi\left(Features_t\right), \tag{3}$$

where σ represents the sigmoid function that maps the gating map to a value between 0 (representing missing regions) and 1 (representing undamaged regions), instead of a boolean value. ϕ is a primitive activation function. Additionally, to ensure that the output pixels are consistent with the actual pixels in the semantic category, we propose a color tuning module. This module adjusts the color properties of the output pixels, which is helpful for leveraging the semantic category information to generate realistic instances. Mathematically, this operation can be represented as:

$$\min_c\left(\|S_t - P_c\|_1\right), \tag{4}$$

where S_t represents the pixel information in the generated semantic video, P_c denotes the actual pixel information in the semantic category, and c indicates the semantic category.

3.3   Video Synthesis Network

The ultimate learning objective of video synthesis is to convert the input semantic segmentation information $\widehat{S}_1^T$ into the output video $\widehat{X}_1^T$, while ensuring that the conditional distribution of the generated results is consistent with the conditional distribution of the actual video. Following the modeling approach in fsvid2vid [26], we construct the generative model F using the simplified Markov hypothesis. Mathematically, this can be expressed as:

$$\widehat{x}_t = F\left(\widehat{X}_{t-\tau}^{t-1}, \widehat{S}_{t-\tau}^t, \{e_1, e_2, \cdots, e_K\}, \{s_1, s_2, \cdots, s_K\}\right) \oplus x_t, \qquad (5)$$

where $\{e_1, e_2, \cdots, e_K\}$ represent the K reference images in the target domain and $\{s_1, s_2, \cdots, s_K\}$ are the corresponding semantic images. The introduction of sample images enables the generative model F to extract useful patterns and synthesize videos that are similar in style to ground truth during testing. In other words, new video frames are obtained through the forward model based on the input video sequence, and the final output video is obtained by concatenating these frames in a continuous sequence. The generative model that produces the video frames in a serialized manner can be represented as follows:

$$F = (1 - \widetilde{m}_t) \odot \widetilde{w}_{t-1}(\widehat{x}_{t-1}) + \widetilde{m}_t \odot \widetilde{h}_t, \qquad (6)$$

where $\odot$ means a element-wise product operator, and symbol 1 represents an image with all ones, $\widetilde{m}_t$ refers to the soft map, $\widetilde{w}_{t-1}$ represents the flow from frame $t-1$ to frame t, and $\widetilde{h}_t$ is the resulting intermediate image. To generate the output image $\widehat{x}_t$, we require a combination of the optical flow distorted version of the previous frame image $\widetilde{w}_{t-1}(\widehat{x}_{t-1})$ and the intermediate image $\widetilde{h}_t$ with the soft map $\widetilde{m}_t$ determining the combination of the two images at each pixel position. These quantities are obtained using the soft map prediction module M, the optical flow prediction module W, and intermediate image generator H (see from Fig. 2(b)).

Intermediate image generator H consists of three sub-networks (see from Fig. 2(c)): SPADE generator E_K, the multi-layer perceptron E_P and the reference feature extractor E_F. The network E_F consists of several convolutional layers and is used to extract the style representation q^i from the reference image e_1. After obtaining the style representation q^i, E_P generates the corresponding weights θ_H^i and adds them to the generator E_K. The network E_K uses completed semantic images $\widehat{s}_t$ to generate the intermediate image $\widetilde{h}_t$. The optical flow prediction network W utilizes FlowNet2 [10] as optical flow estimator and obtains the optical flow $\widetilde{w}_{t-1}$ from frame $t-1$ to frame t. The soft map prediction network M generates occlusion masks with continuous values between zero and one to better handle enlarged areas and recover details.

3.4   Loss Functions

We split the total training loss into three parts, as defined below:

$$L_{total} = L_A + L_B + L_C, \qquad (7)$$

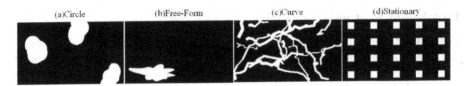

Fig. 3. Four different classes of masks are used, with mask coverage ranging from 5% to 55%. These include circle, free-form, curve and stationary masks.

$$L_A = \lambda_{perc} L_{perc} + \lambda_{style} L_{style}, \tag{8}$$

$$L_B = \lambda_{hole} L_{hole} + \lambda_{valid} L_{valid}, \tag{9}$$

$$L_C = \min_F \left(\max_{D_I} L_I (F, D_I) + \max_{D_V} L_V (F, D_V) \right) + \lambda_w L_w(F), \tag{10}$$

where L_A is style [17] and perceptual [7] differences between overall network outputs and ground truth. L_B is the loss function to train the semantic video completion network, which consists of the L_1 loss for missing regions and valid regions. These are commonly used loss functions in video inpainting. L_C [28] is the loss function of video synthesis network, which consists of image conditioned GAN loss L_I, video conditioned GAN loss L_V and flow loss L_W.

4   Experiments

4.1   Benchmarks and Evaluation Metrics

Datasets. To further demonstrate the effectiveness of the method, we quantitatively evaluate it on two widely recognized urban benchmark datasets, namely Cityscapes dataset and CamVid dataset, both of which exhibit high pedestrian and vehicle diversity. For the Cityscapes dataset, we use the pre-trained network TDNet [9] for semantic segmentation, as continuous semantic video frames are not available. In addition, to evaluate the inpainting performance of our method, we use four different classes of masks(circle, free-form, curve, and stationary masks) as shown in Fig. 3.

Benchmarks. To enable a fair comparison of our models, we select several competitive methods for comparison, including: CPVINet [14], which proposes a new DNN-based framework for video inpainting that utilizes additional information from other video frames. OPN [19], which proposes an attention-based multi-level convolutional neural network for deep video restoration. STTN [34] proposes a novel transformer module for video inpainting which can extract spatio-temporal information from all frames in a video sequence through multiscale attention. FFormer [15] proposes a new fuseformer video inpainting model by introducing soft segmentation and soft composition operations to achieve the

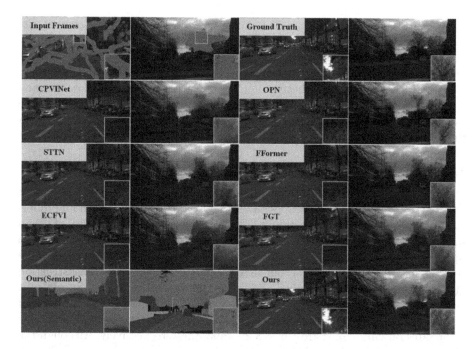

Fig. 4. Qualitative evaluation on the task of video inpainting. We show the inpainting results of the baseline model with our model on two types of masks.

goal of fusing fine-grained information. ECFVI [12] proposes an error compensation framework for video inpainting based on optical flow guidance. FGT [35] proposes a video inpainting model that incorporates optical flow into a transformer-based framework.

Implementation Details. We empirically assign the weights to the different losses as follows: $\lambda_{hole} = \lambda_{valid} = \lambda_{perc} = 1$, $\lambda_{style} = \lambda_w = 10$. The weight parameters for different losses are automatically adjusted during training. We use the ADAM optimizer and set the learning rate to $lr = 1e-5$ and momentum to $(\beta 1, \beta 2) = (0.9, 0.999)$. During training, all video sequences are compressed to 256×128 as input. We first train our semantic video inpainting network and video synthesis network separately on the two datasets. After the loss functions of two networks converge, we conduct joint training for the two networks. Due to limited experimental conditions, we use two blocks of NVIDIA 32 GB 2080ti GPU for training.

Evaluation Metrics. We employ four commonly used metrics in video inpainting to present our results in numerical form: PSNR [33], SSIM [3], E_{warp} [31] and LPIPS [32]. PSNR and SSIM are commonly used metrics to evaluate the quality of a video. The higher PSNR value indicates less distortion in the inpainting result. SSIM, on the other hand, is more in line with human visual characteristics when evaluating image quality. The higher SSIM value indicates a greater similarity between the inpainting result and ground truth. LPIPS is a recently proposed metric that mimics human perception of image similarity. The smaller value of LPIPS indicates a better inpainting effect. The optical flow warping error E_{warp} can measure the temporal consistency of the inpainting results. The smaller value for E_{warp} indicates that the inpainting effect remains consistent in the temporal dimension.

4.2 Results and Discussion

Quantitative Evaluation. We present the quantitative results (see from Table 1) of our model on Cityscapes dataset and CamVid dataset with four different fixed masks. Due to the instability of the masks and the complexity of urban scene, video inpainting is a challenging task. Our model demonstrates high video reconstruction quality and is less affected by lossy compression. Compared to the state-of-the-art (SOTA) model, our model achieves better video reconstruction quality in terms of both per-pixel and overall perceptual measures. Specifically, our model outperforms the SOTA model significantly, particularly in terms of PSNR, optical flow warping error E_{warp}, and LPIPS. The relative improvement rates on Cityscapes are 1.6%, 8.2%, and 8.9%, respectively. Although the CamVid dataset has limited data for model training, our method still outperforms the SOTA model, although all models achieve relatively poor results on this dataset. The superior results demonstrate the effectiveness of our proposed method for video inpainting in complex urban scene.

Table 1. Quantitative comparison with state-of-the-art video inpainting methods. Our model outperforms the baselines in terms of PSNR, SSIM, E_{warp}, and LPIPS.

Methods	Cityscapes				CamVid			
	PSNR↑	SSIM↑	$E_{warp}(\%)$↓	LPIPS↓	PSNR↑	SSIM↑	$E_{warp}(\%)$↓	LPIPS↓
CPVINet	28.208	0.9099	0.2551	0.9295	27.545	0.8777	0.2712	0.9882
OPN	28.425	0.8968	0.2414	0.8801	28.188	0.8795	0.2662	0.9285
STTN	29.217	0.9134	0.1734	0.6441	28.459	0.8832	0.2282	0.7965
FFormer	29.417	0.9182	0.1724	0.6245	29.079	0.8896	0.1949	0.7989
ECFVI	26.814	0.8757	0.3074	1.2416	25.511	0.8605	0.3335	1.3389
FGT	30.268	0.9203	0.1533	0.5172	29.802	0.8946	0.1799	0.6694
Ours	**30.778**	**0.9231**	**0.1466**	**0.4707**	**29.966**	**0.9021**	**0.1660**	**0.5039**

Qualitative Evaluation. To compare the visual results from our model and six baseline models, we follow the common practice in video inpainting literature and evaluate the restoration results on two randomly selected video frames with different mask conditions. When the damaged area is large or the texture is complex, the SOTA methods produce fuzzy inpainting results, whereas our method produces pleasing and consistent results (see from Fig. 4).

4.3   Ablation Experiments

Effectiveness of the Loss Functions. To further evaluate the effectiveness of each loss function in the video inpainting module, we conduct experiments by dividing the overall loss function into five parts: (1) the loss function $L_{hole}\&L_{valid}$ training repair results consistent with groundtruth, (2) the loss function L_I training the conditional image discriminator D_I, (3) the loss function L_v training the conditional video discriminator D_v, (5)the loss function L_w training to generate accurate optical flow, (5) the loss function $L_{perc}\&L_{style}$ training the repair effect to meet human perception. Table 2 shows that the optimized results with the $L_{hole}\&L_{valid}$ loss function have more coherent content compared to the other types of losses.

Table 2. Comparison of the influence of different loss functions on experimental results on cityscapes dataset.

Models	PSNR↑	SSIM↑	$E_{warp}(\%)\downarrow$	LPIPS↓
Our proposal(A)	**30.778**	**0.9231**	**0.1466**	**0.4707**
(A) w/o $L_{hole}\&L_{valid}$	29.586	0.8977	0.1965	0.7124
(A) w/o L_I (F, D_I)	30.025	0.9147	0.1754	0.5873
(A) w/o L_V (F, D_V)	29.682	0.9032	0.1863	0.6843
(A) w/o $L_w(F)$	29.751	0.9074	0.1795	0.6428
(A) w/o $L_{perc}\&L_{style}$	30.425	0.9198	0.1530	0.4957

Effectiveness of Video Synthesis Module. Our model consists of two parts: the semantic video completion network (SVCNet) and the video synthesis network (VSNet). The semantic video has a better inpainting effect than directly inpainting the original video, but using the semantic video as a guide to synthesize the damaged area of the video has a significant impact on the final inpainting result. To demonstrate the effectiveness of the video synthesis module of our model, we replace our module with some popular semantic image/video synthesis modules. Table 3 shows that our model achieves better inpainting results.

Table 3. Comparison of the effects of different video synthesis modules.

Models	PSNR↑	SSIM↑	$E_{warp}(\%)\downarrow$	LPIPS↓
Ours(SVCNet+VSNet)	**30.778**	**0.9231**	**0.1466**	**0.4707**
SVCNet+COVST [4]	28.045	0.8856	0.2481	1.0272
SVCNet+Pix2Pix [11]	28.024	0.8799	0.2567	1.2453
SVCNet+Pix2PixHD [27]	28.571	0.8948	0.2436	0.9574
SVCNet+Vid2Vid [28]	30.565	0.9157	0.1574	0.5428

5   Conclusion

In this paper, we design a novel semantic-guided deep learning model that leverages the semantic information of videos for video inpainting in complex urban scene. Our model achieves state-of-the-art performance on Cityscapes dataset and CamVid dataset with challenging four different classes of masks. Additionally, we propose a new module based on spatial-adaptive normalization(SPADE), which utilizes several actual video frames as the reference to transform the predicted semantic segmentation into the video with the same style as the ground truth. Our ablation experiments validate the effectiveness of our module in achieving good inpainting results.

Although our method outperforms other methods, there is still much room for improvement due to the limited availability of sufficient semantic segmentation video datasets and the precision and accuracy of semantic segmentation. In future work, we aim to incorporate a pixel-level semantic segmentation module to address the issue of missing datasets. Additionally, we will strive to further improve the accuracy of the video compositing module to generate more satisfactory inpainting results.

References

1. Bertalmio, M., Bertozzi, A.L., Sapiro, G.: Navier-stokes, fluid dynamics, and image and video inpainting. In: Proceedings of the 2001 IEEE Computer Society Conference on Computer Vision and Pattern Recognition (CVPR), p. I (2001)
2. Brostow, G.J., Shotton, J., Fauqueur, J., Cipolla, R.: Segmentation and recognition using structure from motion point clouds. In: Forsyth, D., Torr, P., Zisserman, A. (eds.) ECCV 2008. LNCS, vol. 5302, pp. 44–57. Springer, Heidelberg (2008). https://doi.org/10.1007/978-3-540-88682-2_5
3. Chang, Y.L., Liu, Z.Y., Lee, K.Y., et al.: Learnable gated temporal shift module for deep video inpainting. arXiv preprint arXiv:1907.01131 (2019)
4. Chen, D., Liao, J., Yuan, L., et al.: Coherent online video style transfer. In: Proceedings of the IEEE International Conference on Computer Vision, pp. 1105–1114 (2017)
5. Chen, Y., Guo, X., Shen, W.: Robust adaptive spatio-temporal video denoising algorithm based on motion estimation. Comput. Appl. **26**(8), 1882–1887 (2006)
6. Cordts, M., Omran, M., Ramos, S., et al.: The cityscapes dataset for semantic urban scene understanding. In: Proceedings of the IEEE Conference on Computer Vision and Pattern Recognition, pp. 3213–3223 (2016)
7. Gatys, L.A., Ecker, A.S., Bethge, M.: A neural algorithm of artistic style. arXiv preprint arXiv:1508.06576 (2015)
8. Granados, M., Tompkin, J., Kim, K.I., et al.: How not to be seen - object removal from videos of crowded scenes. Comput. Graph. Forum **31**, 219–228 (2012)
9. Hu, P., Caba, F., Wang, O., et al.: Temporally distributed networks for fast video semantic segmentation. In: Proceedings of the IEEE/CVF Conference on Computer Vision and Pattern Recognition, pp. 8818–8827 (2020)
10. Ilg, E., Mayer, N., Saikia, T., et al.: FlowNet 2.0: evolution of optical flow estimation with deep networks. In: Proceedings of the IEEE Conference on Computer Vision and Pattern Recognition, pp. 2462–2470 (2017)

11. Isola, P., Zhu, J.Y., Zhou, T., et al.: Image-to-image translation with conditional adversarial networks. In: Proceedings of the IEEE Conference on Computer Vision and Pattern Recognition, pp. 1125–1134 (2017)
12. Kang, J., Oh, S.W., Kim, S.J.: Error compensation framework for flow-guided video inpainting. In: Avidan, S., Brostow, G., Cissé, M., Farinella, G.M., Hassner, T. (eds) Computer Vision-ECCV 2022: 17th European Conference, pp. 357–390. Springer, Cham (2022). https://doi.org/10.1007/978-3-031-19784-0_22
13. Kim, D., Woo, S., Lee, J.Y.: Deep video inpainting. In: Proceedings of the IEEE/CVF Conference on Computer Vision and Pattern Recognition (CVPR), pp. 5792–5801 (2019)
14. Lee, S., Oh, S.W., Won, D.Y., et al.: Copy-and-paste networks for deep video inpainting. In: Proceedings of the IEEE/CVF International Conference on Computer Vision, pp. 4413–4421 (2019)
15. Liu, R., Deng, H., Huang, Y., et al.: FuseFormer: fusing fine-grained information in transformers for video inpainting. In: Proceedings of the IEEE/CVF International Conference on Computer Vision, pp. 14040–14049 (2021)
16. Liu, R., Li, B., Zhu, Y.: Temporal group fusion network for deep video inpainting. IEEE Trans. Circuits Syst. Video Technol. **32**(6), 3539–3551 (2021)
17. Nazeri, K., Ng, E., Joseph, T., et al.: EdgeConnect: generative image inpainting with adversarial edge learning. arXiv preprint arXiv:1901.00212 (2019)
18. Newson, A., Almansa, A., Fradet, M., et al.: Video inpainting of complex scenes. SIAM J. Imag. Sci. **7**(4), 1993–2019 (2014)
19. Oh, S.W., Lee, S., Lee, J.Y., et al.: Onion-peel networks for deep video completion. In: proceedings of the IEEE/CVF International Conference on Computer Vision, pp. 4403–4412 (2019)
20. Park, T., Liu, M.Y., Wang, T.C., et al.: Semantic image synthesis with spatially-adaptive normalization. In: Proceedings of the IEEE/CVF Conference on Computer Vision and Pattern Recognition, pp. 2337–2346 (2019)
21. Patwardhan, K.A., Sapiro, G., Bertalmio, M.: Video inpainting of occluding and occluded objects. In: IEEE International Conference on Image Processing (ICIP), pp. II-69 (2005)
22. Shang, Y., Duan, B., Zong, Z., Nie, L., Yan, Y.: Lipschitz continuity guided knowledge distillation. In: Proceedings of the IEEE/CVF International Conference on Computer Vision, pp. 10675–10684 (2021)
23. Shang, Y., Xu, D., Zong, Z., Nie, L., Yan, Y.: Network binarization via contrastive learning. In: Avidan, S., Brostow, G., Cissé, M., Farinella, G.M., Hassner, T. (eds) European Conference on Computer Vision. pp. 586–602. Springer, Cham (2022). https://doi.org/10.1007/978-3-031-20083-0_35
24. Shang, Y., Yuan, Z., Xie, B., Wu, B., Yan, Y.: Post-training quantization on diffusion models. In: Proceedings of the IEEE/CVF Conference on Computer Vision and Pattern Recognition, pp. 1972–1981 (2023)
25. Wang, C., Huang, H., Han, X., et al.: Video inpainting by jointly learning temporal structure and spatial details. In: Proceedings of the AAAI Conference on Artificial Intelligence, pp. 5232–5239 (2019)
26. Wang, T.C., Liu, M.Y., Tao, A., et al.: Few-shot video-to-video synthesis. arXiv preprint arXiv:1910.12713 (2019)
27. Wang, T.C., Liu, M.Y., Zhu, J.Y., et al.: High-resolution image synthesis and semantic manipulation with conditional GANs. In: Proceedings of the IEEE Conference on Computer Vision and Pattern Recognition, pp. 8789–8807 (2018)
28. Wang, T.C., Liu, M.Y., Zhu, J.Y., et al.: Video-to-video synthesis. arXiv preprint arXiv:1808.06601 (2018)

29. Wu, Z., Sun, C., Xuan, H., et al.: Deep stereo video inpainting. In: Proceedings of the IEEE/CVF Conference on Computer Vision and Pattern Recognition, pp. 5693–5702 (2023)

30. Wu, Z., Xuan, H., Sun, C., et al.: Semi-supervised video inpainting with cycle consistency constraints. In: Proceedings of the IEEE/CVF Conference on Computer Vision and Pattern Recognition, pp. 22586–22595 (2023)

31. Wu, Z., Zhang, K., Xuan, H., et al.: DAPC-Net: deformable alignment and pyramid context completion networks for video inpainting. IEEE Signal Process. Lett. **28**, 1145–1149 (2021)

32. Wu, Z., Sun, C., Xuan, H., Zhang, K., Yan, Y.: Divide-and-conquer completion network for video inpainting. IEEE Trans. Circuits Syst. Video Technol. **33**(6), 2753–2766 (2023)

33. Xu, R., Li, X., Zhou, B., et al.: Deep flow-guided video inpainting. In: Proceedings of the IEEE/CVF Conference on Computer Vision and Pattern Recognition, pp. 3723–3732 (2019)

34. Zeng, Y., Fu, J., Chao, H.: Learning joint spatial-temporal transformations for video inpainting. In: Vedaldi, A., Bischof, H., Brox, T., Frahm, J.-M. (eds.) ECCV 2020. LNCS, vol. 12361, pp. 528–543. Springer, Cham (2020). https://doi.org/10.1007/978-3-030-58517-4_31

35. Zhang, K., Fu, J., Liu, D.: Flow-guided transformer for video inpainting. In: Avidan, S., Brostow, G., Cissé, M., Farinella, G.M., Hassner, T. (eds.) Computer Vision-ECCV 2022: 17th European Conference, pp. 74–90. Springer, Cham (2022). https://doi.org/10.1007/978-3-031-19797-0_5

Anime Sketch Coloring Based on Self-attention Gate and Progressive PatchGAN

Hang Li[1]📧, Nianyi Wang[2,3]($\boxtimes$)📧, Jie Fang[1], Ying Jia[1], Liqi Ji[1], and Xin Chen[1]

[1] Key Laboratory of China's Ethnic Languages and Information Technology of Ministry of Education, Northwest Minzu University, Lanzhou 730000, China
[2] School of Mathematics and Computer Science, Northwest Minzu University, Lanzhou 730000, China
livingsailor@gmail.com
[3] Key Laboratory of Linguistic and Cultural Computing, Ministry of Education, Northwest Minzu University, Lanzhou, Gansu, China

Abstract. Traditional manual coloring methods require hand-drawn colors to create visually pleasing color combinations, which is both time-consuming and laborious. Reference-based line art coloring is a challenging task in computer vision. However, existing reference-based methods often struggle to generate visually appealing coloring images because sketch images lack texture and training data. To address this, we propose a new sketch coloring network based on the PatchGAN architecture. First, we propose a new self-attention gate (SAG) to effectively and correctly identifying the line semantic information from shallow to deep layers in the CNN. Second, we propose a new Progressive Patch-GAN (PPGAN) to help train the discriminator to better distinguish real anime images. Our experiments show that compared to existing methods, our approach demonstrates significant improvements in some benchmark tests, with Fréchet Inception Distance (FID) improved up to 24.195% and Structural Similarity Index Measure (SSIM) improved up to 14.30% compared to the best values.

Keywords: Anime images · Sketch coloring · Self-attention · GAN

1 Introduction

Anime is a popular art form, and coloring sketches is an important step in creating a finished anime product. In the past, creating excellent anime works required professional artists to spend a lot of time drawing and coloring by hand. However, manually coloring each sketch is very time-consuming and tedious for artists. To address this issue, there is growing interest in using machine learning techniques to develop automatic anime sketch coloring methods.

Anime sketch coloring can be classified as style transfer, whose aim is to extract colors from a style image and transfer them to a content image. Although

Q. Liu et al. (Eds.): PRCV 2023, LNCS 14435, pp. 237–249, 2024.
https://doi.org/10.1007/978-981-99-8552-4_19

|(a)Sketch|(b)Reference|(c) Traditional Style Transfer|(d)Existing method|(e)Our method|(f)Ground Truth|

Fig. 1. The proposed method can produce more accurate coloring and richer details (Fig. 1(e)). Existing deep learning-based methods are prone to color confusion problems (e.g.: the hair part in Fig. 1(d)), while style transfer methods, cannot generate color and texture-rich images due to the lack of content semantic information in the sketch image (Fig. 1(c)). Fig. 1(a) is a sketch image, Fig. 1(b) is a reference image and Fig. 1(f) is a ground truth.

style transfer has made significant progress in the past, there are obvious limitations in sketch coloring: 1) traditional style transfer methods often fail to preserve the structural lines well, resulting in noticeable distortions and artifacts (Fig. 1(c)). This is because of style transfer methods usually deal with content images that are rich in brightness, color, gradient and texture. However, sketch images are very limited in semantic information, and style transfer techniques struggle with the sparse line information and image semantics; and 2) in terms of sketch image coloring, although existing methods have surpassed style transfer methods, they also face serious problems. For example, existing method for densely lined areas, inevitably leads to unsatisfactory coloring results(in Fig. 1(d), the hair part shows significant color confusion). Existing sketch coloring methods such as SV2 [6] and SCFT [7] also fail to generate satisfactory coloring results because of lack of effective network architecture to correctly propagate the semantic information of dense lines in the convolutional neural network (CNN) layers.

In this paper, we propose a novel sketch coloring method for anime sketches. The proposed method consists of two parts: 1) a new self-attention gate (SAG) is proposed for effectively and correctly identifying the line semantic information from shallow to deep layers in the CNN. SAG stabilizes the training of deep networks by correctly propagating the line semantic information by capturing long-range dependencies in the input feature map; and 2) a new Progressive PatchGAN (PPGAN) network is designed to aid the discriminator training strategy, focusing on global and local image detection. We use a progressive discriminator training strategy, starting from a 4×4 sized input image, gradually increasing up to 256×256 size to obtain its local details and global features, making accurate coloring possible.

Experiments show that our method achieves both satisfactory visual performance and outstanding evaluation indexes. In summary, the main contributions of this paper are:

1. A novel self-attention gate for anime sketch coloring is proposed to effectively and correctly identifying the line semantic information from shallow to deep layers in the CNN.
2. A new progressive training strategy based on PatchGAN is designed to generate more realistic anime images.

2   Related Work

Significant research has been conducted on sketch coloring, beyond just style transfer algorithms. Broadly speaking, sketch coloring algorithms can be divided into three categories: 1) automatic sketch coloring; 2) user-guided coloring; and 3) reference-based sketch image coloring.

2.1   Style Transfer

The groundbreaking work of Gatys et al. [21] demonstrated the power of convolutional neural networks (CNNs) in creating artistic images by separating and recombining image content and style. The process of presenting a content image in different styles using CNNs is called neural style transfer (NST). Johnson et al. [16] proposed a perceptual loss for training real-time feedforward networks. However, feedforward methods could only be used for specific style images. Therefore, Huang et al. [25] proposed adaptive instance normalization (AdaIN) in 2017 to achieve real-time arbitrary style transfer. Park et al. [11] considered that simple AdaIN [25] is not sufficient for style transfer and proposed style attention network (SANet) to replace AdaIN for style transfer.

Although these style transfer algorithms have achieved good results, there are still two limitations in their application to sketch coloring: 1) style transfer algorithms mainly focus on the content and style of the input image rather than semantic information, which is severely lacking in sketch images; and 2) sketches are usually more abstract and simplified than actual images, and due to the global nature of style transfer algorithms, colors may be applied to the entire image, as shown in Fig. 1(c).

2.2   Automatic Sketch Coloring

In recent years, deep learning-based automatic sketch coloring methods [1,4,10, 12,18,19,27] have received increasing attention. Liu et al. [10] used feedforward deep neural networks as generators to input sketches and output color images with pixel-level resolution. Frans et al. [12] proposed two concatenated adversarial networks for automatic sketch image coloring. Recent studies [1,19] have improved the use of U-net networks and proposed an architecture for automatic sketch coloring based on U-net.

For automatic sketch coloring methods, there are two main issues: 1) these methods are sensitive to visual artifacts when sketches have complex content and multiple objects; and 2) due to fixed network parameters, existing methods tend to output single-color results and lack multimodality.

2.3 User-Guided Coloring

Ci et al. [20] proposed a deep conditional adversarial architecture to train the network robustly for more natural and realistic synthetic images. Zhang et al. [22] proposed a two-stage colorization framework based on semi-supervised learning, which colors sketches with appropriate colors, textures, and gradients. Yuan et al. [28] proposed a concatenated and U-net framework based on spatial attention modules that can generate more consistent and higher-quality sketch coloring from user-provided clues. Amal et al. [24] proposed a method of image coloring perceptively from raster contour images. Starting from the Delaunay triangulation of the input contour, the triangles are iteratively filled with appropriate colors using the dynamically updated flow values computed from color cues.

However, these methods have limitations: 1) these palette-based coloring methods are susceptible to user aesthetic limitations; and 2) untrained users may have difficulty selecting appropriate points and related colors from the palette.

2.4 Reference-Based Sketch Image Coloring

Coloring sketch images based on reference styles is a user-friendly approach [5–8,17,23]. With the rise of deep neural networks in recent years, Zhang et al. [4] integrated the Residual U-Net into the adversarial conditional generative adversarial networks (AC-GAN) and equipped it with an auxiliary classifier for anime sketch coloring tasks. Due to the limitations of the sketch-reference image pairs dataset, Lee et al. [7] proposed to generate reference images using enhanced self-reference and to color sketches using a pixel-based attention feature transfer module. Li et al. [8] proposed a stop-gradient-attention (SGA) training strategy on the basis of [7] to maintain color consistency while eliminating gradient conflicts in coloring results. Yan et al. [3] proposed a two-step training and spatial latent operation to achieve high-quality and adjustable coloring results using text labels while utilizing reference images.

Although these models have achieved good results, the results are not satisfactory. As shown in Fig. 1, a comparison of existing methods (Fig. 1(d)) and our method (Fig. 1(e)) shows that existing methods have significant color errors and semantic mismatches.

3 Methodology

The proposed method for anime sketch coloring consists of two parts: 1) a self-attention gate (SAG) mechanism to effectively and correctly recognize the line semantics from shallow to deep layer in the CNN; and 2) a novel progressive PatchGAN (PPGAN) network that focuses on global and local detection of the image.

3.1 Overall Workflow

The overall network architecture is shown in Fig. 2. For the reference image, we vertically divide the image into 4 regions. Then, we extract the top 4 colors for each region, using the extracted 16 colors. Finally, we obtain 4 RGB images as shown in Fig. 2(c). The i-th image contains the i-th color for each cropped region. We then concatenate the sketch image and the 4 color histograms (extracted from the reference image) to the generator. Thus, the generator receives an input image of size $512 \times 512 \times 5$ and generates an output image of size $512 \times 512 \times 3$.

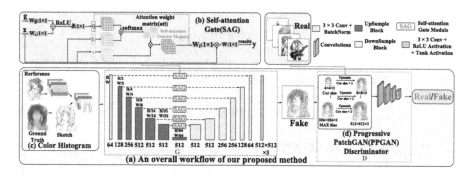

Fig. 2. The proposed colorization method for anime sketches consists of two parts: 1) in order to achieve more precise colorization, a self-attention gate (SAG) is utilized to accurately identify line semantics at different depths of the CNN, starting from the shallow layers and progressing towards the deeper ones; and 2) Progressive PatchGAN (PPGAN), which incrementally increases the input image size to enable the model to balance global and local details, thus improving the discriminative accuracy to enhance the image generation quality.

3.2   Self-attention Gate

Most GAN-based image generation models are built using convolutional layers. Convolution processes local neighborhood information, so using only convolutional layers to simulate long-range dependencies in images is computationally inefficient. In this section, we introduce a self-attention mechanism into the GAN framework to enable the generator to effectively model relationships between spatially distant regions. We proposed a new self-attention method is called self-attention gate (SAG) (see Fig. 2(b)).

Given an input feature map $x \in R^{C_x \times H_x \times W_x}$ and a gated feature map $g \in R^{C_g \times H_g \times W_g}$, first compute their spatial feature maps as follows:

$$q_{att} = \psi \left(\sigma_1 \left(W_x^T x + W_g^T g + b_g \right) \right) + b_\psi \tag{1}$$

where ψ is the learnable mapping function, σ_1 is the element-by-element activation function, W_x and W_g are linear transformations, and b_g and b_ψ are bias terms. The attentional weights are computed using the self-attentive mechanism as follows:

$$att = softmax\left(W_a q_{att} \times (W_a k)^T\right) \tag{2}$$

where W_a is the learned weight matrix, implemented as a 1×1 convolution. k and v denote keys and values, respectively, which are both equal to the input feature map x in this module. The $softmax$ function is normalized along the second dimension. The output of the self-attentive mechanism is the multiplication of the value v and the attention weight element: $c = att * v$. Then, q and c are stitched along the channel dimension and fused by W_f:

$$q' = W_f(concat(q, c)) \tag{3}$$

Finally, the fused feature map q' is multiplied with the input feature map x, and then the final feature map y is obtained by linear transformation W:

$$y = W(q' \odot x) \tag{4}$$

3.3   Progressive PatchGAN

In this section, we propose a new progressive prediction task to enhance the training of the discriminator. Correctly transferring colors to local regions of sketches and balancing global color is a significant challenge in sketch coloring. This is because there are multiple possible results for coloring the same region in a sketch, which can lead to semantic mismatch and color confusion. To overcome this challenge, we establish our model on a generative adversarial network (GAN) [2]. As PatchGAN [26] has achieved great success in the field of image translation, we inherit most of the network structure of the discriminator. As shown in Fig. 2, the discriminator network consists of several layers of standard convolution, but due to the fixed input size of PatchGAN, local and global details are inevitably lost, making it difficult to achieve balance. To tackle this issue, we propose a progressive training approach that balances local and global color.

In practice, the size of the input image is $H_{in} \times W_{in}$, the upsampling factor is f, the current image size is $H_{cur} \times W_{cur}$, and the maximum image size is $H_{max} \times W_{max}$. Therefore, the size of the current image can be expressed as

$$H_{cur}^{(i)} = \min(H_{cur}^{(i-1)} \times f, H_{max}) \tag{5}$$

$$W_{cur}^{(i)} = \min(W_{cur}^{(i-1)} \times f, W_{max}) \tag{6}$$

where $H_{cur}^{(i)}$ and $W_{cur}^{(i)}$ represent the height and width, respectively, of the image sampled at the i-th time step. $H_{cur}^{(0)}$ and $W_{cur}^{(0)}$ represent the initial dimensions of the input image, which are denoted as $H_{in} \times W_{in}$. The min function in the formula is used to ensure that the current image size does not exceed the maximum image size. Then, the final asymptotic training can be expressed as:

$$H_{out}^{(i)} = H_{in} \times \frac{H_{cur}^{(i)}}{H_{cur}^{(i-1)}} \tag{7}$$

$$W_{out}^{(i)} = W_{in} \times \frac{W_{cur}^{(i)}}{W_{cur}^{(i-1)}} \qquad (8)$$

where the dimensions of $H_{out}^{(i)}$ and $W_{out}^{(i)}$ indicate the size of the output image after i sampling steps. The maximum size of the image is represented by H_{max} and W_{max}.

3.4   Loss Function

Most existing methods tend to design and utilize a variety of complex loss functions to achieve multiple constraints in their designs. We only need to employ two widely used loss functions to achieve satisfactory coloring performance.

The adversarial loss for a conditional GAN given a generator G, a discriminator D, and color histogram extractoras C is as follows:

$$\mathcal{L}_{adv} = E_{x,y}[(1 - D(x, G(x, C(y))))] + E_{x,y}[(D(x, y)] \qquad (9)$$

where the generator is D, the discriminator is G, the color histogram extractor is denoted as C, the sketch image is x, and the original coloring image is y.

We adopt smooth L1 loss [13] as the distance metric to avoid the averaging solution in the ambiguous colorization problem. The reconstruction loss can be formulated as:

$$\mathcal{L}_{rec} = E_{x,y}[\|y - G(x, C(y))\|_1] \qquad (10)$$

In summary, the overall loss function for training is defined as:

$$\arg \min_{G} \max_{D} \mathcal{L}_{total} = \mathcal{L}_{adv}(G, D) + \lambda_{rec}\mathcal{L}_{rec}(G) \qquad (11)$$

where λ_{rec} represents the weight of the reconstruction loss.

4   Experimental Results and Analysis

4.1   Implementation Details

To train our network, we used a publicly available dataset [15] consisting of 17,769 pairs of colored comics and corresponding sketches. We used 14,224 pairs for training and 3,545 pairs for evaluation. For our model, we trained for 40 epochs on a single NVIDIA 3090 GPU. We used the Adam solver [14] for optimization with momentum parameters of $\beta_1 = 0.5$ and $\beta_2 = 0.999$. The learning rates for the generator and discriminator were initially set to 0.0001 and 0.0002, respectively. For each dataset, the size of the input image was fixed to 512 × 512.

4.2   Qualitative Evaluation

To showcase the effectiveness of our reference-based coloring approach, we compare its results with those obtained by state-of-the-art baselines, namely, SV2 [6], SCFT [7], SGA [8] and Munit [23]. As shown in Fig. 3, we can easily observe the visual differences between various anime sketch coloring methods. Our method yields more precise colors when compared to the state-of-the-art coloring techniques. Figure 3(g) shows that our method correctly transfers the colors from the reference image to the target sketch and preserves the structural lines of the target sketch well. Munit [23] and SV2 [6] can obtain relatively accurate colors but severely distort the original structural lines (Fig. 3(e,f)), while SGA [8] fails to transfer the colors correctly (Fig. 3(c)). SCFT [7] generates good structure but the generated images have high perceptual color blur and low color consistency (Fig. 3(d)).

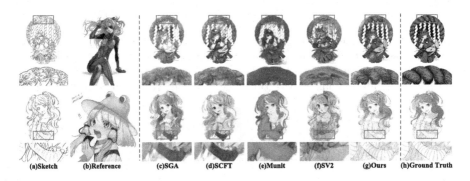

Fig. 3. The comparison of sketch coloring shows that our result (g) achieves better visual performance in terms of image structure and texture than SGA [8] (c), SCFT [7] (d), Munit [23] (e), and SV2 [6] (f). (a) is the anime sketch image. (b) is the reference image. (h) shows the real ground truth.

4.3   Quantitative Evaluation

Fréchet Inception Distance (FID) [9] is a well-known metric for evaluating the performance of generative models. In traditional sketch coloring settings, pixel-level evaluation metrics such as Peak Signal-to-Noise Ratio (PSNR) and contour-preserving evaluation metrics such as Structural Similarity Index (SSIM) are widely used. We employ the above-mentioned three metrics to conduct a quantitative evaluation of our model's performance in this study, with the best result being indicated in bold.

In Table 1, our method outperforms all other methods in terms of both 256 × 256 and 512 × 512 sizes, demonstrating the effectiveness of our approach.

Additionally, we provide more examples of anime sketch coloring generated by our method in Fig. 4. These results further illustrate the capability of our

Table 1. The results of the quantitative evaluation. Our method achieves the best results in all metrics. FID [9] scores: the lower the score the better. PSNR and SSIM scores: the higher the score the better.

Method	256 × 256			512 × 512		
	FID↓	SSIM↑	PSNR↑	FID↓	SSIM↑	PSNR ↑
Munit [23]	36.168	0.703	15.215	39.596	0.699	15.153
SV2 [6]	26.740	0.629	14.999	36.660	0.622	14.861
SCFT [8]	33.609	0.804	17.899	42.747	0.789	17.506
SGA [7]	30.760	0.781	17.034	40.605	0.766	16.664
Ours	**12.044**	**0.934**	**22.530**	**12.465**	**0.932**	**22.173**

method to produce visually appealing and semantically meaningful coloring maps for anime sketches.

Fig. 4. More coloring examples of anime images generated with our method. The proposed method achieves better sketch coloring performance by generating visually satisfying and semantically sound coloring maps for anime images.

4.4 Ablation Study

To evaluate the effectiveness of the proposed SAG and PPGAN training strategies, we conducted two ablation experiments. Figure 5(a) illustrates the results of training that without including the SAG module, which revealed significant semantic mismatch in the generated images. Similarly, Fig. 5(b) shows the results of training without the progressive PPGAN module, which led to poor quality of the generated images.

To provide an objective comparison, we also performed FID, SSIM, and PSNR evaluations, and the results are presented in Table 2. The evaluation metrics demonstrate significant improvements in the quality of the generated images when using the proposed training state.

Table 2. Ablation studies have shown that the proposed SAG and PPGAN are effective. In particular, when both SAG and PPGAN were present in the model (row 3 of Table 2), our model achieved the best results. A lower FID score indicates better quality of the generated image. PSNR and SSIM scores: the higher the score the better.

Method	256×256			512×512		
	FID↓	SSIM↑	PSNR↑	FID↓	SSIM↑	PSNR↑
w/o SGA	13.545	0.930	22.429	14.237	0.928	22.087
w/o PPGAN	13.248	0.932	22.517	13.875	0.931	22.105
FULL	**12.044**	**0.934**	**22.530**	**12.465**	**0.932**	**22.173**

Fig. 5. This is the result of an ablation study. Figure 5(a) shows the results of training without the SAG module, which led to significant semantic mismatch in the generated images. Figure 5(b) illustrates the results of a training strategy without Progressive PatchGAN, which resulted in poor quality of the generated images.

5   Conclusions

We propose a sketch coloring method based on self-attention gate (SAG) and Progressive PatchGAN (PPGAN) for creating anime art. The method consists of two parts: 1) a novel self-attention gate module (SAG) to effectively and correctly identify the line semantic information from shallow to deep layers in the CNN; and 2) a novel Progressive PatchGAN (PPGAN) to help train the discriminator to better distinguish real cartoon images based on global and local color features. In particular, progressive input images help the discriminator balance local and global color combinations. Experiments show that our method can produce visually more reasonable and rich coloring results compared to existing methods. Objective and subjective evaluations have both verified the performance of our method.

Acknowledgment. This work is supported by NSFC (Grant No. 62366047 and No. 62061042).

References

1. Liu, G., Chen, X., Hu, Y.: Anime sketch coloring with swish-gated residual U-Net. In: Peng, H., Deng, C., Wu, Z., Liu, Y. (eds.) ISICA 2018. CCIS, vol. 986, pp. 190–204. Springer, Singapore (2019). https://doi.org/10.1007/978-981-13-6473-0_17
2. Mirza, M., Osindero, S.: Conditional generative adversarial nets. arXiv preprint arXiv:1411.1784 (2014). https://doi.org/10.48550/arXiv.1411.1784
3. Yan, D., Ito, R., Moriai, R., Saito, S.: Two-step training: adjustable sketch colourization via reference image and text tag. In: Computer Graphics Forum, Wiley Online Library (2023). https://doi.org/10.1111/cgf.14791
4. Seo, C.W., Seo, Y.: Seg2pix: few shot training line art colorization with segmented image data. Appl. Sci. **11**, 1464 (2021). https://doi.org/10.3390/app11041464
5. Sato, K., Matsui, Y., Yamasaki, T., Aizawa, K.: Reference-based manga colorization by graph correspondence using quadratic programming. In: SIGGRAPH Asia 2014 Technical Briefs, pp. 1–4 (2014). https://doi.org/10.1145/2669024.2669037
6. Choi, Y., Uh, Y., Yoo, J., Ha, J.-W.: StarGAN v2: diverse image synthesis for multiple domains. In: Proceedings of the IEEE/CVF Conference on Computer Vision and Pattern Recognition, pp. 8188–8197 (2020). https://doi.org/10.1109/CVPR42600.2020.00821
7. Lee, J., Kim, E., Lee, Y., Kim, D., Chang, J., Choo, J.: Reference based sketch image colorization using augmented-self reference and dense semantic correspondence. In: Proceedings of the IEEE/CVF Conference on Computer Vision and Pattern Recognition, pp. 5801–5810 (2020). https://doi.org/10.1109/cvpr42600.2020.00584
8. Z. Li, Z. Geng, Z. Kang, W. Chen, Y. Yang, Eliminating gradient conflict in reference-based line-art colorization, in: Computer Vision- ECCV 2022: 17th European Conference, Tel Aviv, Israel, October 23–27, 2022, Proceedings, Part XVII, Springer, 2022, pp. 579–596. https://doi.org/10.1007/978-3-031-19790-1_35
9. Bynagari, N.B.: GANs trained by a two time-scale update rule converge to a local Nash equilibrium. Asian J. Appl. Sci. Eng. **8**, 25–34 (2019)
10. Liu, Y., Qin, Z., Wan, T., Luo, Z.: Auto-painter: cartoon image generation from sketch by using conditional wasserstein generative adversarial networks. Neurocomputing **311**, 78–87 (2018). https://doi.org/10.1016/j.neucom.2018.05.045
11. Park, D.Y., Lee, K.H.: Arbitrary style transfer with style-attentional networks. In: proceedings of the IEEE/CVF Conference on Computer Vision and Pattern Recognition, pp. 5880–5888 (2019). https://doi.org/10.1109/CVPR.2019.00603
12. Frans, K.: Outline colorization through tandem adversarial networks, arXiv preprint arXiv:1704.08834 (2017). https://doi.org/10.48550/arXiv.1704.08834

13. Huber, P.J.: Robust estimation of a location parameter. In: Kotz, S., Johnson, N.L. (eds.) Breakthroughs in Statistics, pp. 492–518. Springer, New York (1992). https://doi.org/10.1214/aoms/1177703732

14. Tai, Y.-W., Jia, J., Tang, C.-K.: Local color transfer via probabilistic segmentation by expectation-maximization. In: 2005 IEEE Computer Society Conference on Computer Vision and Pattern Recognition (CVPR 2005), vol. 1, pp. 747–754. IEEE (2005). https://doi.org/10.1109/CVPR.2005.215

15. Available at https://www.kaggle.com/ktaebum/animesketch-colorization-pair

16. Johnson, J., Alahi, A., Fei-Fei, L.: Perceptual losses for real-time style transfer and super-resolution. In: Leibe, B., Matas, J., Sebe, N., Welling, M. (eds.) ECCV 2016. LNCS, vol. 9906, pp. 694–711. Springer, Cham (2016). https://doi.org/10.1007/978-3-319-46475-6_43

17. Huang, J., Liao, J., Kwong, S.: Semantic example guided image-to image translation. IEEE Trans. Multimedia **23**, 1654–1665 (2020). https://doi.org/10.1109/TMM.2020.3001536

18. Furusawa, C., Kitaoka, S., Li, M., Odagiri, Y.: Generative probabilistic image colorization. arXiv preprint arXiv:2109.14518 (2021). https://doi.org/10.48550/arXiv.2109.14518

19. Zhang, G., Qu, M., Jin, Y., Song, Q.: Colorization for anime sketches with cycle-consistent adversarial network. Int. J. Perform. Eng. **15**, 910 (2019). https://doi.org/10.23940/ijpe.19.03.p20.910918

20. Ci, Y., Ma, X., Wang, Z., Li, H., Luo, Z.: User-guided deep anime line art colorization with conditional adversarial networks. In: Proceedings of the 26th ACM International Conference on Multimedia, pp. 1536–1544 (2018). https://doi.org/10.1145/3240508.3240661

21. Gatys, L.A., Ecker, A.S., Bethge, M.: Image style transfer using convolutional neural networks. In: Proceedings of the IEEE Conference on Computer Vision and Pattern Recognition, pp. 2414–2423 (2016). https://doi.org/10.1109/CVPR.2016.265

22. Zhang, L., Li, C., Wong, T.-T., Ji, Y., Liu, C.: Two-stage sketch colorization. ACM Trans. Graphics (TOG) **37**, 1–14 (2018). https://doi.org/10.1145/3272127.3275090

23. Huang, X., Liu, M.-Y., Belongie, S., Kautz, J.: Multimodal unsupervised image-to-image translation. In: Ferrari, V., Hebert, M., Sminchisescu, C., Weiss, Y. (eds.) ECCV 2018. LNCS, vol. 11207, pp. 179–196. Springer, Cham (2018). https://doi.org/10.1007/978-3-030-01219-9_11

24. Parakkat, A.D., Memari, P., Cani, M.-P.: Delaunay painting: Perceptual image colouring from raster contours with gaps. In: Computer Graphics Forum, vol. 41, Wiley Online Library, pp. 166–181 (2022). https://doi.org/10.1111/cgf.14517

25. Huang, X., Belongie, S.: Arbitrary style transfer in real-time with adaptive instance normalization. In: Proceedings of the IEEE International Conference on Computer Vision, pp. 1501–1510 (2017). https://doi.org/10.1109/ICCV.2017.167

26. Isola, P., Zhu, J.-Y., Zhou, T., Efros, A.A.: Image-to-image translation with conditional adversarial networks. In: Proceedings of the IEEE Conference on Computer Vision and Pattern Recognition, pp. 1125–1134 (2017). https://doi.org/10.1109/CVPR.2017.632

27. Yan, C., Chung, J.J.Y., Kiheon, Y., Gingold, Y., Adar, E., Hong, S.R.: Flatmagic: improving flat colorization through AI-driven design for digital comic professionals. In: Proceedings of the 2022 CHI Conference on Human Factors in Computing Systems, pp. 1–17 (2022). https://doi.org/10.1145/3491102.3502075
28. Yuan, M., Simo-Serra, E.: Line art colorization with concatenated spatial attention. In: Proceedings of the IEEE/CVF Conference on Computer Vision and Pattern Recognition, pp. 3946–3950 (2021). https://doi.org/10.1109/CVPRW53098.2021.00442

TransDDPM: Transformer-Based Denoising Diffusion Probabilistic Model for Image Restoration

Pan Wei[1,2]([✉])

[1] Institute of Information Engineering, Chinese Academy of Sciences, Beijing, China
weipan@iie.ac.cn
[2] School of Cyber Security, University of Chinese Academy of Sciences, Beijing, China

Abstract. Although diffusion models have achieved impressive success for image generation, its application for image restoration is still underexplored. Following tremendous success in natural language processing, transformers have also shown great success for computer vision. Although several researches indicate that increasing transformer depth/width improves the applicability of diffusion models, application of Transformers in diffusion models is still underexplored due to quadratic complexity with the spatial resolution. In this work, we proposed a Transformer-based Denoising Diffusion Probabilistic Model (TransDDPM) for image restoration. With multi-head cross-covariance attention (MXCA), Trans-DDPM can operates global self-attention with cross-covariance matrix in channel dimension rather than spatial dimension. Another gated feed-forward network (GFFN) is included to enhance the ability to exploit spatial local context. Powered by these designs, TransDDPM is capable for both long-range dependencies and short-range dependencies and flexible for images of various resolutions. Comprehensive experiments demonstrate our TransDDPM achieves state-of-the art performance on several restoration tasks, e.g., image deraining, image dehazing and motion deblurring.

Keywords: DDPM · Cross-Covariance Attention · Image Restoration

1 Introduction

Computer vision applications have been used in various industries such as unmanned driving, video surveillance and so on. However, natural images are often contaminated by various factors, e.g., rain, haze and motion, which is not only unpleasant for human visual perception, but also interferes the performance of other visual applications. Therefore, recovering real images from their degraded versions, i.e., image resto-ration, is a classic task in computer vision.

Early approaches for image restoration mainly used empirical observations or statistical priors to model image degradation under different situations, e.g. rain or haze. These methods are only applicable for certain degradation, while not suitable for others.

© The Author(s), under exclusive license to Springer Nature Singapore Pte Ltd. 2024
Q. Liu et al. (Eds.): PRCV 2023, LNCS 14435, pp. 250–263, 2024.
https://doi.org/10.1007/978-981-99-8552-4_20

Although generative models, such as Generative Adversarial Network (GAN) [9] and Variational Auto-Encoder (VAE) [6], have shown great achievement in image restoration, they still have drawbacks and limitations. GAN suffers from unstable training and model collapse and VAE struggles to capture complex and highly structured image features.

Recently, Denoising Diffusion Probabilistic Model (DDPM) [15] and its variants have achieved impressive results for image generation [19, 20], image super-resolution [24], and other tasks. However, the architecture of DDPM for image restoration is still underexplored, so that most of the diffusion models adopt U-Net architecture [29]. Although Transformer [31] has shown great success for computer vision, its quadratic complexity makes it infeasible to apply in DDPM with high-resolution images.

To address this issue, we propose a Transformer-based Denoising Diffusion Model for image restoration (TransDDPM), which proves to be efficient and effective. Specifically, we introduce multi-head cross-covariance attention block (MXCA) into DDPM architecture inspired by XCiT [35] and TSA [31]. MXCA computes cross-covariance matrix to generate self-attention between keys and queries in channel dimension instead of special dimension. A Gated Feed-Forward Network (GFFN) is proposed to capture local contextual information, as original FFN in Transformer suffers limited capability to leverage local context. With these improvements, TransDDPM shows strong capability for both long-range dependencies and short-range dependencies with linear complexity. Besides, TransDDPM is flexible for images with different resolutions and applicable for several restoration tasks.

The main contributions of this work are summarized below:

- We propose a Transformer-based Denoising Diffusion Probabilistic Model (Trans-DDPM), which achieves state-of-the-art (SOTA) performance for image deraining, image dehazing and motion deblurring.
- We propose a Multi-head Cross-covariance Attention Block (MXCA) that is capable for both global long-range dependencies and local short-range dependencies with linear complexity.
- We propose a gated feed-forward network (GFFN) to control the flow of information with gating mechanism that regulates the information passing through network hierarchy.

2 Related Work

2.1 Image Restoration

Since it is ill-posed to estimate original image from a single image, traditional methods for image restoration aim to model the degradation process by manually designed priors or statistical properties, which is only compatible for certain scenarios.

Convolutional Neural Networks (CNN) have dominated the field of computer vision for image restoration, such as deraining [5, 8, 11, 36], dehazing [6, 10, 14, 17, 25] and deblurring [1, 4, 7]. Among these algorithms, U-net architecture [29] with skip connection has been widely used for various image restoration tasks. Generative models, such as GAN and VAE, have shown SOTA performance for image deblurring [12, 18], image dehazing [38] and deraining [36]. GAN uses adversarial training to generate high fidelity results but suffers from unstable training and model collapse. VAE optimizes the log-likelihood of the data distribution by maximizing Evidence Lower Bound (ELBO), but struggles to capture complex and highly structured image features, such as long-range dependencies or fine-grained texture details.

With strong ability to learn long-range dependencies and the adaptability for given input content, Transformer [39] has become the most popular architecture for image restoration tasks, e.g. image deraining [40, 41] and image deblurring [31, 37]. However, the self-attention in Transformers leads to quadratic complexity, hindering application to high-resolution images. Many methods were proposed to address this issue. For instance, IPT [40] adopt patch-wise self-attention. UFormer [37] and SwinIR [42] adopt local attention based on shifted windows. Restormer [31] and RSFormer [43] adopt transposed self-attention. However, these methods either sacrifice accuracy to reduce computational complexity, or restrict the context aggregation within local neighborhoods, violating the main motivation of self-attention over convolutions.

2.2 Denoising Diffusion Probabilistic Models

Inspired by nonequilibrium thermodynamics, DDPM defines a parameterized Markov chain to corrupt the original data by successively adding Gaussian noise, and then learns to reverse the diffusion process to construct the desired data by sampling from the noise.

For data distribution $x_0 \sim q(x_0)$, the forward process defines a fixed Markov Chain that sequentially corrupts x_0 by injecting Gaussian noise at T diffusion time steps, producing noisy samples x_1 through x_T. The reverse sampling process learns to sample x_0 by reversing the forward process. We can sample x_T firstly and update iteratively with $q(x_t|x_{t-1})$ until we reach x_0 [19]. Because $q(x_{t-1}|x_t)$ cannot be computed exactly as the data distribution q is unknown, we can use a deep neural network to learn $p_\theta(x_{t-1} \mid x_t)$ to approximate $q(x_{t-1}|x_t)$. Ho et al. [25] found that a different objective produces better samples in practice. In particular, the neural network predicts noise $\epsilon_\theta(x_t, t)$ by re-parameterization instead of posterior mean $\mu_\theta(x_t, t)$ in the reversing process.

2.3 Diffusion Models for Image Restoration

Recently, DDPM and its variants have achieved impressive results for image generation [19, 20], image super-resolution [24], and other tasks. However, the architecture of DDPM for image restoration is still underexplored, so that most of current diffusion models adopt U-Net architecture [29]. Although some studies have attempted to apply Transformer [31] in the diffusion models, the quadratic complexity of self-attention still prohibits its application for high-resolution images. Peebles [44] proposed Diffusion Transformers (DiTs) for image generation by replacing U-Net backbone with Transformer. Özdenizci et al. [45] proposed patch-based denoising diffusion models with self-attention layers. Saharia et al. [46] propose image-to-image diffusion models for image inpainting, uncropping and JPEG restoration. However, considering the quadratic complexity of self-attention, these methods apply self-attention within a small patch/window, e.g., at 16×16 resolutions. Although Dhariwal et al. [19] and Saharia et al. [46] demonstrated that using attention at higher resolutions yields better performance than fully convolutional alternatives, using Transformers in diffusion models involving high-resolution images is still underexplored.

3   Transformer-Based Denoising Diffusion Restoration Models

In this section, we first present the overall pipeline of proposed TransDDPM. Then we provide the details of MXCA and GFFN, the core components of TransDDPM. After that, we present how to speed up TransDDPM by implicit sampling.

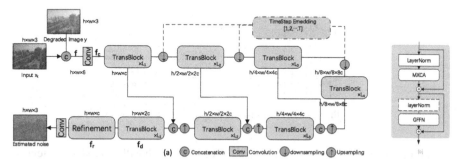

Fig. 1. (a) overview of TransDDPM. (b) Illustration of transBlock.

3.1   Overall Pipeline

For a degraded image $y \in R^{h \times w \times 3}$ and a noisy image $x_t \in R^{h \times w \times 3}$, TransDDPM is trained to predict noise $\epsilon_\theta(x_t, y, t)$ which is used to update input x_t iteratively. As shown in Fig. 1., the overall structure of the proposed TransDDPM is a U-shaped hierarchical network with skip connections between encoders and decoders.

Firstly, we need to build a conditional diffusion model with degraded image y for image restoration. Inspired by previous works [24, 47], input x and degraded image y are concatenated channel-wise, resulting feature $f \in R^{h \times w \times 6}$. The timestep t is also specified into TransDDPM by sinusoidal position embedding [39]. Then TransDDPM extracts low-level features $f_c \in R^{h \times w \times c}$ with a 3×3 convolutional layer, where c is the number of channels. After that, the features f_c are passed into a symmetric encoder-decoder architecture. Each level of the encoder-decoder contains several transBlocks and a downsampling or upsampling layer. The transBlock not only takes advantage of the self-attention mechanism for capturing long-range dependencies, but also cuts the computational cost with Multi-head Cross-covariance Attention (MXCA). From top to bottom, the spatial size decreases while the channel size increases. The number of transBlocks also increases sequentially to maintain model capability. Pixel-unshuffle and pixel-shuffle are applied to each layer. The encoder features are conventionally concatenated with the decoder features via skip connections. Finally, the features are passed into a refinement stage and a convolution layer to obtain the estimated noise $\epsilon_\theta(x_t, y, t)$.

We trained TransDDPM with Charbonnier loss [48], a variant of L1 loss for its robustness and smoothness, which defined as:

$$L_{char} = E_{t \sim [1,T], x_0 \sim q(x_0), \epsilon \sim \mathcal{N}(0,\mathbf{I})} \left[\sqrt{\|\epsilon - \epsilon_\theta(x_t, y, t)\|^2 + \eta^2} \right] \quad (1)$$

where η is empirically set to 10^{-3} for all experiments.

3.2 Multi-Head Cross-Covariance Attention (MXCA)

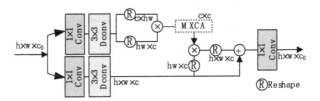

Fig. 2. Illustration of Multi-head Cross-Covariance Attention.

Our main goal is to develop an efficient DDPM that can handle global self-attention for high-resolution image restoration. There are two main challenges to apply transformer in diffusion model for image restoration. Firstly, the calculation of self-attention in Transformer [39] grows quadratically with the spatial resolution of input, making it infeasible to be used in high-resolution images. Secondly, the architecture of Transformer prohibits size-agnostic image restoration [45]. To address these issues, we introduce Multi-head Cross-covariance Attention (MXCA) in diffusion model, shown in Fig. 2. The key component of MXCA is to compute cross-covariance matrix across feature channels rather than spatial dimensions, which is used to encode the global context implicitly.

Furthermore, recent works indicate that self-attention shows a limitation in capturing local dependencies [31, 37, 43], which is important reference for image restoration. To address this issue, we add a 1×1 pixel-wise convolution to aggregate cross-channel context and a 3×3 depth-wise convolution to utilize local context. After reshaping query and key projections, MXCA generates a transposed-attention map A of size $R^{c \times c}$ instead of $R^{hw \times hw}$. The typical self-attention has a time complexity of $O(h^2 w^2 c)$ and memory complexity of $O(h^2 w^2 + hwc)$, while MXCA has a computational cost of $O(hwc^2)$ and a memory complexity of $O(c^2 + hwc)$, which scales linearly with the resolution of images. For input $x \in R^{h \times w \times c_0}$, the output of MXCA $\hat{x}$ is defined as:

$$\hat{x} = \widehat{V} \cdot Softmax(\widehat{K}^T \widehat{Q} / \tau) + x \quad (2)$$

where $\widehat{Q}, \widehat{K}, \widehat{V}$ represents reshaped query, key, value and τ is a learnable scaling parameter. Corresponding to [39], we use multi-head attention by dividing numbers of channels into multiple heads and learn separate attention maps parallelly. Since cross-covariance attention is independent of the input image resolution, MXCA is capable for handling features of different spatial sizes, which is significant for image restoration.

3.3 Gated Feed-Forward Network (GFFN)

Feed-forward network (FFN) was proposed in Transformer [39] to capture non-linear interactions between the features using two 1×1 convolutions at each position. We propose a gated feed-forward network (GFFN) with gating mechanism, which is formulated as the element-wise product of two parallel paths of linear transformation layers. As pointed out by Wu et al. [49], the FFN in the standard Transformer suffers from limited capability to leverage local context. To address this issue, we add a 3×3 depth-wise convolutional layer for capturing local context. We also replace the first linear transformation with Gated Linear Units (GLU) [50]. For input $x \in R^{h \times w \times c_0}$, the GFFN is formulated as:

$$\hat{x} = W_p^0 [\varphi \left(W_d^1 W_p^1(x) \right) \cdot W_d^2 W_p^2(x)] + x \qquad (3)$$

where $W_p^{(\cdot)}$ represents 1×1 pixel-wise convolution, $W_d^{(\cdot)}$ represents 3×3 depth-wise convolution, $\cdot$ denotes element-wise multiplication and φ represents the GELU non-linearity. To keep the number of parameters and the amount of computation constant, we reduce the number of hidden units by a factor of $\frac{2}{3}$ comparing to original FFN [51]. With GFFN, each level of TransDDPM could focus on enriching features with contextual information.

3.4 Accelerated with Implicit Sampling

Despite its high-quality performance, DDPM suffers from slow sampling as it requires thousands of sequential evaluations steps. An implicit sampling approach has been proposed by skipping steps with a certain strategy in sampling process while diffusion process remains the same [52]. Besides, it can correspond to deterministic generative processes that produce high quality samples much faster. With implicit sampling, proposed TransDDPM can be accelerated up to two orders of magnitude. We introduce implicit sampling process with a subsequence $\tau_1, \tau_2, ..., \tau_S$ instead of the entire indices$[1, 2, .., T]$, which is defined as (Fig. 3):

$$\tau_i = (i - 1) \cdot {}^T\!/_S + 1 \qquad (4)$$

which sets $\tau_1 = 1$ at the final step of reverse sampling.

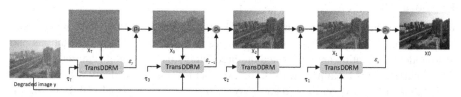

Fig. 3. Illustration of sampling process of TransDDPM.

4 Experiment

In this section, we first discuss the datasets and implementation details of TransDDPM. Then we evaluate TransDDPM quantitatively and qualitatively on three image restoration tasks: (a) deraining, (b) dehazing, and (c) motion deblurring. Finally, we perform ablation studies to evaluate each component of TransDDPM.

4.1 Datasets and Evaluation Metrics

Since not relying on any task-specific architecture or parameters, we trained TransD-DPM separately with the same architecture and parameters for each task. The datasets used in each task are summarized in Table 1. All models are trained for 1M iterations. Quantitative comparisons are performed using the PSNR and SSIM [76] metrics based on the luminance channel Y of the YCbCr color space as in previous work [23, 31].

Table 1. Dataset description for various image restoration tasks.

Task	Deraing			Dehazing		Deblurring		
Train	Rain14000 [53]			BESIDE-ITS [54]	BESIDE-OTS [54]	GoPro [1]		
	14,000			110,000	313,950	2103		
Test	Rain100L [55]	Rain100H [55]	Test100 [56]	SOTS-indoor [54]	SOTS-outdoor [54]	GoPro [1]	HIDE [4]	RealBlur [7]
	100	100	100	500	500	1111	2025	1960

4.2 Implementation Details

We initially sampled 8 images from the training set and randomly cropped patches of size 256×256. Our TransDDPM employs a 4-level Encoder-Decoder U-net architecture. From level 1 to level 4, the number of transBlocks is [4, 6, 8], the number of attention heads in MXCA is [1, 2, 4, 8], and the number of channels is [32, 64, 128, 256]. The refinement stage contains 4 transBlocks. We use Adam optimizer with the initial learning rate of 2×10^{-4}, which is steadily decreased to $1 \times 10 - 6$ using the cosine annealing strategy.

4.3 Image Deraining Experiments

We trained TransDDPM on rain14000 [16],which contains 1,000 clean images and 14,000 corresponding synthesized rainy images with different streak orientations and magnitudes. Evaluation are performed on three widely used synthetic datasets: Test100 [56], Rain100H [55], Rain100L [55]. Table 2 Demonstrate TransDDPM achieves the best PSNR and SSIM on most dataset. We also provide the visual results in Fig. 4. DerainNet [3] and SEMI [5] failed to remove rain streaks. UMRL [11], RESCAN [13] and MSPFN [16] produce blurred details or unnatural artifacts. While MPRNet [23] and Restormer [31] suffer from smooth contents, TransDDPM can completely remove rain streaks while maintaining image details.

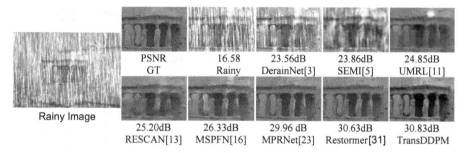

PSNR	16.58	23.56dB	23.86dB	24.85dB
GT	Rainy	DerainNet[3]	SEMI[5]	UMRL[11]
25.20dB	26.33dB	29.96 dB	30.63dB	30.83dB
RESCAN[13]	MSPFN[16]	MPRNet[23]	Restormer[31]	TransDDPM

Fig. 4. Visual results on image deraining.

4.4 Image Dehazing Experiments

We trained TransDDPM separately on ITS and OTS subsets of RESIDE [62] as the training datasets. Evaluation is conducted on SOTS-indoor (500 image pairs) and SOTS-outdoor (500 image pairs). As shown in Fig. 5, DCP [2] and GCANet [17] show color distortion and overexposure in the sky, while MSBDN [25], DeHamer [28] and Dehaze-Former [32] suffer from the details of Ground construction. By utilizing MXCA to capture Long-range Attention, TransDDPM achieves the best performance in the sky area and ground buildings simultaneously (Table 3).

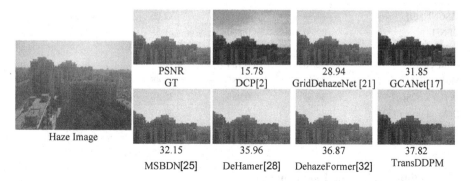

PSNR	15.78	28.94	31.85
GT	DCP[2]	GridDehazeNet [21]	GCANet[17]
32.15	35.96	36.87	37.82
MSBDN[25]	DeHamer[28]	DehazeFormer[32]	TransDDPM

Fig. 5. Visual results on image dehazing.

Table 2. Image deraining results. Best and second scores are **in-bold** and underlined.

Method	Test100		Rain100H		Rain100L	
	PSNR	*SSIM*	*PSNR*	*SSIM*	*PSNR*	*SSIM*
DerainNet [3]	22.77	0.810	14.92	0.592	27.03	0.884
SEMI [5]	22.35	0.788	16.56	0.486	25.03	0.842
DIDMDN [8]	22.56	0.818	17.35	0.524	25.23	0.741
UMRL [11]	24.41	0.829	26.01	0.832	29.18	0.923
RESCAN [13]	25.00	0.835	26.36	0.786	29.80	0.881
MSPFN [16]	27.50	0.876	28.66	0.860	32.40	0.933
MPRNet [23]	30.27	0.897	30.41	0.890	36.40	0.965
SPAIR [27]	30.35	0.909	30.95	0.892	36.93	0.969
Restormer [31]	32.00	0.923	31.46	0.904	**38.99**	0.978
TransDDPM	**32.52**	**0.926**	**31.89**	**0.917**	37.05	**0.979**

Table 3. Image dehazing results. Best and second scores are **in-bold** and underlined.

Method	SOT-indoor		SOTS-outdoor	
	PSNR	*SSIM*	*PSNR*	*SSIM*
DCP[2]	16.62	0.818	19.13	0.815
DehazeNet[6]	19.82	0.821	24.75	0.927
MSCNN[10]	19.84	0.833	22.06	0.908
GFN[14]	22.30	0.880	21.55	0.844
GCANet[17]	30.23	0.980	—	—
GridDehazeNet[21]	32.16	0.984	—	—
MSBDN[25]	33.67	0.985	33.48	0.982
DeHamer[28]	36.63	0.988	35.18	0.986
DehazeFormer [32]	37.84	0.994	34.95	**0.984**
TransDDPM	**37.93**	**0.995**	**35.04**	0.982

4.5 Motion Deblurring Experiments

We trained TransDDPM on GoPro [1], which contains 2,103 image pairs for training and 1,111 pairs for evaluation. We apply GoPro-trained model on HIDE [4] and RealBlur [7] datasets to demonstrate generalizability. To be noted the GoPro and HIDE datasets are synthetically generated while RealBlur are captured in real scene. Table 2. Demonstrate that TransDDPM achieves the best PSNR and SSIM on GoPro. Although TransDDPM only trained with GoPro, it still shows competitive ability with other methods on HIDE [4] and RealBlur [7]. From the visualization results in Fig. 6, TransDDPM are more clear and closer with ground truth compared with other methods (Table 4).

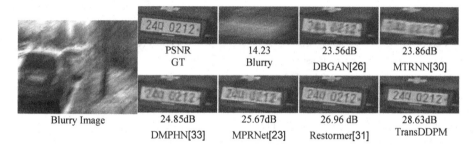

Fig. 6. Visual results on image Deblurring

Table 4. Image deblurring results. Best and second scores are **in-bold** and <u>underlined</u>.

Methods	GoPro [1]		HIDE [4]		RealBlur-R [7]		RealBlur-J [7]	
	PSNR	SSIM	PSNR	SSIM	PSNR	SSIM	PSNR	SSIM
DeblurGAN [12]	28.70	0.858	24.51	0.871	33.79	0.903	27.97	0.834
Nah [1]	28.70	0.858	24.51	0.871	33.79	0.903	27.97	0.834
DeblurGANv2 [18]	29.55	0.934	26.61	0.875	35.26	0.944	28.70	0.866
SRN [22]	30.26	0.934	28.36	0.915	35.66	0.947	28.56	0.867
DBGAN [26]	31.10	0.942	28.94	0.915	33.78	0.909	24.93	0.745
MTRNN [30]	31.15	0.945	29.15	0.918	35.79	0.951	28.44	0.862
DMPHN [33]	31.20	0.940	29.09	0.924	35.70	0.948	28.42	0.860
MIMO-UN++ [34]	32.45	0.957	29.99	0.930	35.54	0.947	27.63	0.837
MPRNet [23]	32.66	0.959	30.96	0.939	35.99	0.952	28.70	0.873
Restomer [31]	<u>32.92</u>	<u>0.961</u>	<u>31.22</u>	0.942	36.19	<u>0.957</u>	28.96	0.879
Uformer [37]	**33.06**	**0.967**	30.90	<u>0.953</u>	<u>36.19</u>	0.956	<u>29.09</u>	<u>0.886</u>
TransDDPM	**33.78**	0.949	**31.25**	**0.957**	**36.22**	**0.958**	**29.17**	**0.887**

4.6 Ablation Experiment

For the ablation experiment, we trained and tested TransDDPM on GoPro [1] for 100K iterations training, with image patches of size 128×128 for efficiency. In Table 5, we start with a simple U-net architecture as baseline, and gradually activate each component of TransDDPM (Table 5a). We observed that the PSNR increases by 0.81 dB after using the original self-attention (SA) at 16×16 resolutions (Table 5b). By replacing SA with MXCA, TransDDPM achieves PSNR gains of 0.43 dB (Table 5c). Table 5d shows that GFFN results in a gain of 0.49 over the original FFN. Finally, TransDDPM yields a significant gain of 2.33 over the baseline.

We empirically study the influence of implicit sampling steps for image deraining. As shown in Table 6, the PSNR gradually increase as the number of steps increases in

Table 5. Ablation experiments.

Network	component	PSNR (GoPro [1])
Baseline	(a) U-net	31.07
MXCA	(b) SA + FFN	31.88
	(c) MXCA + FFN	32.31
GFFN	(d)SA + GFFN	32.37
Overall	(f) MXCA + GFFN	33.40

Table 6. Influence of implicit sampling steps.

steps	Test100		Rain100H		Rain100L	
	PSNR	*SSIM*	*PSNR*	*SSIM*	*PSNR*	*SSIM*
t = 10	31.59	0.879	31.68	0.891	35.85	0.912
t = 25	32.52	0.926	31.89	0.917	37.05	0.979
t = 50	32.64	0.927	31.96	0.916	36.95	0.973
t = 100	32.57	0.935	31.90	0.919	37.02	0.972

the early steps. After $t > 25$, PSNR did not show significant changes under different steps. Therefore, we choose to a fixed choice of $t = 25$ for all the experiments.

4.7 Limitations

Although TransDDPM achieves excellent performance for image restoration, the main limitation is its inference speed, while other end-to-end image restoration networks require only a single forward inference. Although various sampling accelerations have been proposed, it still requires few steps to achieve satisfactory results. Besides, current diffusion models require paired dataset to be effectively trained, which is prohibitively expensive to be obtained in real scene. Therefore, TransDDPM is trained with synthesized datasets and suffers from domain shift between real scene and virtual reality.

5 Conclusion

We present a novel Transformer-based Denoising Diffusion Probabilistic Models (TransDDPM) for image restoration. Firstly, we propose multi-head cross-covariance attention, which can handle global self-attention with linear complexity. Secondly, TransDDPM is flexible to input image with agnostic size. Thirdly, TransDDPM can be conveniently transplanted to other task since not relying any task-specific architecture or parameters. Extensive experiments demonstrate that TransDDPM can handle several image restorations tasks effectively and efficiently, e.g. image deraining, image dehazing and motion deblurring.

References

1. Nah, S., Hyun Kim, T., Mu Lee, K.: Deep multi-scale convolutional neural network for dynamic scene deblurring. In: Proceedings of the IEEE Conference on Computer Vision and Pattern Recognition, pp. 3883–3891 (2017)
2. He, K., Sun, J., Tang, X.: Single image haze removal using dark channel prior. IEEE Trans. Pattern Anal. Mach. Intell. **33**(12), 2341–2353 (2010)
3. Fu, X., et al.: Clearing the skies: a deep network architecture for single-image rain removal. IEEE Trans. Image Process. **26**(6), 2944–2956 (2017)
4. Shen, Z., et al.: Human-aware motion deblurring. In: Proceedings of the IEEE/CVF International Conference on Computer Vision, pp. 5572–5581 (2019)
5. Wei, W., et al.: Semi-supervised transfer learning for image rain removal. In: Proceedings of the IEEE Conference on Computer Vision and Pattern Recognition, pp. 3877–3886 (2019)
6. Cai, B., et al.: DehazeNet: an end-to-end system for single image haze removal. IEEE Trans. Image Process. **25**(11), 5187–5198 (2016)
7. Rim, J., Lee, H., Won, J., Cho, S.: Real-world blur dataset for learning and benchmarking deblurring algorithms. In: Vedaldi, A., Bischof, H., Brox, T., Frahm, J.-M. (eds.) ECCV 2020. LNCS, vol. 12370, pp. 184–201. Springer, Cham (2020). https://doi.org/10.1007/978-3-030-58595-2_12
8. Zhang, H., Patel V.M.: Density-aware single image de-raining using a multi-stream dense network. In: Proceedings of the IEEE Conference on Computer Vision and Pattern Recognition, pp. 695–704 (2018)
9. Goodfellow, I., et al.: Generative adversarial nets. In: Advances in Neural Information Processing Systems, pp. 2672–2680 (2014)
10. Ren, W., Liu, S., Zhang, H., Pan, J., Cao, X., Yang, M.-H.: Single image dehazing via multi-scale convolutional neural networks. In: Leibe, B., Matas, J., Sebe, N., Welling, M. (eds.) ECCV 2016. LNCS, vol. 9906, pp. 154–169. Springer, Cham (2016). https://doi.org/10.1007/978-3-319-46475-6_10
11. Yasarla, R., Patel, V.M.: Uncertainty guided multi-scale residual learning-using a cycle spinning cnn for single image de-raining. In: Proceedings of the IEEE/CVF Conference on Computer Vision and Pattern Recognition, pp. 8405–8414 (2019)
12. Kupyn, O., et al.: DeblurGAN: blind motion deblurring using conditional adversarial networks. In: Proceedings of the IEEE Conference on Computer Vision and Pattern Recognition, pp. 8183–8192 (2018)
13. Li, X., et al.: Recurrent squeeze-and-excitation context aggregation net for single image deraining. In: Proceedings of the European Conference on Computer Vision (ECCV), pp. 254–269 (2018)
14. Ren, W., et al.: Gated fusion network for single image dehazing. In: Proceedings of the IEEE Conference on Computer Vision and Pattern Recognition, pp. 3253–3261 (2018)
15. Ho, J., Jain, A., Abbeel, P.: Denoising diffusion probabilistic models. In: Advances in Neural Information Processing Systems, vol. **33**, pp. 6840–6851 (2020)
16. Jiang, K., et al.: Multi-scale progressive fusion network for single image deraining. In: Proceedings of the IEEE/CVF Conference on Computer Vision and Pattern Recognition, pp. 8346–8355 (2020)
17. Chen, D., et al.: Gated context aggregation network for image dehazing and deraining. In: 2019 IEEE Winter Conference on Applications of Computer Vision (WACV), pp. 1375–1383. IEEE (2019)
18. Kupyn, O., et al.: Deblurgan-v2: Deblurring (orders-of-magnitude) faster and better. In: Proceedings of the IEEE International Conference on Computer Vision, pp. 8878–8887 (2019)

19. Dhariwal, P., Nichol, A.: Diffusion models beat GANs on image synthesis. In: Advances in Neural Information Processing Systems, vol. 34, pp. 8780–8794 (2021)
20. Ho, J., et al.: Cascaded diffusion models for high fidelity image generation. J. Mach. Learn. Res. **23**(47), 1–33 (2022)
21. Liu, X., et al.: GridDehazeNet: attention-based multi-scale network for image dehazing. In: Proceedings of the IEEE/CVF International Conference on Computer Vision, pp. 7314–7323 (2019)
22. Tao, X., et al.: Scale-recurrent network for deep image deblurring. In: Proceedings of the IEEE Conference on Computer Vision and Pattern Recognition, pp. 8174–8182 (2018)
23. Zamir, S.W., et al.: Multi-stage progressive image restoration. In: 2021 IEEE/CVF Conference on Computer Vision and Pattern Recognition, CVPR 2021, pp. 14816–14826 (2021)
24. Saharia, C., et al.: Image super-resolution via iterative refinement. IEEE Trans. Pattern Anal. Mach. Intell. **45**, 4713–4726 (2022)
25. Dong, H., et al.: Multi-scale boosted Dehazing network with dense feature fusion. In: Proceedings of the IEEE/CVF Conference on Computer Vision and Pattern Recognition, pp. 2157–2167 (2020)
26. Zhang, K., et al.: Deblurring by realistic blurring. In: Proceedings of the IEEE/CVF Conference on Computer Vision and Pattern Recognition, pp. 2737–2746 (2020)
27. Purohit, K., et al.: Spatially-adaptive image restoration using distortion-guided networks. In: Proceedings of the IEEE/CVF International Conference on Computer Vision, pp. 2309–2319 (2021)
28. Guo, C.-L., et al.: Image dehazing transformer with transmission-aware 3D position embedding. In: Proceedings of the IEEE/CVF Conference on Computer Vision and Pattern Recognition, pp. 5812–5820 (2022)
29. Ronneberger, O., Fischer, P., Brox, T.: U-net: Convolutional networks for biomedical image segmentation. In: Navab, N., Hornegger, J., Wells, W.M., Frangi, A.F. (eds.) MICCAI 2015. LNCS, vol. 9351, pp. 234–241. Springer, Cham (2015). https://doi.org/10.1007/978-3-319-24574-4_28
30. Park, D., Kang, D.U., Kim, J., Chun, S.Y.: Multi-temporal recurrent neural networks for progressive non-uniform single image deblurring with incremental temporal training. In: Vedaldi, A., Bischof, H., Brox, T., Frahm, J.-M. (eds.) ECCV 2020. LNCS, vol. 12351, pp. 327–343. Springer, Cham (2020). https://doi.org/10.1007/978-3-030-58539-6_20
31. Zamir, S.W., et al.: Restormer: efficient transformer for high-resolution image restoration. arXiv preprint arXiv:2111.09881 (2021)
32. Song, Y., et al.: Vision transformers for single image dehazing. arXiv preprint arXiv:2204.03883 (2022)
33. Zhang, H., et al.: Deep stacked hierarchical multi-patch network for image deblurring. In: Proceedings of the IEEE/CVF Conference on Computer Vision and Pattern Recognition, pp. 5978–5986 (2019)
34. Cho, S.-J., et al.: Rethinking coarse-to-fine approach in single image deblurring. In: Proceedings of the IEEE/CVF International Conference on Computer Vision, pp. 4641–4650 (2021)
35. Ali, A., et al., XCiT: cross-covariance image transformers. In: Advances in Neural Information Processing Systems, vol. 34, pp. 20014-20027 (2021)
36. Qian, R., et al.: Attentive generative adversarial network for raindrop removal from a single image. In: Proceedings of the IEEE Conference on Computer Vision and Pattern Recognition, pp. 2482–2491 (2018)
37. Wang, Z., et al.: Uformer: a general u-shaped transformer for image restoration. In: Proceedings of the IEEE/CVF Conference on Computer Vision and Pattern Recognition, pp. 17683–17693 (2022)

38. Wei, P., et al.: SIDGAN: single image dehazing without paired supervision. In :2020 25th International Conference on Pattern Recognition (ICPR), pp. 2958–2965 IEEE(2021)
39. Vaswani, A., et al.: Attention is all you need. In: Advances in Neural Information Processing Systems, vol. 30 (2017)
40. Chen, H., et al.: Pre-trained image processing transformer. In: Proceedings of the IEEE/CVF Conference on Computer Vision and Pattern Recognition, pp. 12299–12310 (2021)
41. Valanarasu, J.M.J., Yasarla, R., Patel, V.M.: TransWeather: transformer-based restoration of images degraded by adverse weather conditions. In: Proceedings of the IEEE/CVF Conference on Computer Vision and Pattern Recognition, pp. 2353–2363 (2022)
42. Liang, J., et al.: Swinir: Image restoration using swin transformer. In: Proceedings of the IEEE/CVF International Conference on Computer Vision, pp. 1833–1844 (2021)
43. Gao, T., et al.: Towards an effective and efficient transformer for rain-by-snow weather removal. Available at SSRN 4458244 (2023)
44. Peebles, W., Xie, S.: Scalable diffusion models with transformers. arXiv preprint arXiv:2212.09748 (2022)
45. Özdenizci, O., Legenstein, R.: Restoring vision in adverse weather conditions with patch-based denoising diffusion models. IEEE Trans. Pattern Anal. Mach. Intell. (2023)
46. Saharia, C., et al.: Palette: image-to-image diffusion models. In: ACM SIGGRAPH 2022 Conference Proceedings, pp. 1–10 (2022)
47. Whang, J., et al.: Deblurring via stochastic refinement. In: Proceedings of the IEEE/CVF Conference on Computer Vision and Pattern Recognition, pp. 16293–16303 (2022)
48. Lai, W.-S., et al.: Fast and accurate image super-resolution with deep laplacian pyramid networks. IEEE Trans. Pattern Anal. Mach. Intell. 41(11), 2599–2613 (2018)
49. Wu, H., et al.: CVT: introducing convolutions to vision transformers. In: Proceedings of the IEEE/CVF International Conference on Computer Vision, pp. 22–31 (2021)
50. Dauphin, Y.N., et al.: Language modeling with gated convolutional networks. in International conference on machine learning, pp. 933–941. PMLR (2017)
51. Shazeer, N.: GLU variants improve transformer. arXiv preprint arXiv:2002.05202 (2020)
52. Song, J., Meng, C., Ermon, S.: Denoising diffusion implicit models. arXiv preprint arXiv: 2010.02502 (2020)
53. Fu, X., et al.: Removing rain from single images via a deep detail network. In: Proceedings of the IEEE Conference on Computer Vision and Pattern Recognition, pp. 3855–3863 (2017)
54. Li, B., et al.: Benchmarking single-image dehazing and beyond. IEEE Trans. Image Process. 28(1), 492–505 (2019)
55. Yang, W., et al.: Deep joint rain detection and removal from a single image. In: Proceedings of the IEEE Conference on Computer Vision and Pattern Recognition, pp. 1357–1366 (2017)
56. Zhang, H., Sindagi, V., Patel, V.M.: Image de-raining using a conditional generative adversarial network. IEEE Trans. Circ. Syst. Video Technol. 30, 3943–3956 (2019)

One-Stage Wireframe Parsing in Fish-Eye Images

Zhengyang Guo[1,2], Ruqiang Huang[2], Zhongchen Shi[1,2(✉)], Wei Chen[1,2], Liang Xie[1,2], Ye Yan[1,2], and Erwei Yin[1,2]

[1] Defense Innovation Institute, Academy of Military Sciences (AMS), Beijing, China
shizhongchen@buaa.edu.cn

[2] Tianjin Artificial Intelligence Innovation Center (TAIIC), Tianjin, China

Abstract. This paper presents a simple yet efficient algorithm for detecting wireframes in a fish-eye image. Given a fish-eye image, our objective is to extract the semantically and geometrically salient lines and their corresponding junctions directly and without any undistortion processing. To this end, we introduce the circular-arc-based representation, converting the distorted line segments into a part of a circle by drawing inspiration from the projection characteristics of fish-eye lenses. Then a compact and end-to-end trainable model that tailors the redundant multi-module is realized and is used to directly output the vectorized wireframe in a one-stage fashion. Compared with previously considered the state-of-art two-stage method for line segment detection in the fish-eye image, our method achieves an average improvement of **2.8** in structural average precision, and boasts a remarkable increase in inference speed. Moreover, our method demonstrates universality when handling images captured by standard cameras, outperforming existing straight wireframe parsing methods.

Keywords: Line segment detection · Wireframe parsing · Fish-eye camera

1 Introduction

For man-made environments, line segments provide more structural information and robustness to lighting changes than points for the downstream vision tasks, such as image rectification [5], 3D structure reconstruction [8], pose estimation [19] and visual SLAM [23].

The previous studies have demonstrated that a large field-of-view is advantageous for vision-based motion estimation, especially in narrow indoor circumstance [25]. Consequently fish-eye camera has been adopted widely, which results in the emergence of another significantly intricate problem. Considering practical performance and efficiency, a common technique in these point-based approach

This work was supported in part by the grants from the National Natural Science Foundation of China under Grant 62332019 and 62076250.

is that keypoints are first detected in distorted images, followed by a sparse rectification according to pre-calibrated camera distortion parameters. At this aspect, line segments and point features exhibit distinct properties, with the former undergoing significant changes under fish-eye lens, leading to a sharp decline in the performance of general straight line segment detectors. Apart from the time-consuming operation of whole image rectification itself, obtaining an undistorted picture is highly difficult or even impossible for large field cameras. In other words, detecting bending line segments directly in distorted images often provide more relevance for downstream tasks.

On the other hand, deep neural networks have been utilized for line segment detection and achieved significant advancements. Compared to traditional line segment detection rely on underlying information such as image gradients, wireframe parsing based on deep convolutional networks exhibits more applicability, therefore seen as superior approach. Specifically speaking, wireframe refers to salient line segments and their junctions, which possess spatial structural significance, and the line segments demonstrate higher quality and stability under the constraint of junctions. As a quintessential representation of geometric forms, line segments appear to be quite straightforward to expound upon mathematically, nevertheless it has been shown challenging to parse wireframe end-to-end in rasterized images.

Due to the complex and variable nature of distorted wireframes, accurately parsing them remains a challenging task. The objective of our work is to develop a simple and accurate wireframe representation, and apply it to a one-stage detector which outputs directly and real-time for fish-eye cameras. When dealing with distorted wireframe parsing, a key issue that arises is determining how to effectively model and parameterize bending line segments. While the two-endpoint model is commonly used in straight line segment detectors due to its simplicity, it is not suitable for representing curved line segments in distorted images. To address this issue, we utilize arc-based models for representing curved line segments, which has been proven to be an effective approach [11]. Compared to previous two-stage method that use Bezier curve to model arbitrary line segment [14], our method achieves significant performance improvement and faster speedup. Due to the fact that straight lines can be regarded as circular arcs with infinite radii, our approach is also fully applicable to straight wireframe under standard camera imaging. Experimental results have demonstrated that, under the same backbone network, our method outperforms existing approaches on both major straight line datasets.

In summary, the main contributions of this paper are:

(1) We propose a simple and accurate wireframe representation based on the circular arc, where wireframe is represented non-parametrically by their endpoints and midpoint, and symmetry is utilized to improve performance while reducing parameters.

(2) Building upon the aforementioned representation, we design a easy-to-use network architecture, which can simply transform the output feature map of the backbone into curved wireframe. Consequently, we have achieved

a significant enhancement in the efficiency of distorted wireframe parsing, transitioning from a two-stage to a faster one-stage process.

(3) Our method also exhibits exceptional performance even on undistorted images. Compared to other methods that use the same backbone, our approach remains at the forefront as the state-of-the-art solution.

2   Related Work

2.1   Fish-Eye Line

As widely used cameras in industrial settings, fish-eye cameras offer a larger field of view than standard ones. However, they inevitably produce radial distortion. Initially observed, an intuitive and intriguing phenomenon is that the lines passing the center of the image do not appear to bend. Hence, in addition to camera modeling, the fish-eye distortion line itself has also become a research focus. On one hand, researchers strive to model the fish-eye camera as authentically as possible [2,12], so distortion parameters could be estimated by specific form of lines which computed by RANSAC [18] or Hough transform [1], and alternatively, directly output by utilizing deep neural networks [21]. On the other hand, another modeling method involves ignoring the optical characteristics of lenses, uses free-form curves such as Bezier curves to model distortion lines [14]. These methods that rely on free-form curves theoretically allow for the expression of arbitrary high-order curves [16], with accuracy improving as the order increases.

2.2   Wireframe Parsing

Wireframe parsing is an emerging concept that involves modeling and computing line segments and junctions in 2D image, which can be broadly categorized into two-stage and one-stage. The task of deep wireframe parsing was proposed first by Huang et al. [9]. They utilized stacked hourglass network [17] as backbone to extract heatmaps of junctions and lines, and a heuristic function to judge which junctions should be coupled. In addition, they established open datasets providing sufficient data volume for deep learning. L-CNN [24] inspired from the field of object detection [7,13], proposed LoI Pooling to generate fixed-length vectorized straight line representation. They also suggested structural AP, which had become a popular metric of wireframe parsing. To avoid detecting junctions, [20] presents a region-partition based representation for line segment maps, and thus poses wireframe parsing as the region coloring problem. These methods above which use a two-step strategy, relying on extra classifier or verifying module to deal with wireframe proposals, commonly referred to as two-stage ones. The two-stage methods achieve higher accuracy, but often have lower efficiency due to the need for additional sub-modules. Also inspired by detection, TP-LSD [10] suggest one-stage wireframe parsing. The primary issue that needs to be solved for one-stage wireframe parsing is how to parameterize line segments. TP-LSD proposed Tri-Points representation, i.e. the midpoint of a line segment and two displacements to endpoints, which greatly simplifies the network structure and enhances the real-time performance of wireframe parsing. Following this train of thought, some one-stage networks have been successively proposed. In addition to Tri-Points, some representations that utilize symmetry to decrease parameters have emerged, such as the incorporation of unilateral displacement [6] or geometric features like Angle and Length [3,22].

3   Method

3.1   Circular Arc Representation

In contrast to other wireframe parsing methods, we model distorted lines using circular arcs, which has been shown to be reasonable and accurate for fish-eye images [11]. Compared to Bezier curves, circular arcs have several advantages. Firstly, circular arcs or conic sections conform to the properties of the fish-eye model, which makes them simple and effective when there is no inconsistent stretching in the image. Secondly, for a circular arc segment, we only need two endpoints and another point on the arc, usually the midpoint for convenience to represent it accurately, and the extraction of endpoints and midpoints is relatively simple for CNNs. Thirdly, circular arcs have more explicit geometric properties such as curvature, symmetry, perpendicularity between radius and chord, etc., which provide more space for the design of expression methods.

Therefore, we use $l = (\mathbf{p_o}, \mathbf{p_l}, \mathbf{p_r})$ to denote a distorted wireframe, $\mathbf{p} \in \mathbb{R}^2$, where $\mathbf{p_o}$ is the midpoint of modeled arc, while $\mathbf{p_l}$ and $\mathbf{p_r}$ are two endpoints. To output a vectorized arc $(\mathbf{p_o}, \mathbf{p_l}, \mathbf{p_r})$ in a neural network and use a representative point in the feature map to characterize the entire arc, we designed three kinds of representation methods shown in Fig. 1 based on the properties of the arc. After analysis and experimentation, we chose the one in Fig. 1(b).

Tri-Points Arc. Referring to the previous one-stage approach, our first thought was to simply extend the Tri-Points Representation [10] to arcs. As shown in Fig. 1(a), for single wireframe, we predict the midpoint of $\mathbf{p_o}$ and two 2-D vectors $\boldsymbol{\alpha}$ and $\boldsymbol{\beta}$, which are displacements of endpoints $\mathbf{p_l}$ and $\mathbf{p_r}$ relative to $\mathbf{p_o}$, respectively. Regarding this kind of Tri-Points Arc, we have the following relationship:

$$\begin{cases} \mathbf{p_l} = \mathbf{p_o} + \boldsymbol{\alpha} \\ \mathbf{p_r} = \mathbf{p_o} + \boldsymbol{\beta} \end{cases} \tag{1}$$

Symmetry. The statement mentions that the use of symmetry in straight wireframe parsing has been validated to enhance performance [6]. The objective is to introduce axis symmetry (as circular arcs lack central symmetry) into the representation of circular arcs. In Fig. 1(b), we establish indirect connections between the endpoints and midpoint of the arc using the chord: $\boldsymbol{\alpha}$ indicates the

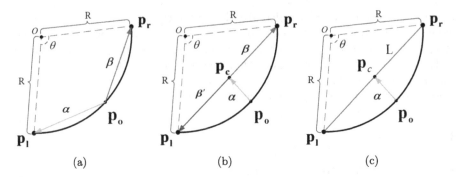

Fig. 1. Illustration shows proposed representations of circular arcs for wireframe parsing. (a): the midpoint $\mathbf{p_o}$ and two displacement vectors α and β are predicted to represent a single wireframe. (b): axis symmetry is introduced into the representation by establishing indirect connections between endpoints and midpoints of arcs using chords. (c): The length of the chord L is used to simplify one of the displacement vectors according to the perpendicular between the radius and chord length. However, this method requires sub-pixel accuracy for the midpoint position and inadequate for situations where $\alpha = 0$.

displacement from the midpoint of the arc to the midpoint of the chord, while β indicates the displacement from the midpoint of the chord to the right endpoint. Based on symmetry, there exists a displacement β' from the midpoint of the chord to the left endpoint, where $\beta' = -\beta$. We have the following relationship:

$$\begin{cases} \mathbf{p_l} = \mathbf{p_o} + \alpha + \beta' \\ \mathbf{p_r} = \mathbf{p_o} + \alpha + \beta \end{cases} \tag{2}$$

Redundant. The above methods require a feature map with 5 channels (1 for the center score map and 2 for each vector) to represent in a neural network, which is actually redundant. In fact, by using some implicit geometric relationships, we can reduce the number of channels. As shown in Fig. 1(c), by using the perpendicular between the radius and chord length, we can simplify β and represent it with scalar L, which denotes the chord length. However, simplification can also cause problems. When the arc tends towards a straight line, α may be particularly small or even zero, which will directly lead to inaccurate determination of the direction of β calculated by the perpendicular. On the other hand, this non-redundant method requires sub-pixel accuracy for the midpoint position, which is also arduous.

3.2   Network

Overall Network Architecture. As shown in Fig. 2, based our proposed Circular Arc Representation, the vectorized arcs can be represented directly in the final feature map. Therefore, we use a deconvolution layer for upsampling and

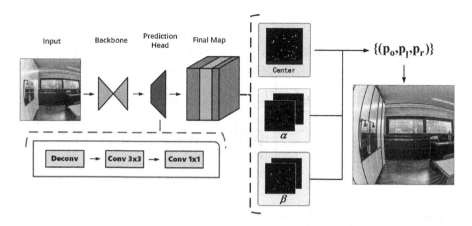

Fig. 2. An overview of our network. Based on the proposed Circular Arc Representation, the architecture is simple and lucid.

two convolution layers to reduce the channels, and then simply use the position of the center in the heatmap as the coordinate of p_o.

Backbone. In order to make comparisons with ULSD [14], we employ the stacked hourglass network [17] as the backbone. Apart from original hourglass network with two stacks (HG2), we employ simplified hourglass with one stack module to get a fast version (HG1). Furthermore, we have devised a lightweight version (Lite) that employs a lightweight backbone and utilizes only one deconvolutional layer to replace the original prediction head for accomplishing the tasks of upsampling and channel reduction. The parameters of the Lite version have been reduced by over 80% compared to the HG2 version. Therefore, it can be deployed on some edge computing devices.

Prediction Head. For an input image of size 512×512, the output size of the hourglass network is $128 \times 128 \times 128$. Given the sub-pixel level α and β in arc representation, complicated offsets were disregarded in our implementation on p_o. To enhance the accuracy of detecting midpoints in modeled arcs, we found using a learnable deconvolution layer provided the best results. Subsequently, we employed a 3×3 convolutional kernel followed by a 1×1 convolutional kernel to reduce the number of channels to 5.

3.3   Training and Inference

Label. For distorted wireframes, we sample hundreds of points on the lines and use method of least squares to fit circles. This provides us with the radius and center coordinates of the fitted circle, then the junctions will be employed to calculate the midpoints circular arcs, so that we can get displacements α and β.

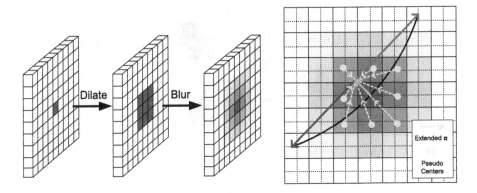

Fig. 3. Illustration of extending positive masks in within the center channel.

As depicted in Fig. 3, we utilize dilation and Gaussian blur techniques on the masks within the center score map. This approach enhances the network's ability to address imbalanced positive/negative data samples while simultaneously reducing errors stemming from manual annotations. In particular, we further extend the concept of alpha, which indicates the displacements of pseudo centers to the midpoints of real chords.

Loss Function. For one-stage wireframe, the problem of prediction centers can be approached as a pixel classification task. So we employ the Focal Loss [15] to the extended centers, which has demonstrated exceptional performance in object detection. The function is defined as follows:

$$
\mathcal{L}_C \doteq -\frac{1}{N} \sum_{i,j} \begin{cases} \alpha \left(1 - \hat{C}_{i,j}\right)^{\gamma} \log \left(\hat{C}_{i,j}\right) & \text{if } C_{i,j} = 1, \\ \left(1 - C_{i,j}\right) \hat{C}_{i,j}^{\gamma} \log \left(1 - \hat{C}_{i,j}\right) & \text{otherwise} \end{cases} \tag{3}
$$

where $\hat{C}_{i,j}$ is the probability of the center of circular arc, N is the total number of pixels of the center score map, while α and γ are the hyper-parameters of Focal Loss.

For regression problems involving geometrical values α and β, we use Smooth L1 Loss:

$$
\mathcal{L}_{\text{smooth L1}}(x) = \begin{cases} 0.5x^2 & \text{if } |x| < 1 \\ |x| - 0.5 & \text{otherwise} \end{cases} \tag{4}
$$

Non-maximum Suppression. During the inference phase, we only need to focus on the essential feature maps, which include the center score map and four displacement maps. Prior to generating the final map, we perform non-maximum suppression on the center score map with a kernel size of 3 to extract more accurate centers.

4   Experiments

4.1   Datasets and Metrics

For comparison purposes, we follow the methodology presented in [14] and train our network on the F-Wireframe datasets. These datasets comprise of 1900 equidistant images spanning 180 °C, which have been manually annotated, as well as 3100 synthetic images with randomly generated Kannala-Brandt parameters. We evaluate the performance of our model on F-Wireframe, which consist of 462 images.

Since the fish-eye model utilized by the aforementioned datasets cannot fully adhere to the requirements of circular arcs, we calculated the average pixel error on the training set to be 0.01 pixels, which is roughly equivalent to a 4-point fitted Bezier curve. We consider such an error to be tolerable.

The structural average precision (sAP) [24] is the mainstream metric for evaluating wireframe parsing, that is based on the sum of squared error between the predicted end-points and the ground truths. The threshold is set to $\vartheta = 5, 10, 15$ while the corresponding results denoted by sAP^5, sAP^{10}, sAP^{15}. To better evaluate the distorted wireframes, we also use multiple-points sAP [14], whose threshold ϑ is calculated by converting the SSE of multiple points into two endpoints.

4.2   Results

For fair comparisons, we employed the same hyper-parameters on backbone as ULSD, while keeping other unique components consistent with its paper. For multiple-points sAP, we followed the approach of ULSD utilizing seven equidistant points. For each input, the top 500 predictions with the highest confidence scores were selected. The Bezier curve order is set to 3 for ULSD because this order is close to our method w.r.t the pixel error. We adopt the representation depicted in Fig. 1(b). All efficiency experiments reported in this paper were conducted on an Nvidia RTX 4090.

Table 1 presents the results. The best scores are highlighted in **bold fonts**. Our method achieves state-of-the-art performance in both efficiency and accuracy. When using the same backbone, our method has a speed advantage that is twice as faster in efficiency. Due to the absence of a heavy prediction head, the advantage will increase continuously as the backbone becomes smaller. Our Lite version modified on the existing lightweight backbone can achieve the effect of a large model with a smaller number of parameters. In terms of accuracy, with the original Stacked Hourglasses model, our network shows substantial superiority on 6 different sAP metrics. Specifically, our method achieved an average improvement of 3.2 in the Multi-points sAP metric, which is more suitable for distorted wireframe. When the backbone is reduced, our method still outperforms ULSD greatly in other 5 kinds of metrics except for 2-points sAP^5.

Moreover, we evaluated our approach on the straight Wireframe datasets, ShanghaiTech [9] and YorkUrban [4]. We compared our method with other

Table 1. Quantitative Results On F-Wireframe Datasets

Method	2-Points sAP			Muilt-points sAP			FPS	Params(M)
	sAP^5	sAP^{10}	sAP^{15}	sAP^5	sAP^{10}	sAP^{15}		
ULSD(HG2)	52.7	57.9	60.2	54.7	59.7	61.7	70.0	10.00
ULSD(HG1)	48.8	54.0	56.4	50.9	55.6	58.0	94.4	6.86
Ours(HG2)	**53.8**	**60.9**	**63.7**	**57.1**	**63.0**	**65.5**	144.7	7.76
Ours(HG1)	47.8	55.6	59.0	51.7	58.4	61.2	235.8	4.62
Ours(Lite)	48.4	55.8	58.9	52.2	58.3	60.8	**276.5**	**1.53**

approaches that utilize similar backbone models such as the Stacked Hourglass, and the results are presented in Table 2. Due to the absence of open-source code for some methods, the reported results from their respective papers were used. Significantly, our method consistently outperformed all other methods, demonstrating superior performance across the board.

Table 2. Quantitative Results On Wirefame and YorkUrban Datasets

Method	ShanghaiTech			YorkUrban		
	sAP^5	sAP^{10}	sAP^{15}	sAP^5	sAP^{10}	sAP^{15}
L-CNN [24]	58.9	62.9	64.9	24.3	26.4	27.5
ULSD [14]	62.6	67.4	69.3	24.9	27.5	29.1
HAWP [20]	62.5	66.5	68.2	26.1	28.5	29.7
TP-LSD [10]	50.9	57.0	–	18.9	22.0	–
ELSD [22]	62.7	67.2	69.0	23.9	26.3	27.9
VLSE [6]	61.1	66.1	68.4	25.1	27.5	29.1
F-Clip [3]	61.3	65.8	–	27.2	29.4	–
Ours	**63.3**	**68.5**	**70.7**	**28.2**	**31.2**	**33.1**

As shown in Fig. 4, we visualized the P-R curves of aforementioned approaches under two sAP^{10} metrics. Scores below 0.1 are not plotted. Our method (HG2) outperforms ULSD in both 2-points sAP and multi-points sAP, with the advantage mainly reflected in recall.

To provide visual clarity, we present the qualitative results of proposed method and the compared methods in Fig. 5. As depicted in the figure, our approach yields results that are more closely aligned with the edges, while ULSD is more prone to extracting erroneous matches between junctions. Which are consistent with the quantitative results in Table 1.

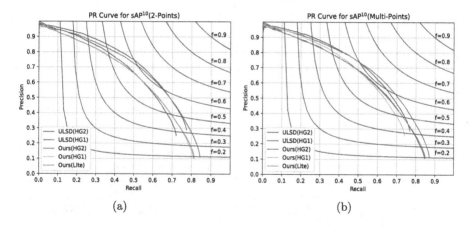

Fig. 4. Precision-Recall Curve for sAP^{10}

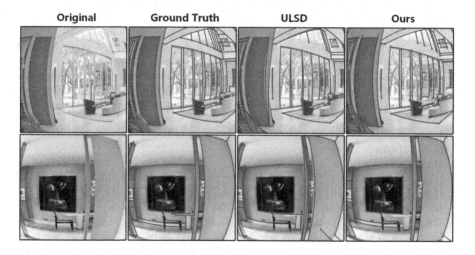

Fig. 5. Qualitative Results and Comparisons on F-Wireframe datasets.

5   Conclusions

In this paper, we propose a method for rapid detection of wireframes in fish-eye images. By employing an arc-based line segment representation, we extend the previous two-stage approach to a one-stage one, which greatly simplifies the model structure. We conducted experiments on both distorted line datasets and standard line datasets. Both quantitative and qualitative results demonstrate that our proposed method is superior in terms of efficiency and accuracy compared to previous methods, suggesting its promising potential for integration into real-time applications. Moving forward, our future vision centers around distorted wireframe parsing, and foster its implementation in practical real-world settings.

References

1. Alemán-Flores, M., Alvarez, L., Gomez, L., Santana-Cedrés, D.: Automatic lens distortion correction using one-parameter division models. Image Process. Line **4**, 327–343 (2014). https://doi.org/10.5201/ipol.2014.106
2. Bräuer-Burchardt, C., Voss, K.: A new algorithm to correct fish-eye- and strong wide-angle-lens-distortion from single images. In: IEEE International Conference on Image Processing, vol. 1, pp. 225–228 (2001). https://doi.org/10.1109/icip.2001.958994
3. Dai, X., Gong, H., Wu, S., Yuan, X., Yi, M.: Fully convolutional line parsing. Neurocomputing **506**, 1–11 (2022). https://doi.org/10.1016/j.neucom.2022.07.026
4. Denis, P., Elder, J.H., Estrada, F.J.: Efficient edge-based methods for estimating Manhattan frames in urban imagery. In: Forsyth, D., Torr, P., Zisserman, A. (eds.) ECCV 2008. LNCS, vol. 5303, pp. 197–210. Springer, Heidelberg (2008). https://doi.org/10.1007/978-3-540-88688-4_15
5. Fan, J., Zhang, J., Tao, D.: SIR: self-supervised image rectification via seeing the same scene from multiple different lenses. IEEE Trans. Image Process. **32**, 865–877 (2023). https://doi.org/10.1109/TIP.2022.3231087
6. Gao, S., et al.: Pose refinement with joint optimization of visual points and lines. In: IEEE International Conference on Intelligent Robots and Systems, pp. 2888–2894 (2022). https://doi.org/10.1109/IROS47612.2022.9981420
7. Girshick, R.: Fast R-CNN. In: 2015 IEEE International Conference on Computer Vision (ICCV), pp. 1440–1448 (2015). https://doi.org/10.1109/ICCV.2015.169
8. Hofer, M., Maurer, M., Bischof, H.: Improving sparse 3D models for man-made environments using line-based 3D reconstruction. In: 2014 2nd International Conference on 3D Vision, pp. 535–542. IEEE, Tokyo (2014). https://doi.org/10.1109/3DV.2014.14
9. Huang, K., Wang, Y., Zhou, Z., Ding, T., Gao, S., Ma, Y.: Learning to parse wireframes in images of man-made environments. In: Proceedings of the IEEE Computer Society Conference on Computer Vision and Pattern Recognition, pp. 626–635 (2018). https://doi.org/10.1109/CVPR.2018.00072
10. Huang, S., Qin, F., Xiong, P., Ding, N., He, Y., Liu, X.: TP-LSD: tri-points based line segment detector. In: Vedaldi, A., Bischof, H., Brox, T., Frahm, J.-M. (eds.) ECCV 2020. LNCS, vol. 12372, pp. 770–785. Springer, Cham (2020). https://doi.org/10.1007/978-3-030-58583-9_46
11. Hughes, C., McFeely, R., Denny, P., Glavin, M., Jones, E.: Equidistant fish-eye perspective with application in distortion centre estimation. Image Vis. Comput. **28**(3), 538–551 (2010). https://doi.org/10.1016/j.imavis.2009.09.001
12. Kannala, J., Brandt, S.: A generic camera model and calibration method for conventional, wide-angle, and fish-eye lenses. IEEE Trans. Pattern Anal. **28**(8), 1335–1340 (2006). https://doi.org/10.1109/TPAMI.2006.153
13. Law, H., Deng, J.: CornerNet: detecting objects as paired keypoints. Int. J. Comput. Vision **128**(3), 642–656 (2020). https://doi.org/10.1007/s11263-019-01204-1
14. Li, H., Yu, H., Wang, J., Yang, W., Yu, L., Scherer, S.: ULSD: unified line segment detection across pinhole, fisheye, and spherical cameras. ISPRS J. Photogramm. Remote. Sens. **178**(June), 187–202 (2021). https://doi.org/10.1016/j.isprsjprs.2021.06.004
15. Lin, T.Y., Goyal, P., Girshick, R., He, K., Dollar, P.: Focal loss for dense object detection. IEEE Trans. Pattern Anal. **42**(2), 318–327 (2020). https://doi.org/10.1109/TPAMI.2018.2858826

16. Mariotti, L., Eising, C.: Spherical formulation of geometric motion segmentation constraints in fisheye cameras. IEEE Trans. Intell. Transp. Syst. **23**(5), 4201–4211 (2022). https://doi.org/10.1109/TITS.2020.3042759

17. Kim, S., Park, K., Sohn, K., Lin, S.: Unified depth prediction and intrinsic image decomposition from a single image via joint convolutional neural fields. In: Leibe, B., Matas, J., Sebe, N., Welling, M. (eds.) ECCV 2016. LNCS, vol. 9912, pp. 143–159. Springer, Cham (2016). https://doi.org/10.1007/978-3-319-46484-8_9

18. Wildenauer, H., Micusik, B.: Closed form solution for radial distortion estimation from a single vanishing point. In: Proceedings of the British Machine Vision Conference 2013, pp. 106.1–106.11. British Machine Vision Association, Bristol (2013). https://doi.org/10.5244/C.27.106

19. Xu, C., Zhang, L., Cheng, L., Koch, R.: Pose estimation from line correspondences: a complete analysis and a series of solutions. IEEE Trans. Pattern Anal. **39**(6), 1209–1222 (2017). https://doi.org/10.1109/TPAMI.2016.2582162

20. Xue, N., et al.: Holistically-attracted wireframe parsing. In: Proceedings of the IEEE Computer Society Conference on Computer Vision and Pattern Recognition, pp. 2785–2794 (2020). https://doi.org/10.1109/CVPR42600.2020.00286

21. Xue, Z., Xue, N., Xia, G.S., Shen, W.: Learning to calibrate straight lines for fisheye image rectification. In: Proceedings of the IEEE Computer Society Conference on Computer Vision and Pattern Recognition, pp. 1643–1651 (2019). https://doi.org/10.1109/CVPR.2019.00174

22. Zhang, H., Luo, Y., Qin, F., He, Y., Liu, X.: ELSD: efficient line segment detector and descriptor. In: Proceedings of the IEEE International Conference on Computer Vision, pp. 2949–2958 (2021). https://doi.org/10.1109/ICCV48922.2021.00296

23. Zhou, L., Huang, G., Mao, Y., Wang, S., Kaess, M.: EDPLVO: efficient direct point-line visual odometry. In: Proceedings - IEEE International Conference on Robotics and Automation, pp. 7559–7565 (2022). https://doi.org/10.1109/ICRA46639.2022.9812133

24. Zhou, Y., Qi, H., Ma, Y.: End-to-end wireframe parsing. In: 2019 IEEE/CVF International Conference on Computer Vision (ICCV), pp. 962–971. IEEE (2019). https://doi.org/10.1109/ICCV.2019.00105

25. Zichao, Z., Rebecq, H., Forster, C., Scaramuzza, D.: Benefit of large field-of-view cameras for visual odometry. In: 2016 IEEE International Conference on Robotics and Automation (ICRA), pp. 801–808. IEEE, Stockholm, Sweden (2016). https://doi.org/10.1109/ICRA.2016.7487210

Out-of-Distribution with Text-to-Image Diffusion Models

Jinglin Tong and Longquan Dai[✉]

Nanjing University of Science and Technology, Nanjing, China
{tongjl,longquandai}@njust.edu.cn

Abstract. Out-of-distribution detection, identifying unexpected data from the known concepts, is essential for reliable machine learning. We present a novel method that explores the application of a text-to-image diffusion model for out-of-distribution detection. Our method is motivated by the fact that the text-to-image diffusion model has shown remarkable capability in generating high-quality images with diverse text descriptions. The text description generates a corresponding text embedding and is injected into the diffusion model to affect image generation. This demonstrates that its internal representation contains semantic information and is highly enhanced by text concepts. This inspires us to apply the diffusion model to extract image representations with suitable text embeddings. In addition, we noticed that describing images directly using native text is often vague and lacking in detail. Thus, we propose an implicit captioner to generate text embeddings for the input images. Subsequently, a compression head is introduced to compress the representations, facilitating easy comparison and removal of noise information. We formulate the proposed text-to-image diffusion model, implicit captioner, and compression head into a network, which we call ODDM: **O**ut-of-distribution **D**etection with Text-to-Image **D**iffusion **M**odels. Several experiments shows that our method can achieved superior performance.

Keywords: out-of-distribution detection · diffusion model · contrastive learning

1 Introduction

Out-of-distribution (OOD) detection is a task that aims to identify samples or instances that come from a distribution different from the training data. The key challenge in OOD detection lies in distinguishing between in-distribution (ID) samples, which are representative of the training data, and OOD samples, which exhibit variations or patterns not seen during training. This differentiation is difficult because OOD samples may introduce novel or rare features that the model has not been exposed to, making them harder to classify accurately. Therefore, developing effective OOD detection methods requires robust generalization capabilities to handle diverse and unforeseen data patterns beyond the ID distribution. One possible approach is to directly apply pre-trained models to obtain the features with rich semantic information.

© The Author(s), under exclusive license to Springer Nature Singapore Pte Ltd. 2024
Q. Liu et al. (Eds.): PRCV 2023, LNCS 14435, pp. 276–288, 2024.
https://doi.org/10.1007/978-981-99-8552-4_22

Recently, diffusion models have revolutionized the field of image synthesis, which are inspired by equilibrium thermodynamics. Numerous studies have explored its applications in many domains, such as inpainting [31], segmentation [2,41], semantic editing [36,39] and super-resolution [11,28]. The text-to-image diffusion models, trained on large-scale data, have demonstrated a remarkable ability to generate a variety of high-quality images controlled via diverse text descriptions. The provided text description is encoded into a corresponding text embedding, which is injected into the diffusion model to affect image generation. It demonstrates that their internal representation contains abundant semantic information and is highly enhanced by text concepts. This inspires us to use the text-to-image diffusion models to extract image representation with suitable text embedding.

However, directly using natural language to describe images often leads to ambiguity and lacks specific details. This is because in OOD detection, the images do not have corresponding text description information. Moreover, OOD samples are unpredictable and there is a semantic gap between text and image. Thus, we propose an implicit captioner to generate text embedding of the input images to enhance the image diffusion representation. Subsequently, due to the large shape and inclusion of redundant background information in the representation, we employ a compression head to compress the representation and get the final image features.

In this paper, we present ODDM: **O**ut-of-distribution **D**etection with Text-to-Image **D**iffusion **M**odels. Our method leverages a combination of a text-image diffusion model, an implicit captioner, and a compression head to effectively implement OOD detection. An overview of our approach is illustrated in Fig. 1. To begin, we utilize the implicit captioner to generate the implicit text embeddings for the input images. The images, along with the corresponding text embeddings, are fed into the diffusion model to obtain a comprehensive representation. Subsequently, we compress the representation using a compression head. Our approach has yielded good results on several datasets.

The contributions of our works are as follows:

1. We propose an OOD detection method based on text-to-image diffusion models. The text-to-image diffusion model is applied to obtain the representation of the input image which contains rich semantic information and can be well-differentiated.
2. Our method uses an implicit captioner to obtain an implicit text embedding to enhance the representation get from diffusion. And a compression head is proposed to compress the representation and eliminate the redundant information to make it easy to be applied.
3. Extensive experiments show that our proposed method outperforms existing methods on several widely-used OOD detection datasets. At the same time we demonstrate the effectiveness of the implicit captioner module through ablation experiments.

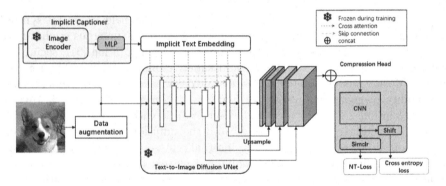

Fig. 1. ODDM Overview. First, we apply data augmentation to the input image, generating multiple variations for contrastive learning. Next, we encode the input image using an implicit captioner to create an implicit text embedding. We then extract the representation from the UNet architecture of a frozen text-to-image diffusion model, using both the image and its implicit text embedding as inputs. The UNet representation is further compressed using a compression head. We obtain a similarity matrix from the output of the SimCLR layer and apply the NT-Xent loss. Additionally, we employ cross-entropy loss between the output of the shift layer and the class of the image's strong data augmentations. The red blocks indicate the blocks where weight updates are needed, while the blue blocks represent the frozen ones.

2   Related Work

2.1   OOD Detection

Classification-based methods [13,24,30] fit the distribution of ID samples with a classification surface, and usually requires building additional negative samples. It is challenging due to the complex boundary separating OOD and ID classes. Reconstruction-based methods attempt to reconstruct the input, then, the difference between the images before and after reconstruction is analyzed to detect OOD [21,25,43,44]. They may suffer from overfitting with a small number of samples, and can significantly impact the generalization ability. Conversely, with high sample diversity, they may exhibit excessive adaptability. Metrics-based methods commonly utilize a deep neural network to extract the features from ID samples normal data, and minimize the intra-class distance of similar data [3,20,27,29,37]. Those methods need to model the representation to better encode ID, which is a challenge to achieve in the absence of OOD samples.

2.2   Diffusion Models

Diffusion models have emerged as the new state-of-the-art family of deep generative models, exhibiting great performance in several applications. Most current research are based on three predominant formulations: denoising diffusion probabilistic models(DDPMs) [17,22] , score-based generative models(SGMs) [34,35], and stochastic differential equations(Score SDEs) [33,36]. DDPMs utilize two

Markov chains, a forward chain to perturb data with noise and a reverse chain to convert noise back to data. For SGMs, data is perturbed with a sequence of intensifying Gaussian noise, and the score functions for all noise data distributions are jointly estimated by training a model conditioned on noise levels. In contrast, Score SDEs perturb data to noise with a diffusion process governed by stochastic differential equation. The stable diffusion model [28] is an application of DDPMs in the field of text-to-image generation. It controls the generation of images by text embedding. Which are obtained through the CLIP [26] model and injected into the intermediate representation of the diffusion model through the cross-attention mechanism.

2.3   Contrastive Learning

Contrastive Learning is self-supervised learning method that enables a model to map similar data samples to a close space and dissimilar data samples to a distant space. Its primary objective is to learn a representation that can capture the characteristics of the data without utilizing any label information. [7] maximizes the similarity between the original image and different, minimizes multiple variations of the same image which are obtained by data augmentation. [8,15] suggest using the momentum contrast mechanism, wherein the query encoder learns representations from a slow key encoder. Other approaches [1,6,19] are based on clustering, enforce consistency when cluster assignments are obtained from different augmented views of the same image. These approaches encourage the model to form clusters of similar representations and facilitate discriminative feature learning.

3   Proposed Method

Our approach utilizes an implicit captioner to encode the input image into embedding space and fed it into diffusion model. Then we extract the representation of the input images from the text-to-image diffusion model and converted into the final image feature by compression head. In this section, we sequentially provide a detailed introduce to OOD scores, loss function and network structure.

3.1   OOD Score

The OOD score is utilized to assign scores to query images during the test phase. A higher score indicates a higher probability of the test image belonging to the in-distribution samples, while OOD samples receive lower scores. Since we know that the training dataset is in-distribution, we determine whether a query image is an OOD sample based on its distance from the training dataset images. Hence, we consider this distance as the OOD score.

We compute the features f_y of the test image y and compare them individually with the features f_x of each training data instances $x \in \{B\}$. The final score of the image is determined by selecting the maximum similarity among these sets of features, as defined in Eq. 1. A lower score indicates a closer resemblance to the OOD image.

$$\text{score}(y) = \max_{x \in \{B\}} \left[\text{sim}(f_y, f_x) \right] \tag{1}$$

Here, the sim function is the cosine similarity between two features.

In addition, we divide the adopted data augmentation methods into a weak data augmentation and a set of strong data augmentations, which is detailed in 3.3. Following [37], we apply each strong data augmentation to test images and training dataset, calculate the test image score respectively, and take the highest score as the final score of the image.

3.2   Training Loss

In our approach, we train our networks by contrastive learning. Given a query image x, $\{X^+\}$ represents a set of positive pair samples, and $\{X^-\}$ represents a set of negative pair samples. We employ the NT-Xent loss [7] as our contrastive learning loss, which is represented as follows:

$$L_{\text{sim}} = -\frac{1}{|\bar{B}|} \sum_{x \in \{\bar{B}\}} \log \frac{\sum_{y \in \{X^+\}} \exp(\text{sim}(f_x, f_y)/\tau)}{\sum_{z \in \{X^+\} \cup \{X^-\}} \exp(\text{sim}(f_x, f_z)/\tau)} \tag{2}$$

Here, $\{\bar{B}\}$ denotes the augmentation dataset and $|\bar{B}|$ denotes the cardinality of $\{\bar{B}\}$. f_x denotes the final feature of the input image x, τ is a temperature hyper-parameter that controls the level of attention given to both positive and negative examples.

At the same time, we introduce a classification subtask to judge the type of strong data augmentation used in the image, which utilize the cross-entropy loss:

$$L_{cls} = -\frac{1}{|\bar{B}|} \sum_{|\bar{B}|}^{i=1} \sum_{M}^{j=1} y_{ij} \log(p_{ij}) \tag{3}$$

where M is the number of augmentation classes, y_{ij} represents the probability of the true label for the i-th sample belonging to class j (ont-hot encoding), and p_{ij} represents the predicted probability of the i-th sample belonging to class j.

We optimize the model by minimizing the overall loss function L:

$$L = L_{\text{sim}} + \lambda L_{\text{cls}} \tag{4}$$

3.3   Network Architecture

Text-to-Image Diffusion Model. The text-to-image diffusion models consists of two Markov chains: a forward chain and a reverse chain. The forward chain is a process of continuously adding Gaussian noise $\epsilon \in \mathcal{N}(0, \mathcal{I})$ to the image, gradually transforming its distribution into a Gaussian distribution. This chain does not involve any learnable parameters and can be formulated as follows:

$$q(x_t|x_0) = \mathcal{N} \left(x_t; \sqrt{\bar{\alpha}_t}x_0, (1 - \bar{\alpha}_t)I\right) \tag{5}$$

$$x_t = \sqrt{\bar{\alpha}_t}x_0 + \sqrt{1 - \bar{\alpha}_t}\epsilon, \quad \epsilon \in N(0,1) \tag{6}$$

In the above equation, the variable $t \in \{0, 1, ..., T\}$ represents the diffusion time step, and the $\bar{\alpha}_t$ is cumulative noise variance schedule.

The reverse process aims to invert the forward process and restore the data from Gaussian noise step by step:

$$p(x_{t-1}|x_t) = \mathcal{N} \left(x_{t-1}; \mu_\theta(x_t, t, z), \widetilde{\beta}_t I\right), \tag{7}$$

$$\mu_\theta(x_t, t, z) = \frac{1}{\sqrt{\alpha_t}}(x_t - \frac{1 - \alpha_t}{\sqrt{1 - \bar{\alpha}_t}}\epsilon_\theta(x_t, t, z)) \quad \text{and} \quad \widetilde{\beta}_t := \frac{1 - \bar{\alpha}_{t-1}}{1 - \bar{\alpha}_t}\beta_t \tag{8}$$

In the reverse process, z is the text embedding obtained from the text corresponding to the image. Instead of directly predicting the mean $\mu_\theta(x_t, t, z)$ of the distribution, a network parameterized by the UNet architecture, denoted as $\epsilon_\theta(x_t, t, z)$, is used to predict the noise component in the image x_t under the influence of the text embedding z.

The UNet architecture comprises multiple convolutional blocks, including skip connections, downsampling and upsampling blocks. The text embedding z is projected onto the intermediate representation of UNet architecture using cross-attentions mechanism, which encourages the intermediate representation to be enhanced with text embedding. In our approach, we obtain the intermediate representation of the input image from the upsampling blocks through a single forward pass using the text-to-image diffusion model. This is different from traditional diffusion models that require multiple steps for generation.

Implicit Captioner. It is important to note that the representations generated by diffusion models exhibit strong correlations with their corresponding textual descriptions. Hence, in cases where there is no available textual description for an image, we employ an implicit captioner to generate the text description.

The implicit captioner consists of a pre-trained image encoder and an MLP block. The image encoder encodes the input image into its embedding space, while the learnable MLP projects the image embedding into the embedding space for implicit text. The resulting implicit text embedding is then fed into the text-to-image diffusion UNet. During the training stage, the parameters of the image encoder and UNet architecture are frozen, and we solely fine-tune the parameters of the MLP.

In brief, the representation r of input image x we get from the text-to-image diffusion model can be expressed by the following formula:

$$r = \text{UNet}(x_t, \text{ImplicitCaptioner}(x)) \tag{9}$$

Table 1. One-class experiment in CIFAR-10 dataset (AUROC%)

Method	Plane	Car	Bird	Cat	Deer	Dog	Frog	Horse	Ship	Truck	Mean
OC-SVM [10]	65.6	40.9	65.3	50.1	75.2	51.2	71.8	51.2	67.9	45.5	58.5
DeepSVDD [29]	61.7	65.9	50.8	59.1	60.9	65.7	67.7	67.3	75.9	73.1	64.8
OCGAN [25]	75.7	53.1	64.0	62.0	72.3	62.0	72.3	57.5	82.0	55.4	65.6
DROCC [13]	81.6	76.7	66.6	67.1	73.6	74.4	74.4	71.3	80.0	76.2	74.2
Uninform [4]	78.9	84.9	73.4	74.8	85.1	79.3	89.2	83.0	86.2	84.8	82.0
IGD [9]	86.8	87.0	73.8	71.6	85.5	76.6	89.0	87.1	89.8	89.9	83.4
MKD [32]	90.5	90.3	79.6	77.0	86.7	91.4	88.9	86.7	91.4	88.9	87.2
GOAD [3]	77.2	96.7	83.3	77.7	87.8	87.8	90.0	96.1	93.8	92.0	88.2
Rot [16]	78.3	94.3	86.2	80.9	89.4	89.0	88.9	95.1	92.3	89.7	88.5
ODDM(ours)	78.7	93.9	85.9	86.0	93.2	88.8	91.6	96.1	90.4	93.4	89.8

Compression Head. The representations r obtained from diffusion model usually have large shape and contain redundant information. We propose a compression head, which employs contrastive learning mechanism to compress the representation and make the representation focus on information common to ID samples.

We treat each image in the training dataset as a separate category, and create multiple images belonging to the same category by applying weak data augmentations W to the original image. It involves making subtle and random changes to an image, with variations applied each time.

We also apply a set of strong data augmentations which are deterministic transformation methods capable of producing more significant changes to the image, denote as $S := \{S_0, S_1, \ldots, S_{M-1}\}$. Each strong data augmentation S_m represents a deterministic transformation method. We consider the output image as a new category and introduce a classification sub-task [37,40,42] to determine which strong data augmentation was used on the image.

Thus, we process the input image in the following formula:

$$\bar{x}_m = W(S_m(x)) \tag{10}$$

In short, when given two images x and y from set $\{B\}$, and we obtain augmented images $\bar{x}_m$ and $\bar{y}_n$ from the set of augmented images $\{\bar{B}\}$. If x and y are identical ($x = y$) and the corresponding augmentation indices are the same ($m = n$), they form a positive pair in contrastive learning. Otherwise, they are considered a negative pair. Formally:

$$(\bar{x}_m, \bar{y}_n) := \begin{cases} \text{positive-pair} & x = y \text{ and } n = m \\ \text{negative-pair} & \text{otherwise} \end{cases} \tag{11}$$

The compression head takes a convolutional neural network module as main body. Then there is a simclr layer responsible for contrastive learning and a shift layer dedicated to the classification sub-task. The resulting feature f_x of the input image x is obtained from the output of the simclr layer.

4 Experiments

4.1 Implementation Details

We use the stable diffusion [28] model pre-trained on LAION-2B dataset as our text-to-image diffusion. We take the output of every third upsampling block in the UNet structure as the representation, which then upsampled to a consistent size and concatenated together. The compression head comprises four blocks, with each block consisting of a convolutional layer using a 3×3 kernel, followed by a ReLU layer and a batch normalization layer. The MLP block consists of linear layer. Notably, the entire model has approximately 3.9 million trainable parameters. For strong data augmentation, we employ the rotation transforms with angle of $0°, 90°, 180°, 270°$, which have been shown to be effective in previous studies [16,37]. For weak data augmentation, we employ a combination of random horizontal flips, random resized crops, random color jitter, random grayscale, and Gaussian blur.

4.2 Results

In order to assess the effectiveness of our proposed model, we evaluate its performance using the widely accepted metric known as the Area Under the Receiver Operating Characteristic Curve (AUROC). This metric provides a threshold-free evaluation of the detection score, making it particularly valuable in our analysis. A higher AUROC value indicates a more favorable performance outcome for the method under consideration. It is worth emphasizing that all AUROC values reported in this article are expressed as percentages.

Table 2. Average (AUROC%) of thirty one-class experiments on the ImageNet-30

Method	Mean
Rot [16]	65.3
Rot+Trans [16]	77.9
Rot+Attn [16]	81.6
Rot+Trans+Attn [16]	84.8
Rot+Trans+Attn+Resize [16]	85.7
CSI [37]	91.6
ODDM(ours)	91.8

One-Class Datasets. We start by considering the one-class setup, here, for a given multi-class dataset, we conduct multiple one-class classification task, and each task chooses one of the classes as in-distribution while the remaining classes being out-of-distribution. There are not any labels or OODabnormal samples data in training stage. We evaluate our method on CIFAR-10 [18] and ImageNet-30 datasets. CIFAR-10 contains ten natural objects is a commonly used dataset in OOD detection. ImageNet-30, following [16], is a subset of ImageNet1k and contains 30 classes. Table. 1 presents the results of 10 experiments conducted on the CIFAR-10 dataset along with the average values obtained. Table. 2 displays the average results of 30 experiments conducted on the ImageNet-30 dataset. We compare ODDM with various priori methods, and its performance is superior to other methods.

Table 3. Mulit-class experiment in ImageNet-30 dataset (AUROC%)

	ImageNet-30					
	cub	flowers-102	food-101	caltech-256	DTD	Mean
Rot [16]	76.5	87.2	72.7	70.9	89.9	79.5
Rot+Trans [16]	74.5	86.3	71.6	70.0	89.4	78.4
GOAD [3]	71.5	82.8	68.7	67.4	87.5	75.6
CSI [37]	90.5	94.7	89.2	87.1	96.9	91.6
ODDM(ours)	85.9	98.2	88.3	77.7	97.6	89.6

Table 4. Ablation results of Implicit Captioner (AUROC%)

	Cifar-10	ImageNet-30
ODDM+Text	82.3	82.7
ODDM+CNN	88.6	87.9
ODDM+CLIP	89.8	91.8

Muliti-class Datasets. In this setup, we assume that in-distribution samples are from a specific multi-class dataset without labels, testing on various external datasets as out-of-distribution. We consider the ImageNet-30 dataset as the in-distribution dataset, and CUB-200 [38], Flowers-102 [23], Food-101 [5], Caltech-256 [14], and DTD [12] as the out-of-distribution datasets. For Food-101, we remove hotdog class to avoid overlap. As show in Table. 3, our method can achieve great results.

4.3 Ablation Study

Implicit Captioner. To validate the functionality of our implicit captioner, we propose three approaches for its construction. The first approach, ODDM+Text, involves utilizing the original pre-trained text conditional module CLIP from stable diffusion. We employ the fixed string "a photo of *" as the text description. In the second approach, ODDM+CNN, we employ a simple convolutional neural network to process the image and generate an implicit text embedding. The third approach, ODDM+CLIP, employs a pre-trained CLIP model, identical to the original text conditional module in stable diffusion, to process the image into embedding space. Additionally, we incorporate an MLP block to project the image embedding to the implicit text embedding, as illustrated in Fig. 1.

As shown in Table 4, we report the mean AUROC values in one-class experiments on CIFAR-10 and ImageNet30 datasets, and the approach using images to generate text embeddings are able to achieve better results, especially in ODDM-CLIP. In our opinion, CLIP is a model that connects text and image, which maps text and image to similar embedding Spaces, and the implicit text embedding generated by CLIP are much closer to the text embeddings used in

Table 5. Ablation results of diffusion time step (AUROC%)

	acorn	airliner	ambulance	alligator	banjo	barn
t = 0	89.4	99.5	99.6	95.7	96.0	99.3
t = 100	86.5	98.0	99.5	93.7	94.2	97.7
t = 200	88.2	99.1	99.1	95.7	93.6	99.0
t = 300	81.1	98.1	98.2	93.4	94.8	98.7
t = 400	78.8	95.5	95.8	89.3	94.4	96.9

the diffusion model. And CLIP as a pre-trained model, it can better represent images than the network trained in limited images.

Diffusion Time Steps. This section we study which diffusion step is more effective for extracting representation. In stable diffusion, t ranges from 0 to 1000, the larger the value of t, the more noise is added to the image. The noise process is defined in Eq. 6. For a fair comparison, we added a fixed noise to each image. We performed one-class experiment on ImageNet-30, and the results on the first 6 classes are shown in the Table 5. When $t = 0$, we can get the best results in most experiments.

5   Conclusion

In this paper, we propose a simple yet effective out-of-distribution detection method named ODDM, which contains of text-to-image diffusion model, implicit captioner and compress head. We find that the internal representation of text-to-image diffusion model contains semantic information. And we apply the implicit captioner to generate the implicit text embedding, which is injected into diffusion model to enhanced the semantic representation. Our work achieves great result and shows the great potential of text-to-image generation models in the representation learning.

Acknowledgments. This work is supported in part by the National Science Foundation of China (62072238 and 62372237), and the Fundamental Research Funds for the Central Universities (30922010911).

References

1. Asano, Y.M., Rupprecht, C., Vedaldi, A.: Self-labelling via simultaneous clustering and representation learning. In: International Conference on Learning Representations (2019)
2. Baranchuk, D., Rubachev, I., Voynov, A., Khrulkov, V., Babenko, A.: Label-efficient semantic segmentation with diffusion models. In: International Conference on Learning Representations (2022)

3. Bergman, L., Hoshen, Y.: Classification-based anomaly detection for general data. In: International Conference on Learning Representations (2020)

4. Bergmann, P., Fauser, M., Sattlegger, D., Steger, C.: Uninformed students: student-teacher anomaly detection with discriminative latent embeddings. In: Computer Vision and Pattern Recognition (2020)

5. Bossard, L., Guillaumin, M., Van Gool, L.: Food-101 – mining discriminative components with random forests. In: Fleet, D., Pajdla, T., Schiele, B., Tuytelaars, T. (eds.) ECCV 2014. LNCS, vol. 8694, pp. 446–461. Springer, Cham (2014). https://doi.org/10.1007/978-3-319-10599-4_29

6. Caron, M., Bojanowski, P., Joulin, A., Douze, M.: Deep clustering for unsupervised learning of visual features. In: Ferrari, V., Hebert, M., Sminchisescu, C., Weiss, Y. (eds.) Computer Vision – ECCV 2018. LNCS, vol. 11218, pp. 139–156. Springer, Cham (2018). https://doi.org/10.1007/978-3-030-01264-9_9

7. Chen, T., Kornblith, S., Norouzi, M., Hinton, G.: A simple framework for contrastive learning of visual representations. In: International Conference on Machine Learning (2020)

8. Chen, X., Fan, H., Girshick, R., He, K.: Improved baselines with momentum contrastive learning. In: Computer Vision and Pattern Recognition (2020)

9. Chen, Y., Tian, Y., Pang, G., Carneiro, G.: Deep one-class classification via interpolated Gaussian descriptor. In: Association for the Advancement of Artificial Intelligence (2022)

10. Chen, Y., Zhou, X.S., Huang, T.S.: One-class SVM for learning in image retrieval. In: International Conference on Image Processing (2001)

11. Chung, H., Sim, B., Ye, J.C.: Come-closer-diffuse-faster: accelerating conditional diffusion models for inverse problems through stochastic contraction. In: Computer Vision and Pattern Recognition (2022)

12. Cimpoi, M., Maji, S., Kokkinos, I., Mohamed, S., Vedaldi, A.: Describing textures in the wild. In: Computer Vision and Pattern Recognition (2014)

13. Goyal, S., Raghunathan, A., Jain, M., Simhadri, H.V., Jain, P.: DROCC: deep robust one-class classification. In: International Conference on Machine Learning (2020)

14. Griffin, G., Holub, A., Perona, P.: Caltech 256 (2007)

15. He, K., Fan, H., Wu, Y., Xie, S., Girshick, R.: Momentum contrast for unsupervised visual representation learning. In: Computer Vision and Pattern Recognition (2020)

16. Hendrycks, D., Lee, K.: Using pre-training can improve model robustness and uncertainty. In: International Conference on Machine Learning (2019)

17. Ho, J., Jain, A., Abbeel, P.: Denoising diffusion probabilistic models. In: Neural Information Processing Systems (2020)

18. Krizhevsky, A., Hinton, G.: Learning multiple layers of features from tiny images. In: Handbook of Systemic Autoimmune Diseases (2009)

19. Li, J., Zhou, P., Xiong, C., Hoi, S.C.: Prototypical contrastive learning of unsupervised representations. In: International Conference on Learning Representations (2020)

20. Liznerski, P., Ruff, L., Vandermeulen, R.A., Franks, B.J., Kloft, M., Müller, K.R.: Explainable deep one-class classification. In: International Conference on Learning Representations (2021)

21. Mei, S., Yang, H., Yin, Z.: An unsupervised-learning-based approach for automated defect inspection on textured surfaces. IEEE Trans. Instrum. Meas. **67**(6), 1266–1277 (2018). https://doi.org/10.1109/TIM.2018.2795178

22. Nichol, A.Q., Dhariwal, P.: Improved denoising diffusion probabilistic models. In: International Conference on Machine Learning (2021)
23. Nilsback, M.E., Zisserman, A.: A visual vocabulary for flower classification. In: Computer Vision and Pattern Recognition (2006)
24. Oza, P., Patel, V.M.: One-class convolutional neural network. IEEE Signal Process. Lett. **26**(2), 277–281 (2019). https://doi.org/10.1109/LSP.2018.2889273
25. Perera, P., Nallapati, R., Xiang, B.: OCGAN: one-class novelty detection using GANs with constrained latent representations. In: Computer Vision and Pattern Recognition (2019)
26. Radford, A., Kim, J.W., Hallacy, C., Ramesh, A., Goh, G., Agarwal, S.: Learning transferable visual models from natural language supervision. In: International Conference on Machine Learning (2021)
27. Reiss, T., Cohen, N., Bergman, L., Hoshen, Y.: PANDA: adapting pretrained features for anomaly detection and segmentation. In: Computer Vision and Pattern Recognition (2021)
28. Rombach, R., Blattmann, A., Lorenz, D., Esser, P., Ommer, B.: High-resolution image synthesis with latent diffusion models. In: Computer Vision and Pattern Recognition (2022)
29. Ruff, L., et al.: Deep one-class classification. In: International Conference on Machine Learning (2018)
30. Sabokrou, M., Fathy, M., Zhao, G., Adeli, E.: Deep end-to-end one-class classifier. IEEE Trans. Neural Netw. Learn. Syst. **32**(2), 675–684 (2021). https://doi.org/10.1109/TNNLS.2020.2979049
31. Saharia, C., Chan, W., Chang, H., Lee, C., Ho, J.: Palette: Image-to-image diffusion models. In: Association for Computing Machinery Special Interest Group on Computer Graphics and Interactive Techniques (2022)
32. Salehi, M., Sadjadi, N., Baselizadeh, S., Rohban, M.H., Rabiee, H.R.: Multiresolution knowledge distillation for anomaly detection. In: Computer Vision and Pattern Recognition (2021)
33. Song, Y., Durkan, C., Murray, I.: Maximum likelihood training of score-based diffusion models. In: Neural Information Processing Systems (2021)
34. Song, Y., Ermon, S.: Generative modeling by estimating gradients of the data distribution. In: Neural Information Processing Systems (2019)
35. Song, Y., Ermon, S.: Improved techniques for training score-based generative models. In: Neural Information Processing Systems (2020)
36. Song, Y., Sohl-Dickstein, J., Kingma, D.P., Kumar, A., Ermon, S., Poole, B.: Score-based generative modeling through stochastic differential equations. In: International Conference on Learning Representations (2020)
37. Tack, J., Mo, S., Jeong, J., Shin, J.: CSI: novelty detection via contrastive learning on distributionally shifted instances. In: Neural Information Processing Systems (2020)
38. Wah, C., Branson, S., Welinder, P., Perona, P., Belongie, S.: The caltech-UCSD birds-200-2011 dataset. Tech. Rep. CNS-TR-2011-001, California Institute of Technology (2011)
39. Wang, T., Zhang, T., Zhang, B., Ouyang, H., Chen, D., Chen, Q.: Pretraining is all you need for image-to-image translation. In: arXiv (2022)
40. Wang, X., Qi, G.J.: Contrastive learning with stronger augmentations. IEEE Trans. Pattern Anal. Mach. Intell. **45**(5), 5549–5560 (2023). https://doi.org/10.1109/TPAMI.2022.3203630

41. Xu, J., Liu, S., Vahdat, A., Byeon, W., Wang, X., De Mello, S.: Open-vocabulary panoptic segmentation with text-to-image diffusion models. In: Computer Vision and Pattern Recognition (2023)

42. Zhang, L., Yu, M., Chen, T., Shi, Z., Bao, C., Ma, K.: Auxiliary training: towards accurate and robust models. In: Computer Vision and Pattern Recognition (2020)

43. Zhou, K., et al.: Encoding structure-texture relation with P-Net for anomaly detection in retinal images. In: Vedaldi, A., Bischof, H., Brox, T., Frahm, J.-M. (eds.) ECCV 2020. LNCS, vol. 12365, pp. 360–377. Springer, Cham (2020). https://doi.org/10.1007/978-3-030-58565-5_22

44. Zimmerer, D., Petersen, J., Kohl, S.A., Maier-Hein, K.H.: A case for the score: Identifying image anomalies using variational autoencoder gradients. In: Neural Information Processing Systems (2019)

Generalized Zero-Shot Learning with Noisy Labeled Data

Liqing Xu[1], Xueliang Liu[1,2(✉)], and Yishun Jiang[2]

[1] School of Computer and Information, Hefei University of Technology, Hefei, China
liuxueliang@hfut.edu.cn
[2] Institute of Artificial Intelligence, Hefei Comprehensive National Science Center, Hefei, China
ysjiang@iai.ustc.edu.cn

Abstract. Generalized zero-shot learning (GZSL) is a challenging task that aims to classify samples with both seen and unseen categories. In practice, noisy labels can significantly degrade the performance of GZSL classifiers, which has received little attention in previous research. When noisy labeled data exists, the class-level semantic representation may fail to accurately describe some samples of the corresponding class. At the same time, noisy data can also disturb the original distributions of the corresponding classes, leading to estimation errors in modeling the sample distributions. To address these issues, we propose a novel method that aims to alleviate the influence of noisy samples in GZSL. Specifically, we propose a sample-level semantic generation method to ensure an accurate description of the corresponding sample. Furthermore, we introduce an unbalanced learning framework to address the sample distribution estimation with noisy labels to make the estimation error on each class and dimension balanced and dynamically mitigate the negative effects of the error distributions from multiple classes. Experimental results on benchmark datasets demonstrate that our approach effectively mitigates the influence of noisy samples and outperforms other advanced methods.

Keywords: Zero-Shot Learning · Noisy Labels

1 Introduction

Different from conventional machine learning tasks, Zero-Shot Learning (ZSL) could train a recognition model on seen objects to recognize and classify objects that are unseen during training [14]. It simulates the process of human recognition by exploiting semantic descriptors to understand new concepts instead of learning from a large number of labeled samples.

Based on the test samples that the recognition model sees at test time, ZSL could be categorized into conventional and generalized settings. In conventional ZSL setting [26,32], the test samples only come from unseen classes. For generalized zero-shot learning setting [2,33], the test samples are from not only unseen classes but also seen classes.

© The Author(s), under exclusive license to Springer Nature Singapore Pte Ltd. 2024
Q. Liu et al. (Eds.): PRCV 2023, LNCS 14435, pp. 289–300, 2024.
https://doi.org/10.1007/978-981-99-8552-4_23

In most conventional ZSL methods [11,16,21,34], it is all assumed that a high-quality dataset from seen classes is available for the training process, which is often hard to obtain in practice. However, the training set often contains noisy labeled samples that could degrade the performance of GZSL classifiers. This problem is never been studied in previous work of ZSL/GZSL.

In zero-shot learning, the semantic descriptors refer to the attribute representations to describe the characteristics or properties of different classes [18], which plays an important role in transferring knowledge from seen to unseen categories during training. Actually, due to the noisy labels in training set, the class-level semantic descriptors can not match all samples of the corresponding classes precisely, which makes the ZSL with noisy labeled data challenging. In addition, since the samples of each class follow a specific distribution in the ZSL scenario, sample distributions of different classes are different. The noisy labeled data also disturb the original sample distributions, which brings a problem in sample distributions estimation.

To solve these issues, we propose a novel generalized zero-shot learning method to alleviate the inference of noisy labeled data. In our solution, we proposed a method to generate a sample-level semantic descriptor for each sample by modeling the discriminative regions of the visual feature. The sample-level semantic descriptors enhance the correlation between visual features and class-level semantic representations to accurately describe each sample, which reduces the impact of noisy samples on fitting joint distribution between image features and semantic descriptors.

Furthermore, inspired by the principle that treating every semantic feature fairly is helpful to visual recognition systems [22,34], we design an image features re-weighting method when estimating the class distribution, and introduce a re-balancing strategy to balance estimation errors corresponding to multiple classes. Specifically, we use two weight factors to rebalance the errors from both class-wise and dimension-wise perspectives, and obtain balanced errors to constrain modeling the sample distributions. This approach can effectively reduce the estimation errors caused by noisy labeled data, mitigating the influence of noisy samples during the generator training process.

To the best of our knowledge, it is the first paper that studies the problem of noisy labels in generalized zero-shot learning. To summarize, we make the contributions as follows.

- For the first time, we study the problem of noisy labels in generalized zero-shot learning and propose a solution to make the model robust.
- To alleviate the inference of noisy labels and make the semantic descriptors, we propose a method to generate the sample-level semantic descriptors, which makes the semantic descriptors describe each sample precisely.
- To suppress the errors caused by the noisy labeled data in class distribution estimation, we propose a rebalancing method to make each class and dimension of estimation errors equally, which obtains the balanced errors to constrain the deviations of modeling class distributions.
- We have conducted a large number of experiments to show that our method has achieved significant performance on two benchmark datasets.

2   Related Work

2.1   Zero-Shot Learning

Zero-Shot Learning [4,11,16] aims to understand examples of unseen classes depending on semantic descriptions during inference, while Generalized Zero-Shot Learning could recognize not only unseen classes but also seen classes. In general, ZSL (GZSL) can be broadly categorized into embedding-based methods, generative methods, and common space learning-based methods. The embedding-based methods learn to project the visual features into embedding space with the mapping between visual features and semantic description [1,3], which may overfit to seen classes in the GZSL scenario. The generative methods are to synthesize the visual features of unseen classes by a generative model that has learned the mapping of semantic description to visual features [4,15,29]. In these methods, the generative model could be designed through variational autoencoders (VAEs) [29], generative adversarial networks (GANs) [4] or generative flows [24]. Common space learning-based methods also focus on learning a common representation space for interaction between visual and semantic spaces [5,7] for knowledge transferring.

2.2   Learning with Noisy Labeled Data

The existing robust learning methods for noisy labels include label correction, training strategy optimization, and robust objective function. Label correction is designed to correct wrong labels from raw datasets. These methods [17,27,28] require support from extra clean data or a potentially expensive detection process, which is often unavailable in real-world applications. Training strategy optimization adjusts the training strategy to make the model more adaptive to noisy labels, such as MentorNet [13] and Co-teaching [10], which usually involves sophisticated training procedures and takes much time for tuning. Another possible direction for modeling noisy labeled data is designing robust loss functions to make the optimization schemes robust to noisy samples [9]. Theoretically, some loss functions such as Mean Absolute Error (MAE), and Generalized Cross Entropy (GCE) [36] are more robust to noisy labels, than the commonly used Cross Entropy (CE) loss. The active and passive losses (APLs) [19] are designed recently and have a promising performance to improve the model robustness. Inspired by the work on robust loss function, we introduce a re-balance loss function in our work, to reduce the estimation errors of class distributions caused by noisy labeled data.

3   Our Method

In this work, to alleviate the inference of noisy labeled data in ZSL, we propose a generalized zero-shot learning with noisy labels (NLGZSL) method in the generative framework. Specifically, we generate the sample-level semantic descriptors for describing each sample precisely. We also design a rebalancing method to reduce errors in sample distribution estimation caused by noisy samples, which is also beneficial for the algorithm to model the sample distributions of all seen classes in a balanced manner. In this section, we will introduce the details of model structure and optimization process.

3.1   Problem Definition

Mathematically, we define the images of seen and unseen classes as X^s and X^u, and we also have two disjoint sets of classes: K seen classes in Y^s and L unseen classes in Y^u, where K and L are the numbers of seen classes and unseen classes. All semantic representations which encode the relationships between seen classes and unseen classes are partitioned into non-overlapping sets as A^s and A^u. The training set is defined as $D_{train} = \{(\boldsymbol{x}_i, y_i, \boldsymbol{a}_i) | \boldsymbol{x}_i \in X^s, y_i \in Y^s, \boldsymbol{a}_i \in A^s\}$, where $\boldsymbol{x}_i$ is the ith seen sample, y_i is its class label and $\boldsymbol{a}_i$ is its semantic representation. ZSL is to fit a classifier $f_{ZSL} : X^u \rightarrow Y^u$ and GZSL is to learn a classifier $f_{GZSL} : X^s \cup X^u \rightarrow Y^s \cup Y^u$. In this paper, we try to learn the classifiers with noisy labeled data, which means that the Y^s may not be labeled correctly.

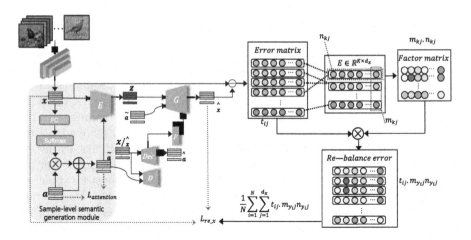

Fig. 1. The main framework of the proposed method.

3.2   Preliminaries: TF-VAEGAN

As shown in Fig. 1, the architecture of the proposed method is inspired by the TF-VAEGAN structure [21]. The network includes a VAEGAN network, a semantic embedding decoder, and a feedback module, where the VAEGAN includes a VAE and a GAN. The VAE includes an encoder $E(\boldsymbol{x}, \boldsymbol{a})$ and a generator $G(\boldsymbol{z}, \boldsymbol{a})$ where $\boldsymbol{z}$ is output of $E(\boldsymbol{x}, \boldsymbol{a})$. The model can be optimized by minimizing the loss: $L_V = KL(E(x, a) \| p(z \mid a)) - \mathbb{E}_{E(x,a)} [\log G(z, a)]$. The GAN designed for feature synthesis comprises a generator $G(z, a)$ and a discriminator $D(x, a)$, which is optimized with WGAN loss: $L_W = \mathbb{E}[D(x, a)] - \mathbb{E}[D(\hat{x}, a)] - \lambda \mathbb{E}\left[(\|\nabla D(\tilde{x}, a)\|_2 - 1)^2\right]$. In addition, the semantic embedding decoder Dec is used to reconstruct semantic embedding $\hat{a}$, which is constrained by $L_R = \mathbb{E}[\| Dec(x) - a\|_1] + \mathbb{E}[\| Dec(\hat{x}) - a\|_1]$. Therefore, the overall loss function of TF-VAEGAN is:

$$L_{tf} = L_V + \alpha L_W + \beta L_R. \tag{1}$$

The TF-VAEGAN model works well on well-labeled data. But in practice, the noisy labeled data existing in the training data may degrade its performance. In this work, we address the problem of generalized zero-shot learning with noisy labeled data and propose a solution to alleviate the inference of noisy labeled data. The details are described in Sect. 3.3 and Sect. 3.4.

3.3 Sample-Level Semantic Descriptor Generation

With noisy labeled data, the class-level semantic representation a_i used in conventional ZSL/GZSL approaches fails to accurately capture the characteristics of all samples with the same label. To address this limitation, we propose a sample-level semantic representation generation module to enrich the semantic representations for each individual sample.

As illustrated in Fig. 1, the sample-level semantic representation generation module takes the image feature x_i and its related class-level semantic vector a_i as input, while the semantic descriptor a_i is utilized to guide the generation of the sample-level semantic descriptors. And the sample-level semantic representation could be obtained by:

$$\tilde{a} = a + a \bigodot softmax\left(f\left(x\right)\right) \tag{2}$$

where f denotes a linear layer and $\bigodot$ represents Hadamard product. The module is based on bilinear pooling mechanism [8,35] which learns the attention on discriminating regions of visual features, then maps them to the semantic domain to obtain discriminative semantics.

Meanwhile, the center loss and softmax loss are employed to optimize the semantic representation generation module. The center loss is used to minimize the differences between intra-class features, inspired by the fact that the semantic vectors $\tilde{a}$ of noisy and clean samples in each class are clustered as close to a central point as possible. The softmax loss is used to enlarge the distance of samples from different classes, reducing the similarity of $\tilde{a}$ among different classes. The loss function formula is defined as follows:

$$
\begin{aligned}
L_{attention} &= L_S + \lambda_1 L_C \\
&= \frac{1}{N} \sum_{i=1}^{N} \left[-\log \frac{e^{\widetilde{a_i} C_{y_i}}}{\sum_{k=1}^{K} e^{\widetilde{a_i} C_k}} + \lambda_1 \|\widetilde{a}_i - C_{y_i}\|_2^2 \right]
\end{aligned} \tag{3}
$$

where N is the number of samples, $\boldsymbol{C}_{y_i}$ is class center point corresponding to the sample $\tilde{a}_i$.

3.4   Re-balanced Sample Distribution Modeling

Noisy labeled data can also disturb the semantic modeling dominated by clean samples and makes the model optimization collapsed. To address this issue, we propose a novel rebalanced MSE loss function for minimizing the estimation error. The loss ensures that our algorithm not only estimates class distribution reliably and accurately, but also models the multiple class distributions in a balanced manner. To the best of our knowledge, our approach is the first to apply rebalanced MSE for constraining the estimated distribution of image features in ZSL to alleviate the influence of noisy labels.

Mathematically, we define the class-level feature generation error as the Euclidean distance between the synthesized features and the real features. And the k_{th} class of j_{th} dimension generation error could be defined as follows:

$$E_{kj} = \frac{1}{|D_k|} \sum_{(\boldsymbol{x}_i, y_i) \in D_k} (\hat{x_{ij}} - x_{ij})^2 \tag{4}$$

where $|D_k|$ is the number of samples in D_k and x_{ij} is the jth dimension of the sample $\boldsymbol{x}_i$.

Notably, the error E may vary on different classes or dimensions. Therefore, to effectively balance the error associated with each class and feature dimension, we second define two balancing factors: the class-level weighting factor denoted as m, and the dimension-level weighting factor denoted as n. The former is to balance the weights of the same dimension across different classes, while the latter is utilized to balance the weights of the same class across different dimensions. The balancing factors ensure that our algorithm not only effectively utilizes each dimension of the samples to model the corresponding distribution, but also treats all seen classes equally. The two factors are mathematically defined as follows:

$$m_{kj} = \left(\log \frac{E_{kj}}{min_{y \in Y^s} E_{yj}} + 1\right)^{\alpha_1}, n_{kj} = \left(\log \frac{E_{kj}}{min_{1 \leq w \leq dx} E_{kw}} + 1\right)^{\beta_1} \tag{5}$$

where α_1 and β_1 are the reweighted hyper-parameters, m_{kj} is the class-level factor on the k_{th} class for the j_{th} dimension.

Then, we use the balancing factors to balance the estimation error in T, and finally, the re-balanced error is leveraged to compute the loss L_{re_x}, which acts as a constraint in modeling class distribution. The loss L_{re_x} facilitates that the generated features $\hat{\boldsymbol{x}}$ are close to the distribution of real features $\boldsymbol{x}$, and yields a balanced algorithm for modeling multiple class distributions. The rebalanced loss for image features can be calculated as follows:

$$L_{re\_x} = \frac{1}{N} \sum_{i=1}^{N} \sum_{j=1}^{dx} m_{y_ij} n_{y_ij} (\hat{x_{ij}} - x_{ij})^2 \tag{6}$$

Therefore, the loss for training the proposed NLGZSL is finally defined as follows:

$$L_{total} = L_{tf} + L_{attention} + \lambda_2 L_{re\_x} \tag{7}$$

where λ_2 is a hype-parameter to adjust the different components.

4 Experiments

4.1 Datasets and Configuration

We evaluate our method on two challenging benchmark datasets: CUB (Caltech-UCSD-Birds) [31] and SUN (SUN Attribute) [23]. On the two datasets, the 2048-dim visual features extracted by ResNet-101 [12] pre-trained with ImageNet-1K [6] are fed as input to the model. The class-level attributes of CUB (312-d) and SUN (102-d) are used as the semantic representations. On the other hand, the noisy labels are generated according to the methods in previous works [20].

In ZSL, the top-1 accuracy of unseen classes is the evaluation metric. In the GZSL scenario, we evaluate not only the top-1 accuracy of unseen classes but also seen classes, which are denoted as U and S respectively. Furthermore, the harmonic mean is also used to evaluate the performance of GZSL, which is defined as: $H = (2 \times S \times U)/(S + U)$.

Table 1. The accuracy (%) of different methods on benchmark datasets with different label noise rate and the top 1 best results and second-best results are marked with red and blue.

	Noise Rate	Methods	ZSL	S	U	H		Noise Rate	Methods	ZSL	S	U	H
CUB	0	LisGAN [15]	58.80	57.90	46.50	51.60	SUN	0	LisGAN [15]	61.70	37.80	42.90	40.20
		LsrGAN [30]	60.30	59.10	48.10	53.00			LsrGAN [30]	62.50	37.70	44.80	40.90
		TF-VAEGAN [21]	64.90	64.70	52.80	58.10			TF-VAEGAN [21]	66.00	40.70	45.60	43.00
		CNZSL [25]	54.53	50.70	49.90	50.30			CNZSL [25]	60.69	41.60	44.70	43.10
		Ours	65.59	63.40	54.01	58.33			Ours	66.04	39.73	46.18	42.71
	0.2	LisGAN [15]	56.28	50.67	43.11	46.59		0.2	LisGAN [15]	57.85	26.82	42.92	33.01
		LsrGAN [30]	57.11	42.60	48.21	45.23			LsrGAN [30]	59.58	28.57	45.35	35.05
		TF-VAEGAN [21]	61.71	53.41	49.69	51.48			TF-VAEGAN [21]	62.15	33.02	42.71	37.25
		CNZSL [25]	52.57	39.55	48.59	43.61			CNZSL [25]	56.94	38.53	36.67	37.57
		Ours	62.47	49.00	54.06	51.41			Ours	63.13	35.16	42.78	38.59
	0.4	LisGAN [15]	52.47	40.37	39.54	39.95		0.4	LisGAN [15]	56.04	20.39	37.64	26.45
		LsrGAN [30]	54.42	30.10	44.42	35.88			LsrGAN [30]	56.25	22.98	42.57	29.85
		TF-VAEGAN [21]	56.80	42.39	44.18	43.27			TF-VAEGAN [21]	57.64	25.54	39.17	30.92
		CNZSL [25]	50.21	34.69	47.39	40.06			CNZSL [25]	52.36	33.45	35.76	34.57
		Ours	57.56	42.14	44.94	43.49			Ours	58.12	26.90	38.40	31.64
	0.6	LisGAN [15]	46.74	27.36	35.84	31.03		0.6	LisGAN [15]	48.19	14.34	21.32	17.15
		LsrGAN [30]	48.56	20.72	33.76	25.68			LsrGAN [30]	49.93	16.67	34.44	22.46
		TF-VAEGAN [21]	46.63	33.48	32.56	33.01			TF-VAEGAN [21]	48.40	17.79	35.42	23.68
		CNZSL [25]	42.12	22.45	39.56	28.64			CNZSL [25]	44.38	29.34	24.51	26.71
		Ours	48.02	34.27	33.27	33.76			Ours	51.53	20.54	29.79	24.32
	0.8	LisGAN [15]	31.02	13.93	18.08	15.74		0.8	LisGAN [15]	33.13	7.44	12.78	9.41
		LsrGAN [30]	31.76	8.80	18.09	11.84			LsrGAN [30]	39.24	9.34	19.93	12.72
		TF-VAEGAN [21]	28.48	19.96	16.42	18.02			TF-VAEGAN [21]	47.85	12.56	24.44	16.59
		CNZSL [25]	25.02	9.12	24.34	13.27			CNZSL [25]	36.81	17.67	18.33	18.00
		Ours	29.49	18.90	18.76	18.83			Ours	46.74	13.18	24.24	17.07

Moreover, we use the Adam optimizer with a learning rate of 10^{-4} for training, while the batch size of all datasets is set to 64, and the number N of input during each iteration is 64.

4.2 Comparision with State-of-the-Arts

Conventional Zero-Shot Learning. We first compare our proposed method with the state-of-the-art methods in the ZSL setting, which includes embedding-based [25] and generative methods [15,21,30]. The accuracy of them with the noise rate of 0 come from the original papers, others come from our replication. Table 1 shows the results of ZSL on two benchmark datasets. Our method achieves the best accuracy under low noise rates of 0.2, 0.4 for both datasets. In the absence of noise, our method also achieves the best results 65.59% and 66.04% respectively for CUB and SUN. On the other hand, our method still achieves competitive performance under high noise rates for both datasets. From the result, we can observe that as the noise rate increases, our method demonstrates superior performance compared to traditional ZSL and GZSL methods. On the CUB dataset, the difference in accuracy between our method and TF-VAEGAN [21] was 0.69%, 0.76%, 0.76%, and 1.39% respectively, for noise rates of 0, 0.2, 0.4, and 0.6. Similarly, on the SUN dataset, our method achieved an accuracy difference of 0.04%, 0.98%, 0.48%, and 3.13% respectively. Overall, our method demonstrated promising performance on both benchmark datasets, and as the noise rates increase, the advantages of our proposed method over the baseline become increasingly evident.

Generalized Zero-Shot Learning. Table 1 also shows the results of different methods in GZSL settings on two datasets in terms of S, U, H. In most cases, our

Table 2. Ablation studies of ZSL/GZSL for different components of our method on two datasets and the top 1 best results are boldfaced.

	Noise Rate	Methods	ZSL	S	U	H		Noise Rate	Methods	ZSL	S	U	H
CUB	0	B	64.90	64.70	52.80	58.10	SUN	0	B	66.00	40.70	45.60	**43.00**
		B+I	65.02	61.24	55.04	57.97			B+I	65.49	39.65	45.90	42.55
		Full	**65.59**	63.40	54.01	**58.33**			Full	**66.04**	39.73	46.18	42.71
	0.2	B	61.71	53.41	49.69	**51.48**		0.2	B	62.15	33.02	42.71	37.25
		B+I	62.28	50.83	51.04	50.94			B+I	62.22	33.26	42.43	37.29
		Full	**62.47**	49.00	54.06	51.41			Full	**63.13**	35.16	42.78	**38.59**
	0.4	B	56.80	42.39	44.18	43.27		0.4	B	57.64	25.54	39.17	30.92
		B+I	57.04	41.87	43.73	42.78			B+I	57.71	27.56	37.15	31.64
		Full	**57.56**	42.14	44.94	**43.49**			Full	**58.12**	26.90	38.40	**31.64**
	0.6	B	46.63	33.48	32.56	33.01		0.6	B	48.40	17.79	35.42	23.68
		B+I	47.22	33.67	33.51	33.59			B+I	50.76	21.67	29.65	**25.04**
		Full	**48.02**	34.27	33.27	**33.76**			Full	**51.53**	20.54	29.79	24.32
	0.8	B	28.48	19.96	16.42	18.02		0.8	B	**47.85**	12.56	24.44	16.59
		B+I	29.32	17.83	18.82	18.31			B+I	45.90	13.41	22.50	16.81
		Full	**29.49**	18.90	18.76	**18.83**			Full	46.74	13.18	24.24	**17.07**

method can achieve competitive results in H for two benchmark datasets. On the CUB, our method has achieved the best performance of H When the noise rate is 0, 0.4, 0.6, 0.8. On the SUN, when the noise rate is 0.2, our method achieves the best result in H, while the H increases by 1.02% and 1.34% respectively compared with the second-best result and baseline. Even at other noise rates, competitive results can also be obtained.

4.3   Ablation Study

We conduct a comparison of our proposed method with each of its decomposed parts on two benchmark datasets. As reported in Table 2, B, B+I, and Full are the result of baseline, baseline with semantic generation structure and our whole method respectively.

When adding semantic generation to the baseline, on the CUB, the accuracy improves significantly under the ZSL setting. When the noise rate is 0, 0.2, 0.4, 0.6 and 0.8, the accuracy increases by 0.12%, 0.57%, 0.24%, 0.59% and 0.84% respectively. It is also helpful under the GZSL setting. The same result could be found on the SUN dataset. From these results, we can obtain that the semantic generation module can lead to better visual feature to semantic embedding and achieves better performance in ZSL/GZSL.

We also find that the re-balancing strategy is also helpful to improve the performance in ZSL/GZSL. On the CUB, when the noisy ratio is 0.4 and 0.6, the difference between our method and baseline with semantic generation is 0.52%, 0.8% respectively in ZSL. And when the noisy ratio is 0.4, the difference reaches 0.71% on H. On the SUN, when the noise rate is 0.2, the difference between both is 0.91% for ZSL and 1.3% for H. Our method is also effective at high noise rates, with a difference of 0.77% and 0.84% at 0.6 and 0.8 respectively. From the above comparison and evaluation, we have found that our proposed method demonstrates superior robustness on noisy samples.

4.4   Hyper-parameter Analysis

We conduct experiments to determine the effectiveness of the hype-parameter λ_2 in Fig. 2. We find that the best parameters are also different for different datasets and different noise rates. The inconsistency in optimal hyper-parameter values can be attributed to the unique characteristics of different datasets and noisy samples, which has a significant impact on the process of modeling class distribution.

In addition, we evaluate the impact of the re-balance factors α_1 and β_1 in the re-balance loss function, as shown in Fig. 3. We find that when $\alpha_1 = 0.5$ and $\beta_1 = 0.5$, the H values corresponding to all noise rates on both datasets can almost achieve the best performance. The equal contribution of both α_1 and β_1 ensures effective handling of noisy data, and leads to improved performance in terms of the H values. This balanced approach achieves a favorable equilibrium and consistent performance, across different noisy rates and datasets.

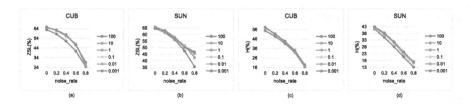

Fig. 2. The effectiveness of λ_2 in our method on two datasets

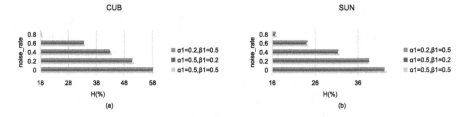

Fig. 3. The result with different re-weighting hyper-parameters α_1 and β_1.

5   Conclusion

In this paper, we address the problem of generalized zero-shot learning with noisy labeled samples, and propose a novel method to suppress their influence by semantic descriptor correction and feature re-balancing. The descriptors correction refines the class-level semantic representations of corresponding samples by highlighting intra-class variations to obtain sample-level descriptors, so that the semantic representations can accurately describe the corresponding samples. The re-balance strategy forces the model to treat each class and each dimension of the class equally, so that the deviations of estimated distributions could be reduced and the robustness of the model to noise is improved. We evaluate our method on two widely used benchmark datasets and the results demonstrate its superiority over existing methods in both ZSL and GZSL scenarios with noisy data.

Acknowledgement. This work was supported in part by the National Key R&D Program of China (No. 2022ZD0118201), in part by the National Natural Science Foundation of China under grants (No. 61976076, 72188101), and by in part the Fundamental Research Funds for the Central Universities (No. JZ2022HGTB0250 and PA2023IISL0096).

References

1. Akata, Z., Perronnin, F., Harchaoui, Z., Schmid, C.: Label-embedding for image classification. IEEE Trans. Pattern Anal. Mach. Intell. **38**(7), 1425–1438 (2015)
2. Chao, W.-L., Changpinyo, S., Gong, B., Sha, F.: An empirical study and analysis of generalized zero-shot learning for object recognition in the wild. In: Leibe, B.,

Matas, J., Sebe, N., Welling, M. (eds.) ECCV 2016. LNCS, vol. 9906, pp. 52–68. Springer, Cham (2016). https://doi.org/10.1007/978-3-319-46475-6_4

3. Chen, L., Zhang, H., Xiao, J., Liu, W., Chang, S.F.: Zero-shot visual recognition using semantics-preserving adversarial embedding networks. In: Proceedings of the IEEE Conference on Computer Vision and Pattern Recognition, pp. 1043–1052 (2018)

4. Chen, S., et al.: Free: feature refinement for generalized zero-shot learning. In: Proceedings of the IEEE/CVF International Conference on Computer Vision, pp. 122–131 (2021)

5. Chen, S., et al.: HSVA: hierarchical semantic-visual adaptation for zero-shot learning. Adv. Neural. Inf. Process. Syst. **34**, 16622–16634 (2021)

6. Deng, J., Dong, W., Socher, R., Li, L.J., Li, K., Fei-Fei, L.: Imagenet: a large-scale hierarchical image database. In: 2009 IEEE Conference on Computer Vision and Pattern Recognition, pp. 248–255. IEEE (2009)

7. Frome, A., et al.: Devise: a deep visual-semantic embedding model. In: Advances in Neural Information Processing Systems, vol. 26 (2013)

8. Fukui, A., Park, D.H., Yang, D., Rohrbach, A., Darrell, T., Rohrbach, M.: Multimodal compact bilinear pooling for visual question answering and visual grounding. arXiv preprint arXiv:1606.01847 (2016)

9. Han, B., et al.: Masking: a new perspective of noisy supervision. In: Advances in Neural Information Processing Systems, vol. 31 (2018)

10. Han, B., et al.: Co-teaching: robust training of deep neural networks with extremely noisy labels. In: Advances in Neural Information Processing Systems, vol. 31 (2018)

11. Han, Z., Fu, Z., Chen, S., Yang, J.: Contrastive embedding for generalized zero-shot learning. In: Proceedings of the IEEE/CVF Conference on Computer Vision and Pattern Recognition, pp. 2371–2381 (2021)

12. He, K., Zhang, X., Ren, S., Sun, J.: Deep residual learning for image recognition. In: Proceedings of the IEEE Conference on Computer Vision and Pattern Recognition, pp. 770–778 (2016)

13. Jiang, L., Zhou, Z., Leung, T., Li, L.J., Fei-Fei, L.: Mentornet: learning data-driven curriculum for very deep neural networks on corrupted labels. In: International Conference on Machine Learning, pp. 2304–2313. PMLR (2018)

14. Lampert, C.H., Nickisch, H., Harmeling, S.: Learning to detect unseen object classes by between-class attribute transfer. In: 2009 IEEE Conference on Computer Vision and Pattern Recognition, pp. 951–958. IEEE (2009)

15. Li, J., Jing, M., Lu, K., Ding, Z., Zhu, L., Huang, Z.: Leveraging the invariant side of generative zero-shot learning. In: Proceedings of the IEEE/CVF Conference on Computer Vision and Pattern Recognition, pp. 7402–7411 (2019)

16. Li, Y., Zhang, J., Zhang, J., Huang, K.: Discriminative learning of latent features for zero-shot recognition. In: Proceedings of the IEEE Conference on Computer Vision and Pattern Recognition, pp. 7463–7471 (2018)

17. Li, Y., Yang, J., Song, Y., Cao, L., Luo, J., Li, L.J.: Learning from noisy labels with distillation. In: Proceedings of the IEEE International Conference on Computer Vision, pp. 1910–1918 (2017)

18. Liu, M., Li, F., Zhang, C., Wei, Y., Bai, H., Zhao, Y.: Progressive semantic-visual mutual adaption for generalized zero-shot learning. In: Proceedings of the IEEE/CVF Conference on Computer Vision and Pattern Recognition, pp. 15337–15346 (2023)

19. Ma, X., Huang, H., Wang, Y., Romano, S., Erfani, S., Bailey, J.: Normalized loss functions for deep learning with noisy labels. In: International Conference on Machine Learning, pp. 6543–6553. PMLR (2020)

20. Ma, X., et al.: Dimensionality-driven learning with noisy labels. In: International Conference on Machine Learning, pp. 3355–3364. PMLR (2018)
21. Narayan, S., Gupta, A., Khan, F.S., Snoek, C.G.M., Shao, L.: Latent embedding feedback and discriminative features for zero-shot classification. In: Vedaldi, A., Bischof, H., Brox, T., Frahm, J.-M. (eds.) ECCV 2020. LNCS, vol. 12367, pp. 479–495. Springer, Cham (2020). https://doi.org/10.1007/978-3-030-58542-6_29
22. Park, S., Lee, J., Lee, P., Hwang, S., Kim, D., Byun, H.: Fair contrastive learning for facial attribute classification. In: Proceedings of the IEEE/CVF Conference on Computer Vision and Pattern Recognition, pp. 10389–10398 (2022)
23. Patterson, G., Hays, J.: Sun attribute database: discovering, annotating, and recognizing scene attributes. In: 2012 IEEE Conference on Computer Vision and Pattern Recognition, pp. 2751–2758. IEEE (2012)
24. Shen, Y., Qin, J., Huang, L., Liu, L., Zhu, F., Shao, L.: Invertible zero-shot recognition flows. In: Vedaldi, A., Bischof, H., Brox, T., Frahm, J.-M. (eds.) ECCV 2020. LNCS, vol. 12361, pp. 614–631. Springer, Cham (2020). https://doi.org/10.1007/978-3-030-58517-4_36
25. Skorokhodov, I., Elhoseiny, M.: Class normalization for (continual)? generalized zero-shot learning. arXiv preprint arXiv:2006.11328 (2020)
26. Socher, R., Ganjoo, M., Manning, C.D., Ng, A.: Zero-shot learning through cross-modal transfer. In: Advances in Neural Information Processing Systems, vol. 26 (2013)
27. Vahdat, A.: Toward robustness against label noise in training deep discriminative neural networks. In: Advances in Neural Information Processing Systems, vol. 30 (2017)
28. Veit, A., Alldrin, N., Chechik, G., Krasin, I., Gupta, A., Belongie, S.: Learning from noisy large-scale datasets with minimal supervision. In: Proceedings of the IEEE Conference on Computer Vision and Pattern Recognition, pp. 839–847 (2017)
29. Verma, V.K., Arora, G., Mishra, A., Rai, P.: Generalized zero-shot learning via synthesized examples. In: Proceedings of the IEEE Conference on Computer Vision and Pattern Recognition, pp. 4281–4289 (2018)
30. Vyas, M.R., Venkateswara, H., Panchanathan, S.: Leveraging seen and unseen semantic relationships for generative zero-shot learning. In: Vedaldi, A., Bischof, H., Brox, T., Frahm, J.-M. (eds.) ECCV 2020. LNCS, vol. 12375, pp. 70–86. Springer, Cham (2020). https://doi.org/10.1007/978-3-030-58577-8_5
31. Wah, C., Branson, S., Welinder, P., Perona, P., Belongie, S.: The Caltech-UCSD birds-200-2011 dataset (2011)
32. Xian, Y., Lampert, C.H., Schiele, B., Akata, Z.: Zero-shot learning-a comprehensive evaluation of the good, the bad and the ugly. IEEE Trans. Pattern Anal. Mach. Intell. **41**(9), 2251–2265 (2018)
33. Xian, Y., Schiele, B., Akata, Z.: Zero-shot learning-the good, the bad and the ugly. In: Proceedings of the IEEE Conference on Computer Vision and Pattern Recognition, pp. 4582–4591 (2017)
34. Ye, Z., Yang, G., Jin, X., Liu, Y., Huang, K.: Rebalanced zero-shot learning. arXiv preprint arXiv:2210.07031 (2022)
35. Yu, Z., Yu, J., Fan, J., Tao, D.: Multi-modal factorized bilinear pooling with co-attention learning for visual question answering. In: Proceedings of the IEEE International Conference on Computer Vision, pp. 1821–1830 (2017)
36. Zhang, Z., Sabuncu, M.: Generalized cross entropy loss for training deep neural networks with noisy labels. In: Advances in Neural Information Processing Systems, vol. 31 (2018)

Q-TrHDRI: A Qurey-Based Transformer for High Dynamic Range Imaging with Dynamic Scenes

Bin Chen[1], Jia-Li Yin[1], Bo-Hao Chen[2](✉), and Ximeng Liu[1](✉)

[1] Fuzhou University, Fuzhou, China
snbix@gmail.com
[2] Yuan Ze University, Taoyuan City, Taiwan
hd840207@gmail.com

Abstract. In the absence of well-exposed contents in images, high dynamic range image (HDRI) provides an attractive option that fuses stacked low dynamic range (LDR) images into an HDR image. Existing HDRI methods utilized convolutional neural networks (CNNs) to model local correlations, which can perform well on LDR images with static scenes, but always failed on dynamic scenes where large motions exist. Here we focus on the dynamic scenarios in HDRI, and propose a Query-based Transformer framework, called Q-TrHDRI. To avoid ghosting artifacts induced by moving content fusion, Q-TrHDRI uses Transformer instead of CNNs for feature enhancement and fusion, allowing global interactions across different LDR images. To further improve performance, we investigate comprehensively different strategies of transformers and propose a query-attention scheme for finding related contents across LDR images and a linear fusion scheme for skillfully borrowing complementary contents from LDR images. All these efforts make Q-TrHDRI a simple yet solid transformer-based HDRI baseline. The thorough experiments also validate the effectiveness of the proposed Q-TrHDRI, where it achieves superior performances over state-of-the-art methods on various challenging datasets.

1 Introduction

High-level computer vision techniques, such as segmentation, detection, and tracking, have brought impressive advances to the state-of-the-art across a wide variety of outdoor vision-based applications. At the same time, however, these leaps in performance come only when captured images and videos with full contents are available, i.e., no under/over-exposed regions. Due to the limited dynamic range of camera sensors, it is hard to achieve an image with fully well-exposed contents under outdoor scenarios. At the moment, for problems lacking well-exposed content, it can be still possible to obtain multiple LDR images with different exposure levels that are enough for generating an HDR image.

Learning to generate well-exposed content from multiple LDR images is known as HDRI. In recent years, deep CNNs have yielded considerable success in HDRI. Typically, conventional HDRI approaches are proposed in the context

Q. Liu et al. (Eds.): PRCV 2023, LNCS 14435, pp. 301–312, 2024.
https://doi.org/10.1007/978-981-99-8552-4_24

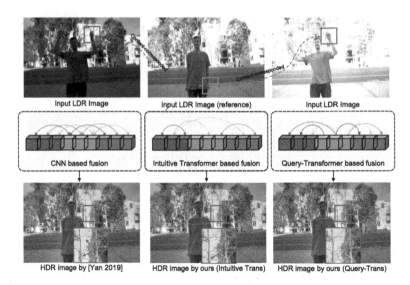

Fig. 1. Comparison of different HDRI schemes. **Left:** The CNN-based fusion models local correlations and aggregates the position-corresponded regions in each input image. **Middle:** Intuitive Transformer based fusion can model long-range correlations across LDR images but lacks explicit alignment to the reference image. **Right:** Our Query-based fusion performs query-attention in each transformer layer, which ensures the alignment before the fusion process.

of shallow learning, i.e., in a situation when the multiple LDR images possess only static and fixed contents. The approaches then devise some end-to-end fusion processes to convert multiple LDR images into one HDR image [14,20]. Despite its effectiveness, a more practical scenario is neglected: the presence of moving objects, background motions, or small shifts by hand-held cameras is unavoidable. Applying conventional HDRI methods can generate severe ghosting artifacts in their resulting images because different contents are fused during the fusion process. Thus a method specifically dedicated to HDRI under dynamic scenarios is encouraged. We therefore focus on the challenging case in HDRI where large-scale moving objects and occlusions exist.

Researchers have recently looked for effective fusions on moving contents with attention mechanisms [21] or non-local operations [22] by assuming the contents among the LDR images have shared only short-range correlations, which is too strong in practical application. Here we go much further. We first observe that there are two key tasks under dynamic scenarios: (i) suppressing the artifacts caused by moving objects and (ii) recovering details in the occlusion regions. This inspires us to satisfy the following two requirements during the HDRI process: (i) avoid aggregation of different content components across LDR images and (ii) skillfully borrow complementary contents from non-reference images. To meet these requirements, conventional position-corresponded aggregation in CNN-based models should be redesigned and long-range correlations across LDR images are crucially needed. However, this is obviously beyond the ability of CNN models, even if additional modules are equipped.

In advancements of machine learning, Transformer, which designs a self-attention scheme to model global interactions of inputs, has shown great promise in modeling long-range correlations. Motivated by this, recent studies [1,9] have applied Transformers in various vision tasks. The Transformers in these works embedded images into patch sequences, linking the non-local patches to explore long-range correlations and model global interactions of all pixels. In this paper, we thus turn to Transformer for HDRI with dynamic scenes. However, we may not train an intuitive Transformer out-of-the-box as the existing Transformers are generally designed for single-image tasks. A direct way is to first concatenate the images and then consider the concatenated images as the input for transformer layers. Indeed, such architecture can explore long-range correlations across input images. Yet, it lacks explicit alignment with the reference image, thus will lead to unsatisfied results, as shown in Fig. 1. To overcome this issue and make Transformers more applicable for HDRI tasks, we propose a variational framework: Query-Transformer HDRI (Q-TrHDRI). Instead of using self-attention for each image, we consider the reference image as the query to find consistent contents in the non-reference images and thus introduce a query-attention scheme. Moreover, we design a linear-based feature fusion for the case, in which the features are mapped into latent space by a linear operation. Compared with stacked convolution operations in previous works, such a fusion strategy can well assist our Q-TrHDRI in finding complementary contents throughout the whole tokens across the LDR images.

The main contributions of this work are:

- We provide a Query-based Transformer architecture, namely Q-TrHDRI, to HDRI on dynamic scenes. This architecture opens up a new frontier in the HDRI process.
- We introduce a query-attention scheme in transformer layers to further ensure the alignment with reference image, and a linear-based fusion to find complementary contents across whole LDR images.
- Extensive experiments are conducted to validate the superiority of our proposed method, we believe the proposed Q-TrHDRI established a new baseline for Transformer-based HDRI approaches.

2   Related Work

2.1   High Dynamic Range Imaging

Numerous HDRI methods have been proposed in recent years, and here we focus on the most related works on HDRI with dynamic scenes.

HDRI with "alignment-merging". Multiple methods perform HDRI with dynamic scenes by first aligning the input images and then merging the aligned images. An important aspect of this "alignment-merging" strategy is the way the input images are aligned. Some approaches perform this by detecting the motions in each image and then refusing the motion pixels during the merging process [7,11], while other works seek an explicit motion correction that would align each input image before the merging process [5,15]. However, the pixel-level alignment process is not always reliable, especially for image sequences with different exposure levels. The errors induced in the alignment process can be amplified during the merging process, leading to even more ghosting artifacts in the resulting images.

HDRI with Direct Merging. Several approaches skip the alignment process and count on the CNNs to directly merge the misaligned images. [19] first proposed an encoder-decoder framework, where each input LDR image is encoded into latent features and then these features are concatenated to feed into a decoding process. To avoid harmful features from LDR images, following works further incorporated attention module [21–23], residual connections [13], and exposure-share block to blend different features [3] to pick up useful features before the feature concatenation. Since the merging process in these network architectures aggregates the features from each LDR image through feature concatenation followed by convolution operations, the features can only receive local information from different images, lacking enough exploration of long-range correlations which thus restricts reconstruction ability of the network.

Transformers. After the introduction of Transformers in [17], which designs a new attention-based modeling architecture for machine translation, it has shown great influence in DNNs. [1] first proposed the Vision Transformer (ViT) for image classification task. [9] further proposed to combine local and global attention in Transformers by window shifting, named Swin Transformer. Following works further improve ViT in terms of either the effectiveness [4,9] or efficiency [12,18]. In the low-level vision tasks, [2] designed a pre-trained image processing transformer (IPT) for the image restoration. [8] adopted the Swin Transformer architecture as the backbone to form features for image super-resolution tasks. However, these architectures can execute self-attention only for a single image and are limited to building correlations across multiple images.

3   Method

3.1   Overview

Given a set of misaligned LDR images, our goal is to fuse them into a HDR image that is aligned to one of the input images, i.e., reference image. Following previous works [13,22], we set the image with middle exposure level as the reference image. For example, we have three LDR images as the input, denoted as $\{I_1, I_2, I_3\}$. Particularly, I_2 is set to be the reference image, i.e., $I_r = I_2$, and the HDRI task can be represented as:

$$I_h = \mathcal{F}_\theta(I_1, I_2, I_3), \tag{1}$$

where $\mathcal{F}_\theta$ is the HDRI network paremeterized with θ. Note that I_h is aligned to I_r. An overview of our Q-TrHDRI network is shown in Fig. 2. We discuss the details of each part in the following subsections.

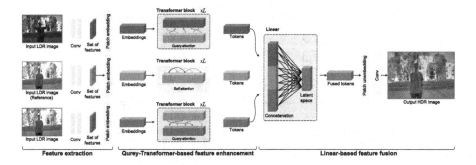

Fig. 2. Architecture of Q-TrHDRI, which consists of three parts including a feature extraction part, a Query-Transformer-based feature enhancement part, and a linear-based feature fusion part.

3.2   Feature Extraction

We first encode LDR images into feature representations via the usage of CNNs, as shown in the left part of Fig. 2. We use a 3×3 convolution layer to transfer each LDR image into feature space as:

$$\mathbf{F}_n = \mathrm{Conv}([\mathbf{I}_n, \mathcal{H}(\mathbf{I}_n)]), \tag{2}$$

where $\mathrm{Conv}(\cdot)$ represents the convolutional operation. Note that we extract feature representations $\mathbf{F}_n$ of image $\mathbf{I}_n$ from the concatenation of $\mathbf{I}_n$ in both LDR domain and HDR domain as previous works [10,21] suggested that images in LDR domain help to detect saturated regions and encourage the detection of moving areas in HDR domain. The LDR to HDR mapping is conducted by using gamma correction [19]:

$$\mathcal{H}(\mathbf{I}_n) = \frac{\mathbf{I}_n^\gamma}{e_n}, \tag{3}$$

where γ is the gamma correction factor, which is fixed to 2.2, and e_n denotes exposure time of the image $\mathbf{I}_n$.

Why CNN for Feature Extraction? HDRI is an image-to-image task, thus we need more contextual information during feature processing to eventually form an HDR image. As pointed out in previous works [4], spatially transferring images into features can help augment more contextual information, thus can be helpful in our task.

3.3   Query-Trans-based Feature Enhancement

After the features are obtained, Transformer layers are first applied to enhance the useful components while avoiding harmful ones. A transformer layer is formed by a multi-head self-attention (MSA) layer and a multi-layer perceptron (MLP). LayerNorm (LN) layer and skip connections are added after both MSA and

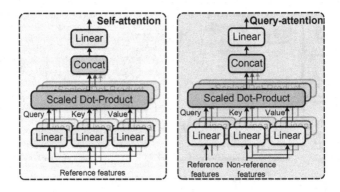

Fig. 3. An illustration of self-attention and query-attention in our Q-TrHDRI.

MLP. Inspired by the recent success of Swin Transformer [9], which incorporates local attention and shifted window mechanism in the original transformer layer, here we use Swin transformer layer as the base architecture in our feature enhancement process.

Given the features from each image of size $\mathbf{F}_n \in \mathbb{R}^{H \times W \times C}$, $n = \{1, 2, 3\}$, patch embedding is first applied to reshape the input into a $\frac{HW}{M^2} \times M^2 \times C$ feature sequence by using a non-overlapping $M \times M$ local window. Then the attention schemes are computed on these local windows. In self-attention, there are three basic elements: *query* Q, *key* K, and *value* V, and its goal is to find the *key* K that is most related to *query* Q and obtain *value* V based on the relations. Correspondingly, we propose a query-attention which uses the reference image as *query* Q and find the *key* K of non-reference images to obtain *value* V. An illustration is shown in Fig. 3. For a local window feature $X_n \in \mathbb{R}^{M^2 \times C}$, the *query* Q, *key* K, and *value* V are first computed as:

$$Q_n = X_n \mathbf{W}_q, K_n = X_n \mathbf{W}_k, V_n = X_n \mathbf{W}_v, \tag{4}$$

where $\mathbf{W}_q, \mathbf{W}_k$, and $\mathbf{W}_v \in \mathbb{R}^{C \times d}$ denotes the learnable parameters for obtaining Q, K, and V, respectively. Next, we apply the self-attention for the reference features X_r and use query-attention for non-reference features X_n to enhance related and consistent contents.

Self-attention. For reference features X_r, the self-attention is then formulated as:

$$\text{SA}(X_r) = \text{softmax}(\frac{Q_r K_r^T}{\sqrt{d}} + B)V_r, \tag{5}$$

where B denotes the learnable relative positional encoding [8], and d is a scaling factor which can be typically set to 64.

Query-attention. For non-reference images X_n, $n = \{1, 3\}$, the query-attention imports Q_r as the *query* vector and we have:

$$\text{QA}(X_n, X_r) = \text{softmax}(\frac{Q_r K_n^T}{\sqrt{d}} + B)V_n. \tag{6}$$

Table 1. Quantitative comparison of different variants of Q-TrHDRI architecture. Note that bold number in each column indicates the best performance.

Variants	PSNR-μ ↑	PSNR-L ↑	HDR-VDP-2 ↑
w/o Query-attention	41.4954	41.414	62.7116
w/o Linear-based fusion	41.0741	41.3224	62.7019
Ours	**42.2613**	**42.0813**	**63.3611**

Following previous works, we perform the above attention functions for h times in parallel and concatenate the results. Note that a residual skip is then applied to the attention layer following by a norm layer before feeding into MLP block:

$$X_r = X_r + \text{LN}(\text{MSA}(X_r)), \quad s.t. \quad r = 2,$$
$$X_n = X_n + \text{LN}(\text{MQA}(X_n)), \quad s.t. \quad n = \{1, 3\}. \tag{7}$$

Then MLP further transforms the result by two linear transformations:

$$\text{MLP}(X_n) = \max(0, X_n \mathbf{W}_1 + b_1)\mathbf{W}_2 + b_2, \tag{8}$$

where $\mathbf{W}_1$, b_1, $\mathbf{W}_2$, and b_2 are the learnable parameters in MLP blocks. Moreover, to build connections across windows, we adopt the shifted window mechanism in Swin Transformer architecture in [9], where the partitioning will be shifted by $(\lfloor \frac{M}{2} \rfloor, \lfloor \frac{M}{2} \rfloor)$ pixels for different layers.

3.4 Linear-Based Feature Fusion

Instead of conventional fusion by stacked convolutional layers, here we adopt linear operation to directly fuse the transformer tokens. In this way, complementary contents can be found from different images to generate the final HDR image. We can formulate the process as:

$$X = \text{LN}(X_1 \oplus X_2 \oplus X_3), \tag{9}$$

where $\oplus$ denotes the concatenation. Finally, an unembedding layer and a convolution layer are employed to turn the fused tokens back into an image.

Why Linear-Based Feature Fusion? Conventional methods first transform the tokens into feature maps and then concatenate the feature maps to form the final image by convolutional layers. Here we use linear-based fusion instead of CNN-based fusion for two reasons: (i) The linear-based fusion globally finds components contributing to the fused image but the CNN performs locally; (ii) the linear layer can be analogous to the extension of the last transformer layer in the feature enhancement process, which can better ensure the feature consistency across the whole framework. Further experimental analysis is shown in Sec. 4.2.

3.5   Training Loss

To learn the HDR image from our entire network with respect to ground-truth $\mathbf{I}_{gt}$, we use the $L-2$ loss over the aligned output $\mathbf{I}_h$:

$$\mathcal{L}(\mathbf{I}_{gt}, \mathbf{I}_h) = \|\mathcal{T}(\mathbf{I}_{gt}) - \mathcal{T}(\mathbf{I}_h)\|_2^2, \tag{10}$$

Table 2. Quantitative comparison between our method and other HDRI models on various datasets. We denote the best result in bold.

Method	Kalantari's dataset					Sen's dataset	Tursun's dataset
	PSNR-μ ↑	PSNR-L ↑	SSIM-μ ↑	SSIM-L ↑	HDR-VDP-2 ↑	NIQE ↓	NIQE ↓
[15]	28.4380	31.5896	0.7596	0.8711	51.7446	5.3321	5.3424
[5]	40.3253	40.8113	0.9871	0.9866	62.0067	5.3955	5.4087
[19]	41.6500	40.8800	0.9860	0.9858	60.4955	5.4703	5.0231
[21]	42.2192	40.8471	0.9899	0.9866	62.3044	6.1618	5.3171
[22]	35.2843	39.0639	0.9826	0.9826	61.2518	5.5813	5.8041
[13]	41.2189	41.5845	0.9815	0.9877	62.0913	5.2810	5.0752
Ours	**42.2613**	**42.0813**	**0.9914**	**0.9879**	**63.4171**	**5.0498**	**4.9784**

where $\mathcal{T}(\cdot)$ is a tonemapping function considering the HDR images are usually displayed after tonemapping in practical. Following previous works [19,22], we adopt the differentiable μ-law function as the tonemapping function:

$$\mathcal{T}(\mathbf{I}) = \frac{\log(1 + \mu\mathbf{I})}{\log(1 + \mu)}, \tag{11}$$

where μ is a compression parameter which is set to $5,000$ according to [5].

4   Experiments

4.1   Experimental Setup

Datasets. We use the most popular Kalantari's dataset [5] for training and testing. It consists of 74 training samples and 15 testing samples. Each sample contains three differently exposed input LDR images with large foreground motions and an HDR image as ground truth. To avoid overfitting during training, we crop the original image to 128×128 with a stride of 64 and apply data augmentations, i.e., horizontal flip, and rotation, to produce abundant data. Finally, the size of training data is expanded from 74 to $27,000$. We also use Sen's dataset [15] and Tursun's dataset [16] without ground truth for testing.

Implementation Details. In our implementation, We set the number and depth of Transformer block, windows size, and number of attention head in MSA as 4, 6, 8, and 9, respectively. For training, we use Adam optimizer [6] with default parameter setting and set the batch size as 16. We used $6,000$ warm-up steps and set the maximum value as $1e^{-3}$ starting from an initial learning rate of 0. The network weights are initialized using Xavier method. We implement our model using PyTorch and train the model on a NVIDIA A100 GPU.

4.2   Ablation Studies

Impact of Query-Attention. We compare our Q-TrHDRI with intuitive self-attention Transformers for feature enhancement, to verify the merits of introducing query-attention in HDRI task. The results are shown in the second and fourth rows of Table 1. It can be seen that introducing the query-attention improves the HDRI performance. We believe the reason is that our query-attention directly finds correlation and complementary contents in the non-reference images and can thereby benefit fusion performance. The qualitative results in Fig. 4 (d) and (f) also show that using the query-attention can more effectively suppress the ghosting artifacts than that of without it.

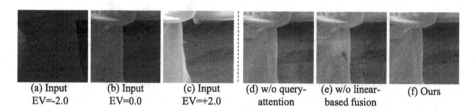

| (a) Input EV=-2.0 | (b) Input EV=0.0 | (c) Input EV=+2.0 | (d) w/o query-attention | (e) w/o linear-based fusion | (f) Ours |

Fig. 4. Visual results of different variants of our Q-TrHDRI architecture. (a)-(c): Zoom-in regions cropped from input images which are misaligned; (d)-(f): Fusion results produced by variants of our proposed method.

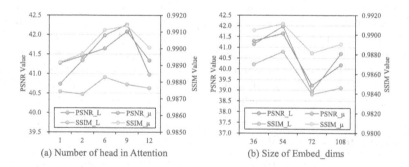

Fig. 5. Ablation study on different parameter settings in the proposed Q-TrHDRI. Results are tested on Kalantari's dataset.

Impact of Linear Fusion. The third and fourth rows of Table 1 validate the effectiveness of linear fusion in our framework. We perform this validation by replacing linear fusion component in our framework with convolution-based fusion. In CNN-based fusion, the transformer tokens of each image are directly reshaped into feature maps and then stacked convolution layers are applied to fuse them. As we can see, our linear-based fusion outperforms the CNN-based fusion in terms of all the metrics. Moreover, we show a qualitative example in Fig. 4 (e) and (f), where the CNN-based fusion failed to maintain the brightness consistency in its result.

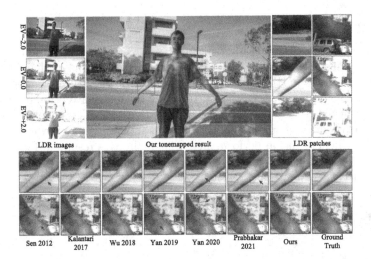

Fig. 6. Visual comparison with HDRI models on Kalantari's dataset. The upper part depicts the three input LDR images, our tonemapped HDR result, and the zoom-in regions cropped from each input image, respectively. The bottom part shows the resulting zoom-in regions produced by each method.

Impact of Hyper-parameters. We also study the sensitivity of our Q-TrHDRI to the parameter setting in Transformer layers, including the dimensions of patch embedding and the number of head h in the MSA layers. We show the results in Fig. 5. Generally, our model is less sensitive to the change of head number in the MSA layers. As h gets larger, the PSNR and SSIM values increase steadily before decreasing. Compared to h, we find that the size of embedding dimensions have deeper influence on the overall performance of our framework. The best and worst PSNR$-\mu$ values have a margin of 3 dB. We use the parameter value that indicates the best results throughout the whole experiments.

4.3   Comparison with SOTAs

We compare our Q-TrHDRI with six HDRI models, including "alignment-merging" based HDRI approaches [5,15], and merging-based HDRI [19,21,22] and [13]. Note that we re-train the model in [22] since the authors did not release their pre-trained model weights. This might result in the results being slightly different from [22].

Evaluation on Kalantari's Dataset. The second to sixth columns in Table 2 show the quantitative comparison of different methods on Kalantari's dataset with ground truth. We can see that our Q-TrHDRI achieves SOTA performance with 42.0813 dB in PSNR-L, 0.9879 score in SSIM-L, and 63.4171 score in HDR-VDP-2, which surpasses the second best method with a large margin. We also show a qualitative comparison in Fig. 6. As we can see, compared to other HDRI

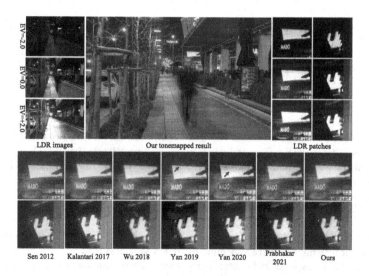

Fig. 7. Visual comparison with HDRI models on Tursun's dataset. The upper part depicts the three input LDR images, our tonemapped HDR result, and the zoom-in regions cropped from each input image, respectively. The bottom part shows the resulting zoom-in regions produced by each method.

methods, our Q-TrHDRI can produce ghosting-free results in the regions containing moving object (see the red bounding boxes), and can more effectively recover details covered by some occlusions (see the blue bounding boxes).

Evaluation on Sen's and Tursun's Datasets. To evaluate the generalization of our Q-TrHDRI, we further test the model on Sen's and Tursun's datasets that are without ground truth images. The seventh to eighth columns of Table 2 illustrate the NIQE values of each method. Our method outperforms all other compared methods on both datasets. Moreover, we show an example in Fig. 7. As can be observed, our method can better generate textures and recover colors under challenging scenarios with over-exposed regions.

5   Conclusions

In this paper, we propose a strong baseline Q-TrHDRI for HDRI. Specially, we propose query-attention schemes to enhance image features for avoiding motion features, and a linear-based fusion to find complementary components from each image. We extensively evaluate our Q-TrHDRI on various scenarios and achieve the state-of-the-art results, which makes us strongly believe that Q-TrHDRI can serve as a solid baseline for further research on Transformer-based HDRI, or misaligned image fusion tasks. One interesting thing we discovered is that, for the first time, long-range correlations between contents matter in a highly accurate HDRI process with dynamic scenarios.

References

1. Alexey, D., et al.: An image is worth 16 ×16 words: transformers for image recognition at scale. In: ICLR (2019)
2. Chen, H., et al.: Pre-trained image processing transformer. In: CVPR, pp. 12299–12310, June 2021
3. Chi, Y., Zhang, X., Chan, S.H.: HDR imaging with spatially varying signal-to-noise ratios. In: CVPR, pp. 5724–5734, June 2023
4. Hassani, A., Walton, S., Shah, N., Abuduweili, A., Li, J., Shi, H.: Escaping the big data paradigm with compact transformers. arXiv:2104.05704 (2021)
5. Kalantari, N.K., Ramamoorthi, R.: Deep high dynamic range imaging of dynamic scenes. ACM TOG **36**(4), 144 (2017)
6. Kingma, D.P., Ba, J.: Adam: a method for stochastic optimization. In: ICLR (2015)
7. Lee, C., Li, Y., Monga, V.: Ghost-free high dynamic range imaging via rank minimization. IEEE SPL **21**(9), 1045–1049 (2014)
8. Liang, J., Cao, J., Sun, G., Zhang, K., Van Gool, L., Timofte, R.: SwinIR: image restoration using Swin transformer. In: ICCVW (2021)
9. Liu, Z., et al.: Swin transformer: hierarchical vision transformer using shifted windows. In: ICCV, pp. 10012–10022 (2021)
10. Niu, Y., Wu, J., Liu, W., Guo, W., Lau, R.W.H.: HDR-GAN: HDR image reconstruction from multi-exposed LDR images with large motions. IEEE TIP **30**, 3885–3896 (2021)
11. Oh, T., Lee, J., Tai, Y., Kweon, I.S.: Robust high dynamic range imaging by rank minimization. IEEE TPAMI **37**(6), 1219–1232 (2015)
12. Pan, Z., Zhuang, B., Liu, J., He, H., Cai, J.: Scalable vision transformers with hierarchical pooling. In: ICCV, pp. 377–386, October 2021
13. Prabhakar, K.R., Senthil, G., Agrawal, S., Babu, R.V., Gorthi, R.K.S.S.: Labeled from unlabeled: exploiting unlabeled data for few-shot deep HDR deghosting. In: CVPR, pp. 4875–4885 (2021)
14. Qu, L., Liu, S., Wang, M., Song, Z.: Transmef: a transformer-based multi-exposure image fusion framework using self-supervised multi-task learning. In: AAAI (2022)
15. Sen, P., Kalantari, N.K., Yaesoubi, M., Darabi, S., Goldman, D.B., Shechtman, E.: Robust patch-based HDR reconstruction of dynamic scenes. ACM TOG **31**(6), 203 (2012)
16. Tursun, O.T., Akyüz, A.O., Erdem, A., Erdem, E.: An objective deghosting quality metric for HDR images. In: Eurographics, pp. 139–152 (2016)
17. Vaswani, A., et al.: Attention is all you need. In: NeurIPS, vol. 30 (2017)
18. Wu, G., Zheng, W.S., Lu, Y., Tian, Q.: PSLT: a light-weight vision transformer with ladder self-attention and progressive shift. IEEE TPAMI, pp. 1–16 (2023)
19. Wu, S., Xu, J., Tai, Y.-W., Tang, C.-K.: Deep high dynamic range imaging with large foreground motions. In: Ferrari, V., Hebert, M., Sminchisescu, C., Weiss, Y. (eds.) ECCV 2018. LNCS, vol. 11206, pp. 120–135. Springer, Cham (2018). https://doi.org/10.1007/978-3-030-01216-8_8
20. Xu, H., Ma, J., Zhang, X.: MEF-GAN: multi-exposure image fusion via generative adversarial networks. IEEE TIP **29**, 7203–7216 (2020)
21. Yan, Q., et al.: Attention-guided network for ghost-free high dynamic range imaging. In: CVPR, pp. 1751–1760 (2019)
22. Yan, Q., et al.: Deep HDR imaging via a non-local network. IEEE TIP **29**, 4308–4322 (2020)
23. Yoon, H., Uddin, S.M.N., Jung, Y.J.: Multi-scale attention-guided non-local network for HDR image reconstruction. Sensors **22**(18), 7044 (2022)

FRNet: Improving Face De-occlusion via Feature Reconstruction

Shanshan Du and Liyan Zhang[✉]

Nanjing University of Aeronautics and Astronautics, Nanjing 210016, China
`zhangliyan@nuaa.edu.cn`

Abstract. Face de-occlusion is essential to improve the accuracy of face-related tasks. However, most existing methods only focus on single occlusion scenarios, rendering them sub-optimal for multiple occlusions. To alleviate this problem, we propose a novel framework for face de-occlusion called FRNet, which is based on feature reconstruction. The proposed FRNet can automatically detect and remove single or multiple occlusions through the predict-extract-inpaint approach, making it a universal solution to deal with multiple occlusions. In this paper, we propose a two-stage occlusion extractor and a two-stage face generator. The former utilizes the predicted occlusion positions to get coarse occlusion masks which are subsequently fine-tuned by the refinement module to tackle complex occlusion scenarios in the real world. The latter utilizes the predicted face structures to reconstruct global structures, and then uses information from neighboring areas and corresponding features to refine important areas, so as to address the issues of structural deficiencies and feature disharmony in the generated face images. We also introduce a gender-consistency loss and an identity loss to improve the attribute recovery accuracy of images. Furthermore, to address the limitations of existing datasets for face de-occlusion, we introduce a new synthetic face dataset including both single and multiple occlusions, which effectively facilitates the model training. Extensive experimental results demonstrate the superiority of the proposed FRNet compared to state-of-the-art methods.

Keywords: Face de-occlusion · Image inpainting · Deep learning

1 Introduction

Face de-occlusion technologies aim at automatically detecting and removing occlusions, and inpainting the occluded area simultaneously, which generally serve as a prepossessing step to assist other face-related tasks. The main idea of traditional technologies [1,5] is to inpaint the images according to the existing information. As each part of the face image has its own characteristics,

This work was supported in part by the National Natural Science Foundation of China under Grant 62172212, in part by the Natural Science Foundation of Jiangsu Province under Grant BK20230031.

Fig. 1. The results of removing real-world face occlusions on different datasets.

their results are far from satisfactory. To overcome that, various early face de-occlusion technologies based on deep learning [2,17] have been proposed, making it possible to leverage deep learning techniques to tackle this task. Nevertheless, most of them generate low-resolution images, which may not meet the current demands. Considering the application under real scenes, methods [4,16,20,24,26] designed for high-resolution face image de-occlusion have been proposed. Specifically, Edgeconnect [16] improves detailed information by introducing prior knowledge. Additionally, CTSDG [4] enhances results by fusing structural and texture features. Furthermore, some methods [20,24,26] increase flexibility in handling various-shaped occluded areas through modified convolution mechanisms. These methods have shown promising results in handling face occlusions, however, most of them are not specifically designed for face de-occlusion tasks and usually require manual marking of the occluded area, which can be time-consuming and has certain constraints in practical applications scenarios. Furthermore, there are several face de-occlusion technologies [7] designed for specific occlusions and achieved the expected results. However, they may struggle when generalizing to other types of occlusions that commonly exist in real-world scenarios.

Additionally, face attribute manipulation technologies [11,13] and image translation technologies [21,25] can address face de-occlusion to some extent. However, the presence of diverse types of occlusions poses labeling challenges and disrupts the feature extraction process, usually leading to unsatisfactory outcomes.

In this paper, we take inspiration from two research studies. One of these studies [2] focuses on the utilization of an occlusion-aware stage to enhance the effectiveness of face de-occlusion. The other one [16] highlights the advantages of incorporating image structure as a prior to improve image inpainting outcomes. Building upon these findings, we propose a novel framework for face de-occlusion called FRNet, which is based on feature reconstruction. The proposed de-occlusion model consists of a two-stage occlusion extractor, a two-stage face generator, and a face discriminator. To better reconstruct the global structure of the occluded face images, we introduce an Occlusion Robust Face Segmentation Module based on the PP-LiteSeg [18] network, which is utilized to obtain both occlusion location details and face structure information.

Firstly, to effectively tackle complex occlusion scenarios in the real world, the two-stage occlusion extractor utilizes the input images and the occlusion location details to obtain coarse occlusion masks and then refines the masks to acquire the final occlusion masks. **Secondly**, to address the issues of structural deficiencies and feature disharmony in the generated face images, the two-stage face generator is designed with the idea of "first reconstructing the structural features globally, and then refining important area features locally". In the coarse stage, we adopt the U-Net structure with a large receptive field to reconstruct the global structures of the faces with the guidance of face segmentation maps. As for the refinement stage, we split it into two distinct modules: the Local Areas Refinement Module (LRM) and the Important Areas Refinement Module (IRM). The LRM extracts information from the neighboring areas by a residual network with a small receptive field, thus enhancing the local textures. While the IRM utilizes intra-feature pixel similarity to identify pixels related to the missing pixels from valid pixels of the corresponding feature, then uses them to fill in the occluded area, thereby ensuring feature harmony. We employ an adaptive merging approach to fuse the outputs from the two branches, generating the final refined face images. Meanwhile, to ensure attribute consistency before and after face de-occlusion, assisting face-related tasks, we introduce a gender-consistent loss in the coarse stage and an identity loss in the refinement stage. Both of them encourage the model to pay more attention to attribute features. **Lastly**, to enhance the model's ability to handle multiple occlusions in a single face image during training, we propose a new synthetic face dataset based on the CelebA-HQ. This dataset includes face images with various types and quantities of occlusions in random states, providing a realistic simulation of common occlusion scenarios in real-life situations, which effectively facilitates the training and supervision of the model. Exemplar results are shown in Fig. 1, our method can effectively remove various types of face occlusions in the wild.

Our contributions can be summarized as follows: **(1)** We propose a novel face de-occlusion framework, which can automatically detect and remove single or multiple occlusions from the face images, achieving visually realistic results. **(2)** We propose the idea of "first reconstructing the structural features globally, and then refining important area features locally", and following this idea, propose a two-stage face generator that can efficiently restore face details and preserve face

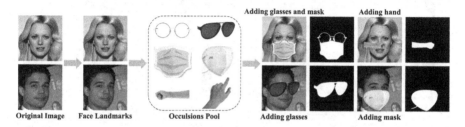

Fig. 2. The process of dataset synthesis. Based on head pose angles and face landmark points, we add occlusions to face images.

attributes. **(3)** We propose a new dataset dedicated to the face de-occlusion task, containing various types and quantities of occlusions. It plays a crucial role in improving the training of our model. **(4)** The experimental results demonstrate the good efficacy of our FRNet in eliminating various face occlusions present in the wild while preserving the essential attribute information of the face images.

2   Synthesis of Face Images with Occlusions

The key to face de-occlusion is to remove occlusions accurately while maintaining the attributes such as gender, skin color, and expression consistent with the input image. Therefore, in this research, we require a contrast dataset that includes both occluded and de-occluded face images to supervise the model's training. However, collecting such a dataset in real life can be challenging, and a more practical solution is to create a dataset with similar characteristics of occluded face images found in the wild. By training on such a dataset, the model can more effectively perceive and remove different types of occlusions and better inpaint the de-occluded face images.

Dataset Preparation. We take the CelebA-HQ as our face image dataset. In addition, we collect 362 glasses images, 324 mask images, and 1,000 hand images as the occlusion data.

Dataset Synthesis. As shown in Fig. 2, we first use the attributes annotations of CelebA-HQ to screen out the occlusion-free face images. Then, we use dlib and OpenCV to get the face pose angles of the face images. According to the pose angles, the face images are divided into eight groups: front looking up, front

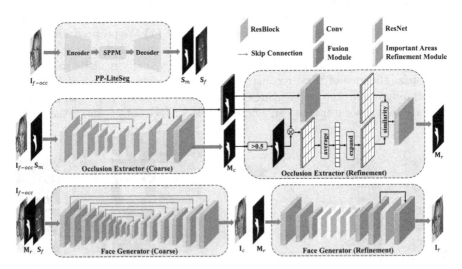

Fig. 3. The overall architecture of FRNet. It has three stages: structure prediction, occlusion extraction, and face inpainting.

looking down, the left deflection angles ranging from 10° to 40°, 40° to 60°, above 60°, and the right deflection angles ranging from 10° to 40°, 40° to 60°, and above 60°. We also classify glasses images and mask images based on the same angle ranges. Subsequently, we use face alignment to extract face features and get 68 face landmark points. Lastly, we combine the occlusion-free face images with glasses and masks based on the head poses and face landmarks. To enhance the realism of the synthetic images, we randomly vary the transparency of the glasses. To simulate real-life scenarios with various types and quantities of occlusions, we randomly add hand occlusions to some face images and both glasses and masks to others, reflecting the complexity of occlusions that may occur in real life. Following the original division of the CelebA, we collect 88,932 images as the training dataset, 11,120 images as the validation dataset, and 10,244 images as the testing dataset. Each set of data includes an original face image, an occluded face image, an occlusion image, and a binary occlusion mask image.

3 FRNet

In this section, we will introduce the overall architecture of FRNet, which is shown in Fig. 3. It has three stages: structure prediction, occlusion extraction, and face inpainting. The structure prediction stage aims to obtain prior information through the segmentation network. The occlusion extraction stage obtains the occlusion mask from the occluded face image with the guidance of occlusion position information. And face inpainting stage reconstructs the global structure with the guidance of face segment image, and refines the features by local information and feature information, to obtain an occlusion-free face image.

3.1 Occlusion Robust Face Segmentation Module

To better remove the occlusions and restore face structure, we propose the Occlusion Robust Face Segmentation Module, which aims to predict the occlusion segmentation map S_m and the face segmentation map S_f from the occluded face image I_{f-occ}. In this paper, we leverage the PP-LiteSeg [18] to realize this module (Fig. 3). To train the module, we fuse the CelebAMask-HQ with occlusion mask images. During the face de-occlusion process, this module provides prior information about the position of occlusion and face structure, which guides occlusion extraction and face reconstruction. This approach enables our model to focus on the deep structural feature of the face.

3.2 Occlusion Extractor

The occlusion extractor consists of a coarse stage and a refinement stage, in which input occluded face image I_{f-occ} and the prior information of occlusion position S_m, output coarse occlusion mask M_c and refinement occlusion mask M_r.

In the coarse stage, we build it based on the U-Net structure and refer to the details of the architecture from [6]. The architectural detail is shown in Fig. 3. Specifically, the encoder firstly increases the number of channels of the feature map to twice by 3×3 Conv-BN-ReLU module. This is followed by using three residual blocks to increase the receptive field of the module and further extract features. Finally, the spatial size of the feature map is reduced by down-sampling through the max pooling layer. The decoder performs the reverse operation. Firstly, the feature map is up-sampled by bilinear interpolation to expand its spatial size. Then the number of channels is reduced by half through the 3×3 Conv-BN-ReLU module, and the result is connected with the output feature map of the corresponding encoder using the skip connection to obtain the concatenated feature map. The concatenated feature map is passed through another 3×3 Conv-BN-ReLU module to reduce the number of channels by half. Finally, the latest feature map passes through three residual blocks to get the new feature map. The output of the last decoder is passed through 1×1 convolution and the sigmoid activation function to obtain the coarse occlusion mask $\mathbf{M}_c$. We optimize it using binary cross-entropy loss:

$$\mathcal{L}_{BCE}^{C} = -\sum_{i,j} \left(\mathbf{M}_{gt_{i,j}} \log \mathbf{M}_{ci,j} + \left(1 - \mathbf{M}_{gt_{i,j}}\right) \log\left(1 - \mathbf{M}_{ci,j}\right) \right), \quad (1)$$

where $\mathbf{M}_{gt_{i,j}}$ (resp., $\mathbf{M}_{ci,j}$) is the (i, j)-th entry in $\mathbf{M}_{gt}$ (resp., $\mathbf{M}_c$).

In the refinement stage, we refer to the Self-calibrated Mask Refinement proposed in [12]. As shown in Fig. 3, we conduct similarity matching between the main feature of the coarse occlusion mask $\mathbf{M}_c$ and features at other positions, thereby fine-tuning the occlusion mask according to its characteristics and enhancing the performance of the occlusion extractor. Through this stage, we obtain the refined occlusion mask $\mathbf{M}_r$. Similarly, we use binary cross-entropy loss $\mathcal{L}_{BCE}^{R}$, which is the same as Eq. (1) except for replacing $\mathbf{M}_{ci,j}$ with $\mathbf{M}_{ri,j}$ in Eq. (1). Additionally, we employ Intersection over Union (IoU) loss. It is defined as:

$$\mathcal{L}_{IoU}^{R} = \frac{\sum_{i,j} \left(\mathbf{M}_{gt_{i,j}} \cdot \mathbf{M}_{ri,j} \right)}{\sum_{i,j} \left(\mathbf{M}_{gt_{i,j}} + \mathbf{M}_{ri,j} - \mathbf{M}_{gt_{i,j}} \cdot \mathbf{M}_{ri,j} \right)}, \quad (2)$$

where $\mathbf{M}_{gt_{i,j}}$ (resp., $\mathbf{M}_{ri,j}$) is the (i, j)-th entry in $\mathbf{M}_{gt}$ (resp., $\mathbf{M}_r$).

Finally, the total loss of the occlusion extractor is

$$\mathcal{L}_{mask} = \lambda_1 \cdot \mathcal{L}_{BCE}^{C} + \lambda_2 \cdot \mathcal{L}_{BCE}^{R} + \lambda_3 \cdot \mathcal{L}_{IoU}^{R}. \quad (3)$$

We set $\lambda_1 = 0.01$, $\lambda_2 = 1$, $\lambda_3 = 0.25$ in the experiment.

3.3   Face Generator

Coarse Face Generator

We use the occluded face image $\mathbf{I}_{f-occ}$ and the predicted occlusion mask $\mathbf{M}_r$ as the input. Moreover, the predicted face segmentation map $\mathbf{S}_f$ is introduced to facilitate the face generation process. Then, the model returns a face image where the occluded area is reconstructed.

Following the [19], the coarse face generator (Fig. 3) uses a U-Net architecture with skip connections and has a large receptive field, which can better pay attention to the global feature of the image. Specifically, the coarse face generator consists of eight encoder-decoder blocks. Each encoder down-samples the feature map through the convolution layer, increasing its number of channels to twice while decreasing its size. The decoder up-samples the feature map to the original size through the transposed convolution layer to obtain coarse face image $\mathbf{I}_{out}^{c}$. According to the mask image $\mathbf{M}_r$, the input image $\mathbf{I}_{f-occ}$ and the generated face image $\mathbf{I}_{out}^{c}$ are combined to obtain a merged face image $\mathbf{I}_{mer}^{c}$,

$$\mathbf{I}_{mer}^{c} = \mathbf{I}_{f-occ} \odot (\mathbb{1} - \mathbf{M}_r) + \mathbf{I}_{out}^{c} \odot \mathbf{M}_r, \tag{4}$$

where $\odot$ is the element-wise product operation.

At this stage, we use the weighted L_1 loss [19] as the pixel-wise reconstruction loss which is defined as:

$$\mathcal{L}_{valid}^{C} = \frac{1}{\text{sum}(\mathbb{1} - \mathbf{M}_r)} \left\| (\mathbf{I}_{out}^{c} - \mathbf{I}_{gt}) \odot (\mathbb{1} - \mathbf{M}_r) \right\|_1, \tag{5}$$

$$\mathcal{L}_{occluded}^{C} = \frac{1}{\text{sum}(\mathbf{M}_r)} \left\| (\mathbf{I}_{out}^{c} - \mathbf{I}_{gt}) \odot (\mathbf{M}_r) \right\|_1, \tag{6}$$

$$\mathcal{L}_1^{C} = \mathcal{L}_{valid}^{C} + \lambda_o \cdot \mathcal{L}_{occluded}^{C}, \tag{7}$$

where $\text{sum}(\mathbf{M}_r)$ (resp., $\text{sum}(\mathbb{1}-\mathbf{M}_r)$) represents the number of non-zero elements in $\mathbf{M}_r$ (resp., $\mathbb{1} - \mathbf{M}_r$). And we set $\lambda_o = 6$.

Meanwhile, we add a patch-based discriminator with spectral normalization [14] and use the least square loss as the adversarial loss. It is defined as:

$$\mathcal{L}_G^{C} = \mathbb{E}_{\mathbf{I_{mer}} \sim p_{\mathbf{I}_{mer}}(\mathbf{I_{mer}})} \left[(D(\mathbf{I}_{mer}^{c}) - 1)^2 \right], \tag{8}$$

$$\mathcal{L}_D = \frac{1}{2} \mathbb{E}_{\mathbf{I} \sim p_{data}(\mathbf{I})} \left[(D(\mathbf{I}_{gt}) - 1)^2 \right] + \frac{1}{2} \mathbb{E}_{\mathbf{I_{mer}} \sim p_{\mathbf{I}_{mer}}(\mathbf{I_{mer}})} \left[(D(\mathbf{I}_{mer}^{c}))^2 \right]. \tag{9}$$

The main challenge of face de-occlusion is to maintain the face attributes before and after de-occlusion, which is crucial for supporting other face-related tasks. To address this, we introduce the gender-consistency loss. We obtain a gender classification model with an accuracy rate of 99.096%, by using the CelebA-HQ to perform transfer learning on the VGG-16. In the training process, the gender $\mathbf{P}_{gen}^{gt}$ of the ground-truth $\mathbf{I}_{gt}$ is used as the target, and the generated face image $\mathbf{I}_{mer}^{c}$ is classified by using the classification model to obtain the classification result $\mathbf{P}_{gen}^{c}$. Finally, we calculate the cross entropy between the classification result and the target. So the gender-consistency loss is defined as:

$$\mathcal{L}_{gen}^{C} = - \left(\mathbf{P}_{gen}^{gt} \log \mathbf{P}_{gen}^{c} + (1 - \mathbf{P}_{gen}^{gt}) \log (1 - \mathbf{P}_{gen}^{c}) \right), \tag{10}$$

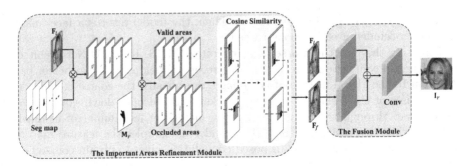

Fig. 4. The architectures of the Important Areas Refinement Module (IRM) and the Fusion Module.

where $\mathbf{P}_{gen}^{gt}$ (*resp.*, $\mathbf{P}_{gen}^{c}$) represents the probability that the image $\mathbf{I}_{gt}$ (*resp.*, $\mathbf{I}_{mer}^{c}$) belongs to male. Finally, the total loss for the coarse face generator is

$$\mathcal{L}_{face}^{C} = \mathcal{L}_1^{C} + \lambda_G \cdot \mathcal{L}_G^{C} + \lambda_{gen} \cdot \mathcal{L}_{gen}^{C}, \tag{11}$$

and we set $\lambda_G = 0.1$, $\lambda_{gen} = 1$.

Refinement Face Generator

Local Areas Refinement Module (LRM). As shown in Fig. 3, the main structure of the LRM is similar to the model proposed in [19]. Specifically, it consists of two down-sampled encoders, four residual blocks, and two up-sampled decoders. Besides, the feature map is padded using the reflection of its boundary before down-sampling and after up-sampling, which makes the edge information better preserved during the convolution processes. We use a shallow neural network to obtain information on surrounding pixels to locally refine the missing areas. In practice, we feed the coarse face image $\mathbf{I}_{mer}^{c}$ and the predicted occlusion mask $\mathbf{M}_r$ into this module to get the local refinement feature map $\mathbf{F}_l$.

Important Areas Refinement Module (IRM). By exploring the similarity of pixels in the same face feature, we propose the IRM, which uses similarity to find valid pixels related to missing areas. This module maintains harmony within the feature and achieves the goal of feature reconstruction. For example, when only one eye is occluded, we can use the features in the occlusion-free eye to inpaint the occluded one, ensuring consistency in features such as the eyeball color. As shown in Fig. 4, we feed the feature map $\mathbf{F}_l$ into the IRM and get the segmentation map of the rough face image $\mathbf{I}_{mer}^{c}$ by using the face segmentation network, then use the segmentation map to get five important areas (brow, eye, nose, lip, mouth) of the feature map. If the current important area contains both a valid area and an occluded area, we calculate cosine similarity between feature points in the valid area and those in the occluded area. Then using the most similar features to fill each feature point in the occluded area. If the current important area is all occluded area, we use the prediction information of LRM to fill them. If the current important area is all valid area, their feature points will not be changed. Through this module, we can acquire the feature map $\mathbf{F}_f$ of feature refinement.

Fusion Module. The LRM inpaints the occluded area using surrounding pixels through the shallow neural network, which enhances the local details. However, the features of face images have particularities, with specific correlations existing between different local areas. For example, the local features of the two eyes should exhibit similarity. However, it is difficult to obtain information from the valid area related to the occluded area in this module, leading to the output result may be discordant within the features. The IRM inpaints the occluded area using valid pixels within the features by calculating the cosine similarity, ensuring harmony within the features. But it cannot inpaint areas without valid information and has difficulty handling images with the large occluded area. Therefore, we adaptively fuse the local refined feature map $\mathbf{F}_l$ with the feature refined feature map $\mathbf{F}_f$, and then up-sample them to the size of the input image to get a refined face image $\mathbf{I}_{out}^r$ and a merged face image $\mathbf{I}_{mer}^r$,

$$\mathbf{I}_{mer}^r = \mathbf{I}_{f-occ} \odot (1 - \mathbf{M}_r) + \mathbf{I}_{out}^r \odot \mathbf{M}_r. \tag{12}$$

Like the coarse stage, we use the weighted L_1 loss as the pixel-wise reconstruction loss $\mathcal{L}_1^R$, which is the same as Eq. (7) except for replacing $\mathbf{I}_{out}^c$ with $\mathbf{I}_{out}^r$ in Eq. (5) and Eq. (6). Meanwhile, following [19], we apply perceptual loss and style loss to the model using the VGG-16 which is pre-trained based on ImageNet. The perceptual loss is defined as:

$$\mathcal{L}_{per}^R = \sum_i \|\phi_i(\mathbf{I}_{out}^r) - \phi_i(\mathbf{I}_{gt})\|_1 + \|\phi_i(\mathbf{I}_{mer}^r) - \phi_i(\mathbf{I}_{gt})\|_1, \tag{13}$$

where ϕ_i means the feature map of i-th layer in pre-trained VGG-16 network ($i \in \{5, 10, 17\}$).

The style loss is defined as:

$$\mathcal{L}_{sty}^R = \sum_i \|\mathcal{G}_i(\mathbf{I}_{out}^r) - \mathcal{G}_i(\mathbf{I}_{gt})\|_1 + \|\mathcal{G}_i(\mathbf{I}_{mer}^r) - \mathcal{G}_i(\mathbf{I}_{gt})\|_1, \tag{14}$$

where $\mathcal{G}_i(\cdot) = \phi_i(\cdot)\phi_i(\cdot)^T$ is the Gram matrix.

Adding total variation (TV) loss as the smoothing penalty, defined as:

$$\mathcal{L}_{tv}^R = \|\mathbf{I}_{mer}^r(i, j+1) - \mathbf{I}_{mer}^r(i,j)\|_1 + \|\mathbf{I}_{mer}^r(i+1, j) - \mathbf{I}_{mer}^r(i,j)\|_1. \tag{15}$$

Like previous work [8], an identity loss is defined on the face recognition network. It is defined as:

$$\mathcal{L}_{id}^R = 1 - \frac{\mathbf{F}_{id}^{gt} \cdot \mathbf{F}_{id}^r}{\max\left(\|\mathbf{F}_{id}^{gt}\|_2 \cdot \|\mathbf{F}_{id}^r\|_2, \epsilon\right)}, \tag{16}$$

where $\mathbf{F}_{id}^{gt}$ and $\mathbf{F}_{id}^r$ are the output vectors of the face recognition network for $\mathbf{I}_{gt}$ and $\mathbf{I}_{mer}^r$. Respectively, ϵ sets to very small values 1e-8.

Finally, the total loss for the refinement face generator is

$$\mathcal{L}_{face}^R = \mathcal{L}_1^R + \lambda_{per} \cdot \mathcal{L}_{per}^R + \lambda_{sty} \cdot \mathcal{L}_{sty}^R + \lambda_{tv} \cdot \mathcal{L}_{tv}^R + \lambda_{id} \cdot \mathcal{L}_{id}^R, \tag{17}$$

and we set $\lambda_{per} = 0.05$, $\lambda_{sty} = 120$, $\lambda_{tv} = 0.1$, $\lambda_{id} = 1$.

Fig. 5. Qualitative results on our synthetic dataset (top) and Gender Occlusion Data (bottom).

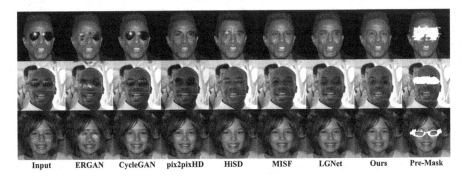

Fig. 6. Qualitative results on CelebA-HQ (top), FFHQ (middle), and MeGlass (bottom).

4   Experiments

4.1   Experimental Settings

Datasets. We train the FRNet using the synthetic dataset proposed in Sect. 2. Furthermore, we also use real-world portrait datasets including CelebA-HQ, FFHQ [9], MeGlass [3], RMFD [22], and the masked face synthesis dataset Gender Occlusion Data [15] to test the model.

Implementation Details. Our method is implemented with PyTorch 1.7.0 using a 24G NVIDIA GTX3090 GPU. And we train the model by the Adam optimizer with $\beta_1 = 0.5$ and $\beta_2 = 0.999$. In practice, we first train the occlusion extractor for 10 epochs, then fix the parameters of the occlusion extractor and train the face generator for 40 epochs. We set the learning rate to 0.0002 when training the occlusion extractor. As for the training of the face generator, we set the learning rate to 0.0002 for the first 20 epochs and linearly decay it to zero for the next 20 epochs.

4.2   Comparisons with State-of-the-Art Methods

We compare our method with state-of-the-art image inpainting methods including CTSDG [4], MADF [26], DSNet [20], WaveFill [23], AOT-GAN [24], MISF [10], LGNet [19], image translation methods including CycleGAN [25], pix2pixHD [21], and glasses removal methods including ERGAN [7], HiSD [11].

Qualitative Comparison. As shown in Fig. 5, Fig. 6 and Fig. 8, our method can restore the global structures and details of the faces more effectively, while maintaining consistency within the features. Furthermore, our model can be directly extended to other datasets, enabling the identification and removal of multiple occlusions while producing visually realistic results without retraining. (More experimental results can be found in https://github.com/dss9964/FRNet.)

Table 1. Quantitative results on our synthetic dataset and Gender Occlusion Data. The best two results are shown in red and blue respectively.

Model	Our Synthetic Dataset			Gender Occlusion Data		
	FID↓	LPIPS↓	SSIM↑	FID↓	LPIPS↓	SSIM↑
pix2pixHD	5.31	0.112	0.843	11.95	0.157	0.803
CTSDG	2.06	0.046	0.938	12.20	0.139	0.853
MADF	3.26	0.049	0.936	14.87	0.137	0.850
DSNet	1.94	0.043	0.939	12.08	0.137	0.855
WaveFill	2.05	0.044	0.935	14.59	0.135	0.854
AOT-GAN	2.13	0.043	0.933	11.55	0.135	0.855
MISF	1.77	0.041	0.942	11.78	0.134	0.858
LGNet	1.87	0.042	0.942	15.49	0.137	0.856
Ours	1.85	0.039	0.945	10.50	0.132	0.858

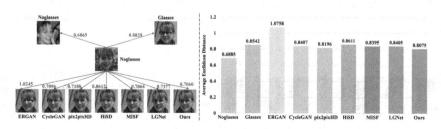

Fig. 7. The results of identity preservation.

Quantitative Comparison

Realism. As shown in Table 1, our method is comparable to MISF and LGNet on the proposed synthetic dataset and achieves the best results on the Gender Occlusion Data. These results indicate that our model possesses better generalization performance and a stronger ability to handle large-scale occlusions.

Identity Preservation. To demonstrate the positive impact of de-occlusion on face recognition, we collected 1,032 sets of face images from MeGlass. Each set consisted of two images without glasses and one image with glasses of the same identity. Various occlusion removal methods were applied to the images with glasses to generate corresponding images without glasses. The Euclidean Distance between the first image without glasses and all other images was then calculated. As shown in Fig. 7, the presence of occlusions significantly increased the Average Euclidean Distance between the occluded face images and the target images. However, our method effectively preserves the identity information of the face images, minimizing the Euclidean Distance. It mitigated the detrimental effects of occlusion to a certain extent, ultimately enhancing the accuracy of face recognition.

4.3 Ablation Studies

In this subsection, we evaluate the performance of our key contributions in occlusion extraction and face inpainting.

Occlusion Extraction. MS and MR represent occlusion location information and the refinement stage of the occlusion extractor respectively. Table 2 shows that they both have positive effects on occlusion extraction.

Face Inpainting. $\mathcal{L}_{gen}$, $\mathcal{L}_{id}$, and FS represent the gender loss, the id loss, and the face structure information respectively. GFNet represents moving the IRM to the coarse stage. As shown in Table 2, the network with all proposed modules achieves the best performance, indicating the effectiveness of our proposed strategy and loss functions. Figure 9 demonstrates that the quality of images generated by w/o FS is the worst, with noticeable blurriness. The images generated by w/o FR exhibit internal feature inconsistency, such as the eyes in the 2nd row. w/o $\mathcal{L}_{gen}\&\mathcal{L}_{id}$ and w/o GF can effectively inpainting the images, but lack detail in some areas. These qualitative results also reflect the advantages of the FRNet.

Input CycleGAN pix2pixHD MISF LGNet Ours Pre-Mask

Fig. 8. Qualitative results on RMFD.

Table 2. Quantitative comparison of different ablations in occlusion extraction (top) and face inpainting (bottom) on our synthetic dataset.

Input w/o IRM w/o $\mathcal{L}_{gen}\&\mathcal{L}_{id}$ w/o FS GFNet Ours

Fig. 9. Qualitative comparison of different ablations in face inpainting on our synthetic dataset.

Model	FID↓	LPIPS↓	SSIM↑
w/o MR	3.71	0.018	0.987
w/o MS	2.32	0.015	0.987
Ours	**2.26**	**0.014**	**0.990**
w/o IRM	1.94	0.041	0.942
w/o $\mathcal{L}_{gen}\&\mathcal{L}_{id}$	1.89	0.041	0.943
w/o FS	2.17	0.042	0.944
GFNet	1.90	0.041	0.944
Ours	**1.85**	**0.039**	**0.945**

5 Conclusion

In this paper, we propose a new face de-occlusion framework based on feature reconstruction (FRNet), which consists of three stages: structure prediction, occlusion extraction, and face inpainting. In the inpainting process, the global structures are reconstructed with the guidance of the face segment images, and the important areas are refined by local information and feature information. Besides, we build a high-quality synthetic occluded face dataset, which provides supervision for the training of models. Qualitative and quantitative experiment results demonstrate that our model can effectively remove face occlusions and retain attribute information to support face-related tasks.

References

1. Barnes, C., et al.: The patchmatch randomized matching algorithm for image manipulation. Commun. ACM **54**(11), 103–110 (2011)
2. Dong, J., et al.: Occlusion-aware GAN for face de-occlusion in the wild. In: ICME, pp. 1–6. IEEE (2020)
3. Guo, J., Zhu, X., Lei, Z., Li, S.Z.: Face synthesis for eyeglass-robust face recognition. In: Zhou, J., et al. (eds.) CCBR 2018. LNCS, vol. 10996, pp. 275–284. Springer, Cham (2018). https://doi.org/10.1007/978-3-319-97909-0_30
4. Guo, X., et al.: Image inpainting via conditional texture and structure dual generation. In: ICCV, pp. 14134–14143 (2021)
5. He, K., et al.: Computing nearest-neighbor fields via propagation-assisted KD-trees. In: CVPR, pp. 111–118. IEEE (2012)
6. Hertz, A., et al.: Blind visual motif removal from a single image. In: CVPR, pp. 6858–6867 (2019)
7. Hu, B., et al.: Unsupervised eyeglasses removal in the wild. TCYB **51**(9), 4373–4385 (2020)
8. Ju, Y.J., et al.: Complete face recovery GAN: unsupervised joint face rotation and de-occlusion from a single-view image. In: WACV, pp. 3711–3721 (2022)
9. Karras, T., et al.: A style-based generator architecture for generative adversarial networks. In: CVPR, pp. 4401–4410 (2019)
10. Li, X., et al.: MISF: multi-level interactive siamese filtering for high-fidelity image inpainting. In: CVPR, pp. 1869–1878 (2022)
11. Li, X., et al.: Image-to-image translation via hierarchical style disentanglement. In: CVPR, pp. 8639–8648 (2021)
12. Liang, J., et al.: Visible watermark removal via self-calibrated localization and background refinement. In: ACM MM, pp. 4426–4434 (2021)
13. Liu, M., et al.: STGAN: a unified selective transfer network for arbitrary image attribute editing. In: CVPR, pp. 3673–3682 (2019)
14. Miyato, T., et al.: Spectral normalization for generative adversarial networks. arXiv preprint arXiv:1802.05957 (2018)
15. Modak, G., et al.: A deep learning framework to reconstruct face under mask. In: CDMA, pp. 200–205. IEEE (2022)
16. Nazeri, K., et al.: Edgeconnect: structure guided image inpainting using edge prediction. In: ICCV (2019)
17. Pathak, D., et al.: Context encoders: feature learning by inpainting. In: CVPR, pp. 2536–2544 (2016)

18. Peng, J., et al.: PP-LiteSeg: a superior real-time semantic segmentation model. arXiv preprint arXiv:2204.02681 (2022)
19. Quan, W., et al.: Image inpainting with local and global refinement. TIP **31**, 2405–2420 (2022)
20. Wang, N., et al.: Dynamic selection network for image inpainting. TIP **30**, 1784–1798 (2021)
21. Wang, T.C., et al.: High-resolution image synthesis and semantic manipulation with conditional GANs. In: CVPR, pp. 8798–8807 (2018)
22. Wang, Z., et al.: Masked face recognition dataset and application. arXiv preprint arXiv:2003.09093 (2020)
23. Yu, Y., et al.: Wavefill: a wavelet-based generation network for image inpainting. In: ICCV, pp. 14114–14123 (2021)
24. Zeng, Y., et al.: Aggregated contextual transformations for high-resolution image inpainting. TVCG (2022)
25. Zhu, J.Y., et al.: Unpaired image-to-image translation using cycle-consistent adversarial networks. In: ICCV, pp. 2223–2232 (2017)
26. Zhu, M., et al.: Image inpainting by end-to-end cascaded refinement with mask awareness. TIP **30**, 4855–4866 (2021)

Semantically Guided Bi-level Adaptation for Cross Domain Crowd Counting

Muming Zhao[1(✉)], Weiqing Xu[2], and Chongyang Zhang[2]

[1] Beijing Forestry University, Beijing, China
mumingzhao@bjfu.edu.cn
[2] School of Electronic Information and Electrical Engineering, Shanghai Jiao Tong University, Shanghai, China

Abstract. Visual crowd counting has played an important role in various practical applications. However, domain gap remains a major barrier preventing models trained on the source domain (e.g., training scenes) generalize well to the target domain (e.g., unseen testing scenes). Crowd semantic information are shown to be beneficial to assist crowd counting in supervised training settings, implying the close relationship between crowd density and semantics. Nevertheless, the potential of this powerful cue has bot been fully explored in the unsupervised domain adaptation (UDA) setting. Motivated by the observation that crowd density map share domain-invariant correspondence with the crowd segmentation map, we propose to adapt this correspondence correlation from the source domain to the target domain to address the domain gap. To this end, a semantically guided task correlation layer is introduced to extract the task correspondence map, whose coherence is enforced across domains by adversarial training. To drive the adaption of earlier hidden layers directly, we further align the task correspondence correlation upon intermediate-level outputs. Extensive experiments are conducted on three benchmark datasets. The performances of our method either surpass or are on par with the counterparts, demonstrating the effectiveness of the proposed approach for cross-domain crowd counting.

Keywords: Crowd counting · Task correspondence · Domain adaptation

1 Introduction

Aiming to monitor the crowd distribution and the total amount of pedestrians from images or videos, visual crowd counting has gained much attention nowadays due to its widely applications in public security, resource management and traffic surveillance [1], etc. Although significant progresses have been witnessed for crowd counting in the supervised training setting, it remains challenging to generalize a model trained on one domain to another domain due to the domain gap. Specifically, images in different datasets are usually collected from scenes with different perspective, various crowd appearances and background clutters, resulting in degraded model performances.

Q. Liu et al. (Eds.): PRCV 2023, LNCS 14435, pp. 327–338, 2024.
https://doi.org/10.1007/978-981-99-8552-4_26

Most crowd counting approaches address the domain shift issue by using unsupervised domain adaptation (UDA) techniques [2]. For example, adversarial-based counting methods [3–5] aim to align the source and target feature distributions and learn domain-invariant representations, while generative-based approaches [6,7] translate images between the source and target domain to mitigate the domain gap. Despite their effectiveness, most previous methods merely exploit the conventional density information for adaptation. Actually, semantic information has been shown to be beneficial to improve counting performances under the supervised training setting in various works [8], which indicates the closely coupled relationships between crowd density and semantics. Nevertheless, the potential of this informative cue has not been fully explored for the UDA training. In this work, we aim to exploit the semantic information for the UDA crowd counting.

As illustrated in [8], crowd density and segmentation is closely coupled since density values should reveal both numeric and semantic implications for accurate counting. More specifically, the spatial layout of density values in the crowd density map should match with the foreground layout in the segmentation map. In other words, the predicted density values in background regions should always be suppressed while meaningful density predictions should mostly lie in the foreground regions. The correspondence between these tasks holds regardless of domains. Thus we hypothesize that explicitly adapting the task correspondence correlations between crowd density and segmentation would be beneficial to bridge the domain gap. Compared to the convention method of adapting high-dimensional features that are prone to drift due to the domain gap, the task correspondence correlation is naturally domain-invariant and will be affected less by the domain gap, which could provide reasonable guidance for model adaptation.

Motivated by these observations, in this paper we propose to utilize the task correspondence correlation between crowd density map and segmentation map to address the domain shift problem. Specifically, we present a semantically guided bi-level adaptation framework. The semantic information is embedded into the backbone model using an auxiliary task that predicts crowd semantic segmentation. On top of the multi-task net, we build a task correlation layer to capture the correspondence correlation of the joint predictions at the output space, which is then aligned between the source domain and the target domain for model adaptation. To directly drive the learning of earlier hidden layers, we further impose the task correlation adversarial training on the intermediate outputs as complementary supervisory signal. By enforcing the coherence of task correspondence across different domains, the consistency between the predicted crowd density and segmentation is encouraged, thus improving the domain adaptation performances on the target domain. The semantic segmentation task is supervised via pseudo segmentation masks, which can be easily derived from the original dot annotations in counting datasets and no extra annotation burden is imposed. Moreover, the auxiliary branches added during the UDA training can be painlessly removed once training finishes, empowering the model with cross-domain capacity yet incurring none of extra computations at inference.

In summary, the contribution of this paper are the following:

– We exploit the semantic information in the UDA training for crowd counting and propose to adapt the correspondence correlation between crowd density map and segmentation map to address the domain shift problem.
– We propose a semantically guided bi-level framework, through which the task correlation is extracted and aligned at the final output space as well as the intermediate output space.
– Extensive experiments demonstrate that the performances of the proposed method either surpass or are on par with its counterparts for cross domain crowd counting. We also report ablation studies to provide insights into the proposed approach.

2   Related Work

Crowd Counting. Visual crowd counting refers to the task of estimating the total number of people in an image. Conventional methods mainly rely on hand-crafted features, either combined with detection-based pipelines to delineate individuals or regression functions to learn the feature-to-count mapping [1]. With the revolutionary concept of deep CNNs [9], Zhang *et al.* [10] are among the first ones to train a deep model to estimate the number of people. Following counting approaches mainly focus on the handling of scale variations via the construction of multi-scale features [11] and design effective training loss functions to make fully usage of the ground truth dot annotation [12].

Cross Domain Crowd Counting. Existing methods have achieved significant progress when they train and test in one uniform domain, however the performances usually degraded when large domain variations faced, e.g., datasets with distinct scenes or crowd distributions. To this end, various methods have been proposed to improve cross domain counting accuracy, among which unsupervised domain adaptation (UDA) has received a lot of attention [2]. Wang *et al.* [13] perform adversarial training with pyramid patches of multi-scales from both source- and target-domain to regularize the model training. Zhang *et al.* [14] propose to model domain-related information with attention mechanism during in adversarial feature aligning. Gone *et al.* [5] align both the data and features simultaneously to mitigate the domain gap from synthetic data to realistic data. Beyond the adversarial-based methods, generative techniques [15] are also employed. For instance, SE Cycle-GAN *et al.* [7] transfer synthetic images into real-world scenes to bridge the domain gap. Beyond using the UDA technique, there are works [16,17] aims to build models that only uses data from the source domain yet generalizes well to other domains. Desipte their effectiveness, most existing methods neglect the semantic information which has been demonstrated to be helpful for supervised crowd counting [8]. Pioneering attempts have been made in [4] which uses the task of semantic label prediction to provide additional supervisory signal for both of the source images and the target images. In this paper we also exploit the semantic information however in a way that

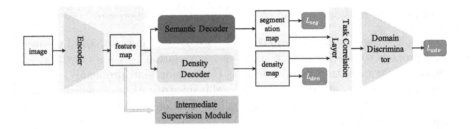

Fig. 1. Illustration of the proposed method. Note that all the auxiliary components except the encoder and the density decoder are only employed to guide the domain adaptation training, and can be removed at inference, without incurring additional computations at test.

considering the task correlation between the crowd density and segmentation, which has not been explored for cross domain crowd counting to the best of our knowledge. The effectiveness of exploiting joint task correlation for domain adaptation has been validated in other related areas such as semantic segmentation [18] with geometric cues e.g., depth is usually adopted for assistance, whose success inspires us to explore its benefits for crowd counting.

3   Semantically Guided Bi-level Adaptation

In the conventional UDA setup for crowd counting, we are given labeled data from the source domain S: $\mathcal{X}_s = \{(X_i^s, Y_i^s)\}_{i=1}^{N_s}$, where X_i^s and Y_i^s denotes the i-th crowd image and its corresponding dot annotation of head center positions, N_s is the total number of labeled source crowd images. For the target domain T, we are only accessible to the unlabeled data $\mathcal{X}_t = \{(X_i^s)\}_{i=1}^{N_t}$, where N_t is the number of unlabeled target training samples.

3.1   Framework Overview

The overall framework of our proposed semantically guided bi-level adaptation approach is illustrated in Fig. 1. The backbone model G is constituted of an encoder G_{enc} extracting high-dimensional features from the input image and a density decoder G_{den} deriving the crowd density map from the features emitted by the encoder. Our aim is to adapt the network G from the source domain to the target domain, taking advantage of semantic information.

Semantically Guided Task Correlation. As described in Sect. 1, the primary hypothesis behind our approach is that the crowd density and segmentation of an image share spatial correspondence no matter which domain the image lies in. Thus we can leverage this consistent information from source samples and transfer it to targets to improve counting performances on the target domain. To this end, we embed semantic information by considering crowd semantic segmentation as an auxiliary task, which is accomplished by an additional segment decoder G_{seg} besides to the density decoder. To capture the correspondence correlation, we further build a task correlation layer F to infer a correspondence map by fusing the joint predictions at the output space. Since semantic information is involved in this process, we dubbed the operation F as *semantically guided*.

Bi-level Adaptation Training. We apply adversarial training to adapt the coherent output-level task correlations across domains. However, this supervisory signal operates on the final predictions, i.e., at the very end of the backbone model, which can be dubbed as the **output level adaptation**. To further sparkle the adaption of earlier hidden layers, we also encourage the coherence of task correlations from intermediate output space with **intermediate-output level adaptation**, anticipating the gradient information directly propagated to earlier layers of the network to improve the training efficiency. To enable the intermediate-output level adaptation, an intermediate supervision module is built on high-dimensional feature maps to generate intermediate outputs for two tasks as well, whose correlation will be further aligned for domain adaptation.

3.2   Semantically Guided Task Correlation

Based on our hypothesis that crowd density map share coherent layout distribution with the crowd segmentation mask, we seek to encode such correspondence between the two predicted outputs. Toward this goal, the task correlation layer F is implemented as the *element-wise product* based on the predicted density map $\widehat{Y}$ and the segmentation map $\widehat{M}$:

$$R = F(\widehat{Y}, \widehat{M}) = \widehat{Y} \odot \widehat{M}, \tag{1}$$

where R is the output task correlation map, $\odot$ denotes the Hadamard matrix product operation with $(A \odot B)_{i,j} = (A)_{i,j}(B)_{i,j}$.

The implication for using this multiplicative operation is illustrated as follows. For the source domain, the two task decoders are well trained with explicit labels, and hence the distribution of crowd density on the predicted density map and that of the foreground/background on the predicted segmentation map are mostly accurate and thus corresponds with each other. In this situation, the multiplicative operation will generate responses with higher values in the foreground regions and lower values in the background regions. While for the target domain, labels are available for none of the two tasks, and the quality of either the density prediction or the segment prediction cannot be guaranteed. Thus situations when the two predictions contradict with each other may occur, e.g., larger density values on the density map probably lie in the background region of the segmentation map. In this situation, the multiplicative operation will degrade values both from the density map and the segmentation mask, resulting to a correspondence map that does not match the patterns as in the source domain. Enforcing the coherence of the task correspondence map will implicitly encourage consistent crowd density and segmentation prediction in the target domain, thus facilitating the domain adaptation efficiency. Despite of its simplicity, the effectiveness of the multiplicative operation is experimentally validated in the ablation study.

3.3   Bi-level Adaptation Training

To address the domain gap more effectively, adversarial-based domain adaptation is performed jointly on two levels, namely output level and intermediate-output level.

Output Level Adversarial Training. During in the output level adaptation, the task decoders G_{seg} and G_{den} concurrently performs semantic segmentation and crowd density estimation for a given input image. Since there is no explicit supervision available for target semantic segmentation, we share the segment decoder for both domains.

We follow [1] and apply 2D same-spread Gaussian kernels on each dotted position to obtain the ground-truth crowd density map $\widetilde{Y}^s$ from the original dotted annotations Y^s provided in the source domain. The ground-truth labelling of crowd foreground are usually not provided in existing crowd counting datasets, and we derive the pseudo ground-truth mask M^s with the binarization transformation on the ground-truth density maps, following the practice in [8]. Subsequently, supervised training can be performed for images of the source domain, with the common practices of using MSE loss L_{den} for density estimation and the standard cross-entropy loss L_{seg} for crowd semantic segmentation:

$$L_{den} = \mathbb{E}_{X^s \sim \mathcal{X}_s} \left[\left\| \widehat{Y}^s - \widetilde{Y}^s \right\|_2^2 \right]$$
$$L_{seg} = \mathbb{E}_{X^s \sim \mathcal{X}_s} \left[CE(\widehat{M}^s, M^s) \right], \tag{2}$$

in which the image width and height dimension are omitted for the sake of simplicity.

To adapt the network to the target domain, a discriminator D is further applied to the outputs. Our work consider the task correlation instead of the direct outputs, as the correspondence between crowd density map and segmentation mask usually holds regardless of domains. In particular, we use the correspondence map R generated by the task correlation layer to train the discriminator D, which learns to distinguish correlation maps between the source and target domain. The loss can be written as:

$$L_d = \mathbb{E}_{X^s \sim \mathcal{X}_s} \left[\log D(R^s) \right] + \mathbb{E}_{X^t \sim \mathcal{X}_t} \left[\log(1 - D(R^t)) \right] \tag{3}$$

At the same time, the parameters of the underlying task prediction network (G and G_{seg}) are learned to fool the discriminator D by maximizing the probability of the target correspondence map being regarded as the correspondence map from the source domain, with an adversarial loss L_{adv}:

$$L_{adv} = \mathbb{E}_{X^t \sim \mathcal{X}_t} \left[\log D(R^t) \right] \tag{4}$$

Intermediate-Output Level Adversarial Learning. Inspired by the Hourglass network [19] that supervises intermediate outputs, we also encourage the coherence of the task correspondence map from intermediate predictions. Specifically, the intermediate supervision module consists two task decoders $\widetilde{G}_{seg}$ and $\widetilde{G}_{den}$, a task correlation layer, and a discriminator $\widetilde{D}$. In the intermediate-output level, the supervised training and the adversarial training are in the same form as for the final output level, and we can rewrite Eq. 2 and Eq. 4 to the overall training objective of the model $\mathcal{G}$, which encompasses the net G, G_{seg}, $\widetilde{G}_{seg}$ and $\widetilde{G}_{den}$:

$$\mathcal{L} = \sum_i \lambda^i_{den} L^i_{den} + \lambda^i_{seg} L^i_{seg} + \lambda^i_{adv} L^i_{adv}, \tag{5}$$

where $i \in (0,1)$ indicates the level that the task predictions are from (0 as the intermediate-output level and 1 as the output level), λ^i_{den}, λ^i_{seg} and λ^i_{adv} denotes the weighting factor for the corresponding loss at the i-th level, respectively. Based on Eq. 5, the prediction model and the discriminators could be optimized using the following min-max criterion:

$$\max_{\mathcal{D}} \min_{\mathcal{G}} \mathcal{L}(X^s, X^t), \tag{6}$$

where $\mathcal{D}$ contains parameters from the two discriminators D and $\widetilde{D}$ on the two output levels. The ultimate goal is to jointly minimize the density estimation loss and the semantic segmentation loss in both output levels for source images, while maximizing the probability of the correspondence map from the target domain being taken as that from the source domain at two output levels to train, respectively.

4   Network Architecture and Implementation Details

We instantiate the backbone crowd counting model G with the commonly used CSRNet [20], which adapts VGG16 network for crowd counting with dilation processing. The architecture of the auxiliary segment decoder G_{seg} at the output level is the same with the density decoder, i.e., the decoder network in the CSRNet. For the intermediate predictions, we use a lightweight architecture with only two convolutional layers interleaved by one ReLU layer. The kernel size of each convolutional layer is 3×3, where the output channel number is 64 and 1, respectively. Still, the two intermediate task decoders $\widetilde{G}_{seg}$ and $\widetilde{G}_{den}$ share the same architecture but are trained separately. For the adversarial learning, we use DC-GAN [21] with 4 convolutional layers with leaky-ReLUs as activation functions.

We implement the network using the Pytorch deep learning framework [27]. The train the prediction network, we use the Stochastic Gradient Descent (SGD) optimization with momentum as 0.9 and weight decay as 5×10^{-4}. The initial learning rate is set to 10^{-4} and is decreased using the polynomial decay with power of 0.9. For training the discriminators, we use the Adam optimizer with

Table 1. Evaluation of cross-domain crowd counting performance: (top) no adaptation methods that does not perform adaptation to specific domains; (bottom) UDA methods using unlabeled data of target domains. The best performance within each category is highlighted bold, with 2nd best underlined.

Type	Method	A→B		A→Q		B→A		B→Q	
		MAE	MSE	MAE	MSE	MAE	MSE	MAE	MSE
No adapt	MCNN (CVPR'16) [11]	85.2	142.3	-	-	221.4	357.8	-	-
	D-ConvNet (CVPR'18) [22]	49.1	99.2	-	-	140.4	226.1	-	-
	SPN + L2SM (ICCV'19) [23]	21.2	38.7	227.2	405.2	126.8	203.9	-	-
	RegNet (ICCV'19) [24]	21.7	37.6	198.7	329.4	148.9	273.9	267.3	477.6
	DetNet (CVPR'19) [25]	55.5	90.0	411.7	731.4	242.8	400.9	411.7	731.4
	D2CNet (TIP'21) [26]	21.6	34.6	126.8	245.5	164.5	286.4	267.5	486.0
	C^{2}MOT (MM'21) [17]	**12.4**	**21.1**	125.7	218.3	120.7	192.0	198.9	368.0
	DICM (AAAI'23) [16]	12.6	24.6	119.4	216.6	121.8	203.1	179.1	316.2
UDA	SE Cycle GAN (CVPR'19) [7]	19.9	28.3	230.4	384.5	123.0	193.4	230.4	384.5
	CODA (ICME'19) [13]	15.9	26.9	-	-	-	-	-	-
	SE+FD (ICASSP'20) [4]	16.9	24.7	221.2	390.2	129.3	187.6	221.2	390.2
	RBT (MM'20) [3]	13.4	29.3	175.0	294.8	112.2	218.2	211.3	381.9
	LDG (TMM'22) [6]	14.2	25.2	179.9	331.3	118.5	190.1	261.1	496.0
	CDANet (TCSVT'22) [14]	13.5	**22.3**	169.2	308.0	**106.5**	**162.5**	232.3	415.8
	Ours	**13.3**	23.9	**141.3**	**241.1**	114.1	190.5	215.3	382.4

the learning rate of 4×10^{-5}. The hyperparameters to balance various loss functions in Eq. 5 are selected experimentally. λ_{den}^1, λ_{seg}^1 and λ_{adv}^1 are set to 1, 1, and 0.001 at the final output level, respectively. At the intermediate-output level, λ_{den}^0, λ_{seg}^0 and λ_{adv}^0 are set to 0.1, 0.1 and 0.0002, respectively.

5 Experiments

5.1 Datasets

We verify the effectiveness of the proposed method for UDA crowd counting using three publicly-available crowd counting datasets, i.e., ShanghaiTech PartA (SHA) [11], ShanghaiTech PartB (SHB) [11] and UCF-QNRF [28]. The ShanghaiTech [11] dataset consists of two parts: SHA and SHB, which respectively contain 300/400 training images and 182/316 testing images. Images in SHB are collected from surveillance cameras on streets, while images in SHA are crawled from the Internet with dense crowd. UCF-QNRF [28] contains 1535 high resolution images with 1,201 for training and 334 for testing, which is a large-scale dataset covering a wide range of crowd densities. Following the convention of existing work [1], metrics of the mean absolute error (MAE) and the mean square error (MSE) are computed for evaluation on these datasets.

5.2   Comparison with State-of-the-Art

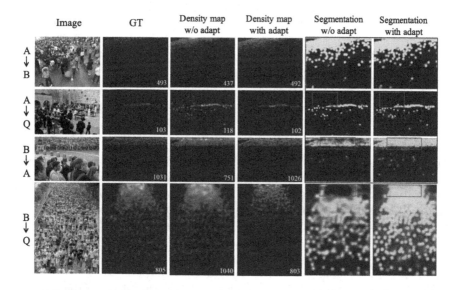

Fig. 2. Visualization and comparison. The first two columns show several sample images and their corresponding ground-truth density maps. The subsequent two columns respectively show the predicted density maps without and with the proposed adaptation technique to the target domain, while the last two columns compare the crowd segmentation maps without and with domain adaptation, respectively. Count estimation are labeled at the bottom right on each density map.

We compare our results to state-of-the-art methods on four pairs of benchmark tasks: from A→B, from A→Q, from B→A, from B→Q, where A, B and Q are respectively the abbreviation for SHA, SHB and UCF-QNRF. The comparing approaches can be categorized mainly into two types: 1)unsupervised domain adaptation (UDA) based methods that perform domain adaptation using the unlabeled data from target domains; 2)methods which only use source data and no adaptation is performed to specific target domains. Our work belongs to the former ones based on UDA. The comparison results are shown in Table 1. Compared with the UDA-based counterparts, our method outperforms others on the sparse-to-dense task of A→Q, and also achieves the best MAE on the sparse-to-dense task of A→B. Besides, we achieve second-best result for the task of B→Q. The overall competitive results demonstrate the effectiveness of the proposed method for cross-domain crowd counting. The latest non-adaptation methods generally achieve better results than UDA-based methods, however they usually rely on integrating extra modules to increase the model capacity [17]. Our method add auxiliary decoders and discriminators to assist the domain adaptation, which could be totally removed at inference and thus will not affect the efficiency of the backbone model. Figure 2 also visualizes the predicted density maps and counts of the backbone model with and without the proposed

Table 2. Ablation study on adapting task correlation at the final output space.

Item	Method	from B→A	
		MAE	MSE
a	baseline (no adapt)	132.5	216.8
b	align density	121.8	198.4
c	align segment	126.4	196.6
d	align den/seg separate	128.4	202.7
e	align den/seg concat	122.5	**191.6**
f	**ours**	**117.0**	191.8

Table 3. Ablation study on the bi-level adaptation.

Dataset	Method	MAE	MSE
B→A	bi-level	**114.1**	**190.5**
	w/o inter-level	117.0	191.8
A→Q	bi-level	**141.3**	**241.1**
	w/o inter-level	152.9	269.1

semantically guided bi-level adaptation approach. It can be observed that overall with the proposed method, the predicted counts are more accurate and also the density distributions become more consistent with the ground-truth, with suppressed background clutters (the 2nd image) and recognition of extremely dense crowd at remote regions (the 3rd and 4th image).

5.3 Ablation Study

In this section, we conduct ablation study to analyze the effects of the proposed approach.

On the Semantically Guided Task Correlation. A key component in our work is the task correlation layer, which extracts task correspondence to be adapted for the UDA training. It is formulated as element-wise product. Is this operator more suitable than others at the similar computational expenses? Is the utilization of two tasks necessary? To answer these questions, we compare several variants at the final output space, including those only align one task i) align density: only adapt the density output for UDA training; ii) align segment: only adapt the segment output, and those align two tasks in different ways iii) align den/seg separate: adapt both of the density and segment separately, with a discriminator for each task respectively; vi) align den/seg concat: encode the task correlation using the concatenation operation. Finally, we compare with the proposed task correlation scheme with element-wise product. Several conclusions could be drawn from Table 2. i) Compared with the baseline results that directly apply CSRNet model trained on SHB to SHA, performing domain adaptation explicitly is beneficial (comparing $b \sim f$ vs a) ii) Different task encoding schemes influence the UDA training efficiency, and improper encoding scheme may be even worse than only adapting one task (comparing d vs b/c). iii) The proposed element-wise product operation based adaptation achieves the lowest MAE, indicating its effectiveness in encoding the task correspondence between density and segment across domains.

On the Bi-level Adaptation. In Table 3 we compare domain adaptation performances with and without the intermediate-output level adaptation. As observed, our bi-level scheme consistently outperforms the one without the intermediate-output level in terms of both MAE and MSE, validating the necessity of adapting networks at the intermediate level.

6   Conclusion

In this paper, we leverage semantic information to assist unsupervised domain adaptation for crowd counting. Instead of adapting high-dimensional features, we propose to adapt the correspondence between the crowd density and segmentation to address the domain shift problem. A task correlation layer is introduced to generate the correspondence map, which is enforced to be consistent between the source and target domains with adversarial training at two levels. With the proposed method, the model is encouraged to generate consistent crowd density and segmentation results for the target domain, thus facilitating the domain adaptation results. We experimentally validate the effectiveness of the proposed approach on three datasets, and achieve competitive results compared with other counterparts.

Acknowledgements. This work was supported in part by the Fundamental Research Funds for the Central Universities under Grant No. BLX202141, the National Natural Science Fund of China under Grant Nos. 62201062 and 61971281, and the foundation of Key Laboratory of Artificial Intelligence, Ministry of Education, P.R. China under Grant No. AI202105.

References

1. Sindagi, V.A., Patel, V.M.: A survey of recent advances in CNN-based single image crowd counting and density estimation. Pattern Recognit. Lett. **107**, 3–16 (2017)
2. Ganin, Y., Lempitsky, V.S.: Unsupervised domain adaptation by backpropagation. In: ICML (2015)
3. Liu, Y., Wang, Z., Shi, M., Satoh, S., Zhao, Q., Yang, H.: Towards unsupervised crowd counting via regression-detection bi-knowledge transfer. In: ACM MM, pp. 129–137 (2020)
4. Han, T., Gao, J., Yuan, Y., Wang, Q.: Focus on semantic consistency for cross-domain crowd understanding. In: ICASSP, pp. 1848–1852. IEEE (2020)
5. Gong, S., Zhang, S., Yang, J., Dai, D., Schiele, B.: Bi-level alignment for cross-domain crowd counting. In: CVPR, pp. 7542–7550 (2022)
6. Zhang, A., Yang, Y., Xu, J., Cao, X., Zhen, X., Shao, L.: Latent domain generation for unsupervised domain adaptation object counting. IEEE Trans. Multimedia (2022)
7. Wang, Q., Gao, J., Lin, W., Yuan, Y.: Learning from synthetic data for crowd counting in the wild. In: CVPR, pp. 8198–8207 (2019)
8. Zhao, M., Zhang, J., Zhang, C., Zhang, W.: Leveraging heterogeneous auxiliary tasks to assist crowd counting. In: CVPR, pp. 12736–12745 (2019)
9. Krizhevsky, A., Sutskever, I., Hinton, G.E.: Imagenet classification with deep convolutional neural networks. In: NIPs, pp. 1097–1105 (2012)
10. Zhang, C., Li, H., Wang, X., Yang, X.: Cross-scene crowd counting via deep convolutional neural networks. In: CVPR, pp. 833–841 (2015)

11. Zhang, Y., Zhou, D., Chen, S., Gao, S., Ma, Y.: Single-image crowd counting via multi-column convolutional neural network. In: CVPR, pp. 589–597 (2016)
12. Ma, Z., Wei, X., Hong, X., Gong, Y.: Bayesian loss for crowd count estimation with point supervision. In: ICCV, pp. 6142–6151 (2019)
13. Li, W., Yongbo, L., Xiangyang, X.: Coda: counting objects via scale-aware adversarial density adaption. In: ICME, pp. 193–198. IEEE (2019)
14. Zhang, A., Jun, X., Luo, X., Cao, X., Zhen, X.: Cross-domain attention network for unsupervised domain adaptation crowd counting. TCSVT **32**(10), 6686–6699 (2022)
15. Zhu, J.-Y., Park, T., Isola, P., Efros, A.A.: Unpaired image-to-image translation using cycle-consistent adversarial networks. In: ICCV, pp. 2223–2232 (2017)
16. Du, Z., Deng, J., Shi, M.: Domain-general crowd counting in unseen scenarios. arXiv preprint arXiv:2212.02573 (2022)
17. Wu, Q., Wan, J., Chan, A.B.: Dynamic momentum adaptation for zero-shot cross-domain crowd counting. In: ACM MM, pp. 658–666 (2021)
18. Wang, Q., Dai, D., Hoyer, L., Van Gool, L., Fink, O.: Domain adaptive semantic segmentation with self-supervised depth estimation. In: CVPR, pp. 8515–8525 (2021)
19. Newell, A., Yang, K., Deng, J.: Stacked hourglass networks for human pose estimation. In: Leibe, B., Matas, J., Sebe, N., Welling, M. (eds.) ECCV 2016. LNCS, vol. 9912, pp. 483–499. Springer, Cham (2016). https://doi.org/10.1007/978-3-319-46484-8_29
20. Li, Y., Zhang, X., Chen, D.: CSRNet: dilated convolutional neural networks for understanding the highly congested scenes. In: CVPR, pp. 1091–1100 (2018)
21. Radford, A., Metz, L., Chintala, S.: Unsupervised representation learning with deep convolutional generative adversarial networks. arXiv preprint arXiv:1511.06434 (2015)
22. Shen, Z., Xu, Y., Ni, B., Wang, M., Hu, J., Yang, X.: Crowd counting via adversarial cross-scale consistency pursuit. In: CVPR, pp. 5245–5254 (2018)
23. Xu, C., Qiu, K., Fu, J., Bai, S., Xu, Y., Bai, X.: Learn to scale: generating multipolar normalized density maps for crowd counting. In: CVPR, pp. 8382–8390 (2019)
24. Liu, L., Qiu, Z., Li, G., Liu, S., Ouyang, W., Lin, L.: Crowd counting with deep structured scale integration network. In: CVPR, pp. 1774–1783 (2019)
25. Liu, W., Liao, S., Ren, W., Hu, W., Yu, Y.: High-level semantic feature detection: a new perspective for pedestrian detection. In: CVPR, pp. 5187–5196 (2019)
26. Cheng, J., Xiong, H., Cao, Z., Hao, L.: Decoupled two-stage crowd counting and beyond. TIP **30**, 2862–2875 (2021)
27. Paszke, A., et al.: Automatic differentiation in pytorch (2017)
28. Idrees, H., et al.: Composition loss for counting, density map estimation and localization in dense crowds. In: ECCV, pp. 532–546 (2018)

Joint Priors-Based Restoration Method for Degraded Images Under Medium Propagation

Hongsheng Chen[1] , Wenbin Zou[1] , Hongxia Gao[1]([✉]) , Weipeng Yang[1] ,
Shasha Huang[1] , and Jianliang Ma[2]

[1] South China University of Technology, Guangzhou, China
hxgao@scut.edu.cn
[2] KUKA Robotics Guangdong Co., Ltd., Foshan, China

Abstract. The deep learning-based methods have shown promising performance in restoring degraded images such as underwater and haze images. However, the majority of existing methods rely on simplified imaging models, which limits their generalization and applicability in real-world scenarios. To address these issues, we incorporate the imaging mechanism in complex underwater environments to redefine the imaging model for degraded images under medium propagation. We then propose a multi-stage restoration framework that combines model-based iterative optimization methods and deep learning methods. At the same time, to tackle the problem of inaccurate parameter estimation in methods relying on a single prior, we introduce a regularization design based on joint priors and develop an attention-based color correction network to correct color distortions in the degraded images. Experimental results on real-world degraded images demonstrate the effectiveness and superiority of our method in both quantitative and subjective evaluations when compared to state-of-the-art methods.

Keywords: Degraded image restoration · Scattering model · Joint priors

1 Introduction

Due to light scattering and absorption during medium propagation, images captured in harsh environments such as haze, sandstorms, and underwater environments often suffer from low visibility. Haze images typically exhibit poor visibility and low contrast due to the presence of suspended particles in the air. Underwater

Supported by the Science and Technology Project of Guangzhou under Grant 202103010003, Science and Technology Project in key areas of Foshan under Grant 2020001006285.

Supplementary Information The online version contains supplementary material available at https://doi.org/10.1007/978-981-99-8552-4_27.

images, on the other hand, suffer from color distortion and blurred details due to the varying absorption of light by water at different wavelengths and the scattering of suspended particles. Therefore, it is essential to restore degraded images under medium propagation for subsequent target recognition and detection tasks.

Scattering imaging model [23] is widely used in haze and underwater image restoration methods, and the expression is:

$$I(x) = J(x)t(x) + A(1 - t(x)) \tag{1}$$

where $I(x)$ is the pixel intensity of the degraded image, and $J(x)$ is the scene radiance. A and $t(x)$ are the ambient light and the transmission map, respectively. Restoring the image with the scattering model (1) is an ill-posed problem, it is necessary to introduce additional prior to estimating the ambient light and the transmission map.

Numerous prior-based approaches, such as dark channel prior (DCP) [13], color-line prior [9], color attenuation prior [39], haze-lines prior [1], and non-local prior [2], have been proposed for haze removal. Although these methods have demonstrated promising results in haze removal, their reliance on a single prior often results in inaccurate parameter estimation and compromises the natural appearance of the restored images. In the context of underwater image restoration, many model-based methods [8,11,21,33] have been proposed for underwater image restoration to address the challenges posed by scattering effects caused by the underwater medium. However, most of these methods rely on simplified scattering models and neglect the forward scattering effect. As a result, they fail to provide comprehensive modeling of the formation process for various types of underwater images, leading to blurry and unnatural restored images, and limiting their applicability.

With the advancement of deep learning techniques, numerous approaches based on convolutional neural networks (CNNs) [3,6,15,19,28,37] have been proposed for the restoration of degraded images, such as haze and underwater images. However, these methods often encounter limitations due to the lack of large-scale datasets containing real-world degraded images. They heavily rely on synthetic degraded images for training, which fail to capture the distinctive characteristics of degradation in real scenarios. Consequently, their performance on real degraded images is relatively subpar.

To enhance the restoration performance on real-world degraded images, we propose a multi-stage restoration framework that combines model-based iterative optimization methods and deep-learning methods. Firstly, we remodel the imaging process for degraded images under medium propagation, taking into account the complex imaging principles in underwater environments. The restoration problem is then formulated as a minimization problem of a variational model. Meanwhile, to address the issue of inaccurate parameter estimation with a single prior, we introduce the regularization design based on joint priors. Furthermore, we employ the Alternating Direction Method of Multipliers (ADMM) to solve the minimization problem of the variational model. Finally, we design a color correction network based on an attention mechanism to effectively correct color distortions in underwater images.

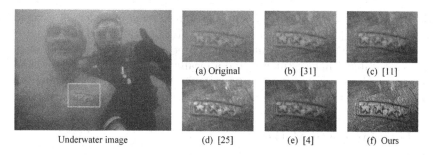

(a) Original   (b) [31]   (c) [11]

Underwater image   (d) [25]   (e) [4]   (f) Ours

Fig. 1. Visual comparisons of underwater image restoration on our test set. Our method outperforms state-of-the-art methods by producing sharper and more detailed image.

The specific contributions are as follows:

- Incorporating the intricate imaging principles in underwater environments, we remodel the imaging model for degraded images under medium propagation. By combining model-based iterative optimization techniques with deep learning methods, we propose a multi-stage restoration framework encompassing an initialization module, an optimization module, and a color correction module.
- To overcome the problem of inaccurate parameter estimation encountered in methods that rely solely on a single prior, we introduce the regularization design that leverages joint priors. Moreover, we develop a color correction network (CO^2-Net) that incorporates the attention mechanism, enabling efficient correction of color distortions present in underwater images.
- Our method exhibits exceptional performance surpassing that of state-of-the-art restoration techniques on a real-world dataset comprising underwater, haze, and sandstorm images. The experimental outcomes are evaluated using objective metrics as well as subjective assessments.

2 Related Work

2.1 Prior-Based Methods

The prior-based image dehazing methods [1,2,10,20] are based on the physical scattering model, which estimates the transmission map and the atmosphere light to recover the haze-free image. Fattal [9] proposed a color-line prior for estimating the optical transmission in hazy scenes based on the observations that the scattered light is eliminated to increase scene visibility. He et al. [13] proposed a dark channel prior (DCP) to estimate transmission map, which is based on the statistics of outdoor haze-free images. Zhu [39] proposed a simple but powerful color attenuation prior by creating a linear model for modeling the scene depth of the hazy image. Due to the similarities between hazed images and underwater images, the DCP-based dehazing approach is widely applied to underwater image restoration [26,38]. Drews et al. [8] proposed underwater dark channel prior (UDCP) to estimate the transmission in underwater environments. Wen et al. [33] proposed a new underwater optical model to estimate the scattering rate and the background light. Lu et al. [21] develop a robust color lines-based ambient light estimator and a locally adaptive filtering algorithm for enhancing underwater images. Galdran et al. [11] proposed an automatic red channel underwater image restoration to restore the underwater image.

2.2 Learning-Based Methods

In recent years, deep learning-based methods [6,15,19,30] have been extensively employed in the restoration of degraded images, including underwater images and haze images, yielding promising restoration outcomes. Li et al. [17] proposed an image dehazing model built with All-in-One Dehazing Network (AOD-Net), which is designed based on a re-formulated atmospheric scattering model. Qin et al. [27] introduced an end-to-end feature fusion attention restoration network (FFA-Net), which consists of three key components: FA module, a basic block, and FFA structure. Ren et al. [29] proposed a multi-scale gated fusion dehazing network (GFN), which learns confidence maps for derived inputs. Hou et al. [14] proposed a residual convolutional neural network to remove the haze effect and increase contrast for the underwater image. Yan et al. [34] formulated a very simple network for learning both the prior and restoration results jointly to guarantee much better results. Wang et al. [32] designed a novel adaptive weighting module to adaptively calibrate priors according to the importance of each prior for the underwater image.

3 Methodology

3.1 Scattering Model

The formation process of degraded images under medium propagation can be expressed as a linear combination [16] of the direct illumination component E_d, the forward scattering component E_{fs}, and the backward scattering component E_{bs}, and is shown below:

$$E_T = E_d + E_{fs} + E_{bs} \tag{2}$$

where E_T represents the total signal energy captured by the camera. The direct component is part of the light emitted from the object that reaches the camera after being lost due to scattering and absorption.

   In previous underwater image restoration methods, the forward scattering component was often overlooked in the modeling process, and simplified imaging models, such as equation (1), were used. However, in complex water bodies, such as those with numerous suspended particles and plankton, the forward scattering effects of underwater optical propagation become particularly significant. Due to the presence of numerous suspended particles, the direct illumination component produces a forward scattering effect that is similar to the direct beam but with a small deviation angle. This leads to the blurring of objects in the captured underwater images (see Fig. 1(a)).

In order to depict the blurring effect present in underwater images, this paper remodels the formation process of underwater images by considering the influence of forward scattering in complex underwater environments. The direct component and the forward scattering component are modeled together as a convolution form of the point spread function of the object's reflected light after absorption attenuation. The expression is as follows:

$$E_d + E_{fs} = Jt * k_s = Je^{-\eta d} * k_s \tag{3}$$

where J and t are the scene radiance and the transmission map, respectively. η is the attenuation coefficient, d is the scene depth, and k_s represents the blur kernel, which describes the blurriness in the forward scattering process. The $*$ denotes the convolution operation.

Additionally, the background scattering component is denoted as:

$$E_{bs} = A(1 - t) = A\left(1 - e^{-\eta d}\right) \tag{4}$$

where the A is the ambient light in the scene.

Therefore, our proposed imaging model for degraded images in medium propagation can be expressed as follows:

$$I_c(x) = J_c(x)t_c(x) * k_s + A_c\left(1 - t_c(x)\right) + N, c \in \{r, g, b\} \tag{5}$$

where $I_c(x)$ is the pixel intensity of the observed image, and $J_c(x)$ is the scene radiance, c corresponds to one of the red, green, and blue channels. N is the noise present in the formation of the degraded image.

3.2 The Joint Priors

To address the limitations of using a single prior for image restoration, we propose the regularization design based on joint priors in the variational model. Our design incorporates both the Dark Channel Prior (DCP) and the Bright Channel Prior (BCP), aiming to enhance the restored image's adherence to the distribution of natural images.

DCP and its variants are widely applied to degraded image restoration such as haze and underwater images. The dark channel of an image J is defined as:

$$D(J)(x) = \min_{y \in \Omega(x)} \left(\min_{c \in \{r, g, b\}} J^c(y) \right) \tag{6}$$

where J^c and $D(J)$ are the scene radiance for one channel among the RGB channels and the dark channel of J, respectively. $\Omega(x)$ is a local patch centered at x. In most of the nonsky patches, the intensity of $D(J)$ is very low and tends to zero $(D(J) \to 0)$.

However, DCP also has its limitations. For example, when restoring hazy images, it can lead to darker restoration results and lower contrast. Moreover, it cannot effectively eliminate haze in the sky areas. Therefore, we introduce a second prior in the regularization term: BCP.

BCP is widely used in various image restoration problems. The bright channel of an image J is defined as:

$$B(J)(x) = \max_{y \in \Omega(x)} \left(\max_{c \in \{r,g,b\}} J^c(y) \right) \tag{7}$$

where $B(J)$ is the bright channel of J and $\Omega(x)$ is a local patch centered at x. In most natural scene patches, the intensity of $B(J)$ is very high and tends to one $(B(J) \to 1)$, except in situations where shadows dominate.

BCP can enhance image contrast and addresses the limitations of the DCP. By incorporating the regularization design based on joint priors, we can overcome the limitation of relying only on a single prior for image restoration, improving the quality of the restored image and enhancing its natural appearance.

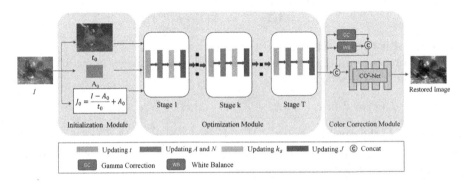

Fig. 2. Proposed multi-stage architecture for restoring degraded images. By passing a degraded image I into the initialization module, the initial transmittance map, initial ambient light value, and initial restored image are generated. Then, the optimization module updates the transmittance map t, ambient light value A, noise value N, blur kernel k_s, and restored image J_{mid} within T iterations. Finally, the color correction module corrects the color of the underwater image.

4   Multi-stage Restoration Framework

As shown in Fig. 2, our multi-stage restoration framework consists of three modules: initialization module, optimization module, and color correction module. The first two stages aim to eliminate the blurring and scattering effects in underwater images, as well as in haze and sandstorm images, while simultaneously enhancing the image contrast. The third stage is dedicated to color correction specifically for underwater images.

4.1 Initialization Module

Considering the crucial influence of initialization on optimization, we employ the methods proposed in [4,25] to estimate the initial values of ambient light A_0 and transmittance map t_0, respectively, for different types of degraded images. More details can be found in the supplementary material.

4.2 Optimization Module

Our objective function for degraded image restoration consists of two main components: the data fidelity term and the regularization term. The data fidelity term is designed based on our newly proposed imaging model. The regularization terms encompass gradient preservation, noise suppression, as well as the L_0-regularization of joint priors: $\|D(J)\|_0 + \|1 - B(J)\|_0$ [35]. These constraints play a role in preserving gradient information, suppressing noise, and ultimately producing a restored image that adheres to the distribution of natural images. The objective function is formulated as follows:

$$\operatorname*{argmin}_{J,A,N,k_s,t} \|Jt * k_s + (1-t)A + N - I\|_F^2 + \delta\|\nabla J - G\|_F^2 + \lambda\|D(J)\|_0$$
$$+\eta\|1 - B(J)\|_0 + \mu\|N\|_F + \gamma\|k_s\|_F^2, \text{ s.t. } A = \widehat{A} \tag{8}$$

where $\|\cdot\|_F$ and $\|\cdot\|_0$ denote Frobenius norm and l_0 norm, respectively, $\nabla(\cdot)$ denotes gradient operation. The first term is the data fidelity term, which ensures that the scene radiance J after degradation closely resembles the observed image I; the second item is the gradient constraint term, which preserves the edge information during the restoration process of the degraded image; the matrix G is represented by: $G = \left(1 + 10e^{-|\nabla I|/10}\right)\nabla I$ [4]; the third term is used to maintain the sparsity of the dark channels; the fifth term is the noise suppression term; the sixth term is used to constrain the stability of the blur kernel k_s; δ, λ, η, μ, and γ are weight parameters. Due to length limitations, we provide a detailed solution of the objective function in the supplementary material.

4.3 Color Correction Module

Considering the complex underwater environment and lighting conditions, we design the CO^2-Net, which utilizes a fusion-based strategy [18], to effectively correct color distortion in underwater images. The network architecture consists of a weight generation module and refinement restoration module that incorporate the attention mechanism, as shown in Fig. 3. Detailed descriptions of the network's architecture, loss functions, and training procedures will be provided in the supplementary material due to space constraints.

5 Experiments

5.1 Comparison with the State-of-the-Art

We assemble a test set comprising a total of 60 real-world underwater images [4,25] and 60 real-world sandstorms, and haze images [4,13]. These images are

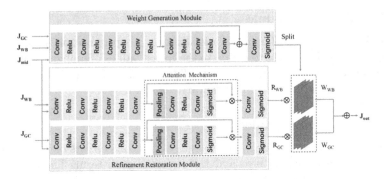

Fig. 3. The illustration about the CO^2-Net of Fig. 2. Here, J_{mid} represents the restored output from the optimization module, while J_{WB} and J_{GC} correspond to the preprocessed results of J_{mid} after applying white balance and gamma correction, respectively.

collected specifically for evaluating the performance of our algorithm and are available in the supplementary material.

For underwater images, we compare our method against several state-of-the-art image restoration methods, including model-based methods described in [11,25], and [4], as well as deep learning methods proposed in [30,31]. For haze and sandstorm images, we compare our method with several image restoration methods, including model-based methods described in [13,39], and [25], as well as deep learning methods proposed in [5,7]. The qualitative evaluation and quantitative evaluation methods are proposed to assess the performance of the above restoration methods.

Qualitative Evaluation. We compare the results of our method with the following state-of-the-art methods, as depicted in Figs. 4 and 5.

(a) Original   (b) [13]   (c) [25]   (d) [39]   (e) [7]   (f) [5]   (g) Ours

Fig. 4. Subjective evaluation of the different dehazing methods on real-world hazy images.

From Fig. 4, we can observe that DCP [13] can effectively remove haze in the foreground area, but it struggles to dehaze the sky regions and introduce severe

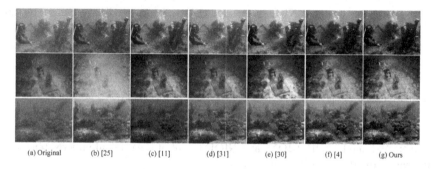

(a) Original   (b) [25]   (c) [11]   (d) [31]   (e) [30]   (f) [4]   (g) Ours

Fig. 5. Subjective evaluation of the different restoration methods on real-world underwater images.

color distortion. [39] results in dark restoration outcomes and fails to effectively enhance contrast. GDCP [25] demonstrates excellent haze removal capabilities, but it may exhibit over-restoration in the sky area. MSBDN [5,7] struggle to remove haze and lack effective color correction, resulting in restored images with low contrast. Our method excels in haze removal and color correction, resulting in restoration results with higher contrast.

From Fig. 5, we can observe that GDCP [25] and ULAP [31] do not compensate enough for the red channel and cause serious color distortion in the restored image. [11] often results in darker restored images and fails to restore texture details effectively. [4,30] excel in color-correcting underwater images. However, they are unable to effectively address the blurring effect caused by forward scattering. Our method effectively performs color correction on underwater images and eliminates blurring effects, resulting in restoration results with enhanced contrast and clear texture details.

Quantitative Evaluation. For haze and sandstorm images, we employ three no-reference image quality assessment indicators: the Natural Image Quality Evaluator (NIQE) [22], e, and r [12]. The NIQE is based on constructing a"quality-aware" collection of statistical features using a space domain natural scene statistic (NSS) model. A lower value of NIQE indicates better image quality. The values of e and r assess the method's ability to restore edges and the quality of contrast restoration, respectively. A larger e and r values indicate higher image quality. As shown in Table 1, our method surpasses the state-of-the-art methods in terms of average e and r.

For underwater images, we utilize three no-reference image quality assessment indicators: NIQE, Underwater Image Quality Measure (UIQM) [24], and Underwater Color Image Quality Evaluation Metric (UCIQE) [36]. UCIQE is a linear combination of contrast, chroma, and saturation in the CIELAB color space. Similarly, UIQM is the linear combination of the underwater image colorfulness measure (UICM), underwater image sharpness measure (UISM), and underwater image contrast measure (UIConM). Hence, a larger value of UIQM

Table 1. Average e, r and NIQE values for the sandstorm and haze images restored by all methods.

	[13]	[25]	[39]	[7]	[5]	Ours
e ↑	0.91	1.06	0.54	0.70	0.21	**1.35**
r ↑	1.37	1.56	1.10	1.36	1.09	**1.83**
NIQE ↓	**3.44**	4.05	3.61	3.74	3.98	3.51

Table 2. Average UCIQE, UIQM, and NIQE values of the original underwater images and their restored results from all methods.

	Original	[25]	[11]	[31]	[30]	[4]	Ours
UCIQE ↑	0.44	0.65	0.59	0.69	0.68	0.76	**0.78**
UIQM ↑	1.02	1.90	2.85	1.87	3.15	3.15	**3.86**
NIQE ↓	4.58	4.32	4.33	4.23	**3.85**	4.18	4.01

and UCIQE indicates better quality of underwater images. As presented in Table 2, our method surpasses other methods in terms of average UIQM and UCIQE.

5.2 Ablation Study

Due to space limitations, the details of the ablation experiments are provided in the supplementary material.

6 Conclusion

In this paper, we propose a multi-stage restoration framework for the degraded images. Firstly, we incorporate the imaging mechanism in complex underwater environments to re-model the imaging model of degraded images under medium propagation. Subsequently, we combine model-based methods with deep learning approaches and introduce the regularization design based on joint priors. Finally, an attention-based color correction network is formulated. Extensive experiments demonstrate the effectiveness of our multi-stage restoration framework in removing medium-induced scattering and blurring effects from degraded images and correcting color distortions.

References

1. Berman, D., Avidan, S., et al.: Non-local image dehazing. In: Proceedings of the IEEE Conference on Computer Vision and Pattern Recognition, pp. 1674–1682 (2016)

2. Berman, D., Treibitz, T., Avidan, S.: Air-light estimation using haze-lines. In: 2017 IEEE International Conference on Computational Photography (ICCP), pp. 1–9. IEEE (2017)
3. Cai, B., Xu, X., Jia, K., Qing, C., Tao, D.: DehazeNet: an end-to-end system for single image haze removal. IEEE Trans. Image Process. **25**(11), 5187–5198 (2016)
4. Cai, Y., Gao, H., Niu, S., Qi, T., Yang, W.: A multi-stage restoration method for degraded images with light scattering and absorption. In: 2022 26th International Conference on Pattern Recognition (ICPR), pp. 407–413. IEEE (2022)
5. Chen, Z., He, Z., Lu, Z.M.: DEA-Net: single image dehazing based on detail-enhanced convolution and content-guided attention. arXiv preprint arXiv:2301.04805 (2023)
6. Dong, H., et al.: Multi-scale boosted dehazing network with dense feature fusion. In: Proceedings of the IEEE/CVF Conference on Computer Vision and Pattern Recognition, pp. 2157–2167 (2020)
7. Dong, H., et al.: Multi-scale boosted dehazing network with dense feature fusion. In: Proceedings of the IEEE/CVF Conference on Computer Vision and Pattern Recognition, pp. 2157–2167 (2020)
8. Drews, P., Nascimento, E., Moraes, F., Botelho, S., Campos, M.: Transmission estimation in underwater single images. In: Proceedings of the IEEE International Conference on Computer Vision Workshops, pp. 825–830 (2013)
9. Fattal, R.: Single image dehazing. ACM Trans. Graph. (TOG) **27**(3), 1–9 (2008)
10. Fattal, R.: Dehazing using color-lines. ACM Trans. Graph. (TOG) **34**(1), 1–14 (2014)
11. Galdran, A., Pardo, D., Picón, A., Alvarez-Gila, A.: Automatic red-channel underwater image restoration. J. Vis. Commun. Image Represent. **26**, 132–145 (2015)
12. Hautiere, N., Tarel, J.P., Aubert, D., Dumont, E.: Blind contrast enhancement assessment by gradient ratioing at visible edges. Image Anal. Stereol. **27**(2), 87–95 (2008)
13. He, K., Sun, J., Tang, X.: Single image haze removal using dark channel prior. IEEE Trans. Pattern Anal. Mach. Intell. **33**(12), 2341–2353 (2010)
14. Hou, M., Liu, R., Fan, X., Luo, Z.: Joint residual learning for underwater image enhancement. In: 2018 25th IEEE International Conference on Image Processing (ICIP), pp. 4043–4047. IEEE (2018)
15. Islam, M.J., Xia, Y., Sattar, J.: Fast underwater image enhancement for improved visual perception. IEEE Robot. Autom. Lett. **5**(2), 3227–3234 (2020)
16. Jaffe, J.S.: Computer modeling and the design of optimal underwater imaging systems. IEEE J. Oceanic Eng. **15**(2), 101–111 (1990)
17. Li, B., Peng, X., Wang, Z., Xu, J., Feng, D.: AOD-Net: all-in-one dehazing network. In: Proceedings of the IEEE International Conference on Computer Vision, pp. 4770–4778 (2017)
18. Li, C., et al.: An underwater image enhancement benchmark dataset and beyond. IEEE Trans. Image Process. **29**, 4376–4389 (2019)
19. Li, R., Pan, J., Li, Z., Tang, J.: Single image dehazing via conditional generative adversarial network. In: Proceedings of the IEEE Conference on Computer Vision and Pattern Recognition, pp. 8202–8211 (2018)
20. Li, Z., Zheng, J.: Edge-preserving decomposition-based single image haze removal. IEEE Trans. Image Process. **24**(12), 5432–5441 (2015)
21. Lu, H., Li, Y., Zhang, L., Serikawa, S.: Contrast enhancement for images in turbid water. JOSA A **32**(5), 886–893 (2015)
22. Mittal, A., Soundararajan, R., Bovik, A.C.: Making a "completely blind" image quality analyzer. IEEE Signal Process. Lett. **20**(3), 209–212 (2012)

23. Narasimhan, S.G., Nayar, S.K.: Vision and the atmosphere. Int. J. Comput. Vis. **48**(3), 233 (2002)
24. Panetta, K., Gao, C., Agaian, S.: Human-visual-system-inspired underwater image quality measures. IEEE J. Oceanic Eng. **41**(3), 541–551 (2015)
25. Peng, Y.T., Cao, K., Cosman, P.C.: Generalization of the dark channel prior for single image restoration. IEEE Trans. Image Process. **27**(6), 2856–2868 (2018)
26. Peng, Y.T., Zhao, X., Cosman, P.C.: Single underwater image enhancement using depth estimation based on blurriness. In: 2015 IEEE International Conference on Image Processing (ICIP), pp. 4952–4956. IEEE (2015)
27. Qin, X., Wang, Z., Bai, Y., Xie, X., Jia, H.: FFA-Net: feature fusion attention network for single image dehazing. In: Proceedings of the AAAI Conference on Artificial Intelligence, vol. 34, pp. 11908–11915 (2020)
28. Ren, W., Liu, S., Zhang, H., Pan, J., Cao, X., Yang, M.-H.: Single image dehazing via multi-scale convolutional neural networks. In: Leibe, B., Matas, J., Sebe, N., Welling, M. (eds.) ECCV 2016. LNCS, vol. 9906, pp. 154–169. Springer, Cham (2016). https://doi.org/10.1007/978-3-319-46475-6_10
29. Ren, W., et al.: Gated fusion network for single image dehazing. In: Proceedings of the IEEE Conference on Computer Vision and Pattern Recognition, pp. 3253–3261 (2018)
30. Sharma, P., Bisht, I., Sur, A.: Wavelength-based attributed deep neural network for underwater image restoration. ACM Trans. Multimed. Comput. Commun. Appl. **19**(1), 1–23 (2023)
31. Song, W., Wang, Y., Huang, D., Tjondronegoro, D.: A rapid scene depth estimation model based on underwater light attenuation prior for underwater image restoration. In: Hong, R., Cheng, W.-H., Yamasaki, T., Wang, M., Ngo, C.-W. (eds.) PCM 2018. LNCS, vol. 11164, pp. 678–688. Springer, Cham (2018). https://doi.org/10.1007/978-3-030-00776-8_62
32. Wang, Z., Shen, L., Yu, M., Lin, Y., Zhu, Q.: Single underwater image enhancement using an analysis-synthesis network. arXiv preprint arXiv:2108.09023 (2021)
33. Wen, H., Tian, Y., Huang, T., Gao, W.: Single underwater image enhancement with a new optical model. In: 2013 IEEE International Symposium on Circuits and Systems (ISCAS), pp. 753–756. IEEE (2013)
34. Yan, K., Liang, L., Zheng, Z., Wang, G., Yang, Y.: Medium transmission map matters for learning to restore real-world underwater images. Appl. Sci. **12**(11), 5420 (2022)
35. Yan, Y., Ren, W., Guo, Y., Wang, R., Cao, X.: Image deblurring via extreme channels prior. In: Proceedings of the IEEE Conference on Computer Vision and Pattern Recognition, pp. 4003–4011 (2017)
36. Yang, M., Sowmya, A.: An underwater color image quality evaluation metric. IEEE Trans. Image Process. **24**(12), 6062–6071 (2015)
37. Zhang, H., Patel, V.M.: Densely connected pyramid dehazing network. In: Proceedings of the IEEE Conference on Computer Vision and Pattern Recognition, pp. 3194–3203 (2018)
38. Zhao, X., Jin, T., Qu, S.: Deriving inherent optical properties from background color and underwater image enhancement. Ocean Eng. **94**, 163–172 (2015)
39. Zhu, Q., Mai, J., Shao, L.: A fast single image haze removal algorithm using color attenuation prior. IEEE Trans. Image Process. **24**(11), 3522–3533 (2015)

GridIIS: Grid Based Interactive Image Segmentation

Pengqi Zhu[1,2], Da-Han Wang[1,2(✉)], and Shunzhi Zhu[1,2]

[1] School of Computer and Information Engineering,
Xiaman University of Technology, Xiamen, China
`wangdh@xmut.edu.cn`
[2] Fujian Key Laboratory of Pattern Recognition and Image Understanding,
Xiamen, China

Abstract. Interactive segmentation enables users to specify the object of interest (OOI) via various interaction strategies to obtain accurate segmentation results. An ideal interactive method should efficiently and accurately express users' segmentation intentions. However, the existing methods can only use a single interactive mode, ignoring the differences in scale and shape between OOIs, resulting in an inflexibility labeling process. In this paper, we propose a grid-based interactive image segmentation method (**GridIIS**). Specifically, GridIIS overlays grids on the image, and users can specify the location and shape of the OOI by selecting the grid areas as the interactive guidance. Users can choose the appropriate grid selection method and size considering the OOI's scale, shape, and boundary clarity to obtain guidance. We accordingly propose a novel grid sampling strategy, that considers the OOI's scale and shapes to adaptively estimate the grid size and area. Experiments on several datasets from different domains (street views, medical images, scene texts, etc.) show that our method achieves superior performance with fewer interaction rounds and exhibits strong generalization ability in cross-domain datasets.

Keywords: Interactive segmentation · Grid-based interactive

1 Introduction

Interactive image segmentation incorporates human assistance into the segmentation process to make the segmentation of OOIs more accurate. In recent years, it has been increasingly popular in various fields, such as medical image analysis [22,30], picture manipulation [7,16], etc. The most prevalent use is low-cost labeling since interactive segmentation techniques can significantly decrease the labor workload in labeling pixel-level annotations.

Existing interactive methods can be mainly categorized into three kinds based on the interaction strategy: clicks [5,13,20,23,28,29,33,35], bounding boxes [27,31,32], and scribbles [1,12,16,18]. Regardless of the interaction method, the aim is to give users well interactive experience through easy operation and precise segmentation. A successful interactive strategy should satisfy the following issues.

Q. Liu et al. (Eds.): PRCV 2023, LNCS 14435, pp. 351–363, 2024.
https://doi.org/10.1007/978-981-99-8552-4_28

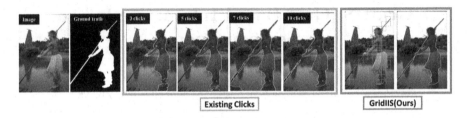

Fig. 1. Comparison of click-based and grid-based methods for labeling the same target. Compared with the results of click-based method with ten clicks, our method has better results in one interaction.

Accurately Express the User's Intent. The interaction should precisely convey the user's segmentation intention [35]. In other words, the user's operation could provide sufficient guidance to help the model correctly segment OOIs in as few interactive rounds as possible.

Easy to Sample and Train. The framework should use simple and effective sampling strategies to simulate the user interaction inputs for network training.

Generalizability. Facing the segmentation challenges of images with different characteristics from various datasets, the interactive segmentation strategy can still complete the high-performance interactive segmentation task.

Existing interaction strategies have their own characteristics. While simple, clicks can only specify a limited number of pixels with little guidance. When the shape of the OOI is complex, or the boundary is fuzzy, it usually requires multiple and dense clicks in the local area. As shown in Fig. 1, the segmentation result after ten clicks is still unsatisfactory. The bounding boxes interaction may be limited due to the loss of details in boundaries when the box contains the other objects, or the OOI is irregular. Since scribbles can mark regions more accurately and flexibly, it is the best choice when the shape is complex. However, the sampling of scribbles is more challenging because scribbles hold a considerable input variation across different users [13]. It is hard to replicate all the potential inputs, resulting in usually insufficiently robust models.

Considering the above analysis and issues, we propose a simple but effective grid-based interactive image segmentation method called GridIIS that can serve as a general framework supporting various interaction strategies and enables users to flexibly label the OOIs. First, overlays a set of grids on the image, then users can 1) click to select a grid. 2) Scribble to select grids. 3) Select any symmetrical corner position to select the entire rectangular area. The Fig. 2 shows the common interaction methods input. It can be seen that the grid area can mark the target more accurately, reflecting the position and approximate shape. The advantages of our GridIIS are as follows.

Accurately Reflect the Users' Intention. By selecting grids to indicate the approximate area and shape of an OOI, the network can be guided more explicitly, achieving high performance with a minimum number of interactions.

Fig. 2. The marking methods of several popular interaction mechanisms. Compared with other mechanisms, the grid area can mark the target more accurately.

As in Fig. 1, compared with the segmentation result of ten clicks, the result of one grid-based interaction is more satisfactory.

Simple and Robust Sampling. Grid-based sampling only needs to simulate the grid area selected by the user.

Strong Generalization. Due to the flexible and accurate guidance, our method can maintain strong generalization performance in different domains.

Experiments on several datasets from different domains, including PASCAL [10], GrabCut [27], Berkeley [26], CityScapes [8], ssTEM [11], MSCMR [38], and CTW1500 [34] show that our method achieves superior performance with a low cost of interactions and exhibits strong generalization ability in cross-domain datasets.

2   Related Work

Since Xu et al. [33] first applied deep learning methods to interactive segmentation in 2016, traditional interactive segmentation methods [2,14] have gradually been replaced by deep learning-based approaches.

Clicks. RIS [17] performs a local refinement branch in the square area formed by positive and negative clicks to refine the segmentation details. EXTER [25] indicate the target with four extreme points. [24,35] attempt to utilize semantic information to enhance understanding of user clicks. Lin et al. [20] proposed a method to strengthen the first click's attention and enhance the first click's guiding role. Ding et al. [9] presented adaptive Gaussian graphs with different variances to encode user annotations, improving sensitivity to details. Our GridIIS selects the grid area as the target area, fundamentally avoiding the problem of target ambiguity. Besides, the precise selection is guaranteed to get the desired result with fewer interaction rounds.

Bounding Boxes. To reduce the influence of the different tightness of the bounding box on the results, Xu et al. [32] turned the rectangular box into a distance guide and a soft constraint. IOG [37] proposed adding internal clicks based on the overall bounding box to indicate the objects to be segmented inside the box.

Scribbles. Bai et al. [1] improved the error tolerance of scribble-based methods by sampling scribble strokes with excess areas. [15] proposes a skeleton extraction method to simulate the user's possible input strokes. The work of Jahanifar et al. [13] improves the sampling method by adding a series of transform cascade operations to generate more complex skeletons. Our method converts the original complex sampling of scribbles trajectories into the selection of grid areas, reducing the sampling difficulty and improves the model robustness.

3   Method

The pipeline of the proposed GridIIS framework is illustrated in Fig. 3. First, users select the appropriate grid area of the OOI. Then user's annotations will be converted into interactive information. The selected grid area is regarded as the positive guidance, while the four corners of the smallest circumscribed rectangle containing the grid area serve as negative guidance. The guidances are concatenated with the RGB cropped images to form the 5-channel input to the network for segmentation.

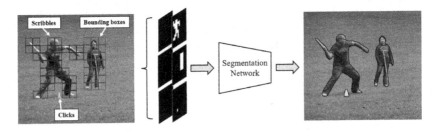

Fig. 3. The pipeline of the proposed GridIIS framework. The green, red, and blue grids are the grid areas selected by clicks, bounding boxes and scribbles, respectively. (Color figure online)

Grid Selection Method: For OOIs with different shapes and appearances, users can choose the most accurate and convenient gird selection method. For example, for the three objects shown in Fig. 3, we can use three different selection methods: 1) The conical plastic pile is simple and small, and we can click the grid where it is located; 2) The woman has a regular shape with clear boundaries. We mark it by dragging from the upper left corner to the lower right corner to draw a rectangular box; 3) The man's shape is more complex, and we can use scribbles to select the grid area.

Adjustment of Grid Size: In addition to different ways of selecting grids, the grid size can be adjusted according to the OOI's scale and segmentation difficulty. Small grids allow users to mark OOI's area more accurately, while large grids allow users to quickly select areas. As shown in Fig. 4, different OOIs marked with different grid sizes. Through experiments and observations, we set the grid size between 10 and 50 pixels, with an adjustment step of 10 pixels. The user's choice of the selection method and the grid size may vary. Still, our model also shows excellent robustness to these individual differences, which we will discuss in Sect. 4.4.

Fig. 4. According to the scale of the target and the difficulty of segmentation, and other attributes, different grid sizes can be selected for marking.

3.1 Strength Analysis

As shown in Fig. 5, to better illustrate the effect of our grid-based interaction mechanism, we visualize the prediction probability maps of the last layer of the network decoder. The advantages of our method can be reflected in the following aspects.

Position and Shape Guidance. The click-based mechanism has a problem that the intent will be ambiguous with a few clicks. It's hard to use several clicks to represent the OOI, which limits the segmentation performance. However, by selecting the grids, we can more clearly mark the position and shape of the target, which is conducive to the subsequent network inference. As shown in the first line of Fig. 5, the OOI is heavily overlapped with the other objects. We can still accurately segment the correct area.

Retention of Details. For high-quality segmentation masks, it's necessary to capture the local details. Nevertheless, the existing methods are easy to omit local details, which requires multiple subsequent interactions to correct the details. Through a precise selection of the grids, the network can notice more details and the boundary trend, which is conducive to restoring more details and reducing local pixel-level misclassification. As shown in the second line of Fig. 5, GridIIS can notice more details of the bird's legs and the tip of its mouth.

Exclude Background Effects. The bounding boxes-based method can't avoid the deterioration of the segmentation results caused by the complex background area or other targets in the box. Since the grid-based method can flexibly select the grids area, users can directly select the grid containing the OOI to avoid

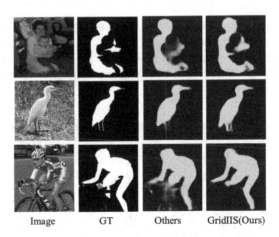

Image   GT   Others   GridIIS(Ours)

Fig. 5. Visualization of the prediction graph of the network decoder in the training stage. The third column is the result after four clicks, and the last column is the result after one interaction of our grid-based method.

redundant background areas, reducing the impact of the background and other objects. As shown in the third line of Fig. 5, our method can shield the network's perception of bicycles, and directly improve the network's prediction process.

3.2   Grid Sampling Strategy

The grid size can be adjusted appropriately for efficient area selection and interaction. As shown Fig. 4, the grid size adjusts according to the specific characteristics of the OOI. We estimate the grid size based on the scale (denoted as s) and the OOI shape represented by the degree of regularity (denoted as t). We denote the short side of the bounding box as w. Then the size of the grid is computed as follows:

$$size = C * w * t * s$$

where $C \in [0.87, 1.0, 1.2]$ is a constant set empirically, we randomly select the value of C in set to enhance the model's robustness and make the value of grid size consistent with the actual interaction needs.

Scale(s): The scale is estimated according to the proportion of the OOI in the whole picture: find the bounding box closely contained in the ground truth, then calculate the proportion of the box area in the original image. We record the proportional value as s'. To facilitate training and model stability, we select a fixed s value according to the calculated s' range. The value of s is as follows:

$$s = \begin{cases} 0.13, & s' \geq 0.5 \\ 0.2, & 0.5 > s' \geq 0.2 \\ 0.4, & s' < 0.2 \end{cases}$$

Degree of Regularity(t)**:** The shape of the OOI is described by the regularity degree, which is computed as the proportion of the ground truth in the bounding box. We record the ratio value as t'. When the segmentation target has an irregular shape, the proportion t' is low, and users tend to choose a smaller grid size, avoiding choosing too many excess regions. Therefore, we take t as the regular coefficient. Similar to s, we re-value t according to the range of t', and the values are as follows:

$$t = \begin{cases} 0.65, & t' \leq 0.3 \\ 1.0, & t' > 0.3 \end{cases}$$

4 Experiments

4.1 Implementation Details

Datasets. We conduct extensive experiments on eight publicly available benchmarks, including PASCAL [10], COCO [19], GrabCut [27], Berkeley [26], CityScapes [8], ssTEM [11], MSCMR [38], and CTW1500 [34] to demonstrate the effectiveness and the generalization capabilities of our GridIIS.

Metrics. Following previous work [25,37], we use the Mean Intersection Over Union (mIoU) as the evaluation metric. After experiments and statistics, we found that the average time for four clicks is 7.4 s, while the average time for one grid selection is about 7.1 s. For fairness, the results of the click-based method presented in the subsequent experiment were obtained after four click interactions.

Training and Testing Details. We adopted the network structure used in IOG [37], which is a cascaded network structure from coarse-to-fine, including two main network modules CoarseNet and FineNet. Accordingly, we trained 100

Table 1. Comparison with the state-of-the-art methods on PASCAL,GrabCut and Berkeley in terms of quality at 4 clicks.* represents the IOU of the grid after one interactive operation.

Method	IoU(%) @ 4 click		
	PASCAL	GrabCut	Berkeley
Graph cut [3]	40.2	59.8	50.0
DOS [33]	75.0	84.3	80.0
RIS-Net [17]	80.7	85.0	83.6
BRS [13]	–	88.1	88.9
DEXTR [25]	91.5	92.8	–
IOG [37]	94.4	95.3	91.6
FocalClick [6]	94.6	95.7	92.7
Grid	96.0*	96.2*	93.8*

epochs on PASCAL 2012 Segmentation, and the batch size is set to 5, whereas for COCO, we trained 10 epochs, and the batch size changed to 10. The learning rate, momentum, and weight decay are set to 10-8, 0.9, and 5×10-4, respectively.

4.2   Comparison with the State-of-the-Arts

As shown in Table 1, we compared with the state-of-the-art on three widely used benchmarks: PASCAL VOC val set, GrabCut, and Berkeley. The proposed GridIIS exceeds the others with 1.4%, 0.5% and 1.1%, respectively. Some qualitative results are shown in Fig. 6. This also means that in most cases, users only need to mark the OOI once to complete the high-precision marking work.

Fig. 6. Qualitative results on PASCAL [10], COCO [19] and COCO-Stuff [4]. Each instance with a simulated girds selection area and corresponding segmentation mask are overlayed on the input image.

4.3   Cross-Domain Evaluation

To demonstrate the powerful performance of GridIIS as an interactive annotation tool and its generalization ability on different datasets, we further conducted experiments on datasets from different fields, including street views (CityScapes [8]), medical images (ssTEM [11], MSCMR [38]), and scene text (CTW1500 [34]).

The results are summarized in Tables 2, 3, 4 to 5. We can see that even without fine-tuning, our results are far superior to other methods. Grid's model trained on the PASCAL can achieve 81.5% accuracy when applied directly to cityscapes. And the application without fine-tuning on MRI images with unclear boundaries can also achieve 83.7% accuracy. Moreover, the performance of GridIIS will be greatly improved after further fine-tuning with a small amount of data (1/10 of the cityscapes training data and 1/5 of the MSCMR training data). Furthermore, we also apply our Grid to a larger-span text detection dataset and still achieve 89%. Figure 7 provides some qualitative examples.

Table 2. Cross domain analysis on Cityscapes [8]. "PASCAL+1/10 CityScapes" indicates that the method is fine-tuned on a small set of the Cityscapes dataset (10%).

Methods	Train	#Clicks	IoU(%)
Curve-GCN [21]	CityScapes	3.6	80.2
DEXTER [25]	CityScapes	4	79.4
IOG [37]	PASCAL	3	79.1
Grid	PASCAL	–	81.5
Grid	PASCAL+1/10 CityScapes	–	83.8

Table 3. Cross domain analysis on MSCMR [38]. "PASCAL+1/5 MSCMR" indicates that the method is fine-tuned on a small set of the MSCMR dataset (20%).

Methods	Train	#Clicks	IoU(%)
IOG [37]	PASCAL	3	72.8
CycleMix [36]	MSCMR(Unsupervised)	–	80.0
Grid	PASCAL	–	83.7
Grid	PASCAL+1/5 MSCMR	–	90.0

Table 4. Cross domain analysis on SSTEM [11].

Methods	Train	#Clicks	IoU(%)
Curve-GCN [21]	CityScapes	2	60.9
IOG [37]	PASCAL	3	83.7
Grid	PASCAL	–	89.1

Table 5. Cross domain analysis on ctw1500 [34].

Methods	Train	#Clicks	IoU(%)
IOG [37]	PASCAL	3	70.1
Grid	PASCAL	–	85.9
Grid	COCO	–	89.0

4.4 Sensitivity to User's Guiding Signals

Users will inevitably have differences in interactive annotation, and these differences will lead to fluctuations in model performance. In the rest of this section, we will verify the robust performance of our GridIIS method from three aspects.

Fig. 7. Cross domain performance. Qualitative results of our Grid on Cityscapes, MSCMR, ctw1500 and ssTEM.

Grid Size. We design an experiment to examine the effect of the grid size on the segmentation performance. As shown in Table 6, based on the grid size obtained by sampling, we adjusted it by 10/20 pixels respectively (the final value is still truncated in the range of 10 to 50). It can be seen that the accuracy only floats within 0.8% at a deviation of 10px. Faced with fluctuations in grid size, our model can maintain stable performance.

Selection Habits. As shown in Table 6, we randomly remove a percent of grids of the central area from the originally selected grid areas. It can be seen that even if 20% of the center area is removed, the accuracy only drops by 0.8%.

Table 6. A series of tests for sensitivity to user guiding signals. None represents the original result of no other operation. Lines 2 and 3 are adjustments to the grid size. Lines 4 and 5 discard the fixed-ratio central area grid. The last row is the result of manual scribble that contain excess grids.

Adjust operation	Test	IoU(%)
None	PASCAL	96.0
Size±10	PASCAL	95.2
Size±20	PASCAL	94.3
Discarded 10%	PASCAL	95.6
Discarded 20%	PASCAL	95.2
None	Grabcut	96.2
Redundant Error	Grabcut	94.9

Tolerate Errors. We also do experiments to verify the effect of the redundant selection of grids. Referring to [1], we roughly scribble on the Grabcut [27] to obtain the grid areas that contain redundant errors. Quantitative results are shown in the last row of Table 6, and it can be found that our algorithm can also tolerate redundant wrong selections outside the OOI region.

5   Conclusion

We propose a simple and effective grid-based interactive method in which the user specifies the OOI by adjusting the grid size and selecting the grid area. The grid selection method can integrate several interaction strategies: clicks, scribbles, and bounding boxes. For different goals, users can flexibly choose a suitable method for fast and accurate annotation, minimizing the number of interactions and improving the annotation efficiency. Furthermore, we unify the grid-based sampling strategy to efficiently predict grid regions. Four public datasets demonstrate the superiority of our algorithm in interactive image segmentation. We show the utility of our method as an annotation tool that can be used across different datasets, even from various domains with widely differing appearances.

Acknowledgements. This work is supported by National Natural Science Foundation of China (No. 61773325), Industry-University Cooperation Project of Fujian Science and Technology Department (No. 2021H6035), Fujian Key Technological Innovation and Industrialization Projects (No. 2023XQ023), and Fu-Xia-Quan National Independent Innovation Demonstration Project (No. 2022FX4).

References

1. Bai, J., Wu, X.: Error-tolerant scribbles based interactive image segmentation. In: CVPR, pp. 392–399 (2014)
2. Bai, X., Sapiro, G.: A geodesic framework for fast interactive image and video segmentation and matting. In: ICCV, pp. 1–8. IEEE (2007)
3. Boykov, Y.Y., Jolly, M.P.: Interactive graph cuts for optimal boundary & region segmentation of objects in ND images. In: ICCV, vol. 1, pp. 105–112. IEEE (2001)
4. Caesar, H., Uijlings, J., Ferrari, V.: Coco-stuff: thing and stuff classes in context. In: CVPR, pp. 1209–1218 (2018)
5. Chen, X., et al.: Conditional diffusion for interactive segmentation. In: ICCV, pp. 7345–7354 (2021)
6. Chen, X., et al.: FocalClick: towards practical interactive image segmentation. In: CVPR, pp. 1300–1309 (2022)
7. Cheng, M.-M., et al.: Intelligent visual media processing: when graphics meets vision. JCST **32**(1), 110–121 (2017)
8. Cordts, M., et al.: The cityscapes dataset for semantic urban scene understanding (2016). arXiv: 1604.01685 [cs.CV]
9. Ding, Z., et al.: A dual-stream framework guided by adaptive gaussian maps for interactive image segmentation. Knowl.-Based Syst. **223**, 107033 (2021)
10. Everingham, M., et al.: The pascal visual object classes (VOC) challenge. IJCV **88**(2), 303–338 (2010)

11. Gerhard, S., et al.: Segmented anisotropic ssTEM dataset of neural tissue. figshare (2013)
12. Grady, L.: Random walks for image segmentation. PAMI **28**(11), 1768–1783 (2006)
13. Jang, W.D., Kim, C.S.: Interactive image segmentation via backpropagating refinement scheme. In: CVPR, pp. 5297–5306 (2019)
14. Kim, T.H., Lee, K.M., Lee, S.U.: Generative image segmentation using random walks with restart. In: Forsyth, D., Torr, P., Zisserman, A. (eds.) ECCV 2008. LNCS, vol. 5304, pp. 264–275. Springer, Heidelberg (2008). https://doi.org/10.1007/978-3-540-88690-7_20
15. Koohbanani, N.A., et al.: NuClick: a deep learning framework for interactive segmentation of microscopic images. MIA **65**, 101771 (2020)
16. Li, Y., et al.: Lazy snapping. ACM Trans. Graph. (ToG) **23**(3), 303–308 (2004)
17. Liew, J.H., et al.: Regional interactive image segmentation networks. In: ICCV. IEEE Computer Society, pp. 2746–2754 (2017)
18. Lin, D., et al.: Scribblesup: scribble-supervised convolutional networks for semantic segmentation. In: CVPR, pp. 3159–3167 (2016)
19. Lin, T.Y., et al.: Microsoft COCO: common objects in context. In: Fleet, D., Pajdla, T., Schiele, B., Tuytelaars, T. (eds.) ECCV 2014. LNCS, vol. 8693, pp. 740–755. Springer, Cham (2014). https://doi.org/10.1007/978-3-319-10602-1_48
20. Lin, Z., et al.: Interactive image segmentation with first click attention. In: CVPR, pp. 13339–13348 (2020)
21. Ling, H., et al.: Fast interactive object annotation with Curve-GCN. In: CVPR, pp. 5257–5266 (2019)
22. Luo, X., et al.: MIDeepSeg: minimally interactive segmentation of unseen objects from medical images using deep learning. MIA **72**, 102102 (2021)
23. Mahadevan, S., Voigtlaender, P., Leibe, B.: Iteratively trained interactive segmentation. arXiv preprint: arXiv:1805.04398 (2018)
24. Majumder, S., Yao, A.: Content-aware multi-level guidance for interactive instance segmentation. In: CVPR, pp. 11602–11611 (2019)
25. Maninis, K.-K. et al.: Deep extreme cut: from extreme points to object segmentation. In: CVPR, pp. 616–625 (2018)
26. McGuinness, K., O'connor, N.E.: A comparative evaluation of interactive segmentation algorithms. Pattern Recogn. **43**, 434–444 (2010)
27. Rother, C., Kolmogorov, V., Blake, A.: "GrabCut" interactive foreground extraction using iterated graph cuts. ACM Trans. Graph. (TOG) **23**(3), 309–314 (2004)
28. Sofiiuk, K., Petrov, I.A., Konushin, A.: Reviving iterative training with mask guidance for interactive segmentation. arXiv preprint: arXiv:2102.06583 (2021)
29. Sofiiuk, K., et al.: F-BRS: rethinking backpropagating refinement for interactive segmentation. In: CVPR, pp. 8623–8632 (2020)
30. Wang, G., et al.: DeepIGeoS: a deep interactive geodesic framework for medical image segmentation. PAMI **41**(7), 1559–1572 (2018)
31. Wu, J., et al.: Milcut: a sweeping line multiple instance learning paradigm for interactive image segmentation. In: CVPR, pp. 256–263 (2014)
32. Xu, N., et al.: Deep GrabCut for object selection. arXiv preprint: arXiv:1707.00243 (2017)
33. Xu, N., et al.: Deep interactive object selection. In: CVPR, pp. 373–381 (2016)
34. Yuliang, L., et al.: Detecting curve text in the wild: new dataset and new solution. arXiv preprint: arXiv:1712.02170 (2017)
35. Zhang, C., et al.: Intention-aware feature propagation network for interactive segmentation. arXiv preprint: arXiv:2203.05145 (2022)

36. Zhang, K., Zhuang, X.: CycleMix: a holistic strategy for medical image segmentation from scribble supervision. In: CVPR, pp. 11656–11665 (2022)
37. Zhang, S., et al.: Interactive object segmentation with inside-outside guidance. In: CVPR, pp. 12234–12244 (2020)
38. Zhuang, X.: Multivariate mixture model for cardiac segmentation from multi-sequence MRI. In: Ourselin, S., Joskowicz, L., Sabuncu, M.R., Unal, G., Wells, W. (eds.) MICCAI 2016. LNCS, vol. 9901, pp. 581–588. Springer, Cham (2016). https://doi.org/10.1007/978-3-319-46723-8_67

ADORE: Adaptive Diffusion Optimized Restoration for AI-Generated Facial Imagery

Junxue Li[1,2], Hong Chen[1,2(✉)], and Guanglei Qi[3]

[1] School of Computer Science (National Pilot Software Engineering School), Beijing University of Posts and Telecommunications, Beijing, China
{lijx,chenhong76}@bupt.edu.cn
[2] Key Laboratory of Interactive Technology and Experience System (BUPT), Ministry of Culture and Tourism, Beijing, China
[3] Century College Beijing University of Posts and Telecommunications, Beijing, China

Abstract. We introduce ADORE (Adaptive Diffusion Optimized Restoration), a pioneering solution that addresses facial distortion issues in diffusion-based, language-guided image generation. ADORE enhances facial quality based on image characteristics and style, improving the visual fidelity of AI-generated images. It also mitigates boundary distortions during the face-background fusion process, offering a novel approach to address instability issues by using generative models for image restoration. Rigorous experiments validate ADORE's proficiency in achieving high-quality, style-consistent facial restorations. ADORE supports text-driven, fine-tuned facial refinement, leveraging the model's open-domain synthesis capability. As the first method tailored to enhance facial generation quality in text-to-image models, with its versatility and innovative solutions, ADORE successfully addresses a pressing issue and paves new avenues in image generation.

Keywords: Generative AI · Diffusion model · Face image restoration · Object detection

1 Introduction

Recent text-to-image generative diffusion models have developed significantly [11–13], resulting in their ability to produce top-notch images based on textual prompts with high quality, diversity, and realism [18]. One crucial aspect of image generation is the generation of human portraits, which holds great significance for various applications. However, existing text-to-image models struggle to meet the demand for high-quality facial generation, often resulting in the omission of crucial facial details. The commonly used approach to address this issue involves employing blind face restoration methods to repair the generated facial images [16,20,23]. Blind face restoration aims to recover high-quality faces from low-quality counterparts affected by unknown degradation, such as low-resolution,

Q. Liu et al. (Eds.): PRCV 2023, LNCS 14435, pp. 364–376, 2024.
https://doi.org/10.1007/978-981-99-8552-4_29

noise, blur, compression artifacts [15]. However, the current pretrained blind face restoration models used in real-world scenarios lack specific learning on the degradation caused by diffusion models, making them less effective in repairing such newly emerged degradation. Moreover, the diverse and ever-changing facial styles present a challenge for blind face restoration methods, raising doubts about their applicability across different styles.

Fig. 1. Face Restoration Results. The top portion presents a comparative display between the output generated by the text-to-image generative diffusion model in response to the textual prompt: 'A painting of ten children on a couch,' and the image restoration by our proposed method. The bottom portion showcases a detailed comparison of facial features.

We attribute this issue to the inefficiency and instability of using blind face restoration neural network models to address the problem of facial detail loss in text-to-image generation. In the era of rapid development of generative models and the increasing demand for AI-generated content [1], there is a pressing need to propose a specialized method that can restore facial details in the generated results of text-to-image models. This method aims to address this challenging problem and cater to the growing requirements in AIGC (artificial intelligence generated content).

In this study, we present ADORE, a pioneering approach to restoring facial details in images produced by text-to-image generative models. Recognizing the potential degradation introduced during the image generation process, ADORE capitalizes on the inherent style of the generated images to conduct effective facial repairs. This results in enhanced image quality, with a distinct improvement in the fidelity of facial details. Our proposed framework provides a unique and robust mechanism for facial feature restoration, striking a balance between preserving the stylistic integrity of the original image and introducing subtle detail improvements. By integrating the output from the generative model, ADORE facilitates a seamless facial restoration process with text-to-image generative models.

In summary, this paper offers the following key contributions:

1. We've pioneered ADORE, the first solution specifically aimed at addressing facial distortion in AI-generated images, thereby breaking the prevailing reliance on transfer learning in this domain.
2. Introduction of a novel face-background fusion algorithm within ADORE that effectively minimizes boundary distortions in diffusion-based, language-guided generated imagery.
3. Empirical validation confirms ADORE's adaptability across diverse styles, effectively bridging the gap between raw outputs of image generation algorithms and high-fidelity AI-generated content.

2   Related Work

2.1   Diffusion Models for Generation

Diffusion models have demonstrated remarkable achievements in generating high-fidelity and diverse images by effectively approximating the data distribution. These models have emerged as a promising approach to image generation, particularly in the context of text-to-image synthesis. Ramesh et al. [11] proposed a two-stage model leveraging contrastive representations from CLIP [10] for diverse and photorealistic image generation. Saharia et al. [13] demonstrated the effectiveness of frozen large pretrained language models as text encoders for text-to-image generation using diffusion models. Rombach et al. [12] pioneered latent diffusion model and cross-attention conditioning mechanism, achieving remarkable results in conditional image synthesis tasks while maintaining high image quality.

2.2   Diffusion Models for Face Restoration and Synthesis

In facial image generation and restoration, significant advancements have been propelled by the adoption of diffusion models. For instance, Yue et al. [22], introduced DifFace, a robust, diffusion-based blind face restoration method. Kim et al. [3] proposed DiffFace, which applies a diffusion model in a high-fidelity face swapping framework. Similarly, the LDM-based face synthesis model, SGLDM by Peng et al. [8], incorporates a diffusion model to handle sketches of varying abstraction levels. Qiu et al. [9] presented DiffBFR, a face restoration model deploying a pure diffusion model to ensure stability and distribution generation.

3   Method

In this section, we introduce our method specifically devised for restoring facial features in generated images. Owing to our astute utilization of text-to-image diffusion models, our method not only excels in discerning the unique style characteristics of the images but also enables us to precisely reconstruct distorted facial details, as skillfully demonstrated in the restoration of multiple child avatars in Fig. 2. We formulate our model as three high-level stages: sketch detection and reference extraction, fine-grained synthesis, fusion and reconstruction.

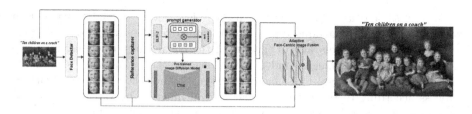

Fig. 2. Overview of the ADORE framework. Our objective is to enhance image quality through the detection and restoration of distorted facial regions in the generated images of the text-to-image model.

3.1 Sketch Detection and Reference Extraction

Upon receiving a generated image $\tilde{x}$ from the text-to-image diffusion model, potentially presenting missing or distorted facial details, we initiate the restoration process with individual facial detection to maintain the uniqueness of each character. For this, we use the state-of-the-art YOLOv8 face detector, renowned for its superior mean average precision (mAP) scores and faster inference speed on the COCO dataset, despite the lack of an official paper. Following face detection, we extract the detected faces' associated features, which we call reference features. These features guide our generative restoration strategy, and their use restricts the generation of irrelevant edge details during restoration.

Algorithm 1. Reference Feature Extraction

Input: Detected face set X_f, entire image $\tilde{x}$.
    for $i = 1$ to $|X_f|$ **do**
        $(h_f, w_f) \leftarrow \text{size}(X_f[i])$
        $s \leftarrow \max(h_f, w_f)$
        $X'_f \leftarrow \text{resize}(X_f[i], s, s)$
        $s = \lambda_i \cdot (1 + \mu_i) \cdot s$
        $R_f \leftarrow \text{computeRegion}(\tilde{x}, X'_f, s)$
        if $\nexists R'_f \in X_{\text{ref}} : R'_f \cap R_f \neq \emptyset$ **then**
            $X_{\text{ref}}[i] \leftarrow \text{extractRegion}(R_f, \tilde{x})$
        end if
    end for
Output: X_{ref}.

---

where, h_f and w_f denote the dimensions of the detected facial image. Two factors, λ and μ, are introduced to adjust reference feature extraction, considering two aspects: λ is the ratio of the size of the facial image to the average size of all detected facial images, and μ is the ratio of the facial image size to the size of the original image. These factors allow us to tune the reference extraction process according to the context of each facial image. The values of λ and μ corresponding to each image are computed using the following formulas:

$$\lambda_i = d_i - \mu_i \cdot \left(\frac{h_f[i] \cdot w_f[i]}{\frac{1}{N} \sum_{j=1}^{N} (h_f[j] \cdot w_f[j])} - 1 \right) \tag{1}$$

$$\mu_i = \left(\frac{\max(h_f[i], w_f[i])}{\frac{h_{\tilde{x}} + w_{\tilde{x}}}{2}} \right) \tag{2}$$

d_i refers to the ratio of the interfacial distance to the currently detected facial image, defining the feasible range for the reference feature. The magnitude of λ is inversely proportional to the ratio between the size of the detected facial feature and the average size of all such detected features, allowing images with prominent facial features to concentrate more effectively on reconstructing local facial details. In contrast, the magnitude of μ is positively proportional to the ratio of the detected facial feature's size to the overall image size, aimed at reducing the undue influence of excessive reference features on the generative reconstruction of smaller-sized detected facial features. *ComputeRegion* and *ExtractRegion* are specifically designed to facilitate our reference feature extraction process. *ComputeRegion* uses the parameters to extract reference features from the original image. *ExtractRegion* eliminates regions containing other facial features after comparing with X_f and then retrieves from the original image the final reference feature, which does not include any irrelevant facial features of the currently detected face.

3.2   Fine-Grained Synthesis

The facial image X_f, reference features X_{ref}, and original image $\tilde{x}$ obtained from the previous module serve as inputs to our generative repair image description process. We employ BLIP-2 [4] to generate image-to-text descriptions $n_f P_f$, $n_{ref} P_{ref}$, and $n_x P_x$ for the images. These descriptions are then combined into a final text representation C_{p_i} using the direct sum operation, as follows:

$$C_{p_i} = \{p_i \mid p_i = n_f P_f[i] \oplus n_{ref} P_{ref}[i] \oplus n_x P_x\} \tag{3}$$

The factors n_f, n_{ref}, and n_x represent the influence of the prompts in decreasing order, allowing the generation details to be reasonably focused on facial repair. We map face image X_f to the latent space representation z_f, and the forward diffusion process is performed to disrupt the original distorted facial details, setting the stage for the generative reverse process:

$$z_n = \sqrt{\overline{\alpha}_n} \cdot z_f + (1 - \overline{\alpha}_n) \cdot \epsilon \tag{4}$$

$$z_{\text{result}} = p_\theta(z_{0:N}|C_{p_i}) = p(z_f) \prod_{n=1}^{N} p_\theta(z_{n-1}|z_n, C_{p_i}) \tag{5}$$

$$x_{\text{result}} = \mathcal{D}(z_{\text{result}}) \tag{6}$$

The parameters α and β are scheduled noise parameters. In the reverse process, a conditional U-Net denoiser ϵ_θ is employed to predict the noise ϵ at each timestep n. We represent the textual guidance as a conditional input by encoding the text using the clip text encoder $\tau = \tau_\theta(y)$. This encoding process generates a text embedding that is subsequently fed into the cross-attention blocks of the U-Net denoiser. Together with the latent space representation z_f of the original facial features, these elements serve as conditioning factors in our generative reconstruction method, facilitating facial image restoration. Finally, we decode the reconstructed latent representation, z_{result}, back into pixel space, thereby obtaining the set of final reconstructed facial images X_{result}.

3.3 Fusion and Reconstruction

A paramount challenge in applying text-to-image diffusion models for image restoration pertains to the ambiguity of the reintroduced details. The absence of a robust strategy to identify and eliminate superfluous details engendered during the restoration process often results in incongruence between the edges of the restored and original images and potential color discrepancies, thus inhibiting successful fusion. Inspired by the forward process formula of the diffusion model [2], we propose an algorithm-based image fusion method to address the issues mentioned earlier. This method involves fusing the unrepaired facial images x_{Sketch} with the repaired images x_{result}, resulting in a fusion photograph R that can be directly embedded into the original image. The expression of the algorithm is as follows:

The fusion region size in the resulting repaired image is determined by φ, which is calculated based on the ratio between the detected face size and the entire image size. The influence factors, p_{sketch} and p_{result}, represent the impact coefficients of x_{sketch} and x_{result}, respectively. Our fusion method mitigates and removes unnecessary edge features from the restored facial image, replacing them with the original image's corresponding features to ensure consistency. Subsequently, we employ Poisson fusion to neutralize color discrepancies between the restored facial image and the original, thus facilitating seamless integration.

4 Experiment

4.1 Implementation Details

In our experiments, we utilized the publicly available stable diffusion model weights[1] as the default in latent diffusion models [12], supplemented by vari-

---

[1] https://huggingface.co/runwayml/stable-diffusion-v1-5.

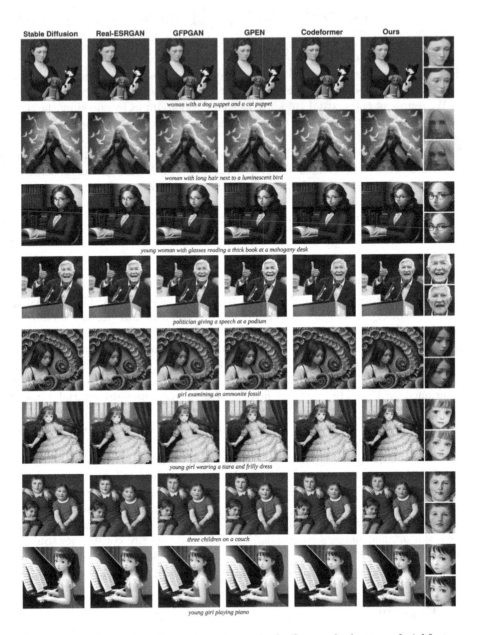

Fig. 3. Comparisons with other restoration methods. Our method ensures facial feature consistency and image style preservation while reconstructing distorted facial details caused by the text-to-image generation process, significantly outperforming other methods in the restoration results.

Algorithm 2. Adaptive Face-Centric Image Fusion

Input: X_{sketch}, X_{result}, $\tilde{x}$

  Initialize R as a copy of image $\tilde{x}$

  for each $(x_{\text{sketch}}, x_{\text{result}})$ in $(X_{\text{sketch}}, X_{\text{result}})$ **do**

    Initialize R_{temp} as an empty image of the same size as x_{sketch} and x_{result}

    $w \leftarrow$ width of x_{sketch}, $h \leftarrow$ height of x_{sketch}

    $t_{\max} \leftarrow w \times h \times \varphi$

    $p_{\text{sketch}} \leftarrow 0$, $p_{\text{result}} \leftarrow 1$

    for $i \leftarrow 1$ to $t_{\max}$ **do**

      $p_{\text{sketch}} \leftarrow i/t_{\max}$, $p_{\text{result}} \leftarrow 1 - p_{\text{sketch}}$

      for $j \leftarrow 1$ to w **do**

        for $k \leftarrow 1$ to h **do**

          $X_{temp_{jk}} \leftarrow p_{\text{sketch}} \cdot x_{\text{sketch}_{jk}} + p_{\text{result}} \cdot x_{\text{result}_{jk}}$

        end for

      end for

    end for

    $R \leftarrow$ Poisson blend (X_{temp})

  end for

Output: R

---

ous style-specific weights to sample different image generation styles[2]. Evaluations were conducted on the "people" category of the PartiPrompts dataset [21], a broad collection of over 1600 English prompts, allowing us to generate and reconstruct facial imagery. For face detection and alignment within the generated images, we deployed YOLOv8, implementing the code from the Ultralytics repository, with a particular focus on tackling our facial recognition task using pre-trained weights trained on the WIDER FACE dataset [19].

During image generation inference, we employed the DDIM sampler [14] with 20 steps and prompts from the PartiPrompts dataset [21] as guidance, applying a guidance scale of 7. For fine-grained synthesis, we found the DPM++ 2 M sampler [5,6] effective with 30 steps and a guidance scale of 11. Comparative methods implemented using their default configurations.

4.2 Main Result

We chose 260 character-descriptive prompts from PartiPrompts to rigorously test the effectiveness of our approach. For each prompt, we generated 2000 images across various resolutions like 512×512, 512×320, and 768×768, among other specifications. In Fig. 3, we present examples comparing our method against state-of-the-art blind face restoration baselines to demonstrate its effectiveness in facial restoration on generated images. We compare our method with a range of baselines commonly employed for enhancing image quality in text-to-image diffusion model image generation, including blind face restoration methods such as

---

[2] https://civitai.com/models/6424/chilloutmix
https://civitai.com/models/3627/protogen-v22-anime-official-release
https://civitai.com/models/4201/realistic-vision-v20.

GFPGAN [16], GPEN [20], CodeFormer [23], and the superresolution restoration method Real-ESRGAN [17]. The comparison results indicate that our method outperforms the aforementioned baselines in terms of both quality and style consistency.

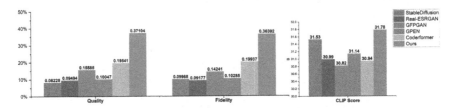

Fig. 4. Quantitative results. The left side illustrates user preference studies for facial restoration in generated images compared to commonly used methods, and the right side presents the evaluation metrics based on CLIP scores.

4.3   Qualitative Evaluation

We compare our method against state-of-the-art blind face restoration baselines and the commonly used superresolution restoration method to demonstrate its effectiveness in facial restoration on images generated by the text-to-image diffusion model. Figure 3 provides a qualitative evaluation of our approach relative to existing methods, namely GFPGAN [16], GPEN [20], CodeFormer [23], and Real-ESRGAN [17]. These established methods demonstrate proficiency in recreating images with precise facial contours and features similar to actual individuals, as highlighted in rows five and six. However, they encounter difficulties when dealing with images exhibiting low fidelity, occluded faces, or unclear facial contours, often due to the denoising generation process inherent in these generative models, as exemplified in rows one, two, three, and four, as shown in the accompanying images. Notably, when confronted with unique styles such as oil paintings or cartoons displayed in rows one and eleven, these methods often fail to capture the distinctive image styles adequately. In contrast, our proposed approach consistently outperforms these methods across all these challenging scenarios, exhibiting superior performance and fidelity.

4.4   Quantitative Evaluation

In our quantitative user preference survey conducted via an online questionnaire, participants from 28 provincial administrative regions in China, representing a uniform age distribution between 18–60 and various professions, evaluated the overall quality and facial feature fidelity of face restoration using our method versus others. The examples used for this assessment were similar to those depicted in Fig. 3. Here, 'quality' refers to the overall excellence of the restored images, while 'fidelity' denotes the accuracy and congruence of facial features with the entire image style. Data collected from 159 random participants, spanning various age groups and professions, were statistically analyzed and visually depicted in Fig. 4. The results clearly demonstrated that our proposed method consistently outperformed the alternatives across all evaluation parameters. The CLIP score measures text-to-image alignment by computing the similarity between the target prompt and the generated image. As visualized in Fig. 4 third column, our approach scores higher than the alternative methods. Notably, automatic metrics such as the CLIP score may not always align with human perception [11] and should be referenced as an imperfect measure.

4.5   Ablation Study

Face Restoration Method. As depicted in Fig. 5, we applied the stable diffusion img2img on the original input image for facial restoration instead. The results validated the superiority of our method. Comparatively, our method yielded higher-quality images with enhanced flexibility in facial expression adjustments when using CodeFormer with varying weights for image restoration shown in the first row. The strategy of employing img2img for restoration, followed by CodeFormer for secondary restoration, fell short in maintaining consistency and quality compared to our method. Figure 5 showcases that our method achieves nuanced facial feature fine-tuning while minimizing unnecessary alterations, thereby maintaining faithfulness to the original image, demonstrating its potential for facial restoration and refinement in generated models. This also substantiates the potential application of ADORE in facilitating text-driven modifications to generated images.

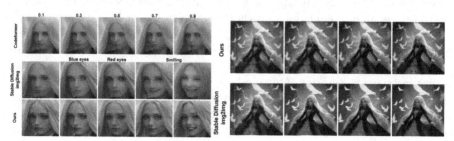

Fig. 5. Ablations on the effectiveness of diffusion-based fine-grained synthesis. Our method not only achieves higher levels of fine-grained results but also allows for the adjustment of facial details in target images, demonstrating the efficacy of the approach.

Image Fusion Method. As depicted in Fig. 6, we removed the image fusion module to directly merge the facial restorations with the original images. We also test devising an inpainting-based approach [7]. We utilized a face detection algorithm to ascertain inpainting mask locations and attempted to repair regenerated image edges for incorporation into source images. Unfortunately, both of

Fig. 6. Ablations on the effectiveness of adaptive face-centric image fusion. Comparisons between inpainting using the diffusion model and image stitching.

these approaches led to improper edge distortion, color differences, and unstable restoration results. Our image fusion module exhibited validated superiority in outcomes.

5   Conclusion

In this work, we introduced ADORE, an innovative technology designed to transcend the constraints of existing image-generation models by yielding intricate facial details. Crucially, our approach leverages generative models' open-domain synthesis capability, enabling zero-shot facial restoration in a variety of styles without the need for additional training. This achievement is primarily due to our strategic application of a diffusion-based image generator and our targeted solution for managing undeserved edge detail generation that often poses challenges in image restoration. Our method, proven effective in facial restoration, offers a versatile paradigm for detail-oriented image generation tasks. Furthermore, its robust framework has significant potential for future exploration and applications in video generation, subject-driven generation, and image editing based on diffusion models.

Limitations. Our method, leveraging pretrained image diffusion model weights, also inherits its limitations. For instance, our method is inherently open-ended, focusing on enhancing the quality of generated images, which may not precisely restore real-world images to their ideal state.

Acknowledgements. This work was supported in part by the project "Digital Twin Application Demonstration for New Museum Public Service Models", a key research topic under the National Key Research and Development Program of China, Grant No. 2022YFF0904305.

References

1. Cao, Y., et al.: A comprehensive survey of AI-generated content (AIGC): a history of generative AI from GAN to ChatGPT. arXiv preprint arXiv:2303.04226 (2023)
2. Ho, J., Jain, A., Abbeel, P.: Denoising diffusion probabilistic models. Adv. Neural. Inf. Process. Syst. **33**, 6840–6851 (2020)
3. Kim, K., et al.: DiffFace: diffusion-based face swapping with facial guidance. arXiv preprint arXiv:2212.13344 (2022)
4. Li, J., Li, D., Savarese, S., Hoi, S.: BLIP-2: bootstrapping language-image pre-training with frozen image encoders and large language models. arXiv preprint arXiv:2301.12597 (2023)
5. Lu, C., Zhou, Y., Bao, F., Chen, J., Li, C., Zhu, J.: DPM-Solver: a fast ode solver for diffusion probabilistic model sampling in around 10 steps. arXiv preprint arXiv:2206.00927 (2022)
6. Lu, C., Zhou, Y., Bao, F., Chen, J., Li, C., Zhu, J.: DPM-Solver++: fast solver for guided sampling of diffusion probabilistic models. arXiv preprint arXiv:2211.01095 (2022)
7. Lugmayr, A., Danelljan, M., Romero, A., Yu, F., Timofte, R., Van Gool, L.: Repaint: inpainting using denoising diffusion probabilistic models. In: Proceedings of the IEEE/CVF Conference on Computer Vision and Pattern Recognition, pp. 11461–11471 (2022)
8. Peng, Y., Zhao, C., Xie, H., Fukusato, T., Miyata, K.: DiffFaceSketch: high-fidelity face image synthesis with sketch-guided latent diffusion model. arXiv preprint arXiv:2302.06908 (2023)
9. Qiu, X., Han, C., Zhang, Z., Li, B., Guo, T., Nie, X.: DiffBFR: bootstrapping diffusion model towards blind face restoration. arXiv preprint arXiv:2305.04517 (2023)
10. Radford, A., et al.: Learning transferable visual models from natural language supervision. In: International Conference on Machine Learning, pp. 8748–8763. PMLR (2021)
11. Ramesh, A., Dhariwal, P., Nichol, A., Chu, C., Chen, M.: Hierarchical text-conditional image generation with clip latents. arXiv preprint arXiv:2204.06125 (2022)
12. Rombach, R., Blattmann, A., Lorenz, D., Esser, P., Ommer, B.: High-resolution image synthesis with latent diffusion models. In: Proceedings of the IEEE/CVF Conference on Computer Vision and Pattern Recognition (CVPR), pp. 10684–10695 (2022)
13. Saharia, C., et al.: Photorealistic text-to-image diffusion models with deep language understanding. Adv. Neural. Inf. Process. Syst. **35**, 36479–36494 (2022)
14. Song, J., Meng, C., Ermon, S.: Denoising diffusion implicit models. arXiv preprint arXiv:2010.02502 (2020)
15. Wang, T., et al.: A survey of deep face restoration: denoise, super-resolution, deblur, artifact removal. arXiv preprint arXiv:2211.02831 (2022)
16. Wang, X., Li, Y., Zhang, H., Shan, Y.: Towards real-world blind face restoration with generative facial prior. In: Proceedings of the IEEE/CVF Conference on Computer Vision and Pattern Recognition, pp. 9168–9178 (2021)
17. Wang, X., Xie, L., Dong, C., Shan, Y.: Real-ESRGAN: training real-world blind super-resolution with pure synthetic data. In: Proceedings of the IEEE/CVF International Conference on Computer Vision, pp. 1905–1914 (2021)

18. Yang, L., et al.: Diffusion models: a comprehensive survey of methods and applications. arXiv preprint arXiv:2209.00796 (2022)
19. Yang, S., Luo, P., Loy, C.C., Tang, X.: Wider Face: a face detection benchmark. In: Proceedings of the IEEE Conference on Computer Vision and Pattern Recognition, pp. 5525–5533 (2016)
20. Yang, T., Ren, P., Xie, X., Zhang, L.: Gan prior embedded network for blind face restoration in the wild. In: Proceedings of the IEEE/CVF Conference on Computer Vision and Pattern Recognition, pp. 672–681 (2021)
21. Yu, J., et al.: Scaling autoregressive models for content-rich text-to-image generation. arXiv preprint arXiv:2206.10789 (2022)
22. Yue, Z., Loy, C.C.: DifFace: blind face restoration with diffused error contraction. arXiv preprint arXiv:2212.06512 (2022)
23. Zhou, S., Chan, K., Li, C., Loy, C.C.: Towards robust blind face restoration with codebook lookup transformer. Adv. Neural. Inf. Process. Syst. **35**, 30599–30611 (2022)

TU-Former: A Hybrid U-Shaped Transformer Network for SAR Image Denoising

Shikang Tian[1,2], Shuaiqi Liu[1,2(✉)], Yuhang Zhao[1,2], Siyuan Liu[1,2], Shuhuan Zhao[1,2], and Jie Zhao[1,2]

[1] College of Electronic and Information Engineering, Hebei University, Baoding 071002, China
sktian_hbu@163.com

[2] Machine Vision Technology Innovation Center of Hebei Province, Baoding 071000, China

Abstract. In order to obtain a better synthetic aperture radar (SAR) image noise suppression effect, we combine Convolutional Neural Network (CNN) and Transformer network to construct a U-shaped hybrid Transformer (TU-former) for SAR image denoising. The encoder of the TU-former network consists of convolution, residual and self-attentive mechanisms. To avoid large-scale training of the model, TU-former first uses convolutional structure and residual structure for shallow feature extraction of images. Subsequently, TU-former performs long-term dependencies by using the self-attention mechanism of the Transformer block to further extract the deep features of the image. Finally, TU-former sums the output of the decoder with the input noise image to obtain the final denoised image. Compared with the state-of-the-art SAR image denoising algorithm, the proposed algorithm not only improves in each objective index but also shows great advantages in the visual effect after denoising.

Keywords: Image denoising · Transformer · Residual

1 Introduction

Synthetic aperture radar (SAR), as a high-resolution imaging radar, has the advantage of being able to image target areas around the clock and in all weather and is therefore widely used in civil and military applications, such as disaster monitoring, target reconnaissance, and surface change detection [1]. Speckle suppression is an important issue in SAR image processing. SAR denoising algorithms can be divided into three main categories: spatial domain filtering algorithms, frequency domain filtering algorithms, and deep learning-based image-denoising algorithms.

The basic principle of spatial domain filtering for noise reduction is to directly perform data operations on the two-dimensional space where the image is located, i.e., to process the gray values of pixels. Because of the simple and feasible implementation of spatial domain filtering, it had been widely used in SAR image denoising, resulting in many classical denoising algorithms, such as Frost filtering [2]. Dabov et al. [3] proposed a block-matching and 3D filtering (BM3D) denoising algorithm based on transform domain image filtering. Parrilli et al. [4] extended the BM3D to the field of SAR image

© The Author(s), under exclusive license to Springer Nature Singapore Pte Ltd. 2024
Q. Liu et al. (Eds.): PRCV 2023, LNCS 14435, pp. 377–389, 2024.
https://doi.org/10.1007/978-981-99-8552-4_30

denoising by proposing a local linear minimum mean square error (LLMMSE) filter and obtained good denoising results for SAR images.

The transform domain denoising method first performed multi-scale transformation on the image, which can transform the image from the spatial domain to the transform domain. Then, the transform coefficients can be shrunk for noise suppression. Finally, the inverse transformation is performed to obtain the denoised image, such as the wavelet transform-based SAR image denoising algorithm [5]. Although the SAR image denoising algorithm based on the transform domain can achieve better denoising effect, there are problems such as large operation and introduction of artifacts in the process.

With the development of deep learning, convolutional neural networks had also been gradually applied to SAR image denoising since 2017. Chierchia et al. [6] first used CNN for SAR image denoising by using residual learning and batch normalization. Liu et al. [7] proposed an algorithm based on a multi-scale CNN, which effectively suppressed the speckle in SAR images. Liu et al. [8] combines generative adversary network (GAN) and total variation (TV) loss to achieve good results in both subjective and objective aspects. Ramesh et al. [9] constructed the AGSDNet for SAR image denoising by fusing the attention mechanism with the gradient of SAR image, which achieved better denoising results. However, this model itself had the disadvantage of high redundancy and required a large number of datasets as support.

In recent years, the Transformer [10] had shown significant performance in natural language and advanced vision tasks. Transformer-based structures excelled in capturing long-range dependencies through global self-attention, which helped the application of Transformer in various vision tasks. Although Transformer achieved good results, it lacks weight sharing and local connection compared with CNN which cannot learn the local features. In order to improve Transformer, SCUNet [11] bring the residual block to Swin Transformer block. SCUNet used UNet as the backbone network, which incorporated the local modeling capability of CNN and the non-local modeling capability of Transformer and could be well applied to various image processing tasks. In [12], the Transformer was first applied to the SAR image denoising by combining the CNN. This method achieved significant improvements over the traditional CNN-based de-noising methods.

In order to solve the problem of introducing boundary artifacts in the denoised image, we propose an effective speckle suppression network: the U-shaped hybrid Transformer denoising network (TU-former). Experimental results show that our algorithm has good denoising effect on both simulated and real SAR images. The main contributions of this paper are as follows: (1) A new network TU-former for speckle suppression is proposed, which can effectively balance the relationship between denoising and texture retention. (2) On the one hand, TU-former uses convolution operation to effectively extract local features and avoid large-scale training of the model; on the other hand, the relationship between adjacent blocks is effectively linked, thus avoiding the generation of boundary artifacts, which is conducive to further improving the effect of image denoising.(3) We use multi-scale decomposition to further improve the performance of the Transformer, and we also use Convolutional Position Embedding (CPE) to enhance the Transformer's local information retention ability.

2   Proposed Method

2.1   The Network Structure

The TU-former constructed is a U-shaped hierarchical network which is shown in Fig. 1. The TU-former network consists of three parts, namely the encoder, the bottleneck module and the decoder. Every part is described in detail as follows.

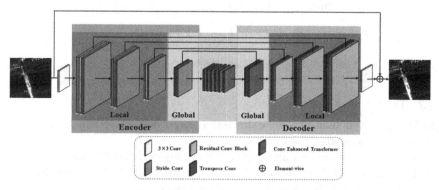

Fig. 1. The overall network structure of TU-former.

2.1.1   The Encoder

The encoder includes local and global encoding modules. The local encoding module consists of the residual convolution block (RCB), while the global encoding module consists of the convolution enhanced Transformer (CETransformer) block.

For the noisy image X, we can obtain the initial features X_0 by a 3 × 3 convolution. The local information of the initial feature X_0 can be obtained by three local encoding modules. Meanwhile, stepwise convolution is used to achieve down-sampling. The above process can be expressed as follows:

$$\begin{cases} X_{E_1} = \mathcal{C}^2_{4\times4}(H_{RCB}(X_0)) \\ X_{E_2} = \mathcal{C}^2_{4\times4}(H_{RCB}(X_{E_1})) \\ X_{E_3} = \mathcal{C}^2_{4\times4}(H_{RCB}(X_{E_2})) \end{cases} \tag{1}$$

where $H_{RCB}(\cdot)$ denotes the mapping function of the RCB structure, and $\mathcal{C}^2_{4\times4}$ denotes a convolutional layer with a step size of 2 and a kernel size of 4 × 4. The RCB structure consists of two convolutional layers with kernel size 3 × 3, ReLU activation function, and one convolutional layer with kernel size 1 × 1 and residual connections.

In this paper, a simple convolution and flattening operation is used to transform the output features into a one-dimensional sequence, that is:

$$X'_{E_3} = LN(\mathcal{C}^1_{3\times3}(X_{E_3})) \tag{2}$$

where $LN(\cdot)$ denotes the layer normalization. The CETransformer takes advantage of the window attention mechanism to capture the long-range dependencies of features, which can be expressed as:

$$X_{E_4} = \mathcal{C}^2_{4\times4}(H_{CET}(X'_{E_3})) \tag{3}$$

where $H_{CET}(\cdot)$ denotes CETransformer.

The X_{E_4} go through the bottleneck module to extract the deeper features of the image. We set six CETransformers to capture longer dependencies, that is:

$$X_5 = H_{CET\times6}(X_{E_4}) \tag{4}$$

where $H_{CET\times6}(\cdot)$ denotes the six CETransformers.

2.1.2 The Decoder

The decoder is composed of a similar structure as the encoder, except that the up-sampling operation uses transposed convolution.

The features X_5 are first up-sampled with transposed convolution, then matched with the output features X_{E_4} for channel concatenation. Finally, the fused features are reconstructed with the global decoding module, that is

$$X_{D_4} = H_{CET}(X_5 \uparrow, X_{E_4}) \tag{5}$$

where $\uparrow$ denotes the transposed convolution of step 2.

We adopted a convolution to reshape the features X_{D_4} into X'_{D_4}. The feature map X'_{D_4} is up-sampled by using transposed convolution, and then X'_{D_4} and X_{E_3} are fed to the local decoding module by channel concatenation, that is

$$\begin{cases} X'_{D_4} = LN(\mathcal{C}^1_{3\times3}(X_{D_4})) \\ X_{D_3} = H_{RCB}(X'_{D_4} \uparrow, X_{E_3}) \end{cases} \tag{6}$$

With the same operation, Feature map X_{D_3} is channel spliced with the outputs of the first two stages in the local encoding module and fed to the corresponding local decoding module for decoding restoration. Finally we obtain the same resolution as the original image size, that is

$$\begin{cases} X_{D_2} = H_{RCB}(X_{D_3} \uparrow, X_{E_2}) \\ X_{D_1} = H_{RCB}(X_{D_2} \uparrow, X_{E_1}) \end{cases} \tag{7}$$

Finally, the X_{D_1} is fed to a 3×3 convolutional. The global residual connection is introduced to sum the input with the output to obtain the final clean image, that is

$$Y = \mathcal{C}^1_{3\times3}(X_{D_1}) + X \tag{8}$$

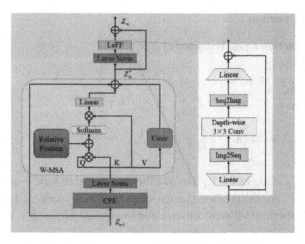

Fig. 2. The structure of CETransformer block.

2.1.3 CETransformer Block

The CETransformer block consists of a window based self-attention (W-MSA) layer, a convolutional layer, and a locally-enhanced feed-forward network (LeFF). The detailed structure of CETransformer module is shown in Fig. 2.

We use CPE to provide position information for Z_{m-1}. The query, key, and value matrices Q, K and V can be obtained by applying LN, that is:

$$\begin{cases} Q = LN(H_{CPE}(Z_{m-1}W^Q)) \\ K = LN(H_{CPE}(Z_{m-1}W^K)) \\ V = LN(H_{CPE}(Z_{m-1}W^V)) \end{cases} \tag{9}$$

where W^Q, W^K, $W^V \in R^{d \times d}$ is the projection matrix, $H_{CPE}(\cdot)$ denotes the CPE.

The Q, K and V are input into the W-MSA, and the self-attentive weight scores are calculated within a non-overlapping window, while relative position encoding B is introduced, and then V is weighted by the softmax. In order to aggregate information in the neighborhood to reduce artifacts, we used an additional convolution on V. The self-attention of each head can be expressed as

$$Attention(Q, K, V) = SoftMax(\frac{QK^T}{\sqrt{d_k}} + B)V + Conv(V) \tag{10}$$

We can obtain Z'_m by summing the obtained attention scores with the input Z_{m-1} like Eq. (12). Finally, we regularized Z'_m by LN, and fed it into LeFF to further enhance the local information, used the residual connection to obtain the final output Z_m. The whole process can be expressed as follows:

$$\begin{aligned} Z'_m &= Attention(Q, K, V) + Z_{m-1} \\ Z_m &= LeFF(LN(Z'_m)) + Z'_m \end{aligned} \tag{11}$$

where Z'_m and Z_m are the outputs of the W-MSA and the LeFF, respectively.

2.2 The Loss Function

The total loss function $\mathcal{L}_{Total}$ is defined as:

$$\begin{cases} \mathcal{L}_{L2} = \left\| \hat{X} - Y \right\|_2^2 \\ \mathcal{L}_{TV} = \sum_{i,j} \left| \hat{X}_{i+1,j} - \hat{X}_{i,j} \right| + \left| \hat{X}_{i,j+1} - \hat{X}_{i,j} \right| \\ \mathcal{L}_{Total} = \lambda_1 \left\| \hat{X} - Y \right\|_2^2 + \lambda_2 \sum_{i,j} \left| \hat{X}_{i+1,j} - \hat{X}_{i,j} \right| + \left| \hat{X}_{i,j+1} - \hat{X}_{i,j} \right| \end{cases} \tag{12}$$

where $\mathcal{L}_{l_2}$ and $\mathcal{L}_{tv}$ denotes the L2 loss and the total variance loss, Y and $\hat{X}$ denote the ground-truth (GT) and the predicted image, respectively, $\hat{X}_{i,j}$ denotes the pixel values in the image. By experimental verification, we set $\lambda_1 = 1$ and $\lambda_2 = 5 \times 10^{-5}$.

3 Experimental Results and Analysis

To verify the effectiveness of the proposed network, TU-former is tested on synthetic data and real image datasets respectively.

3.1 Dataset and Experimental Setup

Training Datasets. We randomly selected 800 from the UC Merced Land Use dataset [13] as the training set where the size of each image is 256×256. The data set is first processed in grayscale, and then the simulated noisy images are obtained by adding multiplicative noise to simulate real images, and the reference images and noisy images are used as the training data of the network.

Test Datasets. nThe test data consists of simulated and real images. We added noise with different Looks (L = 2, 4, 6, 8, 10) in Kodak24 [14] to obtain simulated data. We use real SAR images (as shown in Fig. 3) in four different scenes. These real SAR images are available at the MSAR dataset and the HRSID dataset.

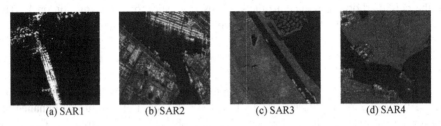

(a) SAR1   (b) SAR2   (c) SAR3   (d) SAR4

Fig. 3. The real SAR images.

In this paper, peak signal-to-noise ratio (PSNR) and structural similarity index (SSIM) [15] are used to evaluate the performance of denoising algorithms. PSNR can effectively estimate the image recovery ability, and the higher, the stronger the denoising ability. SSIM can effectively evaluate the edge preservation ability of the denoised image, and the higher, the better the edge preservation ability.

For real SAR images, since no clean reference image exists, several reference-free metrics are selected for evaluation, namely, equivalent number of looks (ENL) [16], edge preservation degree based on the ratio of average (EPD-ROA) [17], Unassisted Measure (M score) [18], Algorithm running time (TIMES). ENL is used to measure the relative intensity of speckle in an image, and also to measure the performance of the filter, where a larger ENL value in a homogeneous region indicates a better denoising performance. EPD-ROA is used to measure the detail preservation ability of the denoising algorithm, and its value is closer to 1, the image detail is maintained better. M score is used to measure the overall denoising ability of the image, and the smaller the M score, the stronger the overall denoising ability of the algorithm. Times is used to evaluate the algorithm running efficiency.

In the training and testing process, the Pytorch 1.7.1 platform is used to build the deep learning network. The operating system used in this paper is Windows 10, and the GPU of the machine used is NVIDIA GeForce RTX 2080 Ti, and CUDA10.1 are used to enhance the GPU computing power and accelerate the training speed of the network. The network is optimized by using the Adam optimizer [19] with the momentum term of (0.9, 0.999) and the weight decay of 0.02. In the training process, the whole network is iterated 250 times, and the batch size is set to 2, while the initial learning rate is set to 0.0002. In this paper, the learning rate is finally reduced to 0.00006 by using the cosine decay strategy.

3.2 Experiment on Synthetic Speckled Image

For the simulated SAR image test set, four different levels of speckle noise ($L = 2, 4, 8$, and 10) were added to the Kodak24. The average PSNR are given in Table 1, where the bold font indicates the highest PSNR. It can be seen that the proposed algorithm are optimal after denoising the noisy images with four different looks on the Kodak24, which fully illustrates the effectiveness of the proposed algorithm.

Table 1. The average PSNR values of the eight denoising algorithms on the Kodak24

Looks	SAR-BM3D [4]	BSS-SR [20]	Frost [2]	IRCNN [22]	FFDNet [21]	CCSNet [7]	MRDDANet [23]	TU-former
$L = 2$	24.7574	22.0482	19.4576	17.9808	19.9107	17.8555	23.5937	**25.9982**
$L = 4$	26.4657	23.6436	21.5804	20.9961	24.5216	20.8524	24.5831	**27.1575**
$L = 8$	28.0531	24.4495	23.4246	23.7032	26.8213	24.3161	24.2266	**28.4388**
$L = 10$	28.5778	24.6206	23.9599	24.3358	26.9694	25.5600	24.1535	**28.7901**

3.2.1 Experiment on Real Image

In order to fully verify the proposed algorithm, we also tested on the real SAR images—SAR1, SAR2, SAR3 and SAR4. The denoised images are shown in Fig. 4, Fig. 5, Fig. 6 and Fig. 7.

It can be seen from Fig. 4 that SAR-BM3D and BSS-SR have strong denoising ability, but the denoised images are blurred and some texture information is lost. Frost, FFDNet and CCSNet can keep the edges better, but the noise suppression ability is somewhat weakened. Block effect appears in the denoised images of IRCNN. Although MRDDANet has a strong denoising ability, its texture retention ability is poor, and over-smoothing images appear in the denoised images. The TU-former not only has a strong denoising ability, but also can fully restore and maintain the edge texture information of the image, and the overall visual effect of its denoised image is the best.

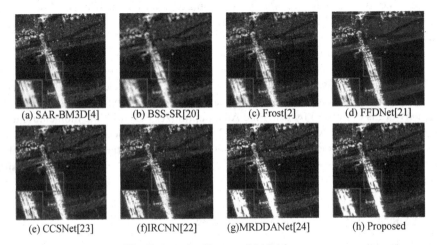

| (a) SAR-BM3D[4] | (b) BSS-SR[20] | (c) Frost[2] | (d) FFDNet[21] |
| (e) CCSNet[23] | (f)IRCNN[22] | (g)MRDDANet[24] | (h) Proposed |

Fig. 4. Denoised image of SAR1 image.

The objective evaluation indexes of the no-reference image are shown in Table 2. The bold font indicates the highest value, and the italic indicates the next highest objective index value. From Table 2, it can be seen that the ENL and M score of the proposed algorithm are slightly low, while the other metrics are the best. Although the ENL is slightly lower than that of BSS-SR and SAR-BM3D, it is higher than that of other deep learning-based denoising algorithms, which indicates that our algorithm can effectively suppress speckle noise. The M score of our algorithm is only second to the BSS-SR and much better than other algorithms. TU-former achieved the optimal EPD-ROA which indicates the strong texture detail retention ability. At last, our algorithm takes the shortest runtime and is more efficient.

Table 2. Objective evaluation index of each denoising algorithm for SAR1.

Method	ENL↑	M score↓	EPD-ROA		TIME(s)↓
			HD↑	VD↑	
SAR-BM3D [4]	*2.3933*	124.3513	0.5997	0.6353	26.7422
BSS-SR [20]	**2.3086**	**2.7526**	0.5765	0.5972	4.7648
Frost [2]	2.1791	155.2209	0.6194	0.6401	1.0802
IRCNN [22]	1.9065	59.2175	0.5856	0.6242	1.2015
FFDNet [21]	1.9576	193.7482	0.5975	0.6274	1.0893
CCSNet [7]	1.9457	96.4688	0.5810	0.6110	1.0750
MRDDANet [23]	2.0928	77.6628	*0.9547*	*0.9600*	*0.3226*
TU-former	2.1808	*35.0717*	**0.9888**	**0.9884**	**0.0647**

From Fig. 5, it can be seen that SAR-BM3D has strong denoising ability and edge retention, but the denoised image still contains some noise. BSS-SR and Frost has strong denoising ability, but the texture details of the denoised image are not well kept. The denoised images of FFDNet, FFDNet-CCS, IRCNN and MRDDANet appear over-smoothed and artificial artifacts appear in the denoised images of IRCNN. The algorithms in this paper are able to achieve a balance between denoising and image detail maintenance (Table 4 and 5).

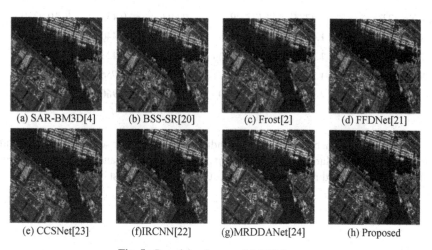

(a) SAR-BM3D[4]	(b) BSS-SR[20]	(c) Frost[2]	(d) FFDNet[21]
(e) CCSNet[23]	(f)IRCNN[22]	(g)MRDDANet[24]	(h) Proposed

Fig. 5. Denoising image of SAR2 image.

Table 3. Objective evaluation index of each denoising algorithm for SAR2.

Method	ENL↑	M score↓	EPD-ROA		TIME(s)↓
			HD↑	VD↑	
SAR-BM3D [4]	3.85	418.0002	0.5889	0.5928	228.0827
BSS-SR [20]	3.1782	**106.2313**	0.3693	0.3793	9.7119
Frost [2]	4.1432	693.0576	0.516	0.5245	6.4732
IRCNN [22]	4.3032	154.4662	0.5213	0.523	10.9635
FFDNet [21]	**4.7461**	499.7736	0.5114	0.5081	6.1965
CCSNet [7]	3.9928	367.3231	0.5128	0.5123	6.2749
MRDDANet [23]	4.0369	475.8186	**0.9615**	**0.9762**	*1.3129*
TU-former	*4.4175*	*134.4626*	*0.9566*	*0.9594*	**0.4075**

Table 4. Objective evaluation index of each denoising algorithm for SAR3.

Method	ENL↑	M score↓	EPD-ROA		TIME(s)↓
			HD↑	VD↑	
SAR-BM3D [4]	1.3735	232.0928	0.5831	0.5234	231.2659
BSS-SR [20]	1.5654	327.1156	0.2129	0.2093	14.0306
Frost [2]	1.4886	607.6411	0.4625	0.4456	6.3508
IRCNN [22]	1.4283	*127.7934*	0.4936	0.4522	10.9250
FFDNet [21]	**1.6119**	232.8575	0.4837	0.4402	6.0656
CCSNet [7]	1.5279	494.1696	0.4648	0.4232	6.3305
MRDDANet [23]	1.5368	546.3707	*0.9672*	*0.9567*	*1.2739*
TU-former	*1.5743*	**35.3240**	**0.9750**	**0.9608**	**0.3743**

From Table 3, it can be seen that all the indicators achieved the second best which indicates the algorithm balances the relationship between denoising and texture maintenance quite well. Our algorithm takes the shortest time and is the most efficient.

Table 5. Objective evaluation index of each denoising algorithm for SAR4.

Method	ENL↑	M score↓	EPD-ROA		TIME(s)↓
			HD↑	VD↑	
SAR-BM3D [4]	3.0660	474.7319	0.5582	0.6136	235.7986
BSS-SR [20]	3.2363	216.5503	0.4387	0.4527	12.3487
Frost [2]	3.1684	1787.2610	0.5174	0.5510	6.5782
IRCNN [22]	2.8732	**89.1502**	0.5052	0.5509	10.9603
FFDNet [21]	3.2557	291.9746	0.4858	0.532	3.0903
CCSNet [7]	3.3025	4239.8280	0.4985	0.5409	6.8226
MRDDANet [23]	*3.3492*	326.3684	*0.9412*	*0.9673*	*1.2903*
TU-former	**3.4142**	*181.7774*	**0.9591**	**0.9690**	**0.3996**

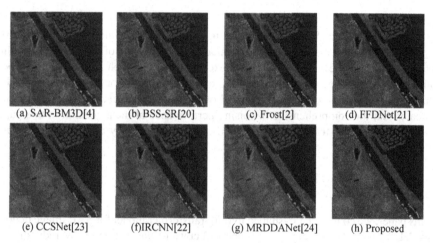

(a) SAR-BM3D[4]   (b) BSS-SR[20]   (c) Frost[2]   (d) FFDNet[21]

(e) CCSNet[23]   (f)IRCNN[22]   (g) MRDDANet[24]   (h) Proposed

Fig. 6. Denoising image of SAR3 image.

The visual effect of SAR3-SAR4 after denoising are given in Fig. 6–Fig. 7. The result of the denoised image is similar to SAR1, and our algorithm outperforms other algorithms in the visual effect of denoising, indicating that our algorithm can effectively remove speckle noise while retaining the edge and texture information of the target image to the maximum extent.

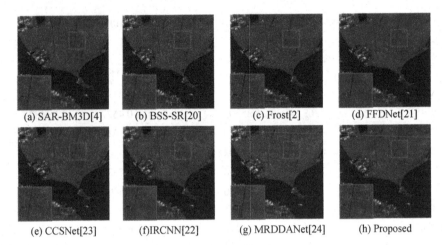

(a) SAR-BM3D[4]   (b) BSS-SR[20]   (c) Frost[2]   (d) FFDNet[21]

(e) CCSNet[23]   (f)IRCNN[22]   (g) MRDDANet[24]   (h) Proposed

Fig. 7. Denoising image of SAR4 image.

In the denoising of SAR3, our algorithm obtains the best value for all the indexes except ENL, which is the secondary best value. In the denoising of SAR4, our algorithm obtains the best value for all the indexes except M score, which is the secondary best value. For SAR3 image denoising, the ENL of our algorithm is only second to FFDNet, which is about 0.03 lower than it. And all other objective indexes are optimal, which indicates that the comprehensive denoising performance of the algorithm in this paper is the best, and has stronger texture detail maintenance ability and higher denoising efficiency high.

4 Conclusion

In this paper, we construct a U-shaped hybrid transformer model for SAR image denoising—TU-former. TU-former reduces the difficulty of early training by reducing the feature map size through local encoder modules. TU-former fully utilizes the global information advantage of the global encoder module to capture features. Therefore, it can effectively suppress noise and the network training time is short. The results show that the proposed algorithm has strong denoising ability and can fully preserve the image details. Therefore, the proposed algorithm contributes to the effective development of high-level image processing in SAR image processing. In future work, we will further improve the network model to perform better in other denoising tasks.

References

1. Liu, F., Wu, J., Li, L., et al.: A hybrid method of SAR speckle reduction based on geometric-structural block and adaptive neighborhood. IEEE Trans. Geosci. Remote Sens. **56**(2), 1–19 (2018)
2. Frost, V.S., et al.: A model for radar images and its application to adaptive digital filtering of multiplicative noise. IEEE Trans. Pattern Anal. Mach. Intell. **4**(2), 157–166 (1982)

3. Dabov, K., Foi, A., Katkovnik, V., Egiazarian, K.: Image denoising with block-matching and 3D fifiltering. In: Image processing: Algorithms and Systems, Neural Networks, and Machine Learning, vol. 6064, p. 606414 (2006)

4. Parrilli, S., Poderico, M., Angelino, C.V., Verdoliva, L.: A nonlocal SAR image denoising algorithm based on LLMMSE wavelet shrinkage. IEEE Trans. Geosci. Remote Sens. **50**(2), 606–616 (2012)

5. Yang, X.J., Chen, P.: SAR image denoising algorithm based on bayes wavelet shrinkage and fast guided filter. J. Adv. Comput. Intell. Intell. Inform. **23**(1), 107–113 (2019)

6. Chierchia, G., Cozzolino, D., Poggi, G., Verdoliva, L.: SAR image despeckling through convolutional neural networks. In: 2017 International Geoscience And Remote Sensing Symposium (IGARSS), pp. 5438–5441 IEEE Press (2017)

7. Liu, S., et al.: SAR speckle removal using hybrid frequency modulations. IEEE Trans. Geosci. Remote Sens. **59**(5), 3956–3966 (2021)

8. Liu, R., et al.: SAR image specle reduction based on a generative adversarial network. In: 2020 International Joint Conference on Neural Networks (IJCNN), pp. 1–6 (2020)

9. Thakur, R.K., Maji, S.K.: AGSDNet: attention and gradient-based SAR denoising network. IEEE Geosci. Remote Sens. Lett. **19**, 1–5 (2022)

10. Ashish, V., et al.: Attention is all you need. In: Neural Information Processing Systems (NIPS), vol. 30 (2017)

11. Zhang, K., Li, Y., Liang, J., et al.: Practical Blind Denoising via Swin-Conv-UNet and Data Synthesis. arXiv preprint arXiv:2203.13278 (2022)

12. Perera, M.V., Bandara, W., Valanarasu, J., et al.: Transformer-based SAR Image Despeckling. In: 2022 IEEE International Geoscience and Remote Sensing Symposium 2022 (2022)

13. Franzen, R.: Lossless ,Kodak.True.Color.Image.Suite, http://r0k.us/plotics/kodak/. Accessed 15 Nov 1999

14. Zhang, L., Wu, X., Buades, A., Li, X.: Color demosaicking by local directional interpolation and nonlocal adaptive thresholding. J. Electron. Imaging **20**(20), 023016–023016 (2011)

15. Wang, Z., Bovik, A.C., Sheikh, H.R., Simoncelli, E.P.: Image quality assessment: From error visibility to structural similarity. IEEE Trans. Image Process. **13**(4), 600–612 (2004)

16. Dellepiane, S.G., Angiati, E.: Quality assessment of despeckled SAR images. IEEE J. Sel. Topics Appl. Earth Observ. Remote Sens. **7**(2), 691–707 (2014)

17. Ma, X., Shen, H., Zhao, X., Zhang, L.: SAR image despeckling by the use of variational methods with adaptive nonlocal functionals. IEEE Trans. Geosci. Remote Sens. **54**(6), 3421–3435 (2016)

18. Gomez, L., Ospina, R., Frery, A.C.: Unassisted quantitative evaluation of despeckling filters. Remote Sens. **9**(4), 389–392 (2017)

19. Ilya, L., Frank, H.: Decoupled Weight Decay Regularization. arXiv preprint arXiv:1711.05101 (2017)

20. Liu, S., et al.: Bayesian Shearlet shrinkage for SAR image de-noising via sparse representation. Multidimension. Syst. Signal Process. **25**(4), 683–701 (2014)

21. Zhang, K., et al.: FFDNet: toward a fast and flexible solution for CNN-based image denoising. IEEE Trans. Image Process. **27**(9), 4608–4622 (2018)

22. Zhang, K., et al.: Learning deep CNN Denoiser prior for image restoration. In: 2017 IEEE Conference on Computer Vision and Pattern Recognition (CVPR), pp. 2808–2817 (2017)

23. Liu, S., et al.: MRDDANet: a multiscale residual dense dual attention network for SAR image denoising. IEEE Trans. Geosci. Remote Sens. **60**(5214213), 1–13 (2022)

New Insights on the Generation of Rain Streaks: Generating-Removing United Unpaired Image Deraining Network

Yu Chen[1,2], Zhaoyong Yan[1], and Liyan Ma[1,2(✉)]

[1] School of Computer Engineering and Science, Shanghai University, Shanghai 200444, China
{Rain_C1715,zyyan,liyanma}@shu.edu.cn
[2] Institute of Artificial Intelligence, Shanghai University, Shanghai 200444, China

Abstract. Most existing deraining methods use synthetic rainy images to train models. They focus on extracting the features to establish mapping models from rainy images to clean background images, while ignoring the domain gap between synthetic and real rainy images. Furthermore, the rain streaks generation is not considered as important as the rain removal. Hence, we propose a novel **G**enerating-**R**emoving **U**nited **U**npaired image deraining **Net**work(GRUUNet) to generate and remove rain streaks unitedly. It helps to reduce the difference between synthetic and real rain streaks to improve the performance of deraining. Specifically, (1)we adopt a dual-way transform strategy between real and synthetic rainy images. There are both rain streaks generation and removal network for real and synthetic rainy images respectively; (2)our model learns the prototypes of rainy degradation images by self-supervised sparse-addressing memory modules; (3)we align the domain gap between real and synthetic rainy images with shared rain streaks generation network. We empirically evaluate our GRUUNet on Cityscape and RainHQ, and the outstanding results prove the promising performance of our method on both real and synthetic rainy images.

Keywords: Image deraining · Rain streaks generation network · Shared memory module

1 Introduction

The rainy/clean image pairs are needed to train image deraining methods based on deep learning [1,2,8,18,19]. However, the real scenes are constantly changing, making it difficult and costly to acquire convincing rainy/clean image pairs. Hence, most current studies rely on synthetic data for training and evaluation. These data are obtained by using realistic rendering techniques [3] or professional photography and manual simulation. It is clear that there are obvious differences between synthetic and real data because the real rainy images contain more

---

Y. Chen and Z. Yan—These authors contributed equally to this work and share first authorship.

Q. Liu et al. (Eds.): PRCV 2023, LNCS 14435, pp. 390–402, 2024.
https://doi.org/10.1007/978-981-99-8552-4_31

complex and diverse types of rain streaks. Due to the bias between the synthetic and the real data, the performance of the model trained on the synthetic datasets will be seriously degraded when evaluating on the real ones, and vice versa.

In order to obtain better deraining performance on real rainy images, some semi-supervised image deraining methods [15,18] adopt the idea of domain alignment and introduce real rainy images to train their models. In this case, the network is able to learn the distribution of real rain streaks, so as to narrow the difference between the features of synthetic and real rain streaks. SRRD [9] used a two-stage distillation method to process unpaired rainy images. In JRGR [19], the rain layer is obtained by decoupling the original images, while the conversion between synthetic and real rain streaks is realized by minimizing the adversarial loss. However, such simple decoupling methods and constraint functions cannot effectively explore the physical characteristics of real rain streaks.

The exploration of the rain streaks generation will be benefited to analyze the characteristics of real rain streaks, so as to improve the performance of deraining. Hence, we propose a novel **G**enerating-**R**emoving **U**nited **U**npaired image deraining **Net**work. We use a unified learning framework to generate and remove rain streaks iteratively to narrow the gap between synthetic and real rain streaks. Besides, we adopt a dual-way transform strategy and there are both rain streaks generation and removal network for real and synthetic rainy images respectively. The shared rain streaks generation network leads to the domain alignment of real and synthetic rain streaks. Since the main difference between the real and the synthetic rainy images is the style of rain streaks, our method focuses on transforming the rain layer between real and synthetic rainy images while preserving the consistency of the image background.

The main contributions of this paper are as follows:

- We propose a novel image deraining method(GRUUNet) to generate and remove rain streaks iteratively within a unified framework.
- We design a dual-way transform strategy to significantly reduce the inter-domain difference between synthetic rainy images and real rainy images.
- A shared sparse addressing memory module is proposed to learn the prototypes of degradation and eliminate the background information unrelated to rain streaks.
- Numerous experiments on both synthetic and real rain datasets show that our GRUUNet outperforms existing deraining methods and maintains consistent superior performance compared with those trained on real rain datasets.

2 Related Work

2.1 Rainy Image Generation

The generation of rainy images has provided crucial support and impetus for the development of image deraining tasks. Luo et al. [10] proposed a non-linear composite model to simulate some visual characteristics of real rainy images to generate visually more realistic rainy images. Yang et al. [17] introduced a comprehensive rain model that incorporates both rain streaks and rain accumulation. Hu et al. [5] incorporated scene depth information and atmospheric lighting intensity into the generation of synthetic rainy images. Although these methods produce various rain degradation effects, they are heuristic and cannot accurately represent the physical characteristics of rain streaks.

2.2   Single Image Deraining

Recently, several image deraining models based on convolutional neural networks(CNNs) have been proposed. They exploit CNNs to extract hierarchical features, allowing them to establish mapping models from rainy images to clean background images. Yang et al. [17] constructed a joint rain detection and removal network. Fu et al. [2] made an attempt to remove rain streaks via a deep detail network(DetailNet). They train models in a fully supervised manner and ignore the domain differences between synthetic and real rainy images, resulting in the degradation of the performance on real rainy images. To address this challenge, researchers have introduced semi/unsupervised methods that leverage real rainy images. Wei et al. [15] introduced a semi-supervised learning approach that utilizes priors from both synthetic paired images and unpaired real images. Ye et al. [20] proposed an unsupervised deraining network that introduced non-local contrastive learning and leveraged the intrinsic self-similarity property within samples. In this work, we aim to narrow the gap between synthetic and real rain streaks to enhance the performance of image deraining.

2.3   Generative Adversarial Networks

Generative adversarial networks(GAN) have been widely studied in recent years in computer vision and machine learning. Zhang et al. [21] directly applied the conditional generative adversarial network(CGAN) for the single image rain removal task. Qian et al. [11] developed an attentive GAN(AttGAN) by injecting visual attention into both the generative and discriminative networks. Ye et al. [19] proposed a decoupled approach to obtain the rain layer independently, enabling mutual transformation of rain streaks through the adversarial loss. However, the mentioned methods, which rely solely on combining clean images with extracted rain streaks for consistency loss, lack of strong constraints. Consequently, they struggle to fully capture the intricate physical characteristics of rain streaks, resulting in suboptimal deraining performance.

3   Methods

3.1   Framework Architecture

The architecture of GRUUNet is shown in Fig. 1. GRUUNet is a dual-way dual-branch model consisting of two branches named Synthetic-to-Real(S2R) and

Real-to-Synthetic(R2S). (1)For S2R, the rain streaks removal network for synthetic data($F_s(\cdot)$) takes the synthetic rainy image as input and outputs the background image $\tilde{B}_s$. Then, $\tilde{B}_s$ is combined with the real rain layer generated by the rain streaks generation network($G(\cdot)$) to obtain the real rainy image. (2)For R2S, the rain streaks removal network for real data($F_r(\cdot)$) takes the real rainy image as input and outputs the background image $\tilde{B}_r$. Then, $\tilde{B}_r$ is combined with the synthetic rain layer generated by $G(\cdot)$ to obtain the synthetic rainy image. The aforementioned dual-way dual-branch is alternately applied to the newly acquired synthetic and real rainy images.

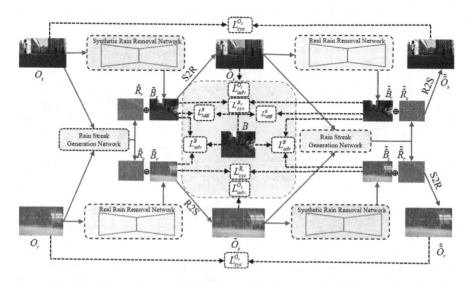

Fig. 1. The overall architecture of our GRUUNet.

3.2 Rain Streaks Generation Network

Memory networks [13,16] use an external storage to store information like feature prototypes and retrieve relevant content to enhance specific tasks. Hence, we take the advantage of memory networks to better improve the performance.

Network Architecture. Followed [6], our $G(\cdot)$ consists of three components: encoder E, decoder D, and memory module M, which is shown in Fig. 2. M adopts a self-supervised update strategy based on the similarity between q and the memory item p. The cosine similarity s_{ij} between the i-th item p_i and the j-th column vector q_j is as follows:

$$s_{ij} = \frac{p_i q_j^T}{\sqrt{\sum_k p_{ik}^2}\sqrt{\sum_k (q_{jk})^2}}. \tag{1}$$

We use Eq. 1 to select $p_{k(j)}$ which is most relevant to q_j:

$$p_{k(j)} = \arg\max_i s_{ij}. \tag{2}$$

Then we use the selected item $p_{k(j)} = p_i$ to update p_i:

$$p_i \leftarrow \tau p_i + (1 - \tau) \frac{\sum_{x \in X} \sum_{j=1}^{n} \mathbb{I}\left(k_{(j)}(x) = i\right) q_j}{\sum_{x \in X} \sum_{j=1}^{n} \mathbb{I}\left(k_{(j)}(x) = i\right)}, \tag{3}$$

where $\tau \in [0, 1]$ is a decay rate. Parameters of the encoder and decoder are iteratively updated with p_i, and X in Eq. 3 represents a batch of rainy images.

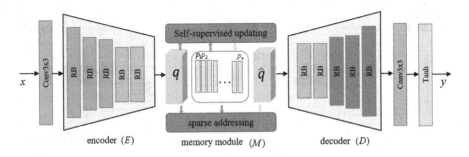

Fig. 2. The shared rain streaks generation network. 'RB' means a basic residual block.

Shared Parameters. We use the self-updating rain streaks generation network $G(\cdot)$ with shared parameters to generate both synthetic and real rain layers. As mentioned before, there is a large domain gap between synthetic and real rainy images. The memory module M trained by synthetic rainy images has the main structural information of rain streaks, but is not rich in details. However, if we train M only with real rainy images, it could not capture the information of rain streaks well because rain streaks and backgrounds are always mixed together in the real rainy images and the structure of rain streaks is not quite clear. Hence, we use shared $G(\cdot)$ for both synthetic and real rainy images to generate rain layers. In this way, M could be trained better with the help of the main structure information of rain streaks provided by synthetic rainy images and the detail information of rain streaks provided by the real ones. We do ablation studies in Sect. 4.3 to verify our points.

Sparse Addressing. After updating M, the reconstruction of the feature $\hat{q}$, which is a memory-based representation, is achieved by retrieving memory entries based on a query q. One intuitive approach to obtain $\hat{q}$ is through hard attention, where the memory entry with the highest similarity to the query is selected. However, this method does not allow gradient back-propagation from the decoder to the encoder. Hence, we follow MOSS [6] to adopt a read strategy based on soft attention. We compute the similarity matrix $S(x) = \{s_{ij}(x)|i = 1, ..., m, j = 1, ..., n\}$ again with the updated memory entries, where m and n are the number of memory entries and queries, respectively. Then we use softmax to obtain the attention weights $W = \{w_{ij}|i = 1, ..., m, j = 1, ..., n\}$:

$$w_{ij} = \frac{\exp(s_{ij})}{\sum_{i=1}^{m} \exp(s_{ij})}, \tag{4}$$

where w_{ij} represents the attention weight between query vector j and memory entry vector i, and s_{ij} is the cosine similarity. It computes the attention distribution over all memory entries for each query vector j. Then we aggregate the attention-weighted memory items to obtain the memory-based representation:

$$\hat{q}_j = \sum_{i=1}^{m} w_{ij} p_i. \tag{5}$$

Using the rain-degraded prototype in the memory to generate the rain layer helps to reduce the domain gap between real and synthetic rain layers. For convenience, we use w_i to represent w_{ij}. However, even with a dense w that has many small elements, some irrelevant values may still affect the rain generation through complex combinations. Inspired by [4], we adopt a hard-threshold operation to promote sparsity in w:

$$\hat{w}_i = h(w_i; \lambda) = \begin{cases} w_i, & \text{if } w_i > \lambda, \\ 0, & \text{otherwise,} \end{cases} \tag{6}$$

where $\hat{w}_i$ represents the i-th element of the memory addressing weight vector $\hat{w}$, and λ is the threshold. But the back-propagation of the discontinuous function in Eq. 6 is not feasible. For convenience, we use the continuous ReLU activation function to approximate the hard-threshold operation:

$$\hat{w}_i = \frac{\max(w_i - \lambda, 0) \cdot w_i}{|w_i - \lambda| + \epsilon}, \tag{7}$$

where $\max(\cdot, 0)$ is known as the ReLU activation function, and ϵ is a very small positive number. In practice, we set λ in $[1/m, 3/m]$, where m is the number of items in the storage module. After that, we re-normalize $\hat{w}$ by $\hat{w}_i = \hat{w}_i / \|\hat{w}\|_1$. The latent representation $\hat{q}$ of the rain layer can be obtained by $\hat{q} = \hat{w} M$.

The sparse addressing mechanism encourages the model to use fewer but more relevant memory items to represent the rain layer feature, thereby learning more meaningful rain streaks prototypes.

3.3   Rain Streaks Removal Network

We utilize two rain streaks removal networks: $F_s(\cdot)$ for synthetic data and $F_r(\cdot)$ for real ones:

$$\tilde{B}_s = F_s(O_s), \quad \tilde{B}_r = F_r(O_r), \tag{8}$$

where $F_s(\cdot)$ takes the original synthetic rainy image O_s as input and produces the background image $\tilde{B}_s$, while $F_r(\cdot)$ takes the original real one O_r as input and produces $\tilde{B}_r$. Then we synthesize new rainy image through an overlaying process represented as follows:

$$\tilde{R}_r = G(O_r), \quad \tilde{R}_s = G(O_s), \tag{9}$$

$$\tilde{O}_r = \tilde{B}_s + \tilde{R}_r, \quad \tilde{O}_s = \tilde{B}_r + \tilde{R}_s, \tag{10}$$

where $G(\cdot)$ is our shared rain streaks generation network. $\tilde{R}_s$ is the rain layer generated based on the synthetic rainy image and $\tilde{R}_r$ is the rain layer generated based on the real one. $\tilde{O}_s$ and $\tilde{O}_r$ represent the newly generated synthetic and real rainy image, respectively. According to the dual-way transform strategy, after obtaining $\tilde{O}_s$ and $\tilde{O}_r$, we perform the direction-reversal propagation:

$$\tilde{\tilde{B}}_s = F_s(\tilde{O}_s), \quad \tilde{\tilde{B}}_r = F_r(\tilde{O}_r), \tag{11}$$

where $\tilde{\tilde{B}}_s$ and $\tilde{\tilde{B}}_r$ represent the final synthetic background and real background images, respectively.

3.4   Objective Function

Here we detail the description of the loss functions each rain streaks removal and generation network applied in this work.

Adversarial Loss. We employ the adversarial loss to assist the generation network to generate more realistic clean images:

$$\begin{aligned} L_{adv}^B(D_B, F_s, F_r) &= \mathbb{E}_{O_s}[\log(1 - D_B(F_s(O_s)))] \\ &+ \mathbb{E}_{O_r}[\log(1 - D_B(F_r(O_r)))] + \alpha \mathbb{E}_B[\log D_B(B)], \end{aligned} \tag{12}$$

where F_s and F_r represent the synthetic and real deraining networks, respectively, and α serves as the balancing weight. In the cycle, the generated synthetic rainy images $\tilde{O}_r$ and $\tilde{O}_s$ are fed into the R2S and S2R modules to form a closed loop. Therefore, the background adversarial loss also applies to the decomposed backgrounds $\tilde{\tilde{B}}_r$ and $\tilde{\tilde{B}}_s$. The overall background adversarial loss is as follows:

$$\begin{aligned} L_{adv}^B(D_B, F_s, F_r, M) &= \alpha \mathbb{E}_B[\log D_B(B)] \\ &+ \mathbb{E}_{O_s}[\log(1 - D_B(F_s(O_s))) + \log(1 - D_B(F_r(\tilde{O}_r)))] \\ &+ \mathbb{E}_{O_r}[\log(1 - D_B(F_r(O_r))) + \log(1 - D_B(F_s(\tilde{O}_s)))]. \end{aligned} \tag{13}$$

For the regenerated rainy images, the model applies two rainy image discriminators to construct the adversarial loss. Taking the real rainy image generation as an example, O_r is used to train the discriminator D_{O_r} and the self-updating rain streaks generation network G in an adversarial manner:

$$L_{adv}^{O_r}(D_{O_r}, F_s, G) = \mathbb{E}_{O_r}[\log D_{O_r}(O_r)] + \mathbb{E}_{O_s}[\log(1 - (D_{O_r}(G(F_s(O_s)))))]. \tag{14}$$

For synthetic rainy images, a similar adversarial loss $L_{adv}^{O_s}$ is applied between M and the discriminator D_{O_s}, aiming to enforce the generation of synthetic rainy images $\tilde{O}_s$.

Cycle Consistency Loss. The cycle consistency loss is used to compute the cycle reconstruction error from the synthetic image to the real image and back to the synthetic image. Specifically, the synthetic rainy image is fed into the S2R generator to generate a real rainy image, and then fed the generated data into the R2S generator to regenerate a synthetic rainy image. We calculate $L1$ loss between the regenerated synthetic rainy image and the input one:

$$L_{cyc}^{O_s}(F_s, F_r, G) = \mathbb{E}_{O_s}[||O_s - G(F_r(G(F_s(O_s))))||_1]. \tag{15}$$

Similarly, $L_{cyc}^{O_r}$ is employed for the regeneration of O_r. Furthermore, the generated real rainy image $\tilde{O}_r$ and the original synthetic one O_s should have consistent backgrounds. The cycle consistency loss can be expressed as follows:

$$L_{cyc}^{B_s}(F_s, F_r, G) = \mathbb{E}_{O_s}[||F_s(O_s) - F_r(G(F_s(O_s)))||_1]. \tag{16}$$

Similarly, the cycle consistency loss for the decomposed backgrounds $\tilde{B}_r$ and $\tilde{\tilde{B}}_r$ is defined in a similar manner as $L_{cyc}^{B_r}$.

MSE Loss. We employ MSE loss to supervise the training of the F_s, which is expressed as follows:

$$L_{mse}^{B_s}(F_s, F_r, G) = [||F_s(O_s) - B||_2^2 + ||F_r(G(F_s(O_s))) - B||_2^2], \tag{17}$$

where B is the corresponding synthesized clean background image and O_s is the synthetic rainy image. It ensures that the generated image closely resembles the clean background image as much as possible. Finally, the total loss is as follows:

$$L(F_s, F_r, M, D_B, D_{O_s}, D_{O_r}) = \lambda_{adv}(L_{adv}^B + L_{adv}^{O_r} + L_{adv}^{O_s})$$
$$+\lambda_{cyc}(L_{cyc}^{O_r} + L_{cyc}^{O_s} + L_{cyc}^{B_r} + L_{cyc}^{B_s}) + \lambda_{mse}L_{mse}^{B_s}. \tag{18}$$

4 Experiments and Results

4.1 Datasets

The synthetic dataset consists of two subsets: RainCityscape [5] contained 1400 pairs of rainy/clean images and Rendering [12] included 1400 unpaired rainy images. The test set consists of 175 rainy images synthesized by [12]. We selected 2000 pairs of synthetic clean/rainy images and 2000 real rainy images from the real rainy dataset SPA-Data [14] for training. All the datasets used for evaluating have corresponding clean images, allowing for quantitative comparisons using PSNR and SSIM metrics.

4.2   Implementation Details

We use PyTorch and train GRUUNet on a NVIDIA GeForce GTX 3090 GPU with 24GB of memory. Images were randomly cropped into 256×256 image patches for training. We use Adam optimizer with weight decay and momentum set to 0.0001 and 0.9, respectively. The model was trained for 200 epochs with a batch size of 4. The initial learning rate was set to 2e-5 and decayed to 1e-5 after 100 epochs. The parameters α, λ_{adv}, λ_{cyc}, and λ_{mse} in Eq. 18 and Eq. 13 were set to 4, 10, 1, and 10, respectively.

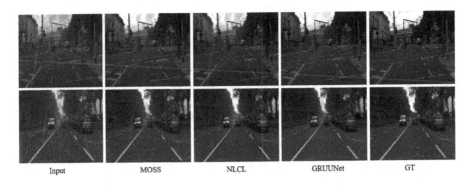

Fig. 3. Comparison of rain steaks removal results on Rendering.

4.3   Comparisons with the State-of-the-Arts

We mainly compare to four supervised methods(DDN [2], JORDER-E [17], RES-CAN [7], and DAF [5]), one unsupervised method(NLCL [20]) and three semi-supervised methods(Syn2Real [18], JRGR [19], and MOSS [6]).

Result on Synthetic Datasets. Table 1 shows PSNR and SSIM for different methods on Rendering [12]. Among them, GRUUNet obtained the best results.

We show the visualization results of various methods on Rendering [12]. It is observed in Fig. 3 that the results of MOSS and NLCL exhibit noticeable rain streaks and blurring. Compared to the aforementioned methods, GRU-UNet obtains more realistic and clear background information. This is mainly attributed to the adoption of a joint approach for rain streaks generation and removal in GRUUNet, which enables it to learn the physical properties of rain.

Result on Real-world Datasets. According to the results in Table 1, our GRUUNet achieves the best performance. Figure 4 visually shows the results obtained from SPA-Data [14]. GRUUNet demonstrates superior performance compared to NLCL and MOSS, showcasing remarkable generalization capability.

Table 1. Comparison of PSNR and SSIM of methods on Rendering and SPA-Data.

Methods	Rendering		SPA-Data	
	PSNR	SSIM	PSNR	SSIM
DDN [2]	24.12	0.878	33.74	0.911
RESCAN [7]	24.89	0.910	33.11	0.925
DAF [5]	25.23	0.882	24.15	0.852
JORDER-E [17]	25.64	0.876	33.28	0.940
Syn2Real [18]	25.32	0.887	33.14	0.918
JRGR [19]	27.51	0.898	33.59	0.939
MOSS [6]	26.21	0.886	33.69	0.942
NLCL [20]	27.43	0.912	33.72	0.945
GRUUNet	**28.78**	**0.917**	**34.46**	**0.951**

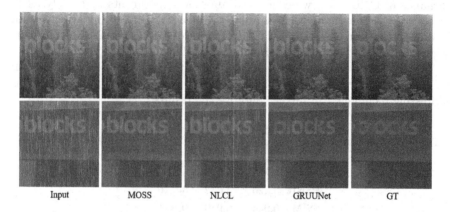

Input   MOSS   NLCL   GRUUNet   GT

Fig. 4. Comparison of rain streaks removal results on SPA-Data.

Ablation Study

Network Architecture. To verify the effect of network structure, we conducted an ablation study on SPA-Data, considering different forms of network architecture: (1)Using a non-shared weights rain streak generation network(D-GRUUNet). (2)Using a shared weights rain streak generation network(GRUUNet).

Table 2. Ablation study results of the network architectures on SPA-Data.

	D-GRUUNet	GRUUNet
PSNR	34.39	**34.46**
SSIM	0.947	**0.951**

It is obvious in Table 2 that GRUUNet performs better than D-GRUUNet, which means that sharing the rain streaks generation network may better capture the distribution of rain streaks in both real and synthetic rainy images, and shows favorable generalization performance.

Basic Module. To verify the impact of the self-updating memory module and sparse addressing method in the rain streaks generation network on the performance of GRUUNet, we conducted ablation experiments and analyzed the effects of these methods. NMNet refers to removing the self-updating memory module in the rain streaks generation network of GRUUNet, NSNet represents using only the soft attention addressing method in GRUUNet. According to the results in Table 3, the deraining network combined with the self-updating memory module achieves better performance. Furthermore, incorporating the sparse addressing technique can further improve the results.

Table 3. Ablation study on the effectiveness of the basic modules on SPA-Data.

	NMNet	NSNet	GRUUNet
PSNR	34.25	34.32	**34.46**
SSIM	0.946	0.949	**0.951**

Loss Function. We remove loss functions in turn to verify the effects of the original loss. According to the results in Table 4, all loss functions we proposed contribute to the performance. Designing suitable loss functions that align with the characteristics of the deraining model is essential, and it also validates the rationality of the employed loss functions in GRUUNet.

Table 4. Ablation study on the effectiveness of loss functions on SPA-Data.

	w/o L_{adv}^{B}	w/o L_{adv}^{O}	w/o L_{cyc}	w/o L_{MSE}	GRUUNet
PSNR	30.53	33.43	25.25	33.70	**34.46**
SSIM	0.854	0.926	0.812	0.936	**0.951**

5   Conclusion

We propose a novel image deraining method called GRUUNet that employs a unified learning framework for rain streaks generation and removal, aiming to narrow the gap between synthetic and real rainy images to improve the deraining performance. We adopt a dual-way transform strategy, where the network of each way consists of two stages: rain streaks generation and removal for real and synthetic rainy images. The shared rain streaks generation network incorporates a memory module to learn and store rain degradation prototypes in an unsupervised manner, achieving domain alignment between real and synthetic rainy images. Also, We focus on transforming the rain layer while preserving the consistency of the image background. Comprehensive experimental results demonstrate that our method significantly improves the generalization performance of the model on real rainy images.

Acknowledgment. This work was supported in part by the National Key R&D Program of China (No. 2021YFA1003004), in part by the Shanghai Municipal Natural Science Foundation (No.21ZR1423300).

References

1. Chen, J., et al.: Robust video content alignment and compensation for rain removal in a CNN framework. In: Proceedings of the IEEE Conference on Computer Vision and Pattern Recognition (CVPR) (2018)
2. Fu, X., et al.: Removing rain from single images via a deep detail network. In: IEEE Conference on Computer Vision & Pattern Recognition (2017)
3. Garg, K., Nayar, S.K.: Vision and rain. Int. J. Comput. Vis. **75**, 3–27 (2007)
4. Gong, D., et al.: Memorizing normality to detect anomaly: memory-augmented deep autoencoder for unsupervised anomaly detection. In: 2019 IEEE/CVF International Conference on Computer Vision (ICCV) (2020)
5. Hu, X., et al.: Depth-attentional features for single-image rain removal. In: 2019 IEEE/CVF Conference on Computer Vision and Pattern Recognition (CVPR) (2019)
6. Huang, H., Yu, A., He, R.: Memory oriented transfer learning for semi-supervised image deraining. In: IEEE/CVF Conference on Computer Vision and Pattern Recognition, pp. 7732–7741 (2021)
7. Li, X., et al.: Recurrent squeeze-and-excitation context aggregation net for single image deraining. In: Proceedings of the European conference on computer vision (ECCV), pp. 254–269 (2018)
8. Li, Y., et al.: Rain streak removal using layer priors. In: 2016 IEEE Conference on Computer Vision and Pattern Recognition (CVPR), pp. 2736–2744 (2016)
9. Lin, H., et al.: Rain O'er Me: synthesizing real rain to derain with data distillation. IEEE Trans. Image Process. **29**, 7668–7680 (2020). https://doi.org/10.1109/TIP.2020.3005517
10. Luo, Y., Xu, Y., Ji, H.: Removing rain from a single image via discriminative sparse coding. In: Proceedings of the IEEE International Conference on Computer Vision, pp. 3397–3405 (2015)

11. Qian, R., et al.: Attentive generative adversarial network for raindrop removal from a single image. In: Proceedings of the IEEE Conference on Computer Vision and Pattern Recognition, pp. 2482–2491 (2018)

12. Sukanta Halder, S., Lalonde, J.-F., de Charette, R.: Physics-based rendering for improving robustness to rain. In: arXiv e-prints, pp. arXiv-1908 (2019)

13. Sukhbaatar, S., et al.: End-to-end memory networks. In: Advances in Neural Information Processing Systems 28 (2015)

14. Wang, T., et al.: Spatial attentive single-image deraining with a high quality real rain dataset. In: Proceedings of the IEEE/CVF Conference on Computer Vision and Pattern Recognition, pp. 12270–12279 (2019)

15. Wei, W., et al.: Semi-supervised transfer learning for image rain removal. In: Proceedings of the IEEE/CVF Conference on Computer Vision and Pattern Recognition (CVPR) (2019)

16. Weston, J., Chopra, S., Bordes, A.: Memory networks. In: arXiv preprint arXiv:1410.3916 (2014)

17. Yang, W., et al.: Deep joint rain detection and removal from a single image. In: Proceedings of the IEEE Conference on Computer Vision and Pattern Recognition, pp. 1357–1366 (2017)

18. Yasarla, R., Sindagi, V.A., Patel, V.M.: Syn2Real transfer learning for image deraining using gaussian processes. In: 2020 IEEE/CVF Conference on Computer Vision and Pattern Recognition (CVPR), pp. 2723–2733 (2020)

19. Ye, Y., et al.: Closing the loop: joint rain generation and removal via disentangled image translation. In: Proceedings of the IEEE/CVF Conference on Computer Vision and Pattern Recognition, pp. 2053–2062 (2021)

20. Ye, Y., et al.: Unsupervised Deraining: where contrastive learning meets self-similarity. In: Proceedings of the IEEE/CVF Conference on Computer Vision and Pattern Recognition, pp. 5821–5830 (2022)

21. Zhang, H., Sindagi, V., Patel, V.M.: Image de-raining using a conditional generative adversarial network. In: IEEE Transactions on Circuits and Systems for Video Technology, vol. 30, issue 11, pp. 3943–3956 (2019)

Knowledge Transfer via Leveraging Teacher-Student Network with Visual Attention to Enhance Atmospheric Sand Image Restoration

Jun Shi[1] and Zhe Li[2(✉)]

[1] School of Information Science and Engineering,
Xinjiang University, Xinjiang, China
jun_shi@stu.xju.edu.cn
[2] Department of Electrical and Electronic Engineering,
The Hong Kong Polytechnic University, Hong Kong, Hong Kong SAR, China
lizhe.li@connect.polyu.hk

Abstract. Deep learning-based methods have recently shown superiority in sand dust image restoration tasks. However, most existing learning-based methods do not pay much attention to the gap between the synthetic image and the real image, making the obtained recovery effect difficult to generalize on the real image. To alleviate this problem, we propose a novel Teacher-Student network (TS-Net) for real-world sand dust image restoration that maximizes the transfer of knowledge learned in the Teacher Network. Thereby improving the recovery ability of the Student Network on real sand dust images. Firstly, the Teacher Network extracts the latent feature information of dust images. Furthermore, use transfer learning to transfer the knowledge learned by the Teacher Network to the faster and more efficient Student Network. We present a Visual Attention (VA) mechanism to couple multi-scale features with contextual features with large kernel convolutions to fully utilize these complementary features to correct the color of sand dust images. Due to the large kernel convolution and feature fusion strategy, the fused features can correct the color cast the problem of dust color, which is very important for high-level tasks. Extensive experiments on the synthetic and real datasets demonstrate that the proposed method significantly improves over the state-of-the-art methods.

Keywords: Sand dust image restoration · Teacher-Student Network · Visual attention

1 Introduction

In dusty weather, the dust particles suspended in the air have absorption and scattering effects on the incident light. As such, the acquired images display low contrast, color shift, and significant impairment of visibility,as shown in Fig. 1. The functioning of outdoor computer vision applications, including video surveillance, autonomous navigation, and intelligent transportation, is detrimentally

© The Author(s), under exclusive license to Springer Nature Singapore Pte Ltd. 2024
Q. Liu et al. (Eds.): PRCV 2023, LNCS 14435, pp. 403–414, 2024.
https://doi.org/10.1007/978-981-99-8552-4_32

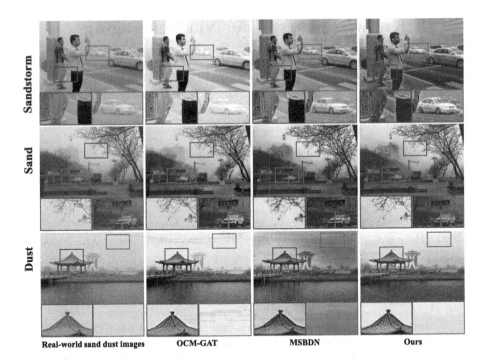

Fig. 1. Results of different dust image restoration methods. From left to right, the leftmost row is a dust image, and OCM-GAT obtains clear images [24], MSBDN [4], and our method. For ease of comparison, both patches are cropped, and their enlarged versions are placed below each image. (Color figure online)

affected by these issues. Thus, exploring the algorithm of sand image restoration and elevating the dependability of outdoor vision tasks has paramount practical implications.

Currently, techniques based on prior and deep learning have been developed for sand image restoration. The prior-based methods estimate the transmission and global atmosphere light value prior knowledge [7] and then put the intermediate variables into the physics model to achieve color constancy and restore clear images. The prior-based methods may lead to an inaccurate estimation of atmospheric scattering model parameters [5,13,18,24]. The other is based on deep learning [19,20]. First, the neural network is used to learn to restore the image color. Then the color-corrected image is compared to the haze image to calculate the restored image using the atmospheric model. The transmission map and atmospheric light estimation greatly influence the performance of these methods.

FitNets [15] is the first to introduce intermediate layer representations. The intermediate layer of the Teacher Network can be used as a hint for the corresponding layer of the student network to improve the performance of the Student

Network model. At its core, it is expected that the Student can directly imitate the feature activation values of the Teacher Network.

Unlike previous sand dust image restoration methods, we designed a Teacher-Student network (TS-Net) for real sand dust images. It generalizes well to the real world, not only on synthetic domains. More specifically, the framework consists of two networks: a Teacher Network that learns the features of sand dust images and a Student Network that acquires knowledge and applies it to the real world. We supervise the intermediate features and use the feature similarity to control the imitation learning from the image reconstruction of the Teacher to the image desand of the Student.

The contributions of this work are summarized as follows:

- We propose a novel trainable Teacher-Student network for single sand dust image restoration. Transfer knowledge of a trained Teacher model to the Student network. The Teacher Network extracts knowledge through the intermediate layers and guides the Student Network to train on the real sand dust image dataset.
- Visual attention by large convolutional kernels and channel shuffling enables the network to aggregate regions of interest. The large convolution kernel has a wider receptive field, which enables the color cast of the dust image to be corrected. Channel shuffling enables information communication between different channel groups and improves accuracy.
- We put forward a new benchmark for synthetic sand dust images. After analyzing the spatial features of the real sand dust images, a large-scale data set that can be used for network training is constructed. Benefiting from it, the realization of deep learning in sand dust image restoration becomes possible.

2   Proposed Method

The overall architecture of the sand dust image restoration network consists of a Teacher and Student network, as shown in Fig. 2. Specifically, the Teacher's goal is to provide intermediate feature representations of real sand dust images (pseudo-ground truth images), and the Student's goal is to recover the blurred image by transforming the intermediate features into the clean image domain.

2.1   Sand Dust Image Dataset

Due to the lack of large-scale sand datasets, there have been no well-known reports of data-driven sand dust image enhancement methods until now. Considering the atmospheric light attenuation effect of the dust floating in the atmosphere on the RGB channel, the atmospheric light model, belonging to the sand dust image, was reconstructed according to the spatial distribution law. The mathematical expression of this model is as follows:

$$\hat{A} = <A_R, k_1 A_R + b_1, k_2 A_R + b_2>. \tag{1}$$

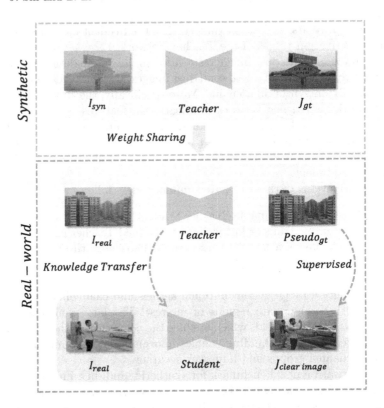

Fig. 2. The overall framework of real sand dust image restoration is proposed in this letter. The teacher network transfers knowledge to the student network, guiding the student network to recover better results on real-world images.

where $A_G = k_1 A_R + b_1$, $A_B = k_2 A_R + b_2$, $\hat{A}$ is the global color deviation value of the sand dust image, A_R, A_G, A_B are the atmospheric light values of each channel of RGB respectively, k is the spatial distribution coefficient of atmospheric light values for the three basic color spectra, b is the perturbation so that $\hat{A}$ fluctuates in the dynamic range.

2.2 Synthetic Training Phase

Visual Attention. Sand dust images suffer from severe color cast, and to enable the network to focus on this important issue, we propose visual attention. Visual attention based on large kernel convolution (LKC) and channel shuffle (CS) [26] is proposed to aim at the shortcomings of standard convolution. First, the input feature $\mathbf{F}$ is convoluted using LKC to obtain an intermediate feature map $\mathbf{M}$.

$$\mathbf{M} = \text{LKC}(\mathbf{F}) \tag{2}$$

Then, this intermediate feature map $\mathbf{M}$ is transformed into an attention map $\mathbf{A}$ by a sigmoid function. The sigmoid function can convert each element value

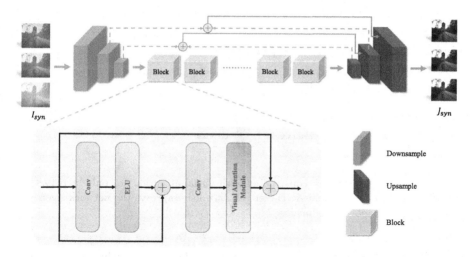

Fig. 3. The Teacher Network uses 6 VA blocks as the restoration network's backbone and skip connections to fuse the shallow network features in upsampling and downsampling.

of $\mathbf{M}$ into a number between 0 and 1, which represents the attention of the corresponding input feature.

$$\mathbf{A} = \text{sigmoid}(\mathbf{M}) = \frac{1}{1 + e^{-\mathbf{M}}} \tag{3}$$

$$\text{Visual Attention} = \mathbf{M} \otimes \mathbf{F} \otimes CS \tag{4}$$

The value in the attention map indicates the importance of each feature. $\otimes$ means element-wise product.

Teacher Network. The Teacher Network consists of a downsampling module, a backbone module, and an upsampling module. As presented in Fig. 3, the Teacher network first adopts 4× downsampling operation (e.g., one regular convolution with stride 1 and two convolution layers, all with stride 2) to make dense Visual Attention(VA) blocks learn the feature representation in the low-resolution space. And then the recovered image is then generated with corresponding 4× upsampling and one regular convolution. We use 6 VA blocks to improve the information flow between layers and more spatially fuse structured information. Reuse local information to capture long-range dependencies and improve the adaptability of Teacher networks in the channel and spatial dimensions. Low-level features (e.g., edges and contours) can be captured in the shallow layers of CNNs [25]. However, with an increase in the network's depth, the shallow features degrade gradually [8]. To deal with this issue, several previous works [8,16] integrate the shallow and deep features to generate new features via skip connections with an addition or concatenation operation. Based on this,

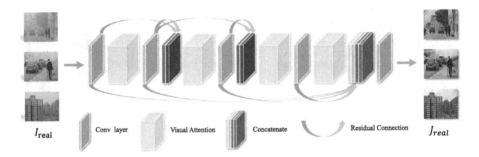

I_{real} Conv layer   Visual Attention   Concatenate   Residual Connection J_{real}

Fig. 4. The Student Network constructed with visual attention not only increases the model capacity of the original AOD-Net [11] but also improves the network's generalization ability on real-world images.

in our Teacher Network, skip connections are applied between the features of the downsampling layer and the upsampling layer to fuse information from shallow layers for feature preservation (see Fig. 3).

2.3   Real-World Training Phase

Pseudo-gt Using Teacher Network. In this work, we argue that real-world image recovery and generation are interrelated and can provide beneficial informative supervision for each other [27]. If we directly apply existing synthetic domain images to real-world image restoration, they tend to suffer severe adaptation problems and yield poor performance [6]. When the real-world image restoration is supervised with real-world images to improve the generalization performance of the network. Specifically, the intermediate latent space is supervised by the error between the pseudo-gt obtained by the Teacher Network. This method learns more efficient intermediate representations by directly forcing the Teacher Network to imitate the features extracted from the synthetic domain.

Student Network. The Student Network's architecture differs from that of the Teacher Network. The Student Network is designed to improve the Teacher's knowledge and reduce the domain gap between the synthetic and real domain to perform the restoration task of real degraded images faster and better. Li et al. [11] proposed that AOD-Net can directly generate dehazing images through a lightweight CNN and is easier to embed into other deep learning models. However, the AOD-Net model has a small capacity, and is difficult to deal with degraded images in complex environments. Moreover, the sand dust image has a serious color cast problem, so a larger receptive field is needed to extract global information. Hence, as shown in Fig. 4, we choose AOD-Net as the primary Student Network and further improve it. The student network introduced with the visual attention module has a larger receptive field, which effectively solves the color cast problem of sand dust images.

Loss for Student Network. In Student Network, the adaptation of the real domain is constrained by several reconstruction losses and perceptual losses [10] to guarantee the effectiveness of knowledge transfer.

Specifically, pseudo-ground truth images are generated through the Teacher Network to reconstruct the original input through the physical scattering model:

$$J_{clear}\boldsymbol{x} = K(\boldsymbol{x}) \odot Pseudo_{gt} - K(\boldsymbol{x}) + b, \tag{5}$$

where $\odot$ indicates element-wise multiplication.

$$K(\boldsymbol{x}) = \frac{\frac{1}{t(\boldsymbol{x})}(I_{real} - A) + (A - b)}{I_{real} - 1}, \tag{6}$$

Since $K(x)$ is dependent on $Pseudo_{gt}$, we then aim to build an input-adaptive deep model, whose parameters will change with input degraded images, that minimizes the reconstruction error between the output $J(x)$ and the ground truth clean image. Then the reconstruction loss L_{Rec} is formulated as:

$$L_{Rec} = \|I_{real} - Pseudo_{gt}\|_1. \tag{7}$$

Unlike the per-pixel loss, the perceptual loss leverages multi-scale features extracted from a pre-trained deep neural network to quantify the visual difference between the estimated image and the ground truth. This paper uses VGG16 [21] as the loss network to guarantee image translation performance. The perceptual loss is defined as:

$$L_P = \sum_{j=1}^{3} \frac{1}{C_j H_j W_j} \|\phi_j(J_{clear}) - \phi_j(Pseudo_{gt})\|_2^2, \tag{8}$$

where $\phi_j(\hat{J}), j = 1, 2, 3$ denote the aforementioned three VGG16 feature maps associated with the corresponding image.

The total loss is defined by combining the reconstruction losses and perceptual losses as follows:

$$L = L_{Rec} + \lambda L_P, \tag{9}$$

where λ is a parameter used to adjust the relative weights on the two loss components. In this paper, λ is set to 0.04 [12].

3   Experiment

3.1   Performance Evaluation

Since the learning-based method is oriented toward haze image reconstruction, it differs from our target task. Under the premise of ensuring fairness, we retrain all learning methods on the synthetic sand dust dataset for testing.

Since clear images corresponding to these synthetic data are available, we calculated PSNR and SSIM for quantitative measurements. Table 1 compares

Table 1. Quantitative results of Prior-based and Learning-based methods using IQA metrics on synthetic and real degraded Images.

Methods	Full-Reference		No-Reference			
	PSNR ↑	SSIM ↑	FADE ↓	NIQE ↓	PIQE ↓	BRISUQE ↓
DCP [7]	17.838	0.840	1.308	4.967	<u>44.509</u>	36.495
MMSP [5]	17.777	0.822	1.060	<u>2.949</u>	44.827	28.145
GDCP [13]	13.629	0.746	0.723	3.513	45.943	<u>27.164</u>
TTFIO [1]	15.159	0.789	1.625	3.185	45.327	39.897
HRDCP [18]	11.931	0.645	1.292	3.423	48.369	32.390
OCM-GAT [24]	15.541	0.789	1.370	3.128	48.208	35.572
RCC [17]	25.549	0.869	0.928	3.223	44.437	27.721
AOD-Net [11]	17.771	0.838	0.786	3.637	50.677	36.637
DehazeNet [2]	19.376	0.851	0.892	4.026	52.725	42.486
GCA-Net [3]	27.945	0.879	0.949	3.528	49.473	29.779
FFA-Net [14]	24.009	0.853	0.892	4.027	52.725	42.486
MSBDN [4]	<u>28.352</u>	<u>0.899</u>	<u>0.698</u>	3.797	49.243	39.085
AECR-Net [23]	26.192	0.898	1.172	4.664	65.512	49.975
Teacher	**32.667**	**0.944**	0.740	2.956	42.506	26.918
Student	21.100	0.868	**0.690**	**2.936**	**39.570**	**26.648**

results based on these metrics. Our proposed method is highly competitive as compared with these top-performing competitors. As shown in Fig. 5, the prior-based method can recover the original color of the scene on the synthetic dataset (such as MMSP [5] and OGM-GAT [24]). Still, there are problems such as over-exposure and loss of details. Since the learning-based methods are oriented to single-image dehazing, it is not easy to achieve ideal results, even if it is retrained on the synthetic sand dust image dataset. However, compared with the prior-based method, the learning-based method can more effectively solve problems such as the color cast of sand dust images utilizing feature extraction. It can be further observed that our results have no color shift and preserve the original details of the image to the greatest extent.

We also evaluate the performance of the proposed method and recent state-of-the-art methods on real-world sand dust test images. The restoration effects for various real-world sand dust images are shown in Fig. 6. DCP [7] cannot correct the color cast of the sand dust image. MMSP [5] can effectively correct the image color, but the scene is too dark. After HRDCP [18] corrects the color cast, there is a problem of overexposure. GCANet [3] effectively removes the color cast of sand dust images, but the color saturation is not enough, and there is a slight distortion. AECR-Net [23] produces blockiness at the edges. From the visual point of view, the method proposed in this paper not only effectively solves the color cast problem of the sand dust image but also preserves the details of

Fig. 5. Qualitative comparisons on the synthetic sand image. (Color figure online)

Table 2. Frames per second for different image sizes.

Methods	720p	1080p	2K
AOD-Net [11]	796.847	748.436	668.855
DehazeNet [2]	543.122	437.217	400.156
GCA-Net [3]	65.484	63.245	59.521
FFA-Net [14]	8.533	4.725	1.076
MSBDN [4]	63.486	61.228	53.349
AECR-Net [23]	87.533	85.244	80.058
Student	283.046	133.333	81.256

the original image to the greatest extent. The main reason is that the proposed method is not only trained on synthetic datasets.

3.2   Ablation Study

Computation Times. Table 2 compares the running times of several state-of-the-art methods at different resolutions. All learning-based methods are computationally more efficient. Compared with these methods, the proposed Student Network can recover the real sand dust images well and achieve a faster time.

Large Kernel Convolution and Channel Shuffle. To further demonstrate the superiority of this structure, an ablation study is performed by considering the different modules of the proposed visual attention and channel shuffle. The following factors will be addressed: 1) The LKC (Large Kernel Convolution). 2) The CS (Channel Shuffle). 3) Different attention mechanisms are applied to

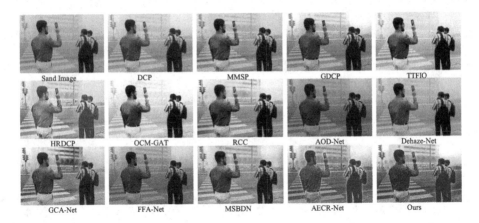

Fig. 6. Qualitative comparisons on the real-world image. (Color figure online)

Table 3. Large kernel convolution and channel shuffle.

Convolution Kernel	LKC	CS	PSNR	SSIM
Stander	$\checkmark$	$\times$	23.100	0.867
Stander	$\checkmark$	$\checkmark$	27.652	0.904
7×7	$\checkmark$	$\times$	30.163	0.911
7×7	$\checkmark$	$\checkmark$	**32.667**	**0.944**
9×9	$\checkmark$	$\times$	<u>30.743</u>	0.902
9×9	$\checkmark$	$\checkmark$	31.765	<u>0.915</u>

Table 4. Quantitative results of different attention mechanisms.

Attention	PSNR	SSIM
CBAM [22]	24.747	0.908
SE [9]	25.417	0.909
FA [14]	27.003	0.910
VA	**32.667**	**0.944**

image restoration. Crop the images to 256×256 as input; the configuration is the same as our implementation details. The results are shown in Tables 3 and 4. The receptive field expansion brought by conventional convolution is still very limited. So we opted to use a larger window (e.g., 7×7) and stack the channel shuffle. The visual attention block forces the features to focus more on the region of interest. More representative features can be obtained when these highlighted features are aggregated together.

4   Conclusions

We propose a Teacher-Student network-based approach to remedy sand dust images' color shift and visibility issues. The framework consists of two networks: a Teacher Network that learns the features of sanddust images and a Student Network that acquires knowledge and applies it to the real world. Furthermore, a synthetic sand dust image dataset is proposed for the long-term development of deep learning methods. Our experiments on many authentic sand dust images have revealed that this method produces better color accuracy and appropriate luminosity. We will endeavor to uncover and develop video sand dust recovery algorithms that are more effective and fitting for the various needs of vision application systems in the future.

References

1. Al-Ameen, Z.: Visibility enhancement for images captured in dusty weather via tuned tri-threshold fuzzy intensification operators. Int. J. Intell. Syst. Appl. 8(8), 10 (2016)
2. Cai, B., Xu, X., Jia, K., Qing, C., Tao, D.: Dehazenet: an end-to-end system for single image haze removal. IEEE Trans. Image Process. 25(11), 5187–5198 (2016)
3. Chen, D., et al.: Gated context aggregation network for image dehazing and deraining. In: 2019 IEEE Winter Conference on Applications of Computer Vision (WACV), pp. 1375–1383 (2019)
4. Dong, H., et al .: Multi-scale boosted dehazing network with dense feature fusion. In: Proceedings of the IEEE/CVF Conference on Computer Vision and Pattern Recognition, pp. 2157–2167 (2020)
5. Fu, X., Huang, Y., Zeng, D., Zhang, X.P., Ding, X.: A fusion-based enhancing approach for single sandstorm image. In: 2014 IEEE 16th International Workshop on Multimedia Signal Processing (MMSP), pp. 1–5. IEEE (2014)
6. Guo, Y., et al.: Closed-loop matters: Dual regression networks for single image super-resolution. In: Proceedings of the IEEE/CVF Conference on Computer Vision and Pattern Recognition, pp. 5407–5416 (2020)
7. He, K., Sun, J., Tang, X.: Single image haze removal using dark channel prior. IEEE Trans. Pattern Anal. Mach. Intell. 33(12), 2341–2353 (2010)
8. He, K., Zhang, X., Ren, S., Sun, J.: Deep residual learning for image recognition. In: Proceedings of the IEEE Conference on Computer Vision and Pattern Recognition, pp. 770–778 (2016)
9. Hu, J., Shen, L., Sun, G.: Squeeze-and-excitation networks. In: Proceedings of the IEEE Conference on Computer Vision and Pattern Recognition, pp. 7132–7141 (2018)
10. Johnson, J., Alahi, A., Fei-Fei, L.: Perceptual losses for real-time style transfer and super-resolution. In: Leibe, B., Matas, J., Sebe, N., Welling, M. (eds.) ECCV 2016. LNCS, vol. 9906, pp. 694–711. Springer, Cham (2016). https://doi.org/10. 1007/978-3-319-46475-6_43
11. Li, B., Peng, X., Wang, Z., Xu, J., Feng, D.: An all-in-one network for dehazing and beyond. arXiv preprint arXiv:1707.06543 (2017)
12. Liu, X., Ma, Y., Shi, Z., Chen, J.: Griddehazenet: attention-based multi-scale network for image dehazing. In: 2019 IEEE/CVF International Conference on Computer Vision (ICCV), pp. 7313–7322 (2019)

13. Peng, Y.T., Cao, K., Cosman, P.C.: Generalization of the dark channel prior for single image restoration. IEEE Trans. Image Process. **27**(6), 2856–2868 (2018)
14. Qin, X., Wang, Z., Bai, Y., Xie, X., Jia, H.: FFA-net: feature fusion attention network for single image dehazing. In: Proceedings of the AAAI Conference on Artificial Intelligence, vol. 34, pp. 11908–11915 (2020)
15. Romero, A., Ballas, N., Kahou, S.E., Chassang, A., Gatta, C., Bengio, Y.: Fitnets: hints for thin deep nets. arXiv preprint arXiv:1412.6550 (2014)
16. Ronneberger, O., Fischer, P., Brox, T.: U-net: convolutional networks for biomedical image segmentation. In: Navab, N., Hornegger, J., Wells, W.M., Frangi, A.F. (eds.) MICCAI 2015. LNCS, vol. 9351, pp. 234–241. Springer, Cham (2015). https://doi.org/10.1007/978-3-319-24574-4_28
17. Shi, F., Jia, Z., Lai, H., Song, S., Wang, J.: Sand dust images enhancement based on red and blue channels. Sensors **22**(5), 1918 (2022)
18. Shi, Z., Feng, Y., Zhao, M., Zhang, E., He, L.: Let you see in sand dust weather: a method based on halo-reduced dark channel prior dehazing for sand-dust image enhancement. IEEE Access **7**, 116722–116733 (2019)
19. Shi, Z., Liu, C., Ren, W., Shuangli, D., Zhao, M.: Convolutional neural networks for sand dust image color restoration and visibility enhancement. Chin. J. Image Graph. **27**(5), 1493–1508 (2022)
20. Si, Y., Yang, F., Guo, Y., Zhang, W., Yang, Y.: A comprehensive benchmark analysis for sand dust image reconstruction. arXiv preprint arXiv:2202.03031 (2022)
21. Simonyan, K., Zisserman, A.: Very deep convolutional networks for large-scale image recognition. arXiv preprint arXiv:1409.1556 (2014)
22. Woo, S., Park, J., Lee, J.-Y., Kweon, I.S.: CBAM: convolutional block attention module. In: Ferrari, V., Hebert, M., Sminchisescu, C., Weiss, Y. (eds.) ECCV 2018. LNCS, vol. 11211, pp. 3–19. Springer, Cham (2018). https://doi.org/10.1007/978-3-030-01234-2_1
23. Wu, H., et al.: Contrastive learning for compact single image dehazing. In: 2021 IEEE/CVF Conference on Computer Vision and Pattern Recognition (CVPR), pp. 10546–10555 (2021)
24. Yang, Y., Zhang, C., Liu, L., Chen, G., Yue, H.: Visibility restoration of single image captured in dust and haze weather conditions. Multidimension. Syst. Sig. Process. **31**(2), 619–633 (2020)
25. Zeiler, M.D., Fergus, R.: Visualizing and understanding convolutional networks. In: Fleet, D., Pajdla, T., Schiele, B., Tuytelaars, T. (eds.) ECCV 2014. LNCS, vol. 8689, pp. 818–833. Springer, Cham (2014). https://doi.org/10.1007/978-3-319-10590-1_53
26. Zhang, X., Zhou, X., Lin, M., Sun, J.: Shufflenet: an extremely efficient convolutional neural network for mobile devices. In: 2018 IEEE/CVF Conference on Computer Vision and Pattern Recognition, pp. 6848–6856 (2018)
27. Zhu, Y., Huang, J., Fu, X., Zhao, F., Sun, Q., Zha, Z.J.: Bijective mapping network for shadow removal. In: Proceedings of the IEEE/CVF Conference on Computer Vision and Pattern Recognition, pp. 5627–5636 (2022)

Enhancing Image-to-Image Translation with Contrast Loss Constrained Generators and Selective Neighborhood Sampling

Meifang Li, Xiaoru Wang$^{(\boxtimes)}$, and Dexin Bian

School of Computer Science (National Pilot Software Engineering School),
Beijing University of Posts and Telecommunications, Beijing, China
{limf,wxr,biandexin}@bupt.edu.cn

Abstract. Unpaired image-to-image (I2I) translation has been a popular research topic in the field of computer vision, and researchers have recently obtained the desired images by introducing contrastive loss to constrain the generative process of generators. However, when constructing negative samples for contrastive learning, they use a random sampling strategy, which may affect the performance of the model. In this paper, we address this issue by deliberately selecting image blocks with high overlapping semantic information. We designed an SNS module to compute a similarity matrix by comparing the feature distances between the original image blocks and the generated images, each row of this matrix corresponds to probability distribution between the image patches, and we select the few positions with the highest probability for sampling. In addition, we note that the domain invariant information in the generated images is subject to problems such as distortion and discoloration, and we use the attention mechanism to separate the domain invariant information and use this information as the final result of the auxiliary computation. We validate our proposed method on three different image translation datasets, showing that it improves the quality of the generated images without adding additional learnable parameters.

Keywords: Contrastive learning · Generative model · Attention · Negative example mining

1 Introduction

Image translation aims to migrate the uniform features in the target domain image to the source domain image, and unpaired image translation has become a research hotspot due to the generalization problem of paired image translation. To achieve this goal, an intuitive idea is to extract the unified features in the target domain and separate the features that do not need to change in the source

This work was supported by the National Natural Science Foundation of China under Grant 61976025.

domain image, which is called feature decoupling. Finally, the two features are fused to obtain the result. Gatys et al. [6] proposed the style transfer task and used feature decoupling to solve the problem, which became a pioneering work in this field. They use a neural network to extract the low-level texture features and high-level style features of the image at the same time. Furthermore, they use the output of high-level activation functions in the VGG (Visual Geometry Group) network to represent the content features of the image, mainly including the macroscopic structure and contour information of the image. Then, Gram matrix is used to describe the style features. Finally, the image style transfer is realized by minimizing the difference between the content features and the style features of the generated image and the input image. Since then, there have been many studies on feature decoupling. Gonzalez-Garcia et al. [7] divided. the image attributes in the source domain and the target domain into three different spaces: the shared part, which is used to store the shared attributes of all images, and two exclusive parts, which represent specific attributes in the two domains respectively. Another type of decoupled representation was proposed by Liu et al. [15] and Ren et al. [19] who treated the important features as content features and the remaining features as style features. Meanwhile, Goodfellow et al. [8] proposed GANS (Generative Adversarial Networks), which has received much attention due to its excellent performance on image generation task. Motivated by GANs, researchers have begun to change their thinking, viewing translation tasks as generative task and designing image translation models by improving GANs. Models they proposed generate images that meet the conditions by feeding the original image instead of random noise into the generator to start iterating. However, models based on traditional GAN network architectures require paired images and control the generation process through the loss between the generated image and the target image. However, paired image data are often difficult to obtain in practical applications. In order to improve the generalization ability of the model, researchers began to consider how to build a generative model suitable for image translation without using paired images. Zhu et al. [27] introduced the cycle consistency loss to solve this problem. They input the original image into the generative discriminative network to generate the target image, and then input the generated target image into another generative discriminative network to generate the original image. Finally, they optimize the model by minimizing the loss between the original image and the generated original image, which means that two generators need to be trained, exacerbating the difficulty of training the model.

Park et al. [17] found that the contrastive loss can replace the cycle consistency loss and effectively improve the performance of the model. Based on the fact that a patch in the original image is similar to its corresponding patch in the generated image, they divided the image into blocks, and the patch in the original image forms positive samples with the corresponding patch in the generated image, and forms negative samples with other patches in the generated image. The contrast loss is optimized to maximize the mutual information between positive examples, thereby replacing the cyclic consistency loss. However, in the process of constructing negative examples, most of the negative examples may not be similar to the image patches in the original image at all, and the effect of these negative examples may be negligible. Based on this, when constructing negative examples, we select hard negative examples to improve the performance of the model by improving the performance of contrastive learning. At the same time, in order to solve the problems of deformation and color change in the regions unrelated to the task in the generated image, the attention mechanism is introduced to retain the domain invariant information. Experimental results show that these methods can improve the performance of the generative model.

2   Related Work

2.1   Image Translation

Early image translation models [2,13,18,22] focus on image translation algorithms applied to paired images. These works focus on optimizing the error between the reconstructed image and the real image, and then training a model that can generate high-fidelity images. Isola et al. [13] proposed Pix2Pix, which is based on cGAN (conditional GAN) architecture introduced by Mirza and Osindero [16] to perform image-to-image translation tasks. The original image is first fed into the generator, and the generated image is trained together with the conditional target image to train a discriminative network. Chen and Koltun [2] used a cascading optimization network to generate images, and Park et al. [18] proposed spatially adaptive normalization, which provides rich semantic information for image synthesis by replacing multiple perceptual layers with convolutional layers. Wang et al. [22] propose a novel coarse-to-fine generator and multi-scale discriminator architecture suitable for higher resolution conditional image generation.

However, all of them require pairwise data, which is often difficult to obtain in practical applications. In response to this problem, many unpaired image translation models [1,26,27] have been proposed. The most typical model is CycleGAN proposed by Zhu et al. [27], which does not require paired data. The model is composed of two generators and two discriminators. One set of generators and discriminators is responsible for the generation and discrimination of images from the source domain to the target domain, and the other set is responsible for the generation of images from the target domain to the source domain. With discrimination, the optimization goal of the network is to make the converted source domain image consistent with the original image. In this paper, a loss function, cycle consistency loss, is designed to ensure that the desired image is generated. Beak et al. [1] propose truly unsupervised image-to image translation method (TUNIT), which does not require either input-output pairs or domain labels. An information theoretic method is used to learn to separate image domains, and the estimated domain labels are used to generate the corresponding images. Zhou et al. [26], based on the dense semantic correspondence of two cross-domain images, realized the cross-attention of the feature space calculation of the image to realize the image synthesis.

2.2   Contrastive Learning

The introduction of two generators in CycleGAN increases the difficulty of training the model. To solve this problem, researchers explored the performance of contrastive learning in image translation task. In recent years, there have been a lot of works on contrastive learning [3–5,9,23,24]. Inspired by these works, Park et al. [17] state that image-to-image translation is a decoupled learning problem, as the goal is to convert content to a specific style. For a source domain image and its corresponding target domain image, the images at corresponding spatial locations should have similar information. They start with this common sense and introduce the idea of contrastive learning. They use the infoNCE loss to maximize the mutual information between image patches. Mutual information is one of the most basic and important concepts in contrastive learning, which refers to the same information implied by two feature vectors. By maximizing the mutual information of the corresponding positions of the source domain image and the target domain image, it can not only supervise the generation network to find the common parts of the two corresponding positions, but also supervise the generation network to ignore the characteristic parts of different positions. This idea was further exploited by Han et al. [10] and Zheng et al. [25] in 2021. The former approach uses dual contrastive learning to constrain the generator during translation. The latter method uses contrastive learning to make the generator learn the spatial structure of the image, and uses self-similarity to provide supervision for the structure. Hu et al. [11] found that the image patches in the CUT model are randomly sampled, and this strategy does not achieve optimal results. In view of this problem, they use an attention module to optimize the selection of image patch positions. Jung et al. [14] mentioned that recent methods do not take semantics into account and proposed Semantic Relation Consistency (SRC) to force the model to preserve spatial semantics during translation and negative example mining, further improving the performance of the model by avoiding the selection of simple image patches.

2.3   Negative Example Mining

In the field of deep metric learning, researchers have deeply studied the selection strategy of negative examples. All these works have observed that the use of difficult to distinguish negative samples is helpful to improve the performance of the model. Schroff et al. [21] used some constraints, based on which they observed that some examples are too difficult to distinguish and do not help much to improve the performance of the model, so they suggested choosing "semi-hard" negative samples. The key of contrastive learning is to select appropriate negative and positive examples, so that the positive pairs are as close as possible in the mapping space, and the negative pairs are as far as possible in the mapping space. Motivated by the research results of negative example selection in metric learning, it is reasonable to assume that easy negative examples contain less effective information and have relatively little effect on training the network. On the contrary, those negative examples that can be easily distinguished as positive examples and hard negative examples contain more effective information.

3   Model

3.1   Model Structure

Image Translation aims to transfer the uniform style of several images $Y \in \mathbb{R}^{H \times W \times 3}$ in the target domain to each image $x \in X(X \in \mathbb{R}^{H \times W \times 3})$ in the source domain. To achieve this goal, we propose an image translation model based on contrastive learning and attention mechanism. The experimental results of CUT model proposed by Park et al. [17] showed the effectiveness of contrastive learning in image translation, but during the generation of images, distortions and discoloration occur in the regions that are supposed to be preserved as they are. To address this issue and taking into account the characteristics of image translation tasks, we introduce a global attention mechanism in the generator. In addition, in the process of generating target patches based on contrastive learning, CUT selected negative examples by adopting a random selection strategy, which may affect the performance of the model. We hope to select more relevant image patches to improve the performance of the model. Therefore, in this process, we propose an SNS module to select negative samples with high similarity based on the semantics of image patches. Our model is shown in Fig. 1, which consists of a generator G and a discriminator D. The input image x is fed into the attention module A and the generator G respectively. After the attention module, it is divided into foreground map P and background map B, while the image x is passed through the generator to obtain the preliminary generated image y', and the final output result y is obtained by computing y' with B. The contrast loss is calculated using the feature encoding y_f of the generated image and the feature encoding x_f of the original image, while the generated image is sent to the discriminator to determine whether it is true or false.

3.2   Attention-Based Generator

In the process of image translation, we want to preserve the areas of the original image that do not need to be changed to the maximum extent. Take the transformation of horse and zebra images as an example, as shown in the Fig. 1, we want to convert the horse in the input image to a zebra to obtain the output image. During this process, we need to preserve the background of the image and the shape and pose of the horse. In order to preserve the background in the image, we use the attention mechanism to separate the original image into a foreground image and a background image, and the background image is used to calculate the final result. Firstly, we feed the image $x \in X$ into the image transformation model $G_{x \to y}$ and the attention network A, $y' = G_{x \to y}(x)$ maps the image x in the source domain to the target domain Y to obtain the preliminary results, we then send x to A to obtain the attention mapping map $x_a = A(x)$, then the dot product of x_a and y' to obtain the foreground map x_p. The background image x_b is obtained by dot product operation between $1 - x_a$ and x. Finally, the final output result y is obtained by addition operation: $y = x_a \odot y' + (1 - x_a) \odot x$.

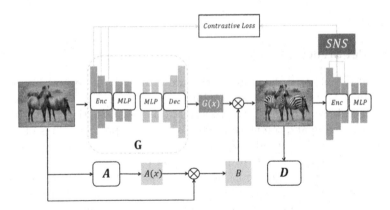

Fig. 1. Model structure. The input image x is sent to the attention module and the generator respectively. After the generator, the preliminary generated image $G(x)$ is obtained, and the final output image y is obtained by computing $G(x)$ with B. y is encoded with the feature code of the original image to calculate the contrast loss. At the same time, the generated image is sent to the discriminator D to determine the authenticity.

3.3 Contrastive Learning with Selective Patches

The key of contrastive learning is to select appropriate positive and negative examples. In CUT, image patches are randomly sampled when the contrastive loss is calculated by maximizing the mutual information of image patches. In other words, for anchor q, CUT randomly selects positive examples k^+ and negative examples k^-, which means that the model may select image patches that are not similar to q, which limits the performance of the model. Robinson et al. [20] show that the performance of contrastive learning models can be improved by selecting hard-to-distinguish negative examples, negative examples that are similar to positive examples. Motivated by this, we select hard negative samples based on the semantic similarity of image patches. Since we want to select the negative example which is similar to the positive example, we first randomly select an anchor q, extract the feature and compare it with the patch features of all other positions to construct the similarity matrix, which can reflect the similarity between the feature of anchor and other features. Using the similarity matrix, we select a number of negative examples with the highest similarity. As shown in Fig. 2, given a patch in a source image as a query q, to obtain its feature $F_x \in \mathbb{R}^{H \times W \times C}$, first, F_x is reconstructed into a two-dimensional matrix $Q \in \mathbb{R}^{HW \times C}$, and Q is multiplied with the global key matrix $K \in \mathbb{R}^{C \times HW}$. The global attention matrix $A_g \in \mathbb{R}^{HW \times HW}$ is obtained. After that, the information entropy H_g can be obtained by Eq. 1 below to measure the importance of features:

$$H_g(i) = - \sum_{j=1}^{HW} A_g(i,j) log A_g(i,j) \tag{1}$$

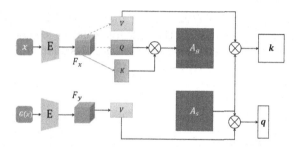

Fig. 2. SNS Module. Given a patch in a source image as a query q, to obtain its feature $F_x \in \mathbb{R}^{H \times W \times C}$, first, F_x is reconstructed into a two-dimensional matrix $Q \in \mathbb{R}^{HW \times C}$, and Q is multiplied with the global key matrix $K \in \mathbb{R}^{C \times HW}$. The global attention matrix $A_g \in \mathbb{R}^{HW \times HW}$ is obtained.

3.4   Loss Function

There are three main loss functions involved in the model, which are adversarial loss, identity loss and contrast loss.

Adversarial Loss. We use adversarial loss to make the generated image as similar as possible to images in target domain. The loss function is as follows Eq. 2:

$$\mathcal{L}_{GAN}(G, D, X, Y) = \mathbb{E}_{y \sim Y} log D(y) + \mathbb{E}_{x \sim X} log(1 - D(G(x))) \qquad (2)$$

where $D(y)$ represents the value returned by feeding the data y in the target domain into the discriminator D, and $G(x)$ represents the value returned by feeding the source domain image into the generator G.

Identity Loss. The identity loss is used to limit some unnecessary changes, backpropagate and update the parameters of G to reduce this loss:

$$\mathcal{L}_{idt} = \mathbb{E}_{x \sim X}[\| G(x) - x \|] + \mathbb{E}_{y \sim Y}[\| G(y) - y \|] \qquad (3)$$

Contrastive Loss. The noise contrast estimation NCE is used to maximize the mutual information between input and output. In general, for image k, the image itself or the image after data enhancement is considered as positive sample k^+, and other images in the same batch or dataset are considered as negative samples k^-. Then the contrast loss is calculated as shown in Eq. 4 below:

$$f(k, k^+, k^-) = -log \left[\frac{exp(k \cdot k^+ / \tau)}{exp(k \cdot k^+ / \tau) + \sum_{n=1}^{N} exp(k \cdot k_n^- / \tau)} \right] \qquad (4)$$

where k_n^- denotes the nth negative sample, and $\tau = 0.07$ is a temperature coefficient that controls how much attention the model pays to the difficult negative samples. The larger the temperature coefficient, the less attention the model pays to the more difficult negative samples and treats all negative samples as close to the same similarity as that sample; the smaller the temperature coefficient, the more attention it pays to the difficult negative samples that are very similar to that sample, giving difficult negative samples a larger gradient in order to separate them from the positive samples.

4   Experiments

4.1   Image Translation

Datasets. Our model is trained and evaluated on three datasets: Horse2Zebra, Cat2Dog, and Cityscapes.

Cat2Dog. The dataset from Animal Faces-HQ (AFHQ) is an animal face dataset, which contains about 15,000 images in three domains: Cat, Dog, and wild animals. Cat2Dog contains 5000 training images and 500 validation images.

Horse2Zebra. The dataset derived from ImageNet dataset and used independently in CycleGAN, contains 2403 training images and 260 zebra images.

Cityscapes. A dataset of 5000 city street view images with 2975 training images, 500 validation images, and 1525 test images.

Evaluation Metric. The Frechet Inception Distance score (FID) is a metric that computes the distance between the feature vectors of a real image and a generated image. Firstly, the mean and variance of the image are calculated. By calculating the mean and covariance of the image, the output of the activation function is summarized as a multivariate Gaussian distribution. These statistics are then used to calculate the activation function in the real image and the generated image set. The distance between these two distributions is then calculated using the Frechet distance. For Cityscapes, the data in the dataset is labeled. So if the generated images accurately correspond to images in the target domain, the model has generated images that can be identified as the correct class. Most researchers use an off-the-shelf network to test the image translation results. We also follow this design, using a pre-trained semantic segmentation network DRN, and compute the average precision (mAP), pixel-by-pixel accuracy (pixAcc), and average Class accuracy (classAcc).

Parameters Setting. In order to verify the improvement effect of model performance, the hyperparameter Settings in our experiment are consistent with those in CUT. The model is composed of a generator, a PatchGAN discriminator, a least squares GAN loss, and an Adam optimizer, where the generator is a ResNet network consisting of 9 residual blocks and the Adam optimizer has a batch size of 1 and a learning rate of 0.002. The encoder Enc is the first half of the CycleGAN generator, which computes the contrastive loss by extracting features from 5 layers, namely the complete image, the first and second downsampled convolution, and the first and fifth residual blocks. For each layer of features, the CUT samples 256 random locations and applies a 2-layer MLP to obtain the final 256-dimensional features. To compute the contrastive loss, the momentum value is 0.999 and the temperature coefficient is 0.07.

4.2   Experimental Results and Analysis

Table 1. Comparison with baselines. The table shows the comparison results of our model with FSeSim, CycleGAN, MUNIT and CUT. The last row shows the numerical metrics of our model, and the best result is shown in bold. Cat2Dog and Horse2Zebra are measured by FID, and Cityscapes are measured by FID, mAP, pixAcc, classAcc.

Method	Cityscapes				Cat2Dog	Horse2Zebra
	mAP	pixAcc	classAcc	FID	FID	FID
CycleGAN [27]	20.4	55.9	25.4	76.3	85.9	77.2
MUNIT [12]	16.9	56.5	22.5	91.4	104.4	133.8
CUT [17]	24.7	68.8	30.7	56.4	76.2	45.5
FSeSim [25]	22.1	**69.4**	27.8	**54.3**	87.8	43.4
our	**25.2**	69.1	**31.6**	55.5	**74.8**	**40.9**

Table 1 shows the numerical comparison results of our model with FSeSim, CUT, CycleGAN, and MUNIT. For the Horse2Zebra and Cat2Dog datasets, our models have smaller FID values and work better than the other methods, with boosts of 4.6 and 1.4, respectively, relative to the CUT model. mAP and classAcc obtain the best results on the Cityscapes dataset, and pixAcc and classAcc, although slightly inferior to the FSeSim model, but the improvement over CUT is 0.3 and 0.9, respectively.

The visual results are shown in Fig. 3 Compared with other methods, our, FSeSim, CUT, CycleGAN, and MUNIT, respectively, our model is able to accurately translate domain specific features. In addition, the consistency of the context is also ensured.

4.3   Ablation Experiments

In order to analyze the influence of global attention module and hard negative example strategy on the model, we refer to the combination of attention module and contrastive learning as Attn, and refer to the combination of hard negative examples mining module and contrastive learning as SNS. Experiments were carried out on the horse2zebra dataset, and the following Table 2 illustrates the experimental results.

The results in Table 2 show that selecting image blocks with high similarity based on their semantics to construct negative samples does improve the performance of the model, with a 4.2 improvement in the FID measure on the

| Input | our | FSeSim | CUT | CycleGAN | MUNIT |

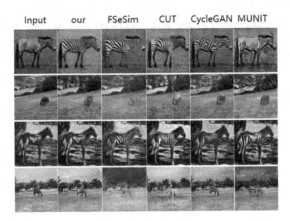

Fig. 3. Results. Our model better retains the background information in the source domain image while performing the horse to zebra transformation task.

Table 2. Ablation The table shows the comparison results of our complete model with Attn, SNS and CUT. The modules of the model are tested on the Horse2Zebra dataset. Horse2Zebra are measured by FID.

Method	Horse2Zebra
	FID
CUT [17]	45.5
Attn	44.9
SNS	41.3
complete model	40.9

Horse2Zebra dataset. While introducing the attention mechanism in the generator improves the performance of the model by 0.6, which proves that the attention mechanism also plays a role in the model.

5   Conclusion

In this paper, we propose an image translation algorithm based on attention mechanism and selective negative sample contrast learning. Instead of randomly selecting sampling points, positive samples and negative samples to calculate the contrast loss, we select the important features in the source domain based on the semantic similarity between image blocks and anchor points. We first compute a similarity matrix using the features of the real image in the source domain, and then compute the entropy of the sampled points based on this matrix. The image blocks corresponding to those sampled points with smaller values are selected to calculate the contrast loss, a strategy that not only expands the acceptance domain of the selected features but also helps the output to preserve the relationships in the input image. In addition, we also introduce attention mechanism force in the model to further preserve the domain invariant information in the original image. We validate the effectiveness of the SNS and attention modules on popular image translation datasets and perform an in-depth ablation study.

References

1. Baek, K., Choi, Y., Uh, Y., Yoo, J., Shim, H.: Rethinking the truly unsupervised image-to-image translation. In: Proceedings of the IEEE/CVF International Conference on Computer Vision, pp. 14154–14163 (2021)
2. Chen, Q., Koltun, V.: Photographic image synthesis with cascaded refinement networks. In: Proceedings of the IEEE International Conference on Computer Vision, pp. 1511–1520 (2017)
3. Chen, T., Kornblith, S., Swersky, K., Norouzi, M., Hinton, G.E.: Big self-supervised models are strong semi-supervised learners. In: Advances in Neural Information Processing Systems, vol. 33, pp. 22243–22255 (2020)
4. Chen, X., Xie, S., He, K.: An empirical study of training self-supervised vision transformers. In: Proceedings of the IEEE/CVF International Conference on Computer Vision, pp. 9640–9649 (2021)
5. Dwibedi, D., Aytar, Y., Tompson, J., Sermanet, P., Zisserman, A.: With a little help from my friends: nearest-neighbor contrastive learning of visual representations. In: Proceedings of the IEEE/CVF International Conference on Computer Vision, pp. 9588–9597 (2021)
6. Gatys, L.A., Ecker, A.S., Bethge, M.: A neural algorithm of artistic style. arXiv preprint arXiv:1508.06576 (2015)
7. Gonzalez-Garcia, A., Van De Weijer, J., Bengio, Y.: Image-to-image translation for cross-domain disentanglement. In: Advances in Neural Information Processing Systems, vol. 31 (2018)
8. Goodfellow, I.J., et al.: Generative adversarial nets. In: Proceedings of the 27th International Conference on Neural Information Processing Systems, vol. 2, pp. 2672–2680 (2014)
9. Grill, J.B., et al.: Bootstrap your own latent-a new approach to self-supervised learning. In: Advances in Neural Information Processing Systems, vol. 33, pp. 21271–21284 (2020)
10. Han, J., Shoeiby, M., Petersson, L., Armin, M.A.: Dual contrastive learning for unsupervised image-to-image translation. In: Proceedings of the IEEE/CVF Conference on Computer Vision and Pattern Recognition, pp. 746–755 (2021)
11. Hu, X., Zhou, X., Huang, Q., Shi, Z., Sun, L., Li, Q.: QS-attn: query-selected attention for contrastive learning in i2i translation. In: Proceedings of the IEEE/CVF Conference on Computer Vision and Pattern Recognition, pp. 18291–18300 (2022)
12. Huang, X., Liu, M.-Y., Belongie, S., Kautz, J.: Multimodal unsupervised image-to-image translation. In: Ferrari, V., Hebert, M., Sminchisescu, C., Weiss, Y. (eds.) ECCV 2018. LNCS, vol. 11207, pp. 179–196. Springer, Cham (2018). https://doi.org/10.1007/978-3-030-01219-9_11
13. Isola, P., Zhu, J.Y., Zhou, T., Efros, A.A.: Image-to-image translation with conditional adversarial networks. In: Proceedings of the IEEE Conference on Computer Vision and Pattern Recognition, pp. 1125–1134 (2017)

14. Jung, C., Kwon, G., Ye, J.C.: Exploring patch-wise semantic relation for contrastive learning in image-to-image translation tasks. In: Proceedings of the IEEE/CVF Conference on Computer Vision and Pattern Recognition, pp. 18260–18269 (2022)

15. Liu, Y., Wang, H., Yue, Y., Lu, F.: Separating content and style for unsupervised image-to-image translation. arXiv preprint arXiv:2110.14404 (2021)

16. Mirza, M., Osindero, S.: Conditional generative adversarial nets. arXiv preprint arXiv:1411.1784 (2014)

17. Park, T., Efros, A.A., Zhang, R., Zhu, J.-Y.: Contrastive learning for unpaired image-to-image translation. In: Vedaldi, A., Bischof, H., Brox, T., Frahm, J.-M. (eds.) ECCV 2020, Part IX. LNCS, vol. 12354, pp. 319–345. Springer, Cham (2020). https://doi.org/10.1007/978-3-030-58545-7_19

18. Park, T., Liu, M.Y., Wang, T.C., Zhu, J.Y.: Semantic image synthesis with spatially-adaptive normalization. In: Proceedings of the IEEE/CVF Conference on Computer Vision and Pattern Recognition, pp. 2337–2346 (2019)

19. Ren, X., Yang, T., Wang, Y., Zeng, W.: Rethinking content and style: exploring bias for unsupervised disentanglement. In: Proceedings of the IEEE/CVF International Conference on Computer Vision, pp. 1823–1832 (2021)

20. Robinson, J., Chuang, C.Y., Sra, S., Jegelka, S.: Contrastive learning with hard negative samples. arXiv preprint arXiv:2010.04592 (2020)

21. Schroff, F., Kalenichenko, D., Philbin, J.: Facenet: a unified embedding for face recognition and clustering. In: Proceedings of the IEEE Conference on Computer Vision and Pattern Recognition, pp. 815–823 (2015)

22. Wang, T.C., Liu, M.Y., Zhu, J.Y., Tao, A., Kautz, J., Catanzaro, B.: High-resolution image synthesis and semantic manipulation with conditional GANs. In: Proceedings of the IEEE Conference on Computer Vision and Pattern Recognition, pp. 8798–8807 (2018)

23. Wang, X., Qi, G.J.: Contrastive learning with stronger augmentations. IEEE Trans. Pattern Anal. Mach. Intell. (2022)

24. Wu, H., et al.: Contrastive learning for compact single image dehazing. In: Proceedings of the IEEE/CVF Conference on Computer Vision and Pattern Recognition, pp. 10551–10560 (2021)

25. Zheng, C., Cham, T.J., Cai, J.: The spatially-correlative loss for various image translation tasks. In: Proceedings of the IEEE/CVF Conference on Computer Vision and Pattern Recognition, pp. 16407–16417 (2021)

26. Zhou, X., et al.: Cocosnet v2: full-resolution correspondence learning for image translation. In: Proceedings of the IEEE/CVF Conference on Computer Vision and Pattern Recognition, pp. 11465–11475 (2021)

27. Zhu, J.Y., Park, T., Isola, P., Efros, A.A.: Unpaired image-to-image translation using cycle-consistent adversarial networks. In: Proceedings of the IEEE International Conference on Computer Vision, pp. 2223–2232 (2017)

Dynamic Neural Networks for Adaptive Implicit Image Compression

Binru Huang[1], Yue Zhang[1], Yongzhen Hu[1], Shaohui Dai[1], Ziyang Huang[1],
and Fei Chao[1,2(✉)]

[1] School of Informatics, Xiamen University, Xiamen, China
fchao@xmu.edu.cn
[2] Key Laboratory of Multimedia Trusted Perception and Efficient Computing,
Ministry of Education of China, Xiamen University, Xiamen 361005,
People's Republic of China

Abstract. Compression with Implicit Neural Presentations (COIN) is a neural network image compression method based on multilayer perceptron (MLP). COIN encodes an image with an MLP that maps pixel positions to RGB values matching, the weights of the MLP are quantized to obtain a code stored as an image. However, this single implicit network structure performs generally when dealing with images of multiple complexities. In this paper, we propose a novel implicit dynamic neural network to process images in a dynamic and adaptive manner. Specifically, this paper uses the Sobel operator to divide the complexity of the images and use it as a criterion to select the network width and depth adaptively. To better fit the image features, this paper concludes with further quantification of the dynamic network parameters and storage matrices. Therefore, only some of the relevant network parameters with their storage matrices are required when storing the images. In training this dynamic network, this paper uses a meta-learning approach for the multi-image compression task. Experimental results show that our method outperforms COIN and JPEG in terms of image reconstruction results for the CIFAR-10 dataset.

Keywords: implicit neural representation · dynamic neural network · multi-level image compression · low-rank matrix synthesis

1 Introduction

The representation of a common picture is a series of discrete two-dimensional spatial pixel points. However, in reality, the information is expressed continuously rather than discrete. Also, the amount of information contained in different pictures varies, which can affect the difficulty and accuracy of target extraction. In contrast, it is more reasonable to use continuous functions to represent the content and complexity of the pictures. However, since we usually have difficulty in obtaining the exact form of continuous functions used to represent pictures, we can resort to advanced multimedia signal representation methods such as implicit neural representation [24]. Implicit neural representation (INR) is an

© The Author(s), under exclusive license to Springer Nature Singapore Pte Ltd. 2024
Q. Liu et al. (Eds.): PRCV 2023, LNCS 14435, pp. 427–443, 2024.
https://doi.org/10.1007/978-981-99-8552-4_34

emerging multimedia signal representation method that uses neural networks to approximate continuous functions that map high-dimensional input coordinates to target scalars (e.g., color, depth, etc.). Applying this method to image compression can enable image reconstruction at arbitrary resolutions. Currently, INR has been widely used in 3D models [5,16,17,19], while in the 2D image domain, it is mainly applied to super-resolution [4,29] and image generation [1,8,22,23] in both areas.

However, the conventional COIN network exhibits significant variance in image fitting of different complexity. The single model structure performs poorly for complex image fitting again, while the deeper model suffers from excessive computational overhead for simple image fitting. Therefore, a single COIN network structure is not suitable for processing images of multiple complexities. In traditional static COIN networks, the test phase maintains the invariance of the parameters and network structure obtained from training [6,11]. However, the inference capability and efficiency of such a single and fixed model can be compromised due to the diversity of samples. Although many existing dynamic neural networks have been successfully applied to tasks such as image classification, target detection, and semantic segmentation [12,31], there are still relatively few applications of dynamic neural networks in the field of image compression. Image compression requires consideration of data compression ratio, distortion control, and decompression speed [15,21,34], which places higher demands on the design of network structures and parameters.

To address the problem that a single network structure cannot effectively handle images of multiple complexities, we propose a dynamic structure approach. This method allows for adaptive selection of network width and depth based on the complexity of the images. This model structure shows better results in image reconstruction compared to a single network structure. Also, we propose a decomposition method of the parameter matrix to address the problem of large redundancy of the actual storage parameters of the pictures. The method decomposes the parameter matrix into two shared low-rank matrices and a complexity-dependent diagonal matrix, which effectively reduces the actual storage volume of the images. In order to make the storage matrices and the network parameters fit better during image reconstruction, we perform class meta-learning [9]. In the outer loop, the base model learns as many dictionary frequencies as possible. In the inner loop, for each image storage matrix, we perform a gradient update to achieve multi-image compression. Such an approach can effectively improve image reconstruction. Finally, we perform the validation of image reconstruction on the "Kodak" dataset and "CIFAR-10" dataset to show the feasibility of our method.

To sum up, our innovations are mainly in two aspects: 1) The innovative combination of dynamic neural network and implicit neural representation to achieve compression of images and show better image reconstruction than COIN and JPEG on the CIFAR-10 dataset. 2) Adaptive dynamic width and depth networks are designed to enable multi-image compression processing for images of different complexity.

The remainder of this paper is organized as follows. Section 2 introduces the basic theories and related work of dynamic neural networks and implicit neural representation. Section 3 constructs a dynamic neural network to realize multi-level image compression of implicit neural representation. Section 4 introduces the setup of the experiment and analyses the experimental results. Section 5 summarises our work and points out our important future directions.

Please note that the appendix has been placed on Github. Please visit https://github.com/ningxiudg/PRCV-appendix for the appendix.

2   Related Work

2.1   Dynamic Neural Networks

A dynamic neural network is an artificial intelligence algorithm that uses dynamic models for training and outcome prediction. Unlike traditional neural networks, dynamic neural networks can adapt the network architecture to changes in the input in order to better fit new data. This flexibility allows dynamic neural networks to show excellent performance in tasks such as processing time-series data [2], image recognition, and natural language processing [25].

In recent years, dynamic convolutional neural networks [30] have been widely studied. The dynamic deep network [3,12,14,27,28] achieves efficient inference in two ways, one way is to exit early [3,12,14] when the classification confidence is high in the shallower subnetworks. The other one is to skip the residual blocks adaptively [27,28]. Whereas the initial research on dynamic width networks focused on controlling the dynamic activation of neurons in linear layers by means of low-rank decomposition [18], for example, the more common approach today is to use multi-expert hybrid systems (MoE) and dynamic channel pruning techniques [11]. In our work, we apply dynamic depth and width neural networks to an image compression reconstruction framework for the first time to solve the storage compression problem under different image complexities.

2.2   Implicit Neural Representation

Implicit neural representation (INR) refers to the use of a neural network to approximate a continuous function of the true state of an image. Thus, INR is able to use neural networks to establish a mapping relationship between coordinates and their signal values for representing images, videos, and other content. For images, the INR function maps two-dimensional coordinates to RGB values, i.e. $I : (x, y) \xrightarrow{\theta} (r, g, b)$; for video, it maps the image 2D coordinates and moments to RGB values, i.e. $I : (x, y, t) \xrightarrow{\theta} (r, g, b)$. In addition, other forms of INR exist, such as NeRF [17] $I : (x, y, z, \gamma, \phi) \xrightarrow{\theta} (r, g, b, \sigma)$. By using the network model, θ, we can map the input data to the underlying model f_θ.

Meta-learning is a method that can integrate features of different learning methods to optimize algorithms and model parameters, with the aim of using

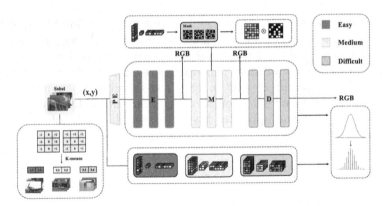

Fig. 1. Dynamic neural network combined with implicit neural representation for multi-image compression. We first divide the image complexity, then store the relevant network parameters and matrixes, and finally carry out further quantization and compression.

a small amount of unlabeled data to achieve fast and effective learning so that the model can quickly adapt to new tasks. The application of meta-learning in INR models can effectively help the model to learn the frequency [7,24]. Through meta-learning techniques, the INR model can better adapt to the requirements of different tasks and improve the performance of the model in various application scenarios.

3   Method

3.1   Overall Framework

Figure 1 shows the main algorithm flow in this paper. The approach uses a dataset of images with varying complexity. The images are evaluated using the Sobel operator and classified into easy, medium, and difficult classes using clustering methods. To implement the dynamic structure, we initialize a storage matrix M_ψ for each image. M_ψ includes two shared matrices ($Q_{d \times r}$ and $V^T_{d \times r}$), as well as l independent diagonal matrices $Q^i_{r \times r}$. The values of l and r are predetermined based on the complexity classes of the images.

During training, we utilize Gaussian encoding to represent the input and propagate it sequentially through each layer. The soft mask for each layer is obtained by multiplying the shared matrices in the storage matrix with the corresponding diagonal matrix of that layer. This mask is generated by applying a sigmoid function and binarizing it with a threshold of 0.5. The model achieves dynamic width by element-wise multiplication of the mask with the layer weights. Depending on the complexity of the images, certain levels of the model can be skipped. To optimize the parameters, we employ the class meta-learning method as described in Appendix B. This involves overfitting the weight mask of each image to the other images and minimizing the fit variance of the base model across the entire dataset. In order to further reduce storage volume, we apply unsigned 8-bit quantization to the storage matrices and network parameters θ. We validate the effectiveness of our approach through experiments conducted on the CIFAR-10 and Kodak datasets. Please refer to the Appendix for the pseudocode of the overall process, specifically "Overall Process Pseudocode" in Sect. A.

3.2   Image Complexity Classification Using the Sobel Operator

Image complexity can be assessed using various features, such as texture and color, which impact the extraction of target content. Simple images are easier to compress and require lighter network structures. Grayscale histograms and co-generation matrices have been used to evaluate image complexity by analyzing grayscale value distribution [20]. However, relying solely on co-generation matrices may not capture image details adequately for effective classification. To address this, one approach is to calculate edge magnitude and orientation using the Sobel operator [32]. In this study, we measure image complexity by computing the mean edge magnitude (G_{mean}), which is obtained by calculating the square root of the sum of squares of the edge computation results (G_x and G_y) in the horizontal and vertical directions at each pixel, and then taking the average of these values. Using G_{mean} values, we categorize images into three classes ("easy", "medium", and "difficult") using k-means clustering. Different image classes are stored with varying information redundancy and represented using distinct shared basis network structures.

3.3   Dynamic Neural Network Structure

SIREN is a structure similar to a structured signal dictionary [33], in which the signal learned at each layer increases exponentially as the number of network layers increases in SIREN. Therefore, it is necessary to construct a dynamic network structure based on different images. The dynamic width of the network is achieved by using a soft mask to dynamically select atomic harmonics based on the characteristics of the input image. In addition, there are differences in the complexity of different images, and therefore the number of frequencies required to generate them varies. Therefore, it makes more sense to use different depths of the underlying network to generate different images. A dynamically architected network not only saves redundant computations but also makes full use of the fundamental frequency signals learned by the underlying model during image reconstruction.

Dynamic Width Network. In order to achieve dynamic width and adaptive activation of neurons in each layer, we employ masks to selectively assign weights, facilitating the dynamic selection of support frequencies for each layer.

Through the superposition of these frequencies, the image signal can be accurately reconstructed. The SIREN model, as described in [33], is represented by the function $f_\theta(x)$, which maps input coordinates x to their corresponding attributes. Notably, we enhance the model by incorporating $\gamma(x)$, which represents the Gaussian encoding [26] of the input x and plays a vital role in capturing high-frequency information during the learning process.

The following describes how the M mask matrix is generated. For each image, we store the low-rank matrices used to generate the M mask matrix and the diagonal matrix equal to the number of layers. Since the direct use of 0-1 mask to filter the weights is crude, we use a soft mask generated with the *sigmoid* function to achieve a dynamic selection of weights.

$$S(x) = \frac{1}{1 + e^{-x}} \tag{1}$$

$$M_{soft}^l = S(Q_{d \times r} \times Q_{r \times r}^l \times V_{d \times r}^T) \tag{2}$$

$$M^l = \begin{cases} 1, M_{soft}^l \geq 0.5 \\ 0, M_{soft}^l < 0.5 \end{cases} \tag{3}$$

$$M_\psi = (Q_{d \times r}, Q_{r \times r}^l, V_{d \times r}^T), l = 1, \ldots, L - 1 \tag{4}$$

where d represents the number of hidden layer nodes in the underlying network, while r is determined based on the picture's complexity type. M_ψ denotes the stored matrices in each layer. The soft mask is generated using shared low-rank matrices $(Q_{d \times r}, V_{d \times r}^T)$ and an independent diagonal matrix $Q_{r \times r}$ for each layer. Binarization of M_{soft}^l is achieved by applying a threshold of 0.5, converting continuous values to binary values. The gradient propagation is retained during this process, resulting in a hard mask with values of 0 or 1. Thus, only the shared low-rank matrix and the diagonal elements of each layer need to be stored for each image. The complexity of the image determines the required complexity of the dictionary signal, which is reflected in the value of r in the low-rank matrix. Images of the same complexity type share the same r value for convenience.

Dynamic Depth Network. Increasing the number of network layers in neural networks improves the model's ability to approximate nonlinear functions. However, excessively deep layers can result in an increased computational burden for simpler samples. On the other hand, if the network is too simple, it may struggle to capture complex image features. Simple images require shallower dictionary depths for image reconstruction since they require fewer initial mappings compared to complex images.

Hence, assigning different depths to images based on their complexity is crucial. To minimize redundant computations, a dynamic depth approach can be employed for image reconstruction. This involves implementing an early retirement mechanism, where images with low complexity exit at shallow levels without further computation at deeper levels.

For image reconstruction with input coordinates x and M_ψ, the output f_θ with L layers can be represented as:

$$y = f_\theta^L \circ f_\theta^{L-1} \circ \cdots \circ f_\theta^1(x, M_\psi) \tag{5}$$

where f_θ^l is denoted as the operation function of the $l(1 \leq l \leq L)^{th}$ layer. Similarly, in dynamic deep networks, l satisfies $l \propto G_{mean}$. Thus, for an image belonging to a certain class of complexity with input coordinates x_i, the storage matrix $M_{\psi i}$ and its type t, its forward propagation can be expressed as

$$y_i = f_\theta^{l_i} \circ f_\theta^{l_i-1} \circ \cdots \circ f_\theta^1(x_i, M_{\psi i}, t), 1 \leq l_i \leq L \tag{6}$$

3.4 Further Compression

Quantization reduces storage and computational costs by converting real-valued parameters into discrete integers. When quantized as k-bit unsigned or signed numbers, the integer values range within specific bounds. To measure the space occupied by each image, we consider the flattened size of storage matrices and the underlying network model. Both the model parameters (θ) and the matrices (M_ψ) are quantized and compressed using 8-bit unsigned quantization. The fully connected layers of the network model are quantized, along with the low-rank matrices (Q and V) in M_ψ. Notably, the diagonal matrix is not quantized due to its smaller size. The space occupied by each image can be represented as a function of the total number of parameter bits (Θ) divided by the number of images in the complexity class (n), plus the size of M_ψ.

$$P_\Theta = \frac{\Theta(f_\theta)}{n} + \Theta(M_\psi) \tag{7}$$

where Θ denotes the total number of bits accounted for by the parameters, and n is the number of images in the complexity class.

4   Experimentation

To verify the effectiveness of applying dynamic neural networks combined with implicit neural representations to multi-image compression, a series of experiments were conducted and the degree of image reconstruction was used as an evaluation metric. All experiments were trained on GPUs to ensure the reliability of the results.

We use two datasets for the evaluation of image reconstruction effectiveness: Kodak [10] and CIFAR-10 [13]. The Kodak dataset contains 24 high-quality color images of different sizes, which meet our requirements for images of different complexity. The CIFAR-10 dataset is a large-scale dataset with labels. We randomly selected 100 images from the CIFAR-10 dataset and used them together with the 24 images in the Kodak dataset to evaluate the image reconstruction effect. We used the mean square error (MSE) as the loss function to train the dynamic network, where the base network parameter is θ, for generating the reconstructed images. We use peak signal-to-noise ratio (PSNR) and bits per pixel (bpp) as evaluation criteria, where PSNR= $10log_{10}(\frac{1}{MSE})[dB]$, bpp= $\frac{1}{n}(\sum\limits_{j=1}^{n} \frac{P_{\Theta}^{(j)}}{WH})[bpp]$.

4.1   Verify the Need for Dynamic Networks

First, the method's effectiveness was tested on dynamic depth networks. In the experiment, we fixed the network width and selected 60 images with varying complexity, with 20 images for each complexity level. We examined the image reconstruction performance with different network depths. Based on the experimental results in the Appendix, particularly in Sect. C.1, it is evident from Fig. 4(a) that increasing the network depth has a detrimental effect on the reconstruction quality for images with easy and medium complexity, beyond a certain range. According to the concept of signal blending learning [33], easy images have relatively fewer frequency components, and excessive network layers may hinder effective learning, leading to a decrease in image reconstruction quality. Additionally, the optimal network depth for different complexity levels varies, with shallow, moderate, and deep depths required for easy, medium, and difficult images, respectively. This highlights the importance of dynamic depth networks when handling images of different complexity.

Second, the validity of dynamic width networks was verified. Similarly, we fixed the network depth and selected 60 images, with 20 images representing each complexity level. The experimental results (see Fig. 4(b) for experimental results) indicate that as the complexity of the image increases, a greater number of nodes is required for the appearance of corresponding inflection points on the image reconstruction effect map, as presented in Sect. C.1 of the Appendix. This implies that easy images necessitate a narrower network width, while difficult images require a relatively wider network width. Hence, dynamic width networks are essential when dealing with images of different complexity.

From the above analysis, we can conclude that it is practical and valuable to consider the dynamic adjustment of network width and depth simultaneously, which allows us to obtain the best reconstruction of images quickly (Fig. 2).

4.2   Performance on Different Datasets

For the Kodak and CIFAR-10 datasets, we tested them separately. We used 100 randomly selected images from the CIFAR-10 dataset for testing. Observing Fig. 3(a), we can see that when the number of network layers is kept constant and the number of nodes in each layer is incremented, the information carried by each pixel gradually increases, while the difference between the compressed image and the original image decreases with the increase of the PSNR value. This implies that the ability of the network to restore image information gradually increases with the increase of the network width within a certain range. The

Original   Ours $\Delta f(x)$ COIN $\Delta f(x)$

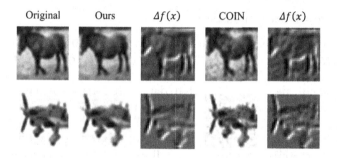

Fig. 2. Qualitative comparison of compression artifacts for models at similar reconstruction quality. Immediately following Ours is the first derivative of Ours image, and immediately following COIN is the first derivative of COIN image. In the image above, COIN achieves 25.9dB at 5.11 bpp while Ours achieves 28.50dB at 4.82 bpp. In the image below, COIN achieves 26.7dB at 5.11 bpp while Ours achieves 28.34dB at 4.82 bpp.

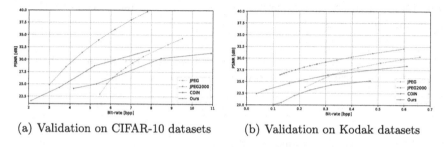

(a) Validation on CIFAR-10 datasets   (b) Validation on Kodak datasets

Fig. 3. Image reconstruction effects of different methods on different datasets. The methods for comparison are JPEG, JPFG2000, COIN and Ours. One hundred images were randomly selected from the CIFAR-10 dataset for testing, and the reconstruction effect was evaluated by PSNR. The whole Kodak data set was selected for testing, and the image reconstruction effect was also evaluated with PSNR.

experimental results show that our method outperforms COIN and JPEG on the CIFAR-10 dataset. Figure 3(a) shows the effect of CIFAR-10 under different compression methods.

We performed a similar test on the Kodak dataset, where we selected all 24 images in the dataset for image reconstruction. The results are shown in Fig. 3(b), and the image reconstruction effect of the Kodak dataset is average. This may be due to the small size of the Kodak dataset and the relatively close image complexity. Some images that may be considered as easy images in the CIFAR-10 dataset are classified as medium or difficult images, which may have some influence on the image reconstruction results.

4.3   Ablation Experiments

In the above experiments conducted on the CIFAR-10 dataset, our method demonstrates strong performance. The dataset's ample collection of images satisfies our requirement for handling multi-complexity images. Ablation experiments were conducted on the CIFAR-10 dataset to further evaluate our method. Two sets of experiments denoted as A and B, were designed for comparison. In Experiment A, the M^l matrix in our method was utilized to select network weights, discarding weights with a value of 0. Conversely, Experiment B randomly discarded weights with the same discard rate as Experiment A. Table 1 displays the image reconstruction results obtained from Experiments A and B. The results indicate that employing the M^l matrix for network weight selection yields significantly superior image reconstruction outcomes compared to randomly discarding network weights. Notably, when using a network structure of 10 layers with a width of 84, the improvement in performance is nearly two-fold. This outcome validates the efficacy of our approach in selecting network weights using the M^l matrix, thereby enhancing the quality of image reconstruction by preserving crucial information.

Table 1. Image reconstruction results table of different weight discard methods. The weight ratio of the method random drop is equal to our mask matrix.

Network structure	Our mask matrix	Random drop
10 layers of width 32	**21.5**	10.47
10 layers of width 48	**25.2**	10.76
10 layers of width 64	**28.7**	10.94
10 layers of width 84	**32.3**	11.08

Secondly, we performed experiments comparing image reconstruction effects for images of varying complexity using networks of the same width but different depths. The detailed operation and results are shown in the Fig. 5(a) and Table 2 in Sect. C.2 of the Appendix.

Finally, we compared the image reconstruction results for images of different complexity with the same network depth and different network widths. Similarly, the corresponding information can be found in Sect. C.3 of the Appendix.

5   Conclusion

Our proposal integrates dynamic neural networks with implicit neural representations. Dynamic width selection is achieved by employing a binarized soft mask, and dynamic depth is proposed in combination with an early retirement mechanism to achieve the goal of efficiency. For multi-image compression, we further apply class element learning and 8-bit quantization to optimize the dynamic neural network. We validate the need for network dynamics and show the results of our method outperforming previous work on the CIFAR-10 dataset. The general performance of the Kodak dataset in our experiments may be related to the overfitting of the underlying model to the learned information. In the future, we will further investigate and address these related issues.

A   Overall Process Pseudocode

As follows, is the pseudo code of the method in this paper.

Algorithm 1: Employ dynamic neural networks to compress images of different complexities with implicit neural representations

---

Data: Set of images

Result: M_ψ and Base model f_θ

1 image set's $G_{mean} \leftarrow$ Sobel operator ;

2 image complexity classification $\leftarrow$ K-means$[G_{mean}]$;

3 initialization Coordinates x and $M_\psi = (Q_{d\times r}, Q^i_{r\times r}, V^T_{d\times r}), i = 1, \ldots, l$
   approximate $r \propto G_{mean}$, $l \propto G_{mean}$;

4 **while** *Training:Taking a complexity level t image y_i set as an example.* **do**

5     $\quad z^0 \leftarrow \gamma(x)$;

6     $\quad$ **for** *Traverse through the layers 1 to l in the base model f_θ* **do**

7     $\quad\quad M^i_{soft} \leftarrow sigmoid(Q_{d\times r} \times Q^i_{r\times r} \times V^T_{d\times r})$

8     $\quad\quad M_i \leftarrow \begin{cases} 1, M^i_{soft} \geq 0.5 \\ 0, M^i_{soft} < 0.5 \end{cases}$

9     $\quad\quad z^i \leftarrow \sin(\omega_0((W^i \odot M^i)z^{i-1} + b^i))$

10    $\quad$ **end**

11    $\quad f_\theta(x, M_\psi, t) \leftarrow W^L_t z^{L-1} + b^L$

12    $\quad L_{mse} \leftarrow ||f_\theta(x, M_\psi, t), y_i||_2$

13    $\quad$ update the parameters using like-meta-learning

14 **end**

15 quantize M_ψ and the base model f_θ

B   Class Meta-Learning M_ψ

Store each image in the dataset explicitly for its M_ψ and evaluate each coordinate point when reconstructing the image. Denote the output y as

$$y = f_\theta(x, M_\psi, t) \qquad (8)$$

Only M_ψ of each image alone needs to be overfitted to each image by the base network. Also, minimize the difference between the reconstruction result of the base dictionary model on the whole dataset and the dataset data.

$$L(x, M_\psi, d) = \sum_{j=1}^{n} ||f_\theta(x_i, M_\psi, t), y_i||_2 \qquad (9)$$

$$\min_{\theta, M_\psi} L(x, M_\psi, d) \tag{10}$$

COIN++ [7], a special MAML is applied to generate well-initialized network parameters. While COIN++ emphasizes the generalization of the model to obtain updated network parameters by several gradient descent, in our experiments, multi-task learning is performed to overfit M_ψ. It is required to meta-learning a θ that over-fits the storage matrix M_ψ at each new data point. Therefore, in the inner loop, M_ψ is learned in the following way:

$$M_\psi^{(j)} = M_\psi^{(j)} - \alpha \nabla_{M_\psi} L\left(\theta, M_\psi, d^{(j)}\right) \tag{11}$$

In the outer loop, the network parameters θ are updated using the errors generated for each data point:

$$\theta \leftarrow \theta - \beta \nabla_\theta \sum_{j=1}^{N} L\left(\theta, M_\psi^{(j)}, d^{(j)}\right) \tag{12}$$

The outer loop base model learns as many dictionary frequencies as possible, while the inner loop performs gradient updates for each image's storage matrix, guiding the generation mask to correctly select the appropriate frequency in the base dictionary to achieve multi-image compression.

C   Experimental Details

C.1   Verify the Need for Dynamic Networks

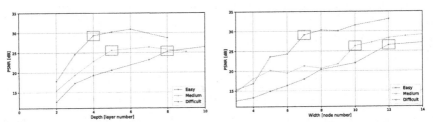

(a) the necessity of dynamic depth of neural network

(b) the necessity of dynamic width of neural network

Fig. 4. Verify the dynamic necessity of neural network. Take twenty pictures of each complexity and change the depth or width to get the image reconstruction effect of each complexity picture in a specific network, which is measured by PSNR.

C.2 Analysis of Experimental Results of Ablation Dynamic Depth

We selected 20 images from each complexity category, maintaining their original width (r) within the category. We then calculated the average image reconstruction effects while increasing the number of network layers.

According to the Figure 5(a), as the complexity of the images gradually increased, we observed that a greater number of network layers were required to achieve a better image reconstruction effect while consuming relatively more resources. For instance, easy images yielded the best effect at 7 layers, medium images at 8 layers, and difficult images required a deeper network with the best effect achieved at 9 layers. This demonstrates that the appropriate network depth varies depending on the complexity of the images. Deeper networks are necessary for more complex images to achieve superior image reconstruction. This further affirms the importance of dynamically adapting the network depth in our approach, allowing for optimal image reconstruction results across images of different complexities.

Table 2. Fixed width and dynamic depth image reconstruction effect table with different complexity

Network Layers	Simple Image	Middle Image	Complex Image
5	34.95	32.64	31.8
6	36.73	33.78	33.45
7	37.8	34.6	34.2
8	38.21	35.8	35.2
9	38.4	36.25	35.9

Our analysis confirms the significance of dynamically adjusting the network depth in our approach, achieving optimal image reconstruction results for images of different complexities.

C.3   Analysis of Experimental Results of Ablation Dynamic Width

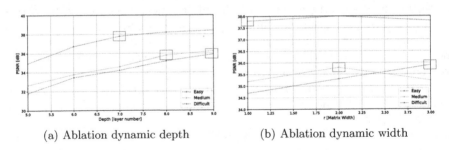

(a) Ablation dynamic depth   (b) Ablation dynamic width

Fig. 5. The results of ablation experiments on dynamic neural networks. During the ablation of dynamic depth, the fixed network width, that is, the fixed R-value, gradually deepens the network depth. Compare the optimal network depth for images of different complexity. When the dynamic width is ablated, the network depth is fixed and the network width is gradually widened, that is, the R-value is increased. Compare the optimal network width for images of different complexity.

For each category, we selected 20 images of each and fixed their optimal depths (verified in previous experiments). Their average image reconstruction effect values were calculated with increasing network width (by adjusting the r value).

According to the Figure 5(b), we can intuitively observe that the best results are achieved with an r value of 2 in medium complexity images and with an r value of 3 in difficult complexity images, both with the same width setting. In the case of simple complexity images, as the r value increases, i.e., the network width becomes wider, the effect of image reconstruction increases slightly and then decreases to the same as the initial one. This is because the simple image requires less frequency composition, while the increase of network width brings more computational consumption, and the optimization effect is not obvious. Therefore the optimal r value for the simple image should be 1, again with the same setting as its width.

Specifically, our set the following r values: r=1 (easy image: 37.8, medium image: 35.2, difficult image: 37.8), r=2 (easy image: 38.03, medium image: 35.88, difficult image: 35.23), r=3 (easy image: 37.8, medium image: 35.3, difficult image: 35.9). From the image reconstruction results with different r values, it can be intuitively seen that the best results are achieved with a r value of 2 in the medium images and with a r value of 3 in the difficult images, which coincides with their width settings. In contrast, in the easy images, as the r value increases (i.e., the network width increases), the image reconstruction effect first slightly improves and then decreases to the same as the initial one. This is because the easy image requires less frequency composition [33], while increasing the network width at the same time adds more computational overhead and is less effective in optimization. Therefore, the optimal r value for simple images should be 1, which is the same as its initial width. And the optimal r values for medium and difficult images are 2 and 3, respectively, and their image reconstruction effects under other r values follow the variation trend.

These results analyzed further validate the necessity of dynamically adapting the network width in our method and being able to adapt to images of different complexity to achieve the best image reconstruction results.

References

1. Anokhin, I., Demochkin, K., Khakhulin, T., Sterkin, G., Lempitsky, V., Korzhenkov, D.: Image generators with conditionally-independent pixel synthesis. In: Proceedings of the IEEE/CVF Conference on Computer Vision and Pattern Recognition, pp. 14278–14287 (2021)
2. Bisoi, R., Dash, P.K.: A hybrid evolutionary dynamic neural network for stock market trend analysis and prediction using unscented kalman filter. Appl. Soft Comput. **19**, 41–56 (2014)
3. Bolukbasi, T., Wang, J., Dekel, O., Saligrama, V.: Adaptive neural networks for efficient inference. In: International Conference on Machine Learning, pp. 527–536. PMLR (2017)
4. Chen, Y., Liu, S., Wang, X.: Learning continuous image representation with local implicit image function. In: Proceedings of the IEEE/CVF Conference on Computer Vision and Pattern Recognition, pp. 8628–8638 (2021)
5. Chen, Z., Zhang, H.: Learning implicit fields for generative shape modeling. In: Proceedings of the IEEE/CVF Conference on Computer Vision and Pattern Recognition, pp. 5939–5948 (2019)
6. Dupont, E., Goliński, A., Alizadeh, M., Teh, Y.W., Doucet, A.: Coin: compression with implicit neural representations. arXiv preprint arXiv:2103.03123 (2021)
7. Dupont, E., Loya, H., Alizadeh, M., Goliński, A., Teh, Y.W., Doucet, A.: Coin++: neural compression across modalities. arXiv preprint arXiv:2201.12904 (2022)
8. Dupont, E., Teh, Y.W., Doucet, A.: Generative models as distributions of functions. arXiv preprint arXiv:2102.04776 (2021)
9. Finn, C., Abbeel, P., Levine, S.: Model-agnostic meta-learning for fast adaptation of deep networks. In: International Conference on Machine Learning, pp. 1126–1135. PMLR (2017)
10. Franzen, R.: Kodak lossless true color image suite (1999). http://r0k.us/graphics/kodak/
11. Han, Y., Huang, G., Song, S., Yang, L., Wang, H., Wang, Y.: Dynamic neural networks: a survey. IEEE Trans. Pattern Anal. Mach. Intell. **44**(11), 7436–7456 (2021)
12. Huang, G., Chen, D., Li, T., Wu, F., Van Der Maaten, L., Weinberger, K.Q.: Multi-scale dense networks for resource efficient image classification. arXiv preprint arXiv:1703.09844 (2017)
13. Krizhevsky, A., Hinton, G., et al.: Learning multiple layers of features from tiny images (2009)
14. Li, H., Zhang, H., Qi, X., Yang, R., Huang, G.: Improved techniques for training adaptive deep networks. In: Proceedings of the IEEE/CVF International Conference on Computer Vision, pp. 1891–1900 (2019)
15. Liu, Z., Karam, L.J., Watson, A.B.: JPEG 2000 encoding with perceptual distortion control. IEEE Trans. Image Process. **15**(7), 1763–1778 (2006)

16. Mescheder, L., Oechsle, M., Niemeyer, M., Nowozin, S., Geiger, A.: Occupancy networks: learning 3D reconstruction in function space. In: Proceedings of the IEEE/CVF Conference on Computer Vision and Pattern Recognition, pp. 4460–4470 (2019)

17. Mildenhall, B., Srinivasan, P.P., Tancik, M., Barron, J.T., Ramamoorthi, R., Ng, R.: NeRF: representing scenes as neural radiance fields for view synthesis. Commun. ACM 65(1), 99–106 (2021)

18. Murata, A., Gallese, V., Luppino, G., Kaseda, M., Sakata, H.: Selectivity for the shape, size, and orientation of objects for grasping in neurons of monkey parietal area AIP. J. Neurophysiol. 83(5), 2580–2601 (2000)

19. Park, J.J., Florence, P., Straub, J., Newcombe, R., Lovegrove, S.: Deepsdf: Learning continuous signed distance functions for shape representation. In: Proceedings of the IEEE/CVF Conference on Computer Vision and Pattern Recognition, pp. 165–174 (2019)

20. Pathak, B., Barooah, D.: Texture analysis based on the gray-level co-occurrence matrix considering possible orientations. Int. J. Adv. Res. Electr. Electron. Instrum. Eng. 2(9), 4206–4212 (2013)

21. de Queiroz, R.L.: Processing JPEG-compressed images and documents. IEEE Trans. Image Process. 7(12), 1661–1672 (1998)

22. Shaham, T.R., Gharbi, M., Zhang, R., Shechtman, E., Michaeli, T.: Spatially-adaptive pixelwise networks for fast image translation. In: Proceedings of the IEEE/CVF Conference on Computer Vision and Pattern Recognition, pp. 14882–14891 (2021)

23. Skorokhodov, I., Ignatyev, S., Elhoseiny, M.: Adversarial generation of continuous images. In: Proceedings of the IEEE/CVF Conference on Computer Vision and Pattern Recognition, pp. 10753–10764 (2021)

24. Strümpler, Y., Postels, J., Yang, R., Gool, L.V., Tombari, F.: Implicit neural representations for image compression. In: Avidan, S., Brostow, G., Cissé, M., Farinella, G.M., Hassner, T. (eds.) Computer Vision – ECCV 2022. ECCV 2022. Lecture Notes in Computer Science, vol. 13686, pp. 74–91. Springer, Cham (2022). https://doi.org/10.1007/978-3-031-19809-0_5

25. Tan, Z., Chen, J., Kang, Q., Zhou, M., Abusorrah, A., Sedraoui, K.: Dynamic embedding projection-gated convolutional neural networks for text classification. IEEE Trans. Neural Networks Learn. Syst. 33(3), 973–982 (2021)

26. Tancik, M., et al.: Fourier features let networks learn high frequency functions in low dimensional domains. Adv. Neural. Inf. Process. Syst. 33, 7537–7547 (2020)

27. Veit, A., Belongie, S.: Convolutional networks with adaptive inference graphs. In: Proceedings of the European Conference on Computer Vision (ECCV), pp. 3–18 (2018)

28. Wang, X., Yu, F., Dou, Z.Y., Darrell, T., Gonzalez, J.E.: SkipNet: learning dynamic routing in convolutional networks. In: Proceedings of the European Conference on Computer Vision (ECCV), pp. 409–424 (2018)

29. Xu, X., Wang, Z., Shi, H.: UltraSR: spatial encoding is a missing key for implicit image function-based arbitrary-scale super-resolution. arXiv preprint arXiv:2103.12716 (2021)

30. Yang, B., Bender, G., Le, Q.V., Ngiam, J.: CondConv: conditionally parameterized convolutions for efficient inference. In: Advances in Neural Information Processing Systems, vol. 32 (2019)

31. Yang, L., Han, Y., Chen, X., Song, S., Dai, J., Huang, G.: Resolution adaptive networks for efficient inference. In: Proceedings of the IEEE/CVF Conference on Computer Vision and Pattern Recognition, pp. 2369–2378 (2020)

32. Yu, H., Winkler, S.: Image complexity and spatial information. In: 2013 Fifth International Workshop on Quality of Multimedia Experience (QoMEX), pp. 12–17. IEEE (2013)

33. Yüce, G., Ortiz-Jiménez, G., Besbinar, B., Frossard, P.: A structured dictionary perspective on implicit neural representations. In: Proceedings of the IEEE/CVF Conference on Computer Vision and Pattern Recognition, pp. 19228–19238 (2022)

34. Zemliachenko, A., Kozhemiakin, R., Vozel, B., Lukin, V.: Prediction of compression ratio in lossy compression of noisy images. In: 2016 13th International Conference on Modern Problems of Radio Engineering, Telecommunications and Computer Science (TCSET), pp. 693–697. IEEE (2016)

Misalignment Insensitive Perceptual Metric for Full Reference Image Quality Assessment

Shunyu Yao, Yue Cao, Yabo Zhang, and Wangmeng Zuo[✉]

School of Computer Science and Technology, Harbin Institute of Technology,
Harbin, China
wmzuo@hit.edu.cn

Abstract. Full-reference (FR) image quality assessment (IQA) is crucial in the evaluation of restored images by comparing them with pristine-quality reference images, offering invaluable insights into the effectiveness of image restoration algorithms. Recently, with the advancement of generative adversarial networks (GANs), the GAN-based restoration algorithms demonstrate excellent restoration capability. Nevertheless, these algorithms introduce local spatial misalignment between the restored and the original reference images, posing a challenge for FR-IQA. To tackle this issue, we present a Misalignment Insensitive Perceptual Metric (MIPM) that strengthens the three components of FR-IQA, namely feature extraction, difference representation and quality regression. Specifically, a Vision Transformer-based network for global feature extraction is employed. Furthermore, MIPM utilizes Local Overlapping Wasserstein Difference (LOWD) and Channel Attention Block (CAB) to provide more accurate difference representation between the features of reference and distorted images in the spatial and channel dimensions, respectively. Lastly, a hybrid loss aids in regressing scores that align better with human subjective perception. Coupled with three key improvements, our MIPM exhibits superior performance over state-of-the-art approaches on five IQA datasets, LIVE, CSIQ, TID2013, KADID-10k, and PIPAL.

Keywords: Image quality assessment · Misalignment insensitive · Wasserstein distance

1 Introduction

Image quality assessment (IQA) is a crucial field of study that aims to objectively evaluate the perceptual quality of images. Within IQA, full-reference image quality assessment (FR-IQA) focuses on evaluating image quality by comparing images to pristine reference images, holding immense significance across various image processing domains, including image restoration, compression, and

---

Supplementary Information The online version contains supplementary material available at https://doi.org/10.1007/978-981-99-8552-4_35.

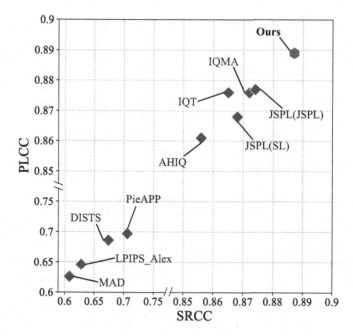

Fig. 1. SRCC v.s PLCC results of FR-IQA methods on validation of NTIRE 2021 IQA challenge.

enhancement. Over time, FR-IQA has undergone rapid development from traditional metrics such as Peak Signal-to-Noise Ratio (PSNR) and Structural Similarity Index (SSIM) [31] to deep learning-based methods, *e.g.*, Learned Perceptual Image Patch Similarity (LPIPS) [38], providing evaluation criteria and perceptual loss function for image restoration algorithms.

In recent years, GAN-based image restoration algorithms have demonstrated remarkable capabilities in restoring images with superior visual quality. However, these techniques often introduce spatial misalignment and pose a great challenge for FR-IQA. Therefore, the evaluation of GAN-based restoration algorithms requires specialized FR-IQA methodologies that can cope with the discrepancies arising from spatial misalignment.

To this end, we rethink the three key components of FR-IQA: feature extraction, difference representation, and quality regression. (i) Feature extraction involves extracting deep features from the reference and distorted images. Most existing learning-based methods utilize the CNNs that heavily rely on local receptive fields, which are susceptible to local spatial misalignment. (ii) Difference representation aims to reflect the feature differences between the distorted and the reference features, while current measurements are mostly based on pixel-wise comparisons, which exaggerate differences due to local misalignment, requiring more suitable difference representation methods. (iii) Quality regression maps the predicted scores to the manually annotated scores, *i.e.*, mean opinion scores (MOS). Existing methods employ mean squared error (MSE) loss, which may struggle to model the nonlinear relationship between visual features and quality scores.

To alleviate the aforementioned concerns, we propose a Misalignment Insensitive Perceptual Metric (MIPM) for FR-IQA. For feature extraction module, we suggest using vision transformers (ViT) [7] rather than convolution networks to obtain global features [35], as the latter are more insensitive to misalignment introduced by GAN-based distortion. Additionally, to enhance robustness to misalignment, we utilize Local Overlapping Wasserstein Difference (LOWD) and Channel Attention Block (CAB) to provide more precise difference representations in spatial and channel dimensions [39], respectively. Finally, a hybrid loss approach combined with mean square error (MSE) loss and Norm-in-Norm (NiN) loss [17] is utilized to achieve nonlinear modeling and promote alignment with human perceptual quality. Experimental evaluations and comparisons with existing methods demonstrate the effectiveness in predicting perceptual image quality accurately and the adaptability to local spatial misalignment of our approach. Figure 1 shows the outstanding performance of our method on PIPAL [10].

Overall, the main innovations of our approach can be summarized as follows:

- Suggesting using ViT-based network architecture to capture global spatial characteristics in feature extraction module of FR-IQA.
- Introducing LOWD and CAB for precise difference representations in spatial and channel dimensions respectively, addressing misalignment issues.
- Employing a combination of MSE loss and NiN loss as a hybrid loss, enabling nonlinear modeling and improved alignment with visual quality.

2   Related Work

FR-IQA networks typically consist of three primary components: feature extraction, difference representation, and quality score regression. Learning-based FR-IQA often leverage pre-trained backbones to extract deep features and design effective modules for difference representation and quality score prediction. To enhance performance in learning-based FR-IQA, several network architectures have been proposed, namely DeepSim [8], LPIPS [38], IQT [4], and AHIQ [15]. These architectures employed pre-trained VGG [28] and AlexNet [14], ResNet [13], and Vision Transformer (ViT) [7] to obtain powerful feature representations, respectively. Regarding difference representation and quality regression, recent research has designed novel metrics to accurately and reliably measure feature difference maps. Unlike the calculation of Euclidean distance at the feature level, DISTS [6] combined correlations of spatial averages with correlations of the feature maps. DeepWSD [18] utilized the Wasserstein distance in the deep feature domain to measure quality degradation from a statistical distribution perspective. Moreover, considering humans' tendency to focus on content-rich image patches, WaDIQaM [2] introduced a weighted average mechanism to recalibrate the difference map, widely employed in current IQA networks [15,34]. In summary, the network structure of FR-IQA still remains worthy of further research for the improvement of IQA performance.

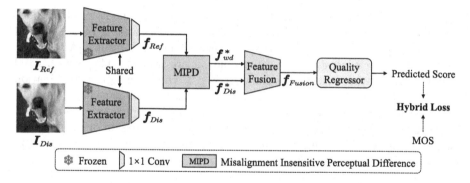

Fig. 2. Illustration of our FR-IQA network. Our network employs a dual feature extraction structure for distorted and reference images, precisely computes perceptual differences via Misalignment Insensitive Perceptual Difference, and generates the final quality score using a quality regressor with spatial attention.

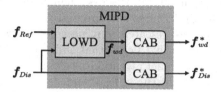

Fig. 3. The proposed Misalignment Insensitive Perceptual Difference (MIPD) for difference representation.

Furthermore, the training losses of IQA networks are also currently hot research topics. Recent studies have enhanced the consistency between predictions of learning-based IQA models and human subjective perception by introducing improvements to optimization objectives, including the adoption of MSE or L1 loss as loss functions, as well as the Norm-in-Norm (NiN) loss [17]. TReS [9] adopted the relative ranking loss to enforce relative ranking among samples, achieving performance enhancement. In this paper, we propose an effective approach to enhance the loss functions in FR-IQA.

3   Method

In the following sections, we propose the Misalignment Insensitive Perceptual Metric (MIPM) for the feature extraction module, difference representation module, quality regression module, and loss function in FR-IQA, aiming to address the issue of local misalignment caused by GAN-based restoration algorithms.

3.1   Feature Extraction Module

CNN-based IQA models extract local spatial features using convolutional layers, whereas ViT-based IQA models capture global context information through self-attention mechanisms. Therefore, ViT-based IQA methods excel at modeling long-range dependencies, enabling a better understanding of GAN-based distorted image, and leading to enhanced performance in certain IQA tasks.

Based on the analysis above, we employ a pre-trained ViT [7] with fixed weights as the feature extractor as shown in Fig. 2, where the output combination of specific ViT layers serves as the raw features for distorted image (I_{Dis}) and the corresponding reference (I_{Ref}) images. To enable further processing, a shared 1×1 convolutional layer is applied to reduce the channel dimension of the reshaped feature maps from ViT output sequences to a suitable dimension d, resulting in processed feature maps denoted f_{Ref} and f_{Dis}.

3.2   Difference Representation Module

By utilizing Wasserstein distance as a loss function, the model can better capture the underlying distributional properties, leading to improved performance in misalignment applications [5,40]. Besides, the channel attention mechanism enhances informative image features and suppresses less relevant ones [39].

Inspired by the previous research, our designed difference representation module, Misalignment Insensitive Perceptual Difference (MIPD) combines the Wasserstein distance and channel attention mechanism to effectively represent the differences between the features f_{Ref} and f_{Dis} obtained by the feature extraction module, enabling a more accurate measurement of the differences that impact visual perception, as depicted in Fig. 3. And a detailed description of the proposed Local Overlapping Wasserstein Difference (LOWD) and Channel Attention Block (CAB) is presented in this section.

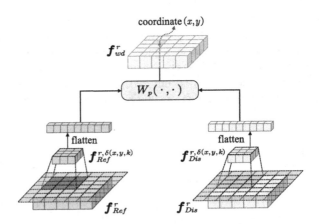

Fig. 4. Illustration of our proposed Local Overlapping Wasserstein Difference which calculates 1-D Wasserstein distance between corresponding patches of f_{Ref} and f_{Dis} for each channel.

Local Overlapping Wasserstein Difference. The Wasserstein distance is a measure of probability distributions. Given two multidimensional random variables X and Y, with corresponding distributions $X \sim P$ and $Y \sim Q$, $\mathcal{J}(P,Q)$ represents joint distributions J for (X,Y) with marginals P and Q. Then the l-Wasserstein distance between X and Y is defined as:

$$W_l(X,Y) = \left(\inf_{J \in \mathcal{J}(P,Q)} \int \|x - y\|^l dJ(x,y) \right)^{1/l}, \tag{1}$$

where l is the order of the l-norm and x and y represent the densities of X and Y respectively. In the case of one-dimensional random variables X and Y, we let F_X denotes the cumulative distribution function of X, and $F_X^{-1}(q) = \inf\{x : F_X(x) \geq q\}, q \in (0,1)$ represents its corresponding inverse function. Then the Wasserstein distance can be conveniently calculated using the following formula:

$$W_l(X,Y) = \left(\int_0^1 |F_X^{-1}(\alpha) - F_Y^{-1}(\alpha)|^l d\alpha \right)^{1/l}. \tag{2}$$

In this paper, we employ the 1-D Wasserstein distance as the metric and utilize a local and overlapping method to calculate patch-wise Wasserstein distance, as shown in Fig. 4. Given $\boldsymbol{f}_{Ref}$ and $\boldsymbol{f}_{Dis}$, LOWD computes the feature distance for each channel, resulting in $\boldsymbol{f}_{wd}$ with the same dimensions as the input features. A $k \times k$ shift window is slid over each channel of the feature maps using the stride length *stride*. The resulting flattened 1-D vectors of length k^2, $\boldsymbol{f}_{Ref}^{r,\delta(i,j,k)}$ and $\boldsymbol{f}_{Dis}^{r,\delta(i,j,k)}$, are fed to $W_l(\cdot,\cdot)$ to compute the value of $\boldsymbol{f}_{wd}^r$ at coordinate (i,j). Zero-padding with the padding size *padding* is performed on $\boldsymbol{f}_{Ref}$ and $\boldsymbol{f}_{Dis}$ to maintain input and output resolution size. Following vanilla 3×3 convolutional layer, we set k, *padding*, and *stride* to 3, 1, and 1, respectively, ensuring overlap and locality in feature distance measurement.

Channel Attention Block. To address interdependencies among feature channels, Channel Attention Blocks are utilized to adjust attention weights across different channel dimensions and prioritize more informative features prior to fusing $\boldsymbol{f}_{wd}$ and $\boldsymbol{f}_{Dis}$. Within our framework, we assign distinct channel-wise weights to $\boldsymbol{f}_{wd}$ and $\boldsymbol{f}_{Dis}$, resulting in two separate Channel Attention Blocks for each branch, thereby generating $\boldsymbol{f}_{wd}^*$ and $\boldsymbol{f}_{Dis}^*$ for subsequent quality regression.

3.3 Quality Regression Module

The quality regression module maps the difference representation to the quality scores that closely align with human subjective perception.

We first concatenate $\boldsymbol{f}_{wd}^*$ and $\boldsymbol{f}_{Dis}^*$ and pass them through a fusion module consisting of three stacked 3×3 Conv layers with ReLU activation functions, yielding the feature maps $\boldsymbol{f}_{Fusion}$. Additionally, inspired by the structure of WaDIQaM [2] and AHIQ [15], we introduce a weighted average patch aggregation module with a spatial attention mechanism to focus on image patches containing more content information. Specifically, we utilize a 1×1 Conv layer ϕ_s to estimate the quality score for each pixel and employ another 1×1 Conv layer ϕ_w followed by a sigmoid function σ to compute the weights associated with each quality prediction. The predicted scores can be calculated as:

$$\hat{q} = \frac{\sum_{i,j} s(i,j) \cdot \omega(i,j)}{\sum_{i,j} \omega(i,j)}, \tag{3}$$

where $s = \phi_s(\boldsymbol{f}_{Fusion}) \in \mathbb{R}^{1 \times h \times w}$, $\omega = \sigma(\phi_w(\boldsymbol{f}_{Fusion})) \in \mathbb{R}^{1 \times h \times w}$, and $\hat{q} \in \mathbb{R}^1$.

3.4 Hybrid Loss Function

In our approach, we employ the mean squared error (MSE) loss to constrain the predicted numerical values, which is defined as:

$$\mathcal{L}_{MSE} = \frac{1}{B} \sum_{i=1}^{B} \| \hat{q}_i - q_i \|^2, \tag{4}$$

where B denotes the number of samples in a batch, $\hat{q}_i$ and q_i indicate the ith predicted quality score and the corresponding MOS, respectively. While the MSE loss improves our algorithm's numerical performance in predicting image quality, alignment between predictions and human subjective perception remains a challenge. To address this issue, we employ Norm-in-Norm (NiN) loss [17]:

$$\mathcal{L}_{NiN} = \frac{1}{2^m B^{1-\frac{m}{n}}} \sum_{i=1}^{B} \| \frac{\hat{q}_i - a_{\hat{q}}}{b_{\hat{q}}} - \frac{q_i - a_q}{b_q} \|^m, \tag{5}$$

where a and b represent the mean and standard deviation respectively. When m and n are all set to 2, NiN loss [17] is equivalent to PLCC-induced loss [17]. As a result, we believe that NiN loss [17] can enhance the correlation between the predicted scores and ground truth MOS to a certain degree. Hence, our method minimizes the loss functions mentioned earlier and the total loss is defined as:

$$\mathcal{L}_{total} = \mathcal{L}_{MSE} + \mathcal{L}_{NiN}. \tag{6}$$

4   Experiments

4.1   Experiment Settings

Datasets. We utilize five widely used IQA datasets, namely LIVE [25], CSIQ [16], TID2013 [22], KADID-10k [19], and PIPAL [10]. PIPAL [10] is a large-scale IQA dataset that includes 250 reference images, 200 of which are used for training. Notably, PIPAL [10] contains restored results from various restoration algorithms, including PSNR-oriented and GAN-based algorithms, in addition to common distortion types. The validation set, used for testing, is partitioned according to the NTIRE 2021 IQA Challenge [11], consisting of 25 reference images and 1000 distorted images. Most of the distorted images are generated by GAN-based image restoration or compression algorithms, which were not seen in the training set. Additional details about datasets can be found in the supplementary material.

Evaluation Criterias. To evaluate the performances of IQA algorithms, we employ two widely used metrics, *i.e.*, Spearman's rank order correlation coefficient (SRCC) and Pearson Linear Correlation Coefficients (PLCC) to measure the prediction monotonicity and the accuracy of the IQA methods, respectively.

Implementation Details. We utilize a ViT-B/16 model pre-trained on ImageNet [24] as the feature extractor, and concatenate the features from its first five layers as the output of this module. The intermediate embedding feature has a dimension of 256, and each CAB has a reduction ratio of 8. Additionally, the hyperparameters m and n in the NiN loss [17], as well as the order of the l-norm l, are empirically set to 1, 2, and 2, respectively.

To ensure a fair comparison with AHIQ [15], we employed same training and testing configurations. For training, we randomly crop the original images into patches of size 224×224 and apply horizontal flip for data augmentation during training. And for testing, we employ uniformly distributed five-point cropping to obtain patch pairs of size 224×224 for each image pair, and then average the quality scores to obtain the final result. More details about implementation information can be found in the supplementary material.

Table 1. Method comparisons on PIPAL-Valid2021. Part of the performances of other methods are borrowed from the NTIRE 2021 IQA challenge report [11].

Type	Methods	SRCC	PLCC	Type	Methods	SRCC	PLCC
Trad. NR	NIQE [21]	0.064	0.102	DL. FR	PieAPP [23]	0.706	0.697
DL. NR	MA [20]	0.201	0.203		LPIPS-Alex [38]	0.628	0.646
	PI [1]	0.169	0.166		DeepWSD [18]	0.397	0.379
Trad. FR	PSNR	0.255	0.292		DISTS [6]	0.674	0.686
	SSIM [31]	0.340	0.398		SWDN [10]	0.661	0.668
	MS-SSIM [32]	0.486	0.563		IQT [4]	0.865	0.876
	UQI [30]	0.548	0.486		IQMA [12]	0.872	0.876
	FSIMc [37]	0.468	0.559		AHIQ [15]	0.856	0.861
	MAD [16]	0.608	0.626		JSPL (SL) [3]	0.868	0.868
	VIF [26]	0.433	0.524		JSPL (JSPL)[1] [3]	0.874	0.877
	IFC [27]	0.594	0.677		Ours	**0.887**	**0.889**

[1] JSPL (JSPL) adopts additional data for training.

4.2 Comparison with State-of-the-Art

Our algorithm is evaluated under two experimental settings: PIPAL [10] evaluation and the cross-dataset evaluation. In PIPAL [10] evaluation, we train the model on PIPAL's training set and test it on the corresponding validation set to assess our method's ability to address local misalignment. In the cross-dataset evaluation, the model is trained on the entire KADID-10k [19] dataset and tested on LIVE [25], CSIQ [16], and TID2013 [22] to evaluate the generalization of our IQA method.

Evaluation on PIPAL Dataset. The comparison results in Table 1 demonstrate the inferior performance of traditional metrics compared to learning-based IQA methods in PIPAL [10] due to the locally misaligned distortion. Addressing the spatial misalignment introduced by GAN-based distortion, SWDN [10] proposed the Space Warping Difference layer to increase robustness but were limited by their feature extractor's capabilities, hindering further improvements. In contrast, JSPL [3] achieved competitive performance by incorporating semi-supervised learning with a specially designed network. To leverage the Vision Transformer's [7] ability to model long-range dependencies, we adopt ViT [7] as the feature extractor to improve representation and mitigate spatial misalignment caused by GAN-based distortion. Our designed difference representation module, consisting of LOWD and CAB, captures feature distributional properties and enhances informative image features, enabling misalignment insensitive feature difference representation.

Cross-Dataset Performance Evaluation. Comparing our proposed model with various IQA algorithms from different categories in Table 2, we observe that while MAD [16] achieves the highest PLCC on LIVE [25], its performance significantly declines when facing more complex datasets like TID2013 [22]. By employing ViT [7] for global feature extraction and NiN loss [17] for nonlinear modeling and prediction-perception alignment enhancement, our method demonstrates superior performance compared to other algorithms, highlighting its strong generalization ability.

4.3 Ablation Study

Ablation studies are conducted to assess the effectiveness of network structure, with supplementary material containing additional ablation studies.

Network Structure. By conducting experiments detailed in Table 3, we evaluate the individual contributions of Local Overlapping Wasserstein Difference (LOWD) and Channel Attention Block (CAB). The baseline method is represented in the first row. The analysis from Table 3 are as follows: (i) Incorporating LOWD into the baseline method significantly improves performance across all datasets, indicating that LOWD effectively captures more accurate differences.

Table 2. Method comparisons on LIVE [25], CSIQ [16], TID2013 [22] databases, while training on the KADID-10k [19], Trad. and DL. represent traditional methods and models based on deep learning respectively.

Type	Methods	LIVE SRCC	LIVE PLCC	CSIQ SRCC	CSIQ PLCC	TID2013 SRCC	TID2013 PLCC
Trad. NR	NIQE [21]	0.914	0.915	–	–	0.317	0.426
	M3 [33]	0.951	0.950	0.795	0.839	0.689	0.771
DL. NR	WaDIQaM-NR [2]	0.855	0.855	0.716	0.750	0.585	0.610
	HyperIQA [29]	0.908	0.903	0.802	0.858	0.686	0.721
Trad. FR	PSNR	0.873	0.865	0.810	0.819	0.687	0.677
	SSIM [31]	0.948	0.937	0.865	0.852	0.727	0.777
	MS-SSIM [32]	0.951	0.940	0.906	0.889	0.786	0.830
	VSI [36]	0.952	0.948	0.942	0.928	0.897	0.900
	MAD [16]	0.967	**0.968**	0.947	0.950	0.781	0.827
DL. FR	LPIPS [38]	0.932	0.934	0.876	0.896	0.670	0.749
	WaDIQaM-FR [2]	0.947	0.940	0.909	0.901	0.831	0.834
	DISTS [6]	0.954	0.954	0.929	0.928	0.830	0.855
	DeepWSD [18]	0.947	0.944	0.962	0.944	0.879	0.878
	IQT [4]	0.970	–	0.943	–	0.899	–
	AHIQ [15]	0.970	0.952	0.951	0.955	0.901	0.899
	JSPL (SL) [3]	**0.973**	–	0.951	–	0.908	–
	Ours	**0.973**	0.957	**0.967**	**0.965**	**0.911**	**0.913**

Table 3. SRCC/PLCC performance with ablation studies about network structure performed on cross-dataset setting and PIPAL [10].

LOWD	CAB_{wd}	CAB_{Dis}	LIVE SRCC	LIVE PLCC	CSIQ SRCC	CSIQ PLCC	TID2013 SRCC	TID2013 PLCC	PIPAL SRCC	PIPAL PLCC
✗	✗	✗	0.968	0.952	0.956	0.956	0.890	0.895	0.862	0.868
✓	✗	✗	0.968	**0.955**	0.957	0.960	0.898	0.902	0.872	0.872
✗	✓	✓	0.961	0.947	0.949	0.952	0.897	0.899	0.864	0.872
✓	✓	✗	**0.971**	0.953	**0.964**	0.962	0.905	0.905	**0.878**	0.876
✓	✓	✓	**0.971**	0.953	**0.964**	**0.963**	0.906	0.906	0.876	**0.877**

(ii) The best performance is achieved when employing the complete version of MIPD, highlighting the importance of simultaneously considering the accurate difference representation in the spatial and channel dimension for predicting quality scores aligned with human perception.

5   Conclusion

In this paper, we propose MIPM, a misalignment insensitive perceptual metric for FR-IQA. By leveraging a pre-trained Vision Transformer, robust feature representations are extracted from distorted and pristine images. We introduce Misalignment Insensitive Perceptual Difference, consisting of Local Overlapping Wasserstein Difference and Channel Attention Block, to effectively represent feature differences despite spatial misalignment. Additionally, the combination of MSE loss and NiN loss enhances the correlation between predicted quality scores and human subjective perception. Extensive experiments demonstrate that MIPM not only exhibits superior generalization ability but also outperforms state-of-the-art methods in FR-IQA.

References

1. Blau, Y., Mechrez, R., Timofte, R., Michaeli, T., Zelnik-Manor, L.: The 2018 PIRM challenge on perceptual image super-resolution. In: Leal-Taixé, L., Roth, S. (eds.) ECCV 2018. LNCS, vol. 11133, pp. 334–355. Springer, Cham (2019). https://doi.org/10.1007/978-3-030-11021-5_21
2. Bosse, S., Maniry, D., Müller, K.R., Wiegand, T., Samek, W.: Deep neural networks for no-reference and full-reference image quality assessment. IEEE Trans. Image Process. **27**(1), 206–219 (2017)
3. Cao, Y., Wan, Z., Ren, D., Yan, Z., Zuo, W.: Incorporating semi-supervised and positive-unlabeled learning for boosting full reference image quality assessment. In: IEEE Conference on Computer Vision and Pattern Recognition, pp. 5851–5861 (2022)
4. Cheon, M., Yoon, S.J., Kang, B., Lee, J.: Perceptual image quality assessment with transformers. In: IEEE Conference on Computer Vision and Pattern Recognition Workshops, pp. 433–442 (2021)
5. Delbracio, M., Talebi, H., Milanfar, P.: Projected distribution loss for image enhancement. arXiv preprint arXiv:2012.09289 (2020)
6. Ding, K., Ma, K., Wang, S., Simoncelli, E.: Image quality assessment: unifying structure and texture similarity. IEEE Trans. Pattern Anal. Mach. Intell. (2020)
7. Dosovitskiy, A., et al.: An image is worth 16x16 words: transformers for image recognition at scale. In: International Conference on Learning Representations (2020)
8. Gao, F., Wang, Y., Li, P., Tan, M., Yu, J., Zhu, Y.: DeepSim: deep similarity for image quality assessment. Neurocomputing **257**, 104–114 (2017)
9. Golestaneh, S.A., Dadsetan, S., Kitani, K.M.: No-reference image quality assessment via transformers, relative ranking, and self-consistency. In: IEEE Winter Conference on Applications of Computer Vision, pp. 1220–1230 (2022)
10. Jinjin, G., Haoming, C., Haoyu, C., Xiaoxing, Y., Ren, J.S., Chao, D.: PIPAL: a large-scale image quality assessment dataset for perceptual image restoration. In: Vedaldi, A., Bischof, H., Brox, T., Frahm, J.-M. (eds.) ECCV 2020. LNCS, vol. 12356, pp. 633–651. Springer, Cham (2020). https://doi.org/10.1007/978-3-030-58621-8_37
11. Gu, J., et al.: NTIRE 2021 challenge on perceptual image quality assessment. In: IEEE Conference on Computer Vision and Pattern Recognition Workshops, pp. 677–690 (2021)

12. Guo, H., Bin, Y., Hou, Y., Zhang, Q., Luo, H.: IQMA network: image quality multi-scale assessment network. In: IEEE Conference on Computer Vision and Pattern Recognition Workshops, pp. 443–452 (2021)

13. He, K., Zhang, X., Ren, S., Sun, J.: Deep residual learning for image recognition. In: IEEE Conference on Computer Vision and Pattern Recognition, pp. 770–778 (2016)

14. Krizhevsky, A., Sutskever, I., Hinton, G.E.: Imagenet classification with deep convolutional neural networks. Commun. ACM **60**(6), 84–90 (2017)

15. Lao, S., et al.: Attentions help CNNs see better: attention-based hybrid image quality assessment network. In: IEEE Conference on Computer Vision and Pattern Recognition Workshops, pp. 1140–1149 (2022)

16. Larson, E., Chandler, D.: Most apparent distortion: full-reference image quality assessment and the role of strategy. J. Electron. Imaging **19**(1), 011006 (2010)

17. Li, D., Jiang, T., Jiang, M.: Norm-in-norm loss with faster convergence and better performance for image quality assessment. In: ACM International Conference on Multimedia, pp. 789–797 (2020)

18. Liao, X., Chen, B., Zhu, H., Wang, S., Zhou, M., Kwong, S.: DeepWSD: projecting degradations in perceptual space to Wasserstein distance in deep feature space. In: ACM International Conference on Multimedia, pp. 970–978 (2022)

19. Lin, H., Hosu, V., Saupe, D.: KADID-10k: a large-scale artificially distorted IQA database. In: IEEE International Conference on Quality of Multimedia Experience, pp. 1–3. IEEE (2019)

20. Ma, C., Yang, C.Y., Yang, X., Yang, M.H.: Learning a no-reference quality metric for single-image super-resolution. Comput. Vis. Image Underst. **158**, 1–16 (2017)

21. Mittal, A., Soundararajan, R., Bovik, A.: Making a "completely blind" image quality analyzer. IEEE Sig. Process. Lett. **20**(3), 209–212 (2012)

22. Ponomarenko, N., et al.: Image database TID2013: peculiarities, results and perspectives. Sig. Process. Image Commun. **30**, 57–77 (2015)

23. Prashnani, E., Cai, H., Mostofi, Y., Sen, P.: PieAPP: perceptual image-error assessment through pairwise preference. In: IEEE Conference on Computer Vision and Pattern Recognition, pp. 1808–1817 (2018)

24. Russakovsky, O., et al.: Imagenet large scale visual recognition challenge. Int. J. Comput. Vision **115**(3), 211–252 (2015)

25. Sheikh, H.: Image and video quality assessment research at live (2003). http://live.ece.utexas.edu/research/quality

26. Sheikh, H., Bovik, A.: Image information and visual quality. IEEE Trans. Image Process. **15**(2), 430–444 (2006)

27. Sheikh, H., Bovik, A., De Veciana, G.: An information fidelity criterion for image quality assessment using natural scene statistics. IEEE Trans. Image Process. **14**(12), 2117–2128 (2005)

28. Simonyan, K., Zisserman, A.: Very deep convolutional networks for large-scale image recognition. arXiv preprint arXiv:1409.1556 (2014)

29. Su, S., et al.: Blindly assess image quality in the wild guided by a self-adaptive hyper network. In: IEEE Conference on Computer Vision and Pattern Recognition, pp. 3667–3676 (2020)

30. Wang, Z., Bovik, A.: A universal image quality index. IEEE Sig. Process. Lett. **9**(3), 81–84 (2002)

31. Wang, Z., Bovik, A., Sheikh, H., Simoncelli, E.: Image quality assessment: from error visibility to structural similarity. IEEE Trans. Image Process. **13**(4), 600–612 (2004)

32. Wang, Z., Simoncelli, E., Bovik, A.: Multiscale structural similarity for image quality assessment. In: Asilomar Conference on Signals, Systems and Computers, vol. 2, pp. 1398–1402. IEEE (2003)

33. Xue, W., Mou, X., Zhang, L., Bovik, A., Feng, X.: Blind image quality assessment using joint statistics of gradient magnitude and Laplacian features. IEEE Trans. Image Process. **23**(11), 4850–4862 (2014)

34. Yang, S., et al.: MANIQA: Multi-dimension attention network for no-reference image quality assessment. In: IEEE Conference on Computer Vision and Pattern Recognition Workshops, pp. 1191–1200 (2022)

35. Zamir, S.W., Arora, A., Khan, S., Hayat, M., Khan, F.S., Yang, M.H.: Restormer: efficient transformer for high-resolution image restoration. In: IEEE Conference on Computer Vision and Pattern Recognition, pp. 5728–5739 (2022)

36. Zhang, L., Shen, Y., Li, H.: VSI: a visual saliency-induced index for perceptual image quality assessment. IEEE Trans. Image Process. **23**(10), 4270–4281 (2014)

37. Zhang, L., Zhang, L., Mou, X., Zhang, D.: FSIM: a feature similarity index for image quality assessment. IEEE Trans. Image Process. **20**(8), 2378–2386 (2011)

38. Zhang, R., Isola, P., Efros, A., Shechtman, E., Wang, O.: The unreasonable effectiveness of deep features as a perceptual metric. In: IEEE Conference on Computer Vision and Pattern Recognition, pp. 586–595 (2018)

39. Zhang, Y., Li, K., Li, K., Wang, L., Zhong, B., Fu, Y.: Image super-resolution using very deep residual channel attention networks. In: Ferrari, V., Hebert, M., Sminchisescu, C., Weiss, Y. (eds.) ECCV 2018. LNCS, vol. 11211, pp. 294–310. Springer, Cham (2018). https://doi.org/10.1007/978-3-030-01234-2_18

40. Zhang, Z., Wang, R., Zhang, H., Chen, Y., Zuo, W.: Self-supervised learning for real-world super-resolution from dual zoomed observations. In: Avidan, S., Brostow, G., Cissé, M., Farinella, G.M., Hassner, T. (eds.) ECCV 2022. LNCS, vol. 13678, pp. 610–627. Springer, Cham (2022). https://doi.org/10.1007/978-3-031-19797-0_35

A Mutual Enhancement Framework for Specular Highlight Detection and Removal

Ge Huang[1,2], Jieru Yao[3], Peiliang Huang[3], and Longfei Han[4,5(✉)]

[1] AHU-IAI AI Joint Laboratory, Anhui University, Hefei, China
[2] Institute of Artificial Intelligence, Hefei Comprehensive National Science Center, Hefei, China
[3] School of Automation, Northwestern Polytechnical University, Xi'an, China
[4] School of Computer Science and Engineering, Beijing Technology and Business University, Beijing, China
[5] School of Information Science and Technology, University of Science and Technology of China, Hefei, China

Abstract. Specular highlights, generated by direct light reflection from surfaces, can significantly reduce the image quality and impair various computer vision applications. Recently, the existing approaches for jointly detecting and removing specular highlights attempted to use the detection as guidance for highlight removal. However, they ignored that this kind of unidirectional enhancement was susceptible to detection tasks. To achieve mutual enhancement, we assume that discriminative features would benefit the highlight detection task which needs to distinguish between highlight areas and highlight-free areas, while coherent features would facilitate the learning of highlight removal since it requires converting highlight areas to highlight-free areas. Specifically, we propose a mutual enhancement framework (MEF-SHDR) that addresses both specular highlight detection and removal in a unified manner. The proposed framework designs a Feature Decomposition and Aggregation Module (FDAM) that separates highlight and highlight-free features explicitly and aggregates them for improved detection and removal performance. Comprehensive experiments are implemented on five widely used datasets, i.e., SHIQ, LIME, SD1, SD2, and RD, demonstrating the superiority of the proposed approach over previous state-of-the-art methods, as well as the effectiveness of jointly detecting and removing highlights. Code is available at https://github.com/drafly/MEF-SHDR.

Keywords: Specular highlights detection · specular highlights removal · mutual enhancement · feature decomposition and aggregation module

Supported by the Institute of Artificial Intelligence, Hefei Comprehensive National Science Center Project under Grant (21KT008), the University Synergy Innovation Program of Anhui Province (GXXT-2022-052), and National Natural Science Foundation of China (No. 62202015).

1   Introduction

Specular highlights are usually generated by direct light reflection on highly reflective surfaces. These highlights can significantly damage the visual quality of images and affect the performance of various computer vision tasks, such as image segmentation [2], object detection [14] and object tracking [5]. Consequently, detecting and removing specular highlights have been regarded as a long-standing challenge in computer vision research.

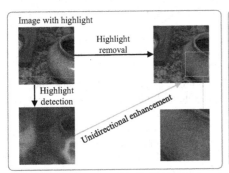

(a) Highlight detection guides highlight removal

(b) Mutual enhancement between highlight detection and removal

Fig. 1. Visual results of the heatmaps of highlight detection and results of removal processes. In contrast to previous methods, the proposed method enhances discriminative features of detection and coherent features of removal by explicitly separating highlight and highlight-free features, enhancing highlight features for detection, and enhancing highlight-free features for removal. Therefore, it makes the detection results more accurate and the removal results more consistent, especially texture details.

Traditional methods for detecting specular highlights employed threshold-based approaches by analyzing chrominance information and highlight characteristics [33]. Similarly, traditional removal methods removed highlights by modeling highlight causes, leveraging contextual information, and finding hidden information under the highlights [1,8,20,28]. However, due to the lack of semantic information, these methods struggle to deal with complex lighting conditions and changeable natural scenes, such as diverse colors, complex textures, and extensive highlight contamination.

Recently, deep learning-based detection methods focused on multi-scale features and used some contextual constraints to adapt to changeable environments [3,23]. Although some removal methods introduced masks to enhance the attention of highlight regions [9,17,19,25,34]. These methods usually treat highlight detection and removal as two separate tasks. Some approaches [4,24,26] have attempted to detect and remove highlights simultaneously, however, they have mainly focused on improving the performance of highlight removal by using information obtained from the highlight detection process, so that the removal result was easily affected by detection. (as shown in Fig. 1(a)).

To achieve mutual enhancement between highlight detection and removal, we propose a novel network framework for these tasks. We assume that highlight detection should learn discriminative features while highlight removal should learn coherent features. Considering that simply combining the two tasks may lead to a conflict, we design a Feature Decomposition and Aggregation Module (FDAM) to explicitly separate and aggregate features. As shown in Fig. 1(b), our method significantly increases the discrimination of highlight and highlight-free features (called inter-class discrimination) and enhances the consistency of removed features (called inter-class consistency) to improve the performance of both tasks. We evaluated the performance of our framework on benchmark datasets and demonstrated its superiority over state-of-the-art methods.

To sum up, our contributions are as follows: (1) For solving the contradiction between highlight detection and removal, we propose a novel framework for establishing the mutual connection between them. (2) To address the above conflict, we leverage Feature Decomposition and Aggregation Module to enhance the inter-class discrimination of detected features and the inter-class consistency of removed features simultaneously. (3) Experimental results on SHIQ [4], LIME [18], and SD1, SD2, RD [9] datasets demonstrate the proposed method can outperform existing state-of-the-art methods in highlight detection and removal.

2 Related Work

2.1 Specular Highlight Detection and Removal

Traditional specular highlight detection methods have made some assumptions and used fixed or automatic thresholds by observing chrominance information and characteristics of specular highlights [33]. However, these methods were heavily reliant on strict premises or assumptions, which may limit their accuracy in complex scenarios. Therefore, researchers focused on developing learning-based methods [3,4,23,24]. For example, Wang et al. [23] used a full-scale with a self-attention module and a refinement strategy to detect highlights on SHIQ [4].

Traditional highlight removal methods were based on physical models, color space analysis [20], and Neighborhood analysis methods [28]. However, these methods were constrained by fixed models and assumptions, making them vulnerable to complex backgrounds and lighting conditions. To solve these problems, researchers proposed deep learning approaches [9,17,19,25,34] and constructed large-scale datasets including real-world, facial, text, and synthetic images [4,9,18,19,25,34] as the foundation. Additional information, such as specular highlight masks, was often used as guidance and constraints to remove highlights, with the expectation that the network would highly focus on the highlighted areas. Compared with the single highlight detection or removal task, the current research focus in the field of highlight is on the multi-task joint learning framework, which is also the focus of our research.

2.2 Enhancement of Highlight Detection and Removal

Existing multi-task framework [4,24,26] tended to leverage highlight detection to improve the effectiveness of highlight removal. Fu et al. [4] presented a multi-task network for joint highlight detection and removal, using a contextual module to accurately detect and remove highlights of varying sizes. They use highlight detection to enhance the performance of highlight removal but ignore the fact that highlight removal is easily affected by one-way detection. Our work aims to improve both highlight detection and removal by collaborative feature learning between the two tasks.

3   Proposed Method

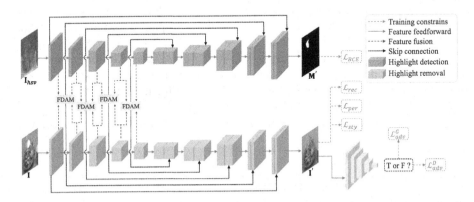

Fig. 2. Illustration of the proposed specular detection and removal framework. The network contains two subtasks, i.e., specular highlight detection (the purple part) and specular highlight removal (the blue and grey part). The two parallel streams decompose and aggregate encoded deep features from each other with a Feature Decomposition and Aggregation Module (FDAM), which is stacked at the former of the decoder. (Color figure online)

3.1   The Overall Framework

As depicted in Fig. 2, we propose a parallel network that is capable of simultaneously performing specular highlight detection and removal (MEF-SHDR). Our principal network is based on GAN [7], where we use an encoder-decoder architecture as the generator and the original GAN discriminator [7] as our discriminator. To be specific, for backbone encoders, we adopt EfficientNet [22] to encoder input highlight images (denoted as $I \in \mathbb{R}^{3 \times h \times w}$) for removal and the corresponding images in the HSV color space (denoted as $I_{hsv} \in \mathbb{R}^{3 \times h \times w}$) for detection. Here, we denote the encoded feature maps as $\mathbf{h}_u, u \in [1,5]$ for detection and $\mathbf{f}_v, v \in [1,5]$ for removal. Highlights of HSV images are characterized by a low saturation and a high value, which served as a weak prior for detection. The decoder in our framework is designed based on the Unet decoder architecture. By progressively combining the features from different layers through skip connections, the decoder is able to capture both low-level and high-level information, facilitating the generation of predicted masks and no-highlight images. Our discriminator is original GAN discriminator. Formally, let $\tilde{I}$ represent the ground-truth no-highlight images, and $\mathbf{I}' = G(\mathbf{I})$ represent the generated no-highlight images. We denote the adversarial loss for the discriminator as:

$$\mathcal{L}_{\text{adv}}^{D} = -\mathbb{E}_{\hat{I} \sim P_d}[log(D(\hat{\mathbf{I}}))] - \mathbb{E}_{I' \sim P_g}[log(1 - D(\mathbf{I}'))], \tag{1}$$

where P_d denotes the real data distribution and P_g denotes the fake data distribution. Correspondingly, the adversarial loss for the generator is denoted as:

$$L_{adv}^{G} = -\mathbb{E}_{I' \sim P_g}[log(D(\mathbf{I}'))]. \tag{2}$$

The principle of choosing optimization objectives for specular highlight detection and removal is to ensure per-pixel accuracy of generated masks and to guarantee both per-pixel reconstruction accuracy and visual fidelity of reconstructed images. For detection, we employ the Binary Cross Entropy(BCE) loss, and for removal, we utilize four optimization objectives: a reconstruction loss, a style loss [6], a perceptual loss [13], and an adversarial loss of the GAN (described in Eq. 2), partial loss formulas providing in 3.3. The entire network is trained by jointly optimizing these objectives, and empirically setting the value of trade-off parameters as $\lambda_{\text{adv}} = 0.01$, $\lambda_{\text{rec}} = 1$, $\lambda_{\text{per}} = 0.1$ and $\lambda_{\text{sty}} = 250$, the overall optimization objective is defined as follows:

$$\mathcal{L} = \mathcal{L}_{\text{BCE}} + \lambda_{\text{adv}}\mathcal{L}_{\text{adv}}^{G} + \lambda_{\text{rec}}\mathcal{L}_{\text{rec}} + \lambda_{\text{per}}\mathcal{L}_{\text{per}} + \lambda_{\text{sty}}\mathcal{L}_{\text{sty}}. \tag{3}$$

3.2   Feature Decomposition and Aggregation Module

In this section, we introduce a core and symmetric module designed to enhance the intersection between the two networks. The overview of our feature decomposition and Aggregation module is depicted in Fig. 3. In the figure, we illustrate an example using the connections of feature maps of detection ($\mathbf{h}_1$ and $\mathbf{h}_2$) and removal ($\mathbf{f}_1$ and $\mathbf{f}_2$) from successive encoder layers. First, we employ the Squeeze-and-Excitation (SE) attention [11] module to disentangle $\mathbf{h}_1$ into two parts: highlight dependent features $\hat{\mathbf{h}}_1$ and highlight independent features $\tilde{\mathbf{h}}_1$. Similarity, $\mathbf{f}_1$ is divided into highlight dependent features $\hat{\mathbf{f}}_1$ and highlight independent features $\tilde{\mathbf{f}}_1$. By using $\mathbf{A}$ to denote SE attention, $\otimes$ to represent pixel-wise multiplication, the formula is shown as:

$$\hat{\mathbf{h}}_1 = \mathbf{h}_1 \otimes \mathbf{A}(\mathbf{h}_1), \quad \tilde{\mathbf{h}}_1 = \mathbf{h}_1 \otimes (1 - \mathbf{A}(\mathbf{h}_1)). \tag{4}$$

Next, we use the combination of standard convolutions, batch normalization, and activation to process the upper results. Finally, to ensure effective utilization of the obtained features, we introduce learnable parameters $\alpha_1^i, i \in [1, 2, 3]$ to control features fusion and obtain the resultant feature for the next layer $\mathbf{h}_2'$. $\mathbf{h}_2'$ is defined as:

$$\mathbf{h}_2' = \alpha_1^1 \mathbf{h}_2 + \alpha_1^2 \hat{\mathbf{h}}_1 + \alpha_1^3 \hat{\mathbf{f}}_1. \tag{5}$$

Finally, we obtain two groups of features, $\mathbf{h}_u', u \in [1, 5]$ and $\mathbf{f}_v', v \in [1, 5]$, where $\mathbf{h}_1'$ and $\mathbf{f}_1'$ are the outputs of backbone, others are the outputs of FDAM.

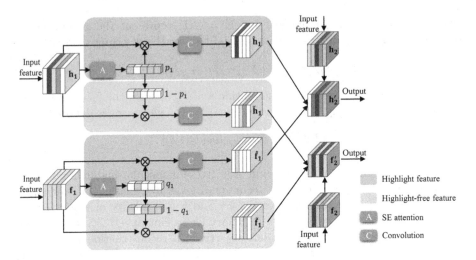

Fig. 3. Feature Decomposition and Aggregation Module. FDAM is helpful for explicit feature separation and enhancement highlight and highlight-free features. Through a series of operations, including attention, pixel-wise multiplication, standard convolution, and pixel-wise addition, we can acquire the new features $\mathbf{h}_2'$ and $\mathbf{f}_2'$ from features of the upper layer and current layer which are provided by the backbone.

3.3   Mutual Training Objective Function

For detection, we set $\hat{\mathbf{M}}$ be the ground-truth highlight masks (with value 1 for existing region, 0 otherwise), and $\mathbf{M}' = G(\mathbf{I}_{hsv})$ be the predicted masks. We utilize the BCE loss to segment the HSV images, and the formula is written as:

$$\mathcal{L}_{\text{BCE}} = -\frac{1}{N} \sum_{i=1}^{N} \left(\hat{\mathbf{M}}_i \log(\mathbf{M}_i') + (1 - \hat{\mathbf{M}}_i) \log(1 - \mathbf{M}_i') \right). \tag{6}$$

For removal, to measure the similarity between $\hat{\mathbf{I}}$ and $\mathbf{I}'$ in terms of pixel values, we employ the l_1 distance as the reconstruction loss, defined as:

$$\mathcal{L}_{\text{rec}}(\hat{\mathbf{I}}, \mathbf{I}') = \|\hat{\mathbf{I}} - \mathbf{I}'\|_1. \tag{7}$$

Besides, we introduce perceptual loss $\mathcal{L}_{\text{per}}$ to evaluate the global structure of the image. It measures the l_1 distance between the feature representations of $\hat{\mathbf{I}}$ and $\mathbf{I}'$ extracted by a pre-trained feature extraction network:

$$\mathcal{L}_{\text{per}}(\hat{\mathbf{I}}, \mathbf{I}') = \sum_{i}^{5} \frac{\left\| \phi_i(\hat{\mathbf{I}}) - \phi_i(\mathbf{I}') \right\|_1}{N_{\phi_i}}, \tag{8}$$

where $\phi_i(\cdot)$ denote the activation map of the i-th layers of the pre-trained network given the input features $\hat{\mathbf{I}}$ and $\mathbf{I}'$, N_{ϕ_i} is the number of elements in $\phi_i(\cdot)$. Lastly, the style loss ensures style consistency between the generated and real images.

Similarly, it calculates the l_1 distance between the Gram matrices (denoted by $\phi_i(\cdot)^T\phi_i(\cdot)$) of deep features extracted from the images:

$$\mathcal{L}_{\text{sty}}(\hat{\mathbf{I}}, \mathbf{I}') = \mathbb{E}_i \left[\left\| \phi_i(\hat{\mathbf{I}})^T\phi_i(\hat{\mathbf{I}}) - \phi_i(\mathbf{I}')^T\phi_i(\mathbf{I}') \right\|_1 \right]. \tag{9}$$

4   Experiments

4.1   Experimental Settings

Datasets. We evaluated the proposed method on several datasets widely used, including SHIQ [4], LIME [18], and SD1, SD2, RD [9]. The SHIQ dataset provides ground truth highlight masks and highlight-free images with a size of 200 × 200, which is used to evaluate both the detection and removal capabilities of our entire method. The LIME dataset with a size of 256 × 256 contains pairs of synthesized images with and without highlights, but they have no explicit division of train and test. Therefore, We follow the random split setting used in JSHDR [4] and employ the highlight removal method individually. The SD1, SD2, and RD datasets consist of text highlight images with a size of 512 × 512. We employ our full method without conducting highlight detection evaluation since they do not provide highlight masks in test sets.

Evaluation Metric. In order to quantitatively evaluate the highlight detection performance for SHIQ, we employ two commonly used metrics [12], namely accuracy (ACC) and balance error rate (BER). For removal, we follow the metrics used in [9], where recall, precision, and f-measure are employed to evaluate the quality of text recovery for SD1, SD2, and RD, while PSNR and SSIM are utilized to assess the visual recovery quality for all datasets.

Implementation Details. We implement our network using PyTorch 1.13.1. For GPU acceleration, we utilize an NVIDIA® GeForce GTX 3090. The learning rate is fixed at $1e-4$ for both the discriminator and the generator training. We employ the Adam optimizer [15] with $\beta_1 = 0.5$ and $\beta_2 = 0.999$ to train our network. In our experiments, all images are trained with their original sizes.

4.2   Comparison to the State-of-the-Art Methods

Table 1 presents the quantitative comparison results of highlight detection on the SHIQ testing dataset. It is evident that our method achieves state-of-the-art performance in terms of both accuracy and BER metrics, which demonstrate our network can accurately detect highlight regions.

Table 1. Quantitative comparison of our method state-of-the-art detection methods on SHIQ dataset. The best results are marked in bold. "↑" means higher value is better.

Method	Li [16]	Zhang [33]	Fu [3]	Fu [4]	Wu [24]	Ours
ACC↑	0.70	0.71	0.91	0.93	0.97	**0.98**
BER↓	18.80	24.40	6.18	5.92	5.92	**5.26**

The quantitative comparison in Table 2 demonstrates that our method has achieved state-of-the-art performance in terms of visual fidelity on the two challenging benchmark datasets, SHIQ and LIME. Take PSNR as an example, our method improves the performance by 1.76(SHIQ) and 1.86(LIME), respectively. We compare the visual performance of our network with current state-of-the-art methods. As shown in Fig. 4, Akashi [1] generates black dots throughout the entire image. Among the three deep-learning methods, Zhang [31] and Hu [10] fail to effectively remove highlights as they only eliminate a portion of them. Fu [4] performs better in highlight removal, but it introduces dark artifacts and fails to preserve the texture details of the original image. In comparison, our method successfully removes the majority of highlights without significant alterations in the brightness, color, and texture of the image.

Table 2. Quantitative comparison of our method state-of-the-art removal methods on SHIQ and LIME dataset. The best results are marked in bold. "↑" means higher value is better.

Method	SHIQ		LIME	
	PSNR ↑	SSIM ↑	PSNR ↑	SSIM ↑
Yang [28]	14.31	0.5	17.64	0.58
Guo [8]	17.18	0.58	18.03	0.60
Shi [21]	18.21	0.61	24.21	0.76
Yamamoto [27]	19.54	0.63	–	–
Yi [29]	21.32	0.72	26.77	0.79
Lin [17]	27.63	0.88	–	–
Wu [25]	28.23	0.94	-	–
Hu [10]	30.94	0.97	32.96	0.98
Xu [26]	31.68	0.97	-	–
Yu [30]	32.19	0.84	–	–
Zhang [32]	33.64	0.94	35.83	0.92
Fu [4]	34.13	0.86	37.01	0.91
Ours	**35.89**	**0.96**	**38.87**	**0.99**

Table 3 presents the performance of text detection and recognition, as well as the visual quality compared to SD1, SD2, and RD. Our method achieves the best performance in all metrics, except for precision. This discrepancy in precision can be attributed to the increased number of recognizable texts after the removal of highlights. As a result, the denominator in the precision calculation becomes larger, which leads to removal methods exhibiting lower precision values compared to the original light images.

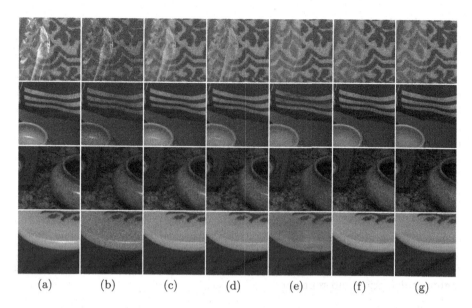

| (a) | (b) | (c) | (d) | (e) | (f) | (g) |

Fig. 4. Qualitative evaluation of specular highlight removal. (a) Highlight images. (b-f) Generated highlight-free images by (b) Akashi [1], (c) Zhang [31], (d) Hu [10], (e) Fu [4], (f) our method, and (g) ground truth.

Table 3. Quantitative comparison of our method state-of-the-art removal methods on SD1, SD2, RD dataset. The best results are marked in bold. "↑" means higher value is better.

Datasets	Method	Metric				
		Recall ↑	Precision ↑	F-measure ↑	PSNR ↑	SSIM ↑
SD1	Highlight Image	85.03	94.70	88.70	17.58	82.37
	Lin [17]	86.28	94.76	89.30	26.29	89.86
	Muhammad [19]	82.39	93.12	86.31	15.61	68.82
	Hou [9]	91.92	96.32	93.57	22.65	88.33
	Ours	**93.71**	**97.16**	**94.92**	**29.53**	**94.78**
SD2	Highlight Image	80.50	**95.89**	87.10	11.79	66.42
	Lin [17]	79.21	93.82	84.88	28.99	91.81
	Muhammad [19]	78.87	95.10	85.55	9.66	53.95
	Hou [9]	83.57	95.00	88.42	29.21	90.67
	Ours	**85.50**	93.63	**89.00**	**31.63**	**95.62**
RD	Highlight Image	64.85	90.60	73.49	17.05	65.04
	Lin [17]	61.58	87.63	70.72	17.17	64.23
	Muhammad [19]	70.59	**91.62**	78.38	14.82	52.49
	Hou [9]	78.50	91.34	83.34	21.62	77.19
	Ours	**80.82**	91.48	**84.67**	**25.26**	**86.85**

4.3   Ablation Study

To validate the effectiveness of the main components in our network, we perform the ablation experiments and design six baselines: "Only D": only train highlight detection component. "Only R": only train highlight removal component. "Ours w/o FDAM": our method without FDAM. "Ours w/o HSV": not use HSV images but use RGB images in highlight detection. "D to R": only performs FDAM in the process of detection to removal, while "R to D" means the opposite.

The results are presented in Table 4. We have some observations: (1) The "Ours w/o FDAM" configuration yields inferior PSNR and SSIM performance compared to "only D" or "only R", indicating that direct combined training of highlight detection and removal leads to poor performance in both tasks. (2) "Ours w/o HSV" baseline, which reveals that utilizing HSV images can enhance the performance of highlight detection and removal. (3) The higher scores obtained by our full method compared to "Ours w/o FDAM", "D to R", and "R to D" baselines demonstrate the effectiveness of FDAM for both highlight detection and removal.

Table 4. Ablation study of the proposed method in SHIQ for highlight detection and removal. The best results are marked in bold. "↑" means higher value is better.

Method	Ours	Only D	Only R	Ours w/o FDAM	Ours w/o HSV	D to R	R to D
BER↓	**5.25**	5.95	–	6.83	5.54	5.93	5.66
PSNR ↑	**35.89**	–	35.25	35.19	35.57	35.49	35.40

5   Conclusion

In this paper, we analyze highlight detection and removal tasks and propose a hypothesis that highlight detection learns discriminant features and highlight removal learns consistent features. Therefore, we introduce a mutual enhancement framework for specular highlight detection and removal. Specifically, we design a Feature Decomposition and Aggregation Module to separate and aggregate features. Through FDAM, we can jointly train the two mutually exclusive tasks and improve the performance in collaborative training. Moreover, we evaluate our method on various datasets, including SHIQ, LIME, SD1, SD2, and RD covering a wide range of scenarios and image types. The results consistently demonstrate the superior performance of our approach in terms of the evaluation metrics of detection and visual effect. Our method leaves highlight-free areas intact, especially when there are white materials in the environment. In the future, we plan to apply our method to the detection and removal of other image degradation with a significant range, such as shadows.

References

1. Akashi, Y., Okatani, T.: Separation of reflection components by sparse non-negative matrix factorization. In: Cremers, D., Reid, I., Saito, H., Yang, M.-H.

(eds.) ACCV 2014. LNCS, vol. 9007, pp. 611–625. Springer, Cham (2015). https://doi.org/10.1007/978-3-319-16814-2_40

2. Arbelaez, P., Maire, M., Fowlkes, C., Malik, J.: Contour detection and hierarchical image segmentation. IEEE Trans. Pattern Anal. Mach. Intell. **33**(5), 898–916 (2010)

3. Fu, G., Zhang, Q., Lin, Q., Zhu, L., Xiao, C.: Learning to detect specular highlights from real-world images. In: Proceedings of the 28th ACM International Conference on Multimedia, pp. 1873–1881 (2020)

4. Fu, G., Zhang, Q., Zhu, L., Li, P., Xiao, C.: A multi-task network for joint specular highlight detection and removal. In: Proceedings of the IEEE/CVF Conference on Computer Vision and Pattern Recognition (CVPR), pp. 7752–7761 (2021)

5. Gao, J., Zhang, T., Xu, C.: Graph convolutional tracking. In: Proceedings of the IEEE/CVF Conference on Computer Vision and Pattern Recognition (CVPR), pp. 4649–4659 (2019)

6. Gatys, L.A., Ecker, A.S., Bethge, M.: Image style transfer using convolutional neural networks. In: Proceedings of the IEEE Conference on Computer Vision and Pattern Recognition (CVPR), pp. 2414–2423 (2016)

7. Goodfellow, I., et al.: Generative adversarial nets. In: Advances in Neural Information Processing Systems 27 (2014)

8. Guo, J., Zhou, Z., Wang, L.: Single image highlight removal with a sparse and low-rank reflection model. In: Ferrari, V., Hebert, M., Sminchisescu, C., Weiss, Y. (eds.) ECCV 2018. LNCS, vol. 11208, pp. 282–298. Springer, Cham (2018). https://doi.org/10.1007/978-3-030-01225-0_17

9. Hou, S., Wang, C., Quan, W., Jiang, J., Yan, D.-M.: Text-aware single image specular highlight removal. In: Ma, H., et al. (eds.) PRCV 2021. LNCS, vol. 13022, pp. 115–127. Springer, Cham (2021). https://doi.org/10.1007/978-3-030-88013-2_10

10. Hu, G., Zheng, Y., Yan, H., Hua, G., Yan, Y.: Mask-guided cycle-GAN for specular highlight removal. Pattern Recogn. Lett. **161**, 108–114 (2022)

11. Hu, J., Shen, L., Sun, G.: Squeeze-and-excitation networks. In: Proceedings of the IEEE Conference on Computer Vision and Pattern Recognition (CVPR), pp. 7132–7141 (2018)

12. Hu, X., Zhu, L., Fu, C.W., Qin, J., Heng, P.A.: Direction-aware spatial context features for shadow detection. In: Proceedings of the IEEE Conference on Computer Vision and Pattern Recognition (CVPR), pp. 7454–7462 (2018)

13. Johnson, J., Alahi, A., Fei-Fei, L.: Perceptual losses for real-time style transfer and super-resolution. In: Leibe, B., Matas, J., Sebe, N., Welling, M. (eds.) ECCV 2016. LNCS, vol. 9906, pp. 694–711. Springer, Cham (2016). https://doi.org/10.1007/978-3-319-46475-6_43

14. Kim, S.-W., Kook, H.-K., Sun, J.-Y., Kang, M.-C., Ko, S.-J.: Parallel feature pyramid network for object detection. In: Ferrari, V., Hebert, M., Sminchisescu, C., Weiss, Y. (eds.) ECCV 2018. LNCS, vol. 11209, pp. 239–256. Springer, Cham (2018). https://doi.org/10.1007/978-3-030-01228-1_15

15. Kingma, D.P., Ba, J.: Adam: a method for stochastic optimization. arXiv preprint arXiv:1412.6980 (2014)

16. Li, R., Pan, J., Si, Y., Yan, B., Hu, Y., Qin, H.: Specular reflections removal for endoscopic image sequences with adaptive-RPCA decomposition. IEEE Trans. Med. Imaging **39**(2), 328–340 (2019)

17. Lin, J., El Amine Seddik, M., Tamaazousti, M., Tamaazousti, Y., Bartoli, A.: Deep multi-class adversarial specularity removal. In: Felsberg, M., Forssén, P.-E., Sintorn, I.-M., Unger, J. (eds.) SCIA 2019. LNCS, vol. 11482, pp. 3–15. Springer, Cham (2019). https://doi.org/10.1007/978-3-030-20205-7_1

18. Meka, A., et al.: LIME: live intrinsic material estimation. In: Proceedings of the IEEE Conference on Computer Vision and Pattern Recognition (CVPR), pp. 6315–6324 (2018)

19. Muhammad, S., Dailey, M.N., Farooq, M., Majeed, M.F., Ekpanyapong, M.: Specnet and spec-CGAN: deep learning models for specularity removal from faces. Image Vis. Comput. **93**, 103823 (2020)

20. Shafer, S.A.: Using color to separate reflection components. Color. Res. Appl. **10**(4), 210–218 (1985)

21. Shi, J., Dong, Y., Su, H., Yu, S.X.: Learning non-lambertian object intrinsics across shapenet categories. In: Proceedings of the IEEE Conference on Computer Vision and Pattern Recognition (CVPR), pp. 1685–1694 (2017)

22. Tan, M., Le, Q.: EfficientNet: rethinking model scaling for convolutional neural networks. In: Proceedings of the 36th International Conference on Machine Learning, pp. 6105–6114. PMLR (2019)

23. Wang, C., Wu, Z., Guo, J., Zhang, X.: Contour-constrained specular highlight detection from real-world images. In: Proceedings of the 18th ACM SIGGRAPH International Conference on Virtual-Reality Continuum and Its Applications in Industry, pp. 1–4 (2022)

24. Wu, Z., Guo, J., Zhuang, C., Xiao, J., Yan, D.M., Zhang, X.: Joint specular highlight detection and removal in single images via unet-transformer. Comput. Vis.Media **9**(1), 141–154 (2023)

25. Wu, Z.: Single-image specular highlight removal via real-world dataset construction. IEEE Trans. Multimedia **24**, 3782–3793 (2021)

26. Xu, J., Liu, S., Chen, G., Liu, Q.: Highlight detection and removal method based on bifurcated-CNN. In: Liu, H., et al. (eds.) Intelligent Robotics and Applications, pp. 307–318. Springer, Cham (2022). https://doi.org/10.1007/978-3-031-13841-6_29

27. Yamamoto, T., Kitajima, T., Kawauchi, R.: Efficient improvement method for separation of reflection components based on an energy function. In: 2017 IEEE International Conference on Image Processing (ICIP), pp. 4222–4226. IEEE (2017)

28. Yang, Q., Tang, J., Ahuja, N.: Efficient and robust specular highlight removal. IEEE Trans. Pattern Anal. Mach. Intell. **37**(6), 1304–1311 (2014)

29. Yi, R., Tan, P., Lin, S.: Leveraging multi-view image sets for unsupervised intrinsic image decomposition and highlight separation. In: Proceedings of the AAAI Conference on Artificial Intelligence, vol. 34, pp. 12685–12692 (2020)

30. Yu, J., Lin, Z., Yang, J., Shen, X., Lu, X., Huang, T.S.: Free-form image inpainting with gated convolution. In: Proceedings of the IEEE/CVF International Conference on Computer Vision (ICCV), pp. 4471–4480 (2019)

31. Zhang, L., Long, C., Zhang, X., Xiao, C.: RIS-GAN: explore residual and illumination with generative adversarial networks for shadow removal. In: Proceedings of the AAAI Conference on Artificial Intelligence, vol. 34, pp. 12829–12836 (2020)

32. Zhang, L., Long, C., Zhang, X., Xiao, C.: Exploiting residual and illumination with gans for shadow detection and shadow removal. ACM Trans. Multimedia Comput. Commun. Appl. **19**, 1–22 (2022)

33. Zhang, W., Zhao, X., Morvan, J.M., Chen, L.: Improving shadow suppression for illumination robust face recognition. IEEE Trans. Pattern Anal. Mach. Intell. **41**(3), 611–624 (2018)

34. Zhu, T., Xia, S., Bian, Z., Lu, C.: Highlight removal in facial images. In: Peng, Y., et al. (eds.) PRCV 2020. LNCS, vol. 12305, pp. 422–433. Springer, Cham (2020). https://doi.org/10.1007/978-3-030-60633-6_35

Pixel-Level Sonar Image JND Based on Inexact Supervised Learning

Qianxue Feng[1], Mingjie Wang[1], Weiling Chen[1(✉)] , Tiesong Zhao[1] ,
and Yi Zhu[2]

[1] Fujian Key Lab for Intelligent Processing and Wireless Transmission of Media
Information, Fuzhou University, Fuzhou 350108, China
weiling.chen@fzu.edu.cn

[2] Key Laboratory of Underwater Acoustic Communication and Marine Information
Technology (Xiamen University), Ministry of Education, Xiamen University,
Xiamen 361005, China

Abstract. The Just Noticeable Difference (JND) model aims to identify perceptual redundancies in images by simulating the perception of
the Human Visual System (HVS). Exploring the JND of sonar images
is important for the study of their visual properties and related applications. However, there is still room for improvement in performance of
existing JND models designed for Natural Scene Images (NSIs), and the
characteristics of sonar images are not sufficiently considered by them.
On the other hand, there are significant challenges in constructing a
densely labeled pixel-level JND dataset. To tackle these issues, we proposed a pixel-level JND model based on inexact supervised learning.
A perceptually lossy/lossless predictor was first pre-trained on a coarse-
grained picture-level JND dataset. This predictor can guide the unsupervised generator to produce an image that is perceptually lossless compared to the original image. Then we designed a loss function to ensure
that the generated image is perceptually lossless and maximally different from the original image. Experimental results show that our model
outperforms current models.

Keywords: Just Noticeable Difference (JND) · Inexact Supervised
Learning · Sonar Images

1 Introduction

The Just Noticeable Difference (JND) measures the threshold at which the
Human Visual System (HVS) can perceive differences in visual stimuli [1]. In
recent years, many researchers have devoted themselves to exploring such thresholds by modeling HVS. However, current achievements are mostly limited to

This work was supported in part by the National Natural Science Foundation of China
under Grant 62171134 and in part by Natural Science Foundation of Fujian Province,
China under Grant 2022J05117 and 2022J02015.

Natural Scene Images (NSIs). Different types of images possess distinct visual redundancy characteristics, thus limiting the applicability of existing JND models to other types of images. In view of this, this paper focuses on the construction of a JND model for sonar images, since JND-guided compression and processing can provide better performance. This is of great importance for sonar images transmitted in the bandwidth-limited underwater acoustic channel. The JND models can be classified into two types based on their design philosophy: bottom-up JND models and top-down JND models.

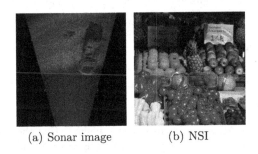

(a) Sonar image   (b) NSI

Fig. 1. Examples of the sonar image and the NSI. They have different characteristics.

The bottom-up JND models involve modeling the primary perceptual factors of JND and then combining them through addition or multiplication to obtain an overall JND. Researchers aim to use this JND threshold to generate a perceptually lossless image. The first pixel-level JND model was proposed by Chou *et al.* [2], which took luminance adaptation and contrast masking as the final pixel-level JND thresholds. Yang *et al.* [3] proposed a model that takes into account the overlapping effects of luminance adaptation and contrast masking. Subsequent studies added new visual perceptual factors such as texture masking [4], pattern masking [5], and salient areas [6]. Pioneering work in subband-domain models began with Ahumada *et al.* [7], they proposed a spatial contrast sensitivity function to obtain the JND model for gray images. Successively, luminance adaptation and contrast masking were incorporated into JND [8–10] and improved by partitioning the image into smooth, edge, and texture blocks [4]. But HVS is too complex to be fully understood now, which hinders the development of bottom-up JND models.

The top-down JND models utilize some methods to obtain a perceptually lossless image and subsequently determine JND thresholds. One prevalent approach is to construct a picture-level JND dataset and take the relevant parameters (*e.g.*, the Quantization parameter, Qp) as JND thresholds. Liu *et al.* [11] proposed a deep-learning-based picture-level JND model. The JND prediction problem was modeled as a binary classification problem, and they proposed a deep-learning-based perceptually lossy/lossless predictor. Given the picture-level JND is still redundant in some regions, Tian *et al.* [12] employed a neural network approach for training to predict the block-level JND. Another approach was recently proposed by Jiang *et al.* [13]. They applied the Karhunen-Loeve Transform (KLT) to the image and then derived its critical point by exploiting the convergence characteristics of the KLT coefficient energy. Using this critical point, the researchers reconstructed the critical perceptually lossless image. However, these methods can only ensure that the generated image is perceptually lossless and do not guarantee that the image has maximum dissimilarity with the original image.

In summary, current JND models face three main challenges. First, existing JND models have shown poor performance when applied to sonar images, possibly due to a lack of investigation into visual redundancy specific to these images. As illustrated in Fig. 1, sonar images exhibit simpler content, limited color information, and are often accompanied by speckle noise compared to NSIs. Yet the influence of factors like speckle noise and regions of interest on JND estimation remains unexplored. Second, when it comes to image compression, pixel-level JND methods offer more accurate and adaptable results. However, existing bottom-up JND models rely heavily on understanding the HVS, while top-down JND models face the impractical task of building a large annotated pixel-level JND dataset. Third, existing top-down methods can only ensure perceptual losslessness of the modified image without guaranteeing the maximization of pixel differences. As a result, there is still a portion of visual redundancy that has not been explored.

To address these challenges, we presented a pixel-level JND model for sonar images based on inexact supervised learning. As depicted in Fig. 2, our model contains three modules and a operation. The information reference module provides information that affects the JND of sonar images. The generation module produces a Critically Distorted Image (CDI) that is perceptually lossless but maximizes the pixel differences compared to the original image. The prediction module assesses the presence of perceptual loss between the CDI and the original image. A subtraction operation is used to obtain the JND map, where each pixel value represents the JND threshold. Our contributions are as follows:

- Proposing a JND model based on inexact supervised learning. A predictor pre-trained on the picture-level JND dataset guides the unsupervised generator to generate the CDI and finally derive the pixel-level JND map. This approach addresses the challenge of building a labeled pixel-level JND dataset.
- Designing a novel loss function that effectively optimizes the network parameters. By adjusting the weights of perceptual loss and pixel differences loss, it ensures that the resulting CDI is perceptually lossless but with maximum pixel differences compared to the original one.
- Incorporating factors (image complexity and edges) that significantly influence the JND of sonar images into the information reference module, due to the simplicity of sonar image content and the importance of edges in target recognition. This ensures that the CDI is more compatible with the perceptual properties of sonar images.

2   Methodology

In this paper, the JND of each pixel is defined as:

$$\text{JND}_{\text{map}}(i,j) = |\mathbf{x}_{\text{ref}}(i,j) - \mathbf{x}_{\text{CDI}_{\text{GT}}}(i,j)|, \tag{1}$$

where $\mathbf{x}_{\text{ref}}$ denotes the reference image, and $\mathbf{x}_{\text{CDI}_{\text{GT}}}$ denotes a ground truth of CDI. We translated the JND threshold estimation into the generation of the CDI, the network-generated CDI is defined as:

$$\mathbf{x}_{\text{CDI}} = \text{G}(\mathbf{x}_{\text{ref}}; \theta), \tag{2}$$

G $(\cdot)$ represents the generator, and θ refers to the generator parameter. The network-generated CDI needs to satisfy the following equation:

$$D(i,j) = |\mathbf{x}_{\text{ref}}(i,j) - \mathbf{x}_{\text{CDI}}(i,j)|, \tag{3}$$

$$\arg\max_{\theta} D(i,j), \ s.t. \ \forall(i,j), D(i,j) \leq \text{JND}_{\text{map}}(i,j), \tag{4}$$

$D(i,j)$ for each pixel difference between the reference image and CDI. It is common to transform the maximization problem into a minimization problem, so we express pixel differences as Peak Signal to Noise Ratio (PSNR):

$$\text{MSE}(\mathbf{x}_{\text{ref}}, \mathbf{x}_{\text{CDI}}) = \sqrt{\frac{1}{H \times W} \sum_{i=0}^{H-1} \sum_{j=0}^{W-1} [D(i,j)]^2}, \tag{5}$$

$$\arg\min_{\theta} \text{PSNR}(\mathbf{x}_{\text{ref}}, \mathbf{x}_{\text{CDI}}) = 20 \cdot \log_{10}(\frac{\text{MAX}_{\mathbf{x}_{\text{ref}}}}{\sqrt{\text{MSE}(\mathbf{x}_{\text{ref}}, \mathbf{x}_{\text{CDI}})}}), \tag{6}$$

$$s.t. \ \forall(i,j), D(i,j) \leq \text{JND}_{\text{map}}(i,j)$$

the perceptual losslessness of $\mathbf{x}_{\text{CDI}}$ is typically assessed through subjective evaluations, which often require significant human resources to conduct subjective experiments. So we used a trained perceptually lossy/lossless predictor to determine:

$$\text{P}(\mathbf{x}_{\text{ref}}, \mathbf{x}_{\text{CDI}}) = \begin{cases} 0, & \forall(i,j), D(i,j) \leq \text{JND}_{\text{map}}(i,j) \\ 1, & \exists(i,j), D(i,j) > \text{JND}_{\text{map}}(i,j) \end{cases}, \tag{7}$$

the question is further organized as:

$$\arg\min_{\theta} \text{PSNR}(\mathbf{x}_{\text{ref}}, \mathbf{x}_{\text{CDI}}), \ s.t. \ \text{P}(\mathbf{x}_{\text{ref}}, \mathbf{x}_{\text{CDI}}) = 0. \tag{8}$$

The overall architecture of the proposed model is shown in Fig. 2. The model has three stages. In the pretraining stage, a picture-level JND dataset is used to train the predictor in the prediction module. In the training stage, the pretrained predictor is used to guide the unsupervised generator within generation module to generate the CDI. In the JND map generation stage, the JND map is generated using the subtraction of the original image and the generated CDI.

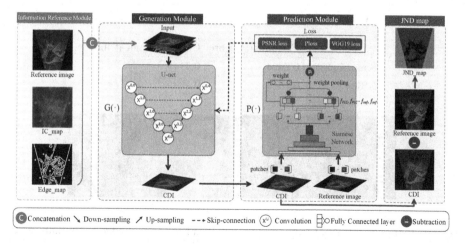

Fig. 2. The architecture of the proposed model. The information reference module provides information that affects the JND of sonar images. The generation module produces the CDI. The prediction module optimizes the generator parameters through a loss function. A subtraction operation is used to obtain the JND map.

2.1 Information Reference Module

The value of the JND thresholds is related to the image's content, and a larger JND threshold represents greater perceptual redundancy. It has been shown that JND thresholds are high in regions with high pattern complexity of the images [5]. Compared to NSIs, sonar images have simpler content, less detail, and little color information. Therefore, image complexity should be a more important factor affecting the JND of sonar images. To verify this, we used the method in [14] to predict the complexity score and complexity heat map of the images. The Pearson Linear Correlation Coefficient (PLCC) was used to measure the correlation between picture-level JND and image complexity. The closer the PLCC value is to 1, the higher the correlation between the two factors (Table 1).

In general, regions with higher attention exhibit increased sensitivity, resulting in lower JND thresholds. For sonar images, the rich information that aids in target detection and recognition is primarily concentrated within contours and edges. However, when dealing with NSIs, additional factors such as color, details, and other image elements also require attention. As depicted in Fig. 3(d), we also need color to distinguish apples from oranges. Therefore, we chose the edges of sonar images as another factor that mainly affects JND. We adopted the Canny operator to extract edges and performed morphological operations to obtain the edge mask map.

In conclusion, we added the image complexity map (IC_map) and the edge mask map (Edge_map) to improve the training performance. This can be seen in the orange dashed box in Fig. 2 and we replaced Eq. 2 with the following:

$$\mathbf{x}_{\text{CDI}} = G(\mathbf{x}_{\text{ref}}, \mathbf{x}_{\text{IC\_map}}, \mathbf{x}_{\text{Edge\_map}}; \theta) \tag{9}$$

Table 1. PLCC value of JND and image complexity.

	Complexity score	Complexity heat map
Sonar images	0.7694	0.7145
NSIs	0.1852	0.1935

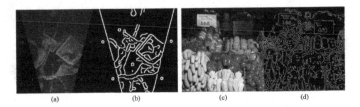

Fig. 3. Images (a)/(c) and (b)/(d) are the original image and edge mask map of the sonar image/NSI. The edges information in sonar images has a greater impact on JND than the NSI.

2.2   Generation Module

We employed U-net [15] as the generator in the generation module, as illustrated in the blue solid line box in Fig. 2. U-net has demonstrated high performance in image segmentation tasks, making it suitable for segmenting different tolerance regions of the human eye. It consists of two parts: the encoder and the decoder. The encoder consists of four convolution blocks, each using two 3×3 convolution layers and a 2×2 max pooling layer. The decoder also consists of four convolution blocks, each containing an up-sampling layer, a 3×3 convolution layer, and a skip connection.

2.3   Prediction Module

The prediction module consists of a perceptually lossy/lossless predictor and a loss calculator. The predictor is exclusively trained on the database constructed in Sect. 3.1 during the pre-training stage, achieving a testing accuracy of 98%. It serves the dual purpose of computing losses and validating the model's effectiveness. The predictor adopted a network structure from the [11], as shown in the green solid line box in Fig. 2. It can predict whether the CDI has perceptual loss for the reference image. A pair of images is initially divided into non-overlapping 32×32 patches. These patches are then processed through a siamese network to extract features. Subsequently, reference image's features, CDI's features, and their differences are concatenated. Finally, the fused features are mapped to either perceptual lossless (0) or lossy (1).

2.4   Loss Function

From Eq. 8 we can see that we turned the JND threshold estimation into an optimization problem. A self-designed loss function was developed in this paper to reduce the perceptual difference between the CDI and the reference image when the generated CDI is perceptually lossy. When the CDI is perceptually lossless, the loss function makes the network maximize pixel differences between the CDI and the reference image. To reduce the perceptual difference, we adopt the L_{VGG19} that is commonly used in style transfer tasks. It calculates semantic differences between the original image and the distorted image. By employing this loss, the network is encouraged to generate CDIs that progressively resemble original images in terms of semantics. After ensuring the CDI is perceptually lossless, pixel differences are maximized using L_{PSNR}. In addition, L_2 regularization is added to prevent overfitting. The loss function is:

$$\text{Loss} = (1 - \rho) \cdot L_{\text{PSNR}} + \rho \cdot L_{\text{VGG19}} + \rho + c'L_2, \tag{10}$$

$$\rho = \frac{1}{\text{batchsize}} \sum_{i=0}^{\text{batchsize}-1} \text{P}(\mathbf{x}_{\text{ref}}^i, \mathbf{x}_{\text{CDI}}^i), \tag{11}$$

$$L_{\text{PSNR}} = 20 \cdot \log_{10}(\frac{\text{MAX}_{\mathbf{x}_{\text{ref}}}}{\sqrt{\text{MSE}(\mathbf{x}_{\text{ref}}, \mathbf{x}_{\text{CDI}})}}), \tag{12}$$

$$L_{\text{VGG19}} = \sum_{l=1}^{n} w_l \cdot \frac{1}{C_l \cdot H_l \cdot W_l} \sum_{c=1}^{C_l} \sum_{h=1}^{H_l} \sum_{w=1}^{W_l} (\mathbf{F}_{\text{ref}}^l(c, h, w) - \mathbf{F}_{\text{CDI}}^l(c, h, w))^2, \tag{13}$$

where ρ indicates the percentage of perceptually lossless images in a batch, $c' = 1 \times 10^{-4}$ is the weight of L_2. $n = 35$ represents the number of layers in the VGG19 model. w_l are weights used to balance the losses across different layers. C_l, H_l, and W_l denote the channel, height, and width dimensions of the l-th layer's feature map, respectively. $\mathbf{F}_{\text{ref}}^l(c, h, w)/\mathbf{F}_{\text{CDI}}^l(c, h, w)$ refers to the feature value at position (c, h, w) in the l-th layer for the reference image/CDI.

3   Experiments

3.1   Sonar Image JND Dataset

Due to significant differences between sonar images and NSIs, it is not appropriate to use the NSIs dataset as training data for the sonar image JND model. So we constructed a picture-level JND dataset for sonar images, named the Sonar Image Distortion Evaluation (SIDE) dataset. The reference images of the dataset were taken from the publicly available SIQD dataset [16], which contains acoustic lens sonar images and side-scan sonar images. 40 reference images were intra-frame compressed using the HEVC/H.265 [17] standard with Qp ranging from 1 to 51, resulting in 2040 compressed images. In this paper, this dataset is only used as training data for the predictor. Details are shown in Table 2.

Subjective Experiment. Thirty volunteers between the ages of 20 and 30 were invited to participate in the subjective experiment. The test was conducted according to the Double-Stimulus Impairment Scale (DSIS) method specified in ITU-R BT.500-13 [18]. They were in a laboratory environment with a mixture of fluorescent and natural light, and a distance from the monitor to the seat of 6 times the image height. A pair of images was displayed side-by-side in their own resolution on a 1920×1080 monitor. The pair of images, called "anchor image" and "comparison image", appeared randomly to the left and right. Then subjects were asked to judge whether the two images were different according to their perception and to make a choice. The "comparison image" for each image was updated using a relaxed binary search algorithm [19].

Table 2. Some details in SIDE dataset.

Characteristics	Information
Dataset	Single-JND for sonar images
Compression	HEVC/H.265 intra (bpg-0.9.8), Qp=[1, 51]
Content	40 original images; 2040 distorted images
Resolution	320×320
Subjective testers	30

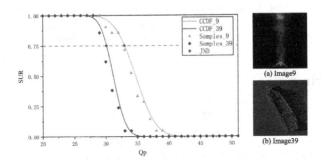

Fig. 4. SUR/Qp curves of image9 and image39. The Qp of 75% user satisfaction is taken as the JND threshold.

After obtaining all the sample data, abnormal subjects and abnormal samples were first screened. Then we modeled the JND distribution as a normal distribution, where the mean and standard deviation values are derived from the subjective test data. The Complementary Cumulative Distribution Function (CCDF) is derived as the Satisfied User Ratio (SUR) function, and the point with 75% SUR is used as the JND threshold. Examples are shown in Fig. 4.

3.2   Training Details

Acoustic lens sonar is a commonly used imaging sonar and we have chosen it as the training data for the generator. Since the sonar material is difficult to obtain, we used 27 acoustic lens sonar videos, extracted at a frame rate of 10, to obtain 6681 320 × 320 acoustic lens sonar images. The corresponding image complexity maps were obtained from the [14] and stitched with the reference images and edge mask maps as the input. The train and test sets were split eight to two, the batch size was set to 4. Adam was taken as the optimizer, with the initial learning rate was 5×10^{-4}. All tests were conducted on the Pytorch software platform on a computer with a 4.90 GHz CPU, 64.00 GB of RAM and an RTX3090.

Noise Injection. We selected 15 images from our test set. Some of these images contain a large number of smooth areas, while others are rich in detail and texture. The content of these images covers several scenarios that are likely to occur in underwater exploration, such as underwater landforms, underwater creatures, aircraft and ship wrecks, swimmers, etc. Equation 14 is used to inject noise into the original image. The more accurate JND model is able to mask as much noise as possible at the same noise level (*i.e.*, the PSNR value is similar). It shows high quality in the objective evaluation and better quality in the subjective evaluation.

$$\mathbf{I}'(i,j) = \mathbf{I}(i,j) + \alpha \cdot \mathbf{N}(i,j) \cdot \text{JND}_{\text{map}}(i,j) \tag{14}$$

where $\mathbf{I}'(i,j)$ is the image after noise injection, $\mathbf{I}(i,j)$ is the reference image, α denotes the noise modulation factor, $\mathbf{N}(i,j)$ represents ×1 bipolar random noise, and $\text{JND}_{\text{map}}(i,j)$ refers to the JND map estimated by the JND model.

3.3   JND Models Comparison

Details Comparison. We selected five JND models for comparison. Classic pixel-level JND model: Wu2013 [20], Wu2017 [5]. Screen image JND model: Scr2021 [21]. Transform domain JND model: Wan2017 [22], KLT2022 [13]. The JND maps of the 15 images were calculated using these models and our model, respectively. Then the same level of noise was injected according to Eq. 14, as shown in Fig. 5.

It can be seen that the Wan2017 noise is very obvious. The KLT2022 model injects noise into the micro-structure, but leaves too much integrity in the macro-structure. Wu2017 introduces more noise into the high pattern complexity region, but obvious noise can be seen in the texture region and the figures. This could be attributed to the speckle noise present in sonar images, leading to an over-estimation of the JND threshold in texture regions. Wu2013 introduces noise into the disordered region with better results, but the speckle noise may affect the segregation of the ordered disordered region, so the distortion is more obvious in some regions, such as the fishing frame. The Scr2021 model introduces noise into areas of low directional sensitivity, and shows better results. However, it does not perform well in some regions. For example, the edge of the fishing frame is a significant region of the sonar image, so the threshold is estimated

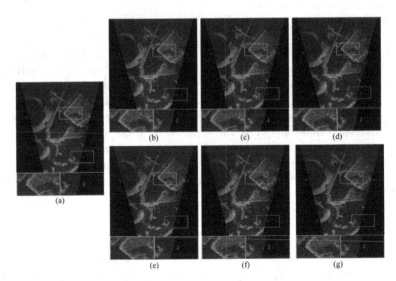

Fig. 5. Visual comparison of different JND models. (a) reference image, (b) KLT2022 [13], (c) Scr2021 [21], (d) Wan2017 [22], (e) Wu2013 [20], (f) Wu2017 [5], (g) the proposed method. Images (b)-(g) have the same noise level PSNR = 30.5 dB.

too high and the burr is obvious. Our method performs best at the same noise level, and can better delineate the region using U-net's powerful segmentation capabilities. And the direction of optimization is constrained by the perceptually lossy/lossless predictor, which can better control the threshold from being overestimated.

Objective Test. We used the full reference sonar image quality assessment algorithm [16] for objective evaluation. The objective results are shown in Table 3. At a noise level of PSNR is about 30.5dB, our model has the highest quality of noise injection images and Scr2021 is the second highest.

Table 3. Objective IQA Test.

Models	KLT2022	Scr2021	Wan2017	Wu2013	Wu2017	Ours
SIQP	59.8444	60.6666	58.7498	60.3887	59.4583	**62.0968**

Subjective Test. The subjective experiment was performed using the DSIS method of ITU-R BT.500-13. Specifically, two noise injection images obtained from the same reference image were displayed side by side on a 1920×1080 monitor. One is the noise injection image derived from our model and the other is the noise injection image derived from one of the other five models. They appeared

Table 4. Scores for Subjective Test.

Score	Description
-3	The right one is much worse than the left one
-2	The right one is worse than the left one
-1	The right one is slightly worse than the left one
0	The right one has the same quality as the left one
1	The right one is slightly better than the left one
2	The right one is better than the left one
3	The right one is much better than the left one

randomly on the left and right. The evaluation criteria and scores are shown in Table 4, subjects were asked to score according to the criteria. Fifteen volunteers were invited to perform the subjective quality test. The results of the subjective test are shown in Table 5. There are two indicators in the table, the "Mean" represents the average score of the subjects and the "Std" represents the degree of agreement between the subjects' opinions. A positive "Avg" means that our model is better, and a negative value means that the comparison model is better. It can be seen that our model shows the best performance.

Table 5. Quality Comparision through Subjective Testing.

Image Id	Ours vs.									
	KLT2022		Scr2021		Wan2017		Wu2013		Wu2017	
	Mean	Std	Mean	Std	Mean	Std	Mean	Std	Mean	Std
1	2.13	0.74	0.60	0.91	2.27	0.59	0.60	0.83	1.80	1.01
2	1.93	0.80	0.13	1.13	2.27	0.70	0.47	0.74	2.07	0.59
3	2.20	0.56	0.73	0.88	1.80	0.68	0.80	0.77	2.07	0.70
4	1.40	0.63	0.40	0.83	2.27	0.59	0.73	0.59	1.80	1.21
5	2.27	0.88	0.20	0.41	2.47	0.64	0.40	0.74	1.53	1.30
6	1.80	0.56	0.13	0.64	2.00	0.76	0.53	0.92	1.53	0.74
7	2.27	0.70	0.13	0.52	2.33	0.62	0.53	0.52	1.73	0.46
8	2.40	0.63	0.13	0.35	2.13	0.74	0.67	0.62	1.87	0.99
9	1.87	0.74	0.13	0.52	1.67	0.62	0.27	0.88	1.00	1.00
10	2.00	0.65	0.27	0.59	2.33	0.62	0.67	0.49	1.33	0.62
11	0.87	1.06	0.47	0.52	2.47	0.52	0.33	1.05	1.67	0.72
12	2.00	0.76	0.47	0.74	2.53	0.52	0.40	0.63	2.20	0.68
13	0.93	1.39	-0.07	0.70	0.73	0.88	0.60	0.63	0.93	0.88
14	1.33	0.82	0.27	0.59	1.13	0.92	0.53	0.83	1.00	0.85
15	0.73	1.10	0.13	0.74	0.40	1.30	0.60	0.63	1.13	0.99
Avg	1.74	0.80	0.28	0.67	1.92	0.71	0.54	0.72	1.58	0.85

4 Conclusion

In this paper, we successfully estimated pixel-level JND thresholds using an inexact unsupervised approach. The loss function designed for the model was able to constrain the model to produce a perceptually lossless CDI that differs the most from the original image. This allowed the estimated JND map to more accurately characterize the visual redundancy. Furthermore, we innovatively investigated the relationship between the JND and the image complexity and edges of sonar images. They were incorporated into the information reference module to make the model more applicable to sonar images. Experimental results indicate that the proposed model can better estimate the perceptual redundancy of sonar images.

References

1. Lin, W., Ghinea, G.: Progress and opportunities in modelling just-noticeable difference (JND) for multimedia. IEEE Trans. Multimedia **24**, 3706–3721 (2021)
2. Chou, C.H., Li, Y.C.: A perceptually tuned subband image coder based on the measure of just-noticeable-distortion profile. IEEE Trans. Circuits Syst. Video Technol. **5**(6), 467–476 (1995)
3. Yang, X., Ling, W., Lu, Z., Ong, E.P., Yao, S.: Just noticeable distortion model and its applications in video coding. Sig. Process. Image Commun. **20**(7), 662–680 (2005)
4. Liu, A., Lin, W., Paul, M., Deng, C., Zhang, F.: Just noticeable difference for images with decomposition model for separating edge and textured regions. IEEE Trans. Circuits Syst. Video Technol. **20**(11), 1648–1652 (2010)
5. Wu, J., Li, L., Dong, W., Shi, G., Lin, W., Kuo, C.C.J.: Enhanced just noticeable difference model for images with pattern complexity. IEEE Trans. Image Process. **26**(6), 2682–2693 (2017)
6. Wang, H., Yu, L., Liang, J., Yin, H., Li, T., Wang, S.: Hierarchical predictive coding-based JND estimation for image compression. IEEE Trans. Image Process. **30**, 487–500 (2020)
7. Ahumada Jr, A.J., Peterson, H.A.: Luminance-model-based DCT quantization for color image compression. In: Human Vision, Visual Processing, and Digital Display III, vol. 1666, pp. 365–374. SPIE (1992)
8. Watson, A.B.: DCT quantization matrices visually optimized for individual images. In: Human Vision, Visual Processing, and Digital Display IV, vol. 1913, pp. 202–216. SPIE (1993)
9. Hontsch, I., Karam, L.J.: Adaptive image coding with perceptual distortion control. IEEE Trans. Image Process. **11**(3), 213–222 (2002)
10. Tong, H.H., Venetsanopoulos, A.N.: A perceptual model for JPEG applications based on block classification, texture masking, and luminance masking. In: Proceedings 1998 International Conference on Image Processing. ICIP98 (Cat. No. 98CB36269), pp. 428–432. IEEE (1998)
11. Liu, H., et al.: Deep learning-based picture-wise just noticeable distortion prediction model for image compression. IEEE Trans. Image Process. **29**, 641–656 (2019)
12. Tian, T., Wang, H., Kwong, S., Kuo, C.C.J.: Perceptual image compression with block-level just noticeable difference prediction. ACM Trans. Multimedia Comput. Commun. Appl. (TOMM) **16**(4), 1–15 (2021)

13. Jiang, Q., Liu, Z., Wang, S., Shao, F., Lin, W.: Toward top-down just noticeable difference estimation of natural images. IEEE Trans. Image Process. **31**, 3697–3712 (2022)

14. Feng, T.: IC9600: a benchmark dataset for automatic image complexity assessment. IEEE Trans. Pattern Anal. Mach. Intell. **45**, 8577–8593 (2022)

15. Ronneberger, O., Fischer, P., Brox, T.: U-Net: convolutional networks for biomedical image segmentation. In: Navab, N., Hornegger, J., Wells, W.M., Frangi, A.F. (eds.) MICCAI 2015. LNCS, vol. 9351, pp. 234–241. Springer, Cham (2015). https://doi.org/10.1007/978-3-319-24574-4_28

16. Chen, W., Gu, K., Lin, W., Yuan, F., Cheng, E.: Statistical and structural information backed full-reference quality measure of compressed sonar images. IEEE Trans. Circuits Syst. Video Technol. **30**(2), 334–348 (2019)

17. Kim, J., Bae, S.H., Kim, M.: An HEVC-compliant perceptual video coding scheme based on JND models for variable block-sized transform kernels. IEEE Trans. Circuits Syst. Video Technol. **25**(11), 1786–1800 (2015)

18. BT.500-13, I.R.: Methodology for the subjective assessment of the quality of television pictures (2012)

19. Wang, H., et al.: VideoSet: a large-scale compressed video quality dataset based on JND measurement. J. Vis. Commun. Image Represent. **46**, 292–302 (2017)

20. Wu, J., Shi, G., Lin, W., Liu, A., Qi, F.: Just noticeable difference estimation for images with free-energy principle. IEEE Trans. Multimedia **15**(7), 1705–1710 (2013)

21. Wang, M., Liu, X., Xie, W., Xu, L.: Perceptual redundancy estimation of screen images via multi-domain sensitivities. IEEE Signal Process. Lett. **28**, 1440–1444 (2021)

22. Wan, W., Wu, J., Xie, X., Shi, G.: A novel just noticeable difference model via orientation regularity in DCT domain. IEEE Access **5**, 22953–22964 (2017)

Discriminative Activation of Information Is What You Need in Image Super-Resolution Transformer

Yixin Qian[✉]

School of Informatics, Xiamen University, Fujian 361005, People's Republic of China
qianyixin@stu.xmu.edu.cn

Abstract. Despite the significant progress made by Transformer in image super-resolution tasks, it has not effectively utilized prior knowledge in the image frequency domain and differentiated the processing of high-frequency and low-frequency information in the image. Previous studies on image super-resolution have shown that the high-frequency and low-frequency regions of the image exhibit distinct differences during the super-resolution process. In this paper, we propose a Discriminative Information Activation Super-Resolution Transformer (DIAST) to further improve the performance of Transformer in SISR tasks by discriminating high-frequency information from low-frequency information in images and discriminating cross-window information from inside-window information efficiently. Our results demonstrate that our method can further utilize the potential of the Transformer. The codes will be available at https://github.com/qyx1999/DIAST.

Keywords: Image Super-resolution · Transformer · Image Restoration

1 Introduction

The problem of Single Image Super-Resolution (SR) has been a long-standing issue in the fields of computer vision and image processing. Its objective is to generate a high-resolution image from a low-resolution input. Despite the advancements in Single Image Super-Resolution (SR), it remains an ill-posed problem due to the infinite number of possible high-resolution images. Numerous deep neural networks have been created to address this challenge [3,4,6–8,13,16,21,23]. In the past few years, Convolutional Neural Networks (CNNs) have become the dominant deep learning technique in this field, with a plethora of methods being proposed. However, with the success of Transformer [15] in natural language processing, it has also gained attention in the computer vision community. By adapting Transformer to vision tasks, it has been successfully applied in image recognition [5,14], object detection [1], and low-level image processing [3,19], including SR [5,12]. One such method, SwinIR [12], has recently been introduced and has achieved a breakthrough improvement in this task.

Q. Liu et al. (Eds.): PRCV 2023, LNCS 14435, pp. 482–493, 2024.
https://doi.org/10.1007/978-981-99-8552-4_38

Despite the significant progress of SwinIR [12] in image super-resolution tasks, it has not yet achieved a good distinction in processing high-frequency and low-frequency information in images. Previous researches on image super-resolution have shown that the high-frequency and low-frequency regions of images exhibit vastly different performance during super-resolution processes [11]. Low-frequency information in smooth regions is easier to recover during super-resolution processes, while high-frequency information in non-smooth regions is difficult to recover effectively, resulting in low-quality image restoration or the appearance of artifacts. Therefore, it is necessary to differentiate between pixels containing high-frequency information and those containing low-frequency information in super-resolution tasks. Additionally, while SwinIR [12] has a stronger ability to model local information, its range of information utilization can be further expanded by expanding its receptive field. When expanding the receptive field outside the window, it is important to consider adjusting the balance between information outside and inside the window adaptively and discriminatively.

To solve the aforementioned problems, a Discriminative Information Activation Super-Resolution Transformer (DIAST) is proposed to enhance the ability of SISR networks to discriminate high-frequency information from low-frequency information in images and discriminate cross-window information from inside-window information. The main contributions are as follows: 1) Our proposed Activating High-frequency Module (AHM) aims to extract high-frequency information and capture texture details of the image while being computationally friendly. 2) We propose Adaptive Double Window Module (ADWM), which takes into account both cross-window information and inside-window information, achieving an adaptive dynamic equilibrium between the two. 3) To enhance the network's representation ability, we have incorporated spatial attention and channel attention in addition to window-based multi-head self-attention.

2   Method

Overall Network Architecture. As depicted in Fig. 1, Discriminative Information Activation Super-Resolution Transformer (DIAST) is composed of three components: shallow feature extraction, deep feature extraction, and image reconstruction. Define I_{LR} as the input and I_{SR} as the output of DIAST. Firstly, a convolutional layer $f_s(\cdot)$ is utilized to extract the shallow feature from I_{LR}.

$$F_0 = f_s(I_{LR}) \tag{1}$$

where F_0 is the extracted shallow feature. By performing shallow feature extraction, the input can be easily mapped from a low-dimensional space to a high-dimensional space, thereby achieving a high-dimensional embedding for each pixel token. Then the extracted shallow feature F_0 is used as the input of deep feature extraction $H_{DF}(\cdot)$ with several Activating High-frequency Modules(AHMs) and Adaptive Double Window Modules (ADWMs)

$$F_{DF} = H_{DF}(F_0) \tag{2}$$

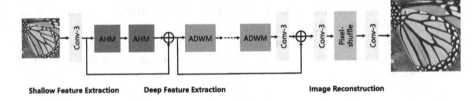

Fig. 1. Overall Network Architecture of DIAST.

where F_{DF} denotes the extracted deep feature. The process of $H_{DF}(\cdot)$ can be decomposed into the following detailed steps

$$F_{high} = H_{AHM}^2(H_{AHM}^1(F_0)) \tag{3}$$

$$F_{DF} = f_{tail}(H_{ADWM}^n(H_{ADWM}^{n-1}(...(H_{ADWM}^1(F_{high} + F_0))))) \tag{4}$$

where H_{AHM}^i represents the i-th AHM, H_{ADWM}^i represents the i-th ADWM, F_{high} denotes the high-frequency information extracted by AHMs and f_{tail} stands for a convolutional layer aiming for better aggregation of information. Then the extracted deep feature F_{DF} is fed into the reconstruction module to get the SR image I_{SR}

$$I_{SR} = f(f_p(f(F_{DF} + F_{high}))) \tag{5}$$

where f and f_p stand for the convolutional layer and Pixel-Shuffle layer, respectively.

Activating High-Frequency Module (AHM). In the process of deep feature extraction, the design purpose of the AHM module is to further activate the high-frequency information in the prior information of the image, so that it can be activated discriminatively from the low-frequency information in the smooth region of the image.

As shown in Fig. 2, the architecture of AHM module mainly consists of a high-frequency Block (HB) and Adaptive Bottleneck Blocks (ABBs). After the use of an ABB to extract the input F_n, the subsequent process can be divided into two branches, one for extracting high-frequency information and the other for extracting low-frequency information. HB is applied to calculate the high-frequency information ($Frequency_{high}$) of the features. At the same time, we reduce the size of the feature map and feed it into three ABB modules to explore the potential information by reducing feature redundancy and preserving low-frequency regions. Then the low-frequency information ($Frequency_{low}$) obtained is upsampled and concatenated with $Frequency_{high}$ processed by a single ABB for feature space alignment. After that, we use a 1×1 convolution layer and a channel attention module to reduce the channel number of concatenated feature and emphasize channels with high activation values, respectively. Finally, an ABB is employed to extract the final features, and a global residual connection is introduced to add the original features F_n to F_{n+1}. This operation can be expressed as

$$Frequency_{low} = \uparrow f_a^3(f_a^2(f_a^1(\downarrow f_a^0(F_n)))) \tag{6}$$

$$Frequency_{high} = f_a(H_{HB}(f_a^0(F_n))) \tag{7}$$

$$F_{n+1} = F_n + f_a(f_{ca}(f([Frequency_{low}, Frequency_{high}]))) \tag{8}$$

where $\uparrow$ and $\downarrow$ denote the upsampling and downsampling operations, respectively. f and f_a stand for the convolutional layer and ABB module. f_{ca} denotes the channel attention module. H_{HB} denotes HB module.

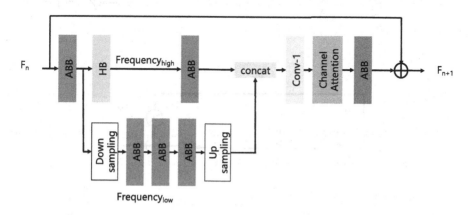

Fig. 2. Activating High-frequency Module (AHM).

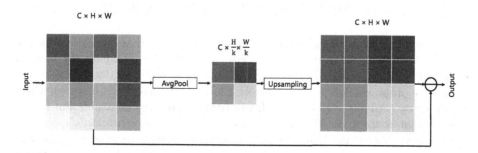

Fig. 3. High-frequency Block (HB).

High-Frequency Block (HB). As depicted in Fig. 3, assuming that the input feature has a dimension of $C \times H \times W$, the HB module first downsampling it using an average pooling layer whose kernel size is $k \times k$, and then upsampling with scaling factor of k. This operation preserves the low-frequency information of the smooth region of the original input feature. Finally, the feature containing low-frequency information is element-wise subtracted from the original input feature to obtain high-frequency information. The efficient process of High-frequency Block can be formulated as

$$F_{output} = F_{input} - \uparrow (avgpool(F_{input}))) \tag{9}$$

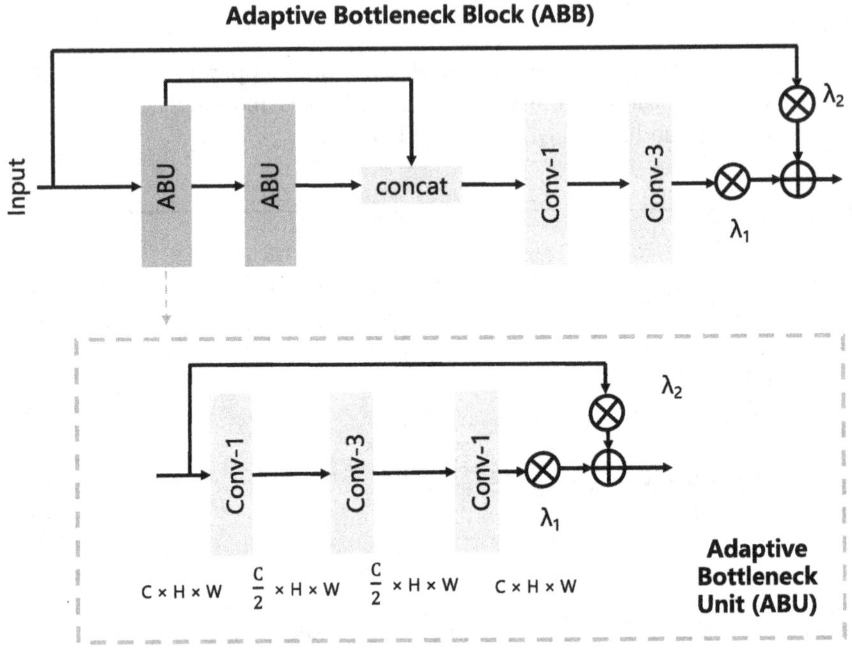

Adaptive Bottleneck Block (ABB)

Fig. 4. Adaptive Bottleneck Block (ABB).

Adaptive Bottleneck Block (ABB). ResNet has shown that by increasing the depth of a model, the residual architecture can effectively address the issue of gradient vanishing and enhance the model's ability to represent data. As shown in Fig. 4, we propose a Adaptive Bottleneck Block (ABB) to extract features, which contains two Adaptive Bottleneck Units (ABUs) and two convolutional layers. The design concept of ABU is to enhance feature representation capabilities without excessively increasing additional computational costs. The ABU module includes two convolutional layers whose kernel size is 1×1 and one convolutional layer whose kernel size is 3×3. Firstly, the first convolutional layer with a kernel size of 1×1 is used to perform a channel down operation on the input features. Then, the features are extracted from the convolutional layer with a kernel size of 3×3. Finally, the second convolutional layer with a kernel size of 1×1 is used to perform a channel up operation. At the same time, two adaptive weights were added to the residual operation for dynamic adjustment of the importance of residual path and identity path. The process of Adaptive Bottleneck Unit can be expressed as

$$Y_{ABU} = \lambda_2 \cdot X_{ABU} + \lambda_1 \cdot f_1(f_3(f_1(X_{ABU}))) \tag{10}$$

where f_1 and f_3 stand for the convolutional layer with a kernel size of 1×1 and 3×3 , respectively. X_{ABU} is input of ABU and Y_{ABU} is the output of ABU. λ_2 and λ_1 denote two adaptive weights added to the residual operation. Meanwhile, the whole process of Adaptive Bottleneck Block can be expressed as

$$Y_{ABB} = \lambda_2 \cdot X_{ABB} + \lambda_1 \cdot f_3(f_1([H_{ABU}(X_{ABB}), H_{ABU}(H_{ABU}(X_{ABB}))])) \quad (11)$$

where H_{ABU} stands for the operation of Adaptive Bottleneck Unit. X_{ABB} is input of ABB and Y_{ABB} is the output of ABB. λ_2 and λ_1 denote two adaptive weights added to the residual operation as well.

Adaptive Double Window Module (ADWM). As shown in Fig. 5, the architecture of Adaptive Double Window Module mainly consists of several Spatial Attention and Channel Attention Blocks (SACABs) and a Adaptive Double Window Block (ADWB), which can be expressed as

$$Y_{ADWM} = X_{ADWM} + f(H_{ADWB}(H_{SACAB}^n(H_{SACAB}^{n-1}(...(H_{SACAB}^1(X_{ADWM})))))) \quad (12)$$

where H_{ADWB} and H_{SACAB}^i denote the operation of ADWB and i-th SACAB. X_{ADWM} is input of ADWM and Y_{ADWM} is the output of ADWM.

Spatial Attention and Channel Attention Block (SACAB). Inspired by many works illustrating that convolution can help Transformer get better visual

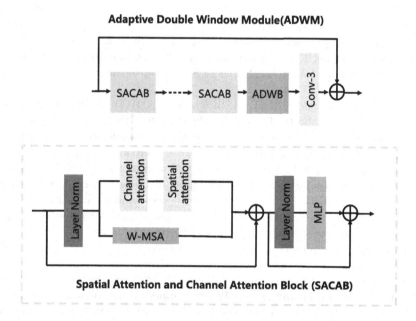

Fig. 5. Adaptive Double Window Module (ADWM) and Spatial Attention and Channel Attention Block (SACAB).

representation [9,17,18,20,24], we modify the standard Transformer block by incorporating a convolution block based on spatial attention and channel attention, in order to further enhance the representation ability of the network. As shown in Fig. 5, a convolution block based on spatial attention and channel attention is inserted into the standard Swin Transformer block [12] in parallel with the window-based multi-head self-attention (W-MSA) module. The whole process of SACAB is computed as

$$F_i = F_{input} + \alpha \cdot SA(CA(LN(F_{input}))) + W - MSA(LN(F_{input})) \qquad (13)$$

$$F_{output} = F_i + MLP(LN(F_i)) \qquad (14)$$

where F_{input}, F_i and F_{output} denote the input feature, intermediate feature and output feature, respectively. LN and MLP represent a LayerNorm layer and a multi-layer perceptron. SA, CA, $W - MSA$ stand for the process of spatial attention, channel attention and W-MSA module. α is a small constant for avoidance of the possible conflict among different attention mechanisms on optimization and visual representation. W-MSA module in swin transformer [12] mainly partition the input feature in to local windows of size $M \times M$ and compute $query$, key and $value$ matrices by linear mappings as Q, K and V, which are used to calculate self-attention inside each window. The window-based self-attention is expressed as

$$Attention(Q, K, V) = SoftMax(QK^T/\sqrt{d} + B)V \qquad (15)$$

where d represents the dimension of $query/key$. B denotes the relative position encoding and is calculated as [15]. In addition, we employ the shifted window partitioning approach [12] to establish connections between adjacent non-overlapping windows. The shift size is set to half of the window size. The detailed process can refer to the original paper of Swin Transformer [12].

Adaptive Double Window Block (ADWB). The ADWB module we propose is an improvement on the traditional W-MSA module to enhance the network's representation ability. The ADWB module mainly includes two sliding windows with inconsistent sizes, one inside and the other outside. The larger window is called the outer window to establish cross-window connections, while the smaller window is called the inner window. The inner window is used to dynamically adjust the balance between the information of cross-window connections and the information inside the window, as well as to compensate for the unevenness of corners in the feature extraction process of the outer window.

Specifically, assuming that the W-MSA module uses a window of size $M \times M$ with the stride of M to partition features and calculate the Q, K and V matrices to obtain self-attention, the outer window of the ADWB module calculates the Q matrix in the same way as the W-MSA module, but double increases the window size as $M_o = 2 \times M$ for the k and v matrices while maintaining the same stride of M and padding the input features to ensure that the number of windows calculating Q, K and V matrices remain consistent.

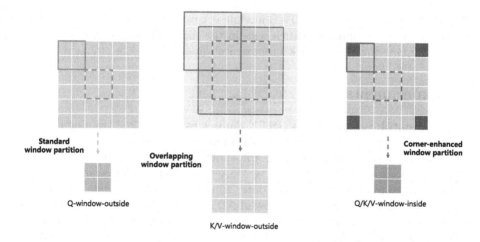

Fig. 6. Adaptive Double Window Block (ADWB).

For a clearer explanation, Fig. 6 demonstrates the calculation process for the Q matrix of the outer window, the K and V matrices of the outer window, and the Q, K, and V matrices of the inner window in a left-to-right sequence. Assuming the size of input feature size is 6×6, following the W-MSA module, the outer window of the ADWB module uses a window size of 2×2 with stride of 2 to partition the input feature into 9 windows and calculate the Q matrix. For the K and V, the size of input feature is transformed to 8×8 after zero-padding operation, and then a window size of 4 with stride of 2 is used to partition the transformed feature into 9 windows and calculate the K and V matrices, thus obtaining self-attention within the outer window.

However, during the partition process of the outer window, the information extraction of the original 6×6 feature is uneven. As can be seen from Fig. 6, although large windows overlap each other, only one window covers four corners of the feature, while the other areas of the feature are covered by two windows. The ignored four corners are marked with white patterns on a blue background in Fig. 6. This imbalance phenomenon is common during the sliding process of large windows, when the window size is twice the step size. Compared to the simplified example in Fig. 6, when the step size is larger, the ignored four corners will contain more feature information. In order to solve this problem, in the calculation of the inner window, the four corners of the 6×6 feature will be enhanced by a constant, and then the calculation of self-attention in the inner window is consistent with the W-MSA module. A window size of 2×2 with stride of 2 to partition the enhanced feature into 9 windows and calculate the Q, K and V matrices. The inner window self-attention and outer window self-attention obtained above will be multiplied by adaptive weights before adding, aiming for dynamically adjustment of the balance between the information of cross-window connections and the information inside the window. It is worth mentioning that although previous work on HAT [2] has also proposed increasing the window size in the W-MSA module, our ADWB module is more comprehensive compared to it, making up for important deficiencies in the sliding process of large windows and also improving the network's representation ability.

3   Experiments

Experimental Setup. We use DF2K dataset as the original training dataset. For the structure of DIAST, the ADWM number and SACAB number are both set to 7. The channel number of the whole network is set to 180. The attention head number are set to 6 for both W-MSA and ADWM. The window size are both set to 16 for W-MSA and the outer window of ADWM, while the window size are set to 32 for the inner window of ADWM. We set the weighting factor α of SACAB as 0.01. To evaluate the quantitative performance, PSNR and SSIM are reported. We use $L1$ loss to optimize the parameters as it can produce more sharp images compared to $L2$ loss.

Comparisons with Advanced SISR Models

Quantitative Results. Quantitative comparison with advanced methods on benchmark datasets is shown in Table 1

It is obvious that our performance improvement is more significant on datasets rich in high-frequency information such as Urban100 dataset.

Table 1. Quantitative comparison with advanced methods on benchmark datasets. Best and second best results are in red and blue colors.

Method	Training Dataset	Scale	Set5	Set14	B100	Urban100	Manga109
Bicubic	–	×2	33.66/0.9299	30.24/0.8688	29.56/0.8431	26.88/0.8403	30.80/0.9339
EDSR [13]	DIV2K	×2	37.99/0.9604	33.58/0.9175	32.16/0.8995	31.99/0.9272	38.54/0.9768
RCAN [22]	DIV2K	×2	38.27/0.9614	34.12/0.9216	32.41/0.9027	33.34/0.9384	39.44/0.9786
SwinIR [12]	DF2K	×2	38.42/0.9623	34.46/0.9250	32.53/0.9041	33.81/0.9427	39.92/0.9797
EDT [10]	DF2K	×2	38.45/0.9624	34.57/0.9258	32.52/0.9041	33.80/0.9425	39.93/0.9800
HAT [2]	DF2K	×2	38.63/0.9630	34.86/0.9274	32.62/0.9053	34.45/0.9466	40.26/0.9809
DIAST	DF2K	×2	38.66/0.9632	34.89/0.9274	32.68/0.9055	34.56/0.9474	40.24/0.9812
Bicubic	–	×3	30.39/0.8682	27.55/0.7742	27.21/0.7385	24.46/0.7349	26.95/0.8556
EDSR [13]	DIV2K	×3	34.39/0.9271	30.29/0.8417	29.08/0.8055	28.16/0.8528	33.44/0.9438
RCAN [22]	DIV2K	×3	34.74/0.9299	30.65/0.8482	29.32/0.8111	29.09/0.8702	34.44/0.9499
SwinIR [12]	DF2K	×3	34.97/0.9318	30.93/0.8534	29.46/0.8145	29.75/0.8826	35.12/0.9537
EDT [10]	DF2K	×3	34.97/0.9316	30.89/0.8527	29.44/0.8142	29.72/0.8814	35.13/0.9534
HAT [2]	DF2K	×3	35.06/0.9329	31.08/0.8555	29.54/0.8167	30.23/0.8896	35.53/0.9552
DIAST	DF2K	×3	34.99/0.9330	31.15/0.8559	29.58/0.8170	30.35/0.8899	35.34/0.9538
Bicubic	–	×4	28.42/0.8104	26.00/0.7027	25.96/0.6675	23.14/0.6577	24.89/0.7866
EDSR [13]	DIV2K	×4	32.09/0.8939	28.59/0.7815	27.56/0.7355	26.09/0.7852	30.34/0.9068
RCAN [22]	DIV2K	×4	32.63/0.9002	28.87/0.7889	27.77/0.7436	26.82/0.8087	31.22/0.9173
SwinIR [12]	DF2K	×4	32.92/0.9044	29.09/0.7950	27.92/0.7489	27.45/0.8254	32.03/0.9260
EDT [10]	DF2K	×4	32.82/0.9031	29.09/0.7939	27.91/0.7483	27.46/0.8246	32.05/0.9254
HAT [2]	DF2K	×4	33.04/0.9056	29.23/0.7973	28.00/0.7517	27.97/0.8368	32.48/0.9292
DIAST	DF2K	×4	33.05/0.9061	29.29/0.7973	27.91/0.7499	28.09/0.8379	32.49/0.9295

Table 2. Study of architecture of DIAST on Urban100(×4).

Case Index	1	2	3	4
HB	✓	✓		
CASAB	✓		✓	
PSNR	28.09 dB	27.96 dB	27.59 dB	27.47 dB

Visual comparison. As is shown in Fig. 7, our method outperforms other models in restoring edge information in high-frequency areas of images.

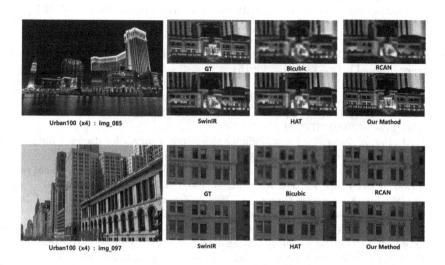

Fig. 7. Visual comparison with advanced methods.

Ablation Study. We explore the effectiveness of HB and CASAB in Table 2. From the table, it can be seen that these two modules can significantly improve the representational ability of the model.

4   Conclusion

In this paper, inspired by prior knowledge in the frequency domain of images, we propose a Discriminative Information Activation Super-Resolution Transformer (DIAST) to further improve the performance of Transformer in SISR tasks by efficiently discriminating high-frequency information from low-frequency information in images and discriminating cross-window information from inside-window information. We also incorporate spatial attention and channel attention in addition to window-based multi-head self-attention to the network's representation ability. The results demonstrate that our method can better utilize the potential of the Transformer.

References

1. Carion, N., Massa, F., Synnaeve, G., Usunier, N., Kirillov, A., Zagoruyko, S.: End-to-end object detection with transformers. In: Vedaldi, A., Bischof, H., Brox, T., Frahm, J.-M. (eds.) ECCV 2020. LNCS, vol. 12346, pp. 213–229. Springer, Cham (2020). https://doi.org/10.1007/978-3-030-58452-8_13
2. Chen, X., Wang, X., Zhou, J., Dong, C.: Activating more pixels in image super-resolution transformer. arXiv e-prints (2022)
3. Dai, T., Cai, J., Zhang, Y., Xia, S.T., Zhang, L.: Second-order attention network for single image super-resolution. In: Proceedings of the IEEE/CVF Conference on Computer Vision and Pattern Recognition, pp. 11065–11074 (2019)
4. Dong, C., Loy, C.C., He, K., Tang, X.: Image super-resolution using deep convolutional networks. IEEE Trans. Pattern Anal. Mach. Intell. **38**(2), 295–307 (2016)
5. Dosovitskiy, A., et al.: An image is worth 16×16 words: transformers for image recognition at scale. arXiv preprint arXiv:2010.11929 (2020)
6. Kim, J., Lee, J.K., Lee, K.M.: Accurate image super-resolution using very deep convolutional networks. IEEE (2016)
7. Li, J., Fang, F., Mei, K., Zhang, G.: Multi-scale residual network for image super-resolution. In: Ferrari, V., Hebert, M., Sminchisescu, C., Weiss, Y. (eds.) ECCV 2018. LNCS, vol. 11212, pp. 527–542. Springer, Cham (2018). https://doi.org/10.1007/978-3-030-01237-3_32
8. Li, J., Pei, Z., Zeng, T.: From beginner to master: a survey for deep learning-based single-image super-resolution. arXiv preprint arXiv:2109.14335 (2021)
9. Li, K., et al.: UniFormer: unifying convolution and self-attention for visual recognition. IEEE Trans. Pattern Anal. Mach. Intell. **45**, 12581–12600 (2023)
10. Li, W., Lu, X., Lu, J., Zhang, X., Jia, J.: On efficient transformer and image pre-training for low-level vision. arXiv e-prints (2021)
11. Liang, J., Zeng, H., Zhang, L.: Details or artifacts: a locally discriminative learning approach to realistic image super-resolution. In: Proceedings of the IEEE/CVF Conference on Computer Vision and Pattern Recognition, pp. 5657–5666 (2022)
12. Liang, J., Cao, J., Sun, G., Zhang, K., Timofte, R.: SwinIR: image restoration using swin transformer. IEEE (2021)
13. Lim, B., Son, S., Kim, H., Nah, S., Lee, K.M.: Enhanced deep residual networks for single image super-resolution. IEEE (2017)
14. Touvron, H., Cord, M., Douze, M., Massa, F., Sablayrolles, A., Jégou, H.: Training data-efficient image transformers & distillation through attention. In: International Conference on Machine Learning, pp. 10347–10357. PMLR (2021)
15. Vaswani, A., et al.: Attention is all you need. In: Advances in Neural Information Processing Systems 30 (2017)
16. Wang, Y., Wang, L., Wang, H., Li, P.: Resolution-aware network for image super-resolution. IEEE Trans. Circuits Syst. Video Technol. **29**(5), 1259–1269 (2018)
17. Wu, H., et al.: CvT: introducing convolutions to vision transformers. In: Proceedings of the IEEE/CVF International Conference on Computer Vision, pp. 22–31 (2021)
18. Xiao, T., Singh, M., Mintun, E., Darrell, T., Dollár, P., Girshick, R.: Early convolutions help transformers see better. Adv. Neural. Inf. Process. Syst. **34**, 30392–30400 (2021)
19. Yang, F., Yang, H., Fu, J., Lu, H., Guo, B.: Learning texture transformer network for image super-resolution. In: Proceedings of the IEEE/CVF Conference on Computer Vision and Pattern Recognition, pp. 5791–5800 (2020)

20. Yuan, K., Guo, S., Liu, Z., Zhou, A., Yu, F., Wu, W.: Incorporating convolution designs into visual transformers. In: Proceedings of the IEEE/CVF International Conference on Computer Vision, pp. 579–588 (2021)

21. Zewei, H., Siliang, T., Jiangxin, Y., Yanlong, C., Ying, Y.M., Yanpeng, C.: Cascaded deep networks with multiple receptive fields for infrared image super-resolution. IEEE Trans. Circ. Syst. Video Technol. 1 (2018)

22. Zhang, Y., Li, K., Li, K., Wang, L., Zhong, B., Fu, Y.: Image super-resolution using very deep residual channel attention networks. In: Ferrari, V., Hebert, M., Sminchisescu, C., Weiss, Y. (eds.) ECCV 2018. LNCS, vol. 11211, pp. 294–310. Springer, Cham (2018). https://doi.org/10.1007/978-3-030-01234-2_18

23. Zhang, Y., Tian, Y., Kong, Y., Zhong, B., Fu, Y.: Residual dense network for image super-resolution. In: Proceedings of the IEEE Conference on Computer Vision and Pattern Recognition, pp. 2472–2481 (2018)

24. Zhao, Y., Wang, G., Tang, C., Luo, C., Zeng, W., Zha, Z.J.: A battle of network structures: an empirical study of CNN, transformer, and MLP. arXiv preprint arXiv:2108.13002 (2021)

SentiImgBank: A Large Scale Visual Repository for Image Sentiment Analysis

Yazhou Zhang[1,2,3] , Yu He[3], Rui Chen[3], and Lu Rong[4(✉)]

[1] Artificial Intelligence Laboratory, China Mobile Communication Group Tianjin Co., Ltd., Tianjin 300456, China
[2] Key Laboratory of Dependable Service Computing in Cyber Physical Society (Chongqing University), Ministry of Education, Chongqing, China
[3] Software Engineering College, Zhengzhou University of Light Industry, Zhengzhou 450001, China
[4] Human Resources Office, Zhengzhou University of Light Industry, Zhengzhou 450001, China
lurong2013@outlook.com

Abstract. In contrast to existing methods which detect sentiment directly from low/high-level features, we construct a large-scale visual repository, namely SentiImgBank, with the aim of providing a benchmarking visual sentiment lexicon. More specially, the SentiImgBank consists of 24 categories, 5,487 adjective-noun pairs (ANPs), in total of 648,946 images that are collected from social media such as Twitter. In view that ANPs might express different sentiments in different contexts, i.e., contextuality, SentiImgBank annotates the discrete sentiment and emotion scores instead of directly defining the golden label. Hence, each image is associated with ten numerical scores, where three of them are sentiment scores, the remaining seven scores denote different emotions. To alleviate the manually annotation cost, a committee of 15 pre-trained language models based classifiers is proposed to automatically produce the sentiment and emotion scores. Finally, the strong baselines are proposed to evaluate the potential of SentiImgBank. We hope this study provides a publicly available resource for visual sentiment analysis. The full dataset will be publicly available for research (https://github.com/anonymity2024/SentiImgBank).

Keywords: Visual sentiment analysis · Emotion recognition · Opinion mining · Dataset

1 Introduction

The widespread popularity of social media has enriched the ways of human communication, which provides more comprehensive and diversified data sources,

---

Supported by National Science Foundation of China under grant No. 62006212, 61702462, Fellowship from the China Postdoctoral Science Foundation (2023M733907), Foundation of Key Laboratory of Dependable Service Computing in Cyber-Physical-Society (Ministry of Education), Chongqing University (PJ.No: CPSDSC202103).

Q. Liu et al. (Eds.): PRCV 2023, LNCS 14435, pp. 494–505, 2024.
https://doi.org/10.1007/978-981-99-8552-4_39

e.g., natural language, facial expressions, etc., for sentiment analysis, and increases the difficulty and challenge due to the challenge of mining sentimental knowledge from multiple modalities.

The recent studies have mainly focused on exploring the correlation between different modalities or extracting refined features for sentiment analysis and emotion recognition. Meanwhile, there are a few researchers who attempt to incorporate the sentimental knowledge into their models. Although periodic achievement has been made, we argue that the potential of visual and multi-modal sentiment analysis has not been fully explored. The lack of publicly available benchmark dataset of visual affective knowledge leads to the helplessness on understanding and interpreting the emotions conveyed in visual content.

To this day, there are only two visual sentiment ontologies, i.e., VSO [1] and MVSO [2]. For example, Borth et al. constructed a large-scale visual sentiment ontology (VSO), a set of 3,000 mid-level concepts using structured semantics called adjective-noun pairs (ANPs). The noun portion of the ANP allows for computer vision detectability and the adjective serves to polarize the noun toward a positive or negative sentiment. In addition, MVSO is an extension of VSO. The MVSO dataset contains more than 7.36M image, constructing 4,342 English adjective-noun pairs. Although their works have made significant contributions, they also expose a few limitations. First, all of their images are constructed based on the adjective-noun pairs (ANPs), instead of categories, failing to satisfy the rule of "representative". Second, the contextuality of ANPs has been neglected. They only assign one sentiment label to each ANP. For example, they assigned a positive value (sentiment: 1.77) to the ANP *"beautiful flower"*. However, this ANP can express opposite sentiments in different contexts. Hence, the contextuality of ANPs decides the fact that only one sentiment polarity cannot serve different affective scenarios. Finally, similar to texts, images should also exhibit polysemy, meaning that each ANP should correspond to images with varying sentiment scores. As demonstrated in Fig. 1, the SentiImgBank dataset is better able to capture the subtle differences in sentiment and emotion conveyed in images compared to the VSO.

To overcome this limitation, we create a large-scale visual sentiment resource across 24 different categories. Our category selection criteria aims to satisfy that: (1) it should cover various images that are closely related to human life; (2) it should have strong associations with sentiments and emotions; (3) the images are frequently used in practice. In line with VSO and MVSO, we adopt *Plutchik's Wheel of Emotions* [3] as the guiding principle to obtain ANPs from label words, extending the ANPs in VSO. Through all ANPs, we have collected qualified images from a wide range of social media resources such as Twitter, Getty Images and Flickr. The reason we employ ANPs is that they can transform a neutral noun (such as *"day"*) to an ANP with strong sentiment polarity (such as *"wonderful day"*) by adding an adjective term. In addition, a graded (as opposed to "hard") evaluation of the sentimental attributes of images will deepn the understanding of visual sentiment, and help the extraction of visual knowledge in different contexts. For example, images containing *"beautiful flowers"*

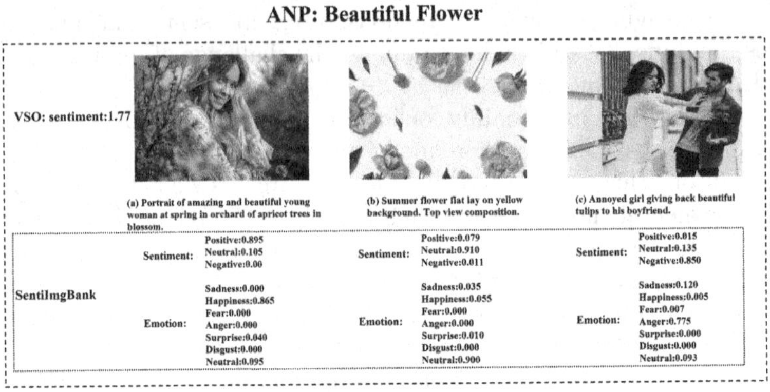

ANP: Beautiful Flower

Fig. 1. An example shows the difference between VSO and our SentiImgBank.

may express different sentiments in different scenarios, where a graded visual affective database like SentiImgBank may provide enough information to capture these nuances.

Then, in the labelling phase, a committee of 15 multi-task multi-class classifiers is proposed to automatically produce the sentiment and emotion scores. Finally, the state-of-the-art baselines are proposed to evaluate the potential of SentiImgBank. We hope this study provides a publicly available resource for visual sentiment analysis, and also benefits other multi-modal sentiment analysis communities. In summary, the major contributions of the work are:

- The first visual sentiment repository annotated with three sentiment scores and seven emotion score is created and released publicly, providing a rich knowledge base for related research such as multi-modal sentiment and emotion analysis.
- This work proposes fine-grain and graded annotations of sentiment and emotion categories. The quality control and agreement analysis are also described.
- The visual sentiment repository covers 24 different domains and themes, such as scenery, portrait, plant, animal, etc.
- The strong baselines are reported to show the potential of SentiImgBank and the need of using visual sentiment knowledge.

2 Related Work

We focus on investigating related resources covering visual and multi-modal sentiment and emotion.

The most representative work, Borth et al. constructed VSO by understanding the visual concepts that are strongly related to sentiments. VSO and MVSO are the only two datasets that cover ANPs and sentimental images with the largest scale. But there are also other multi-modal sentiment datasets, such as Multi-ZOL [4], MOSEI, IEMOCAP, MELD [5], etc. And some datasets focus on

Table 1. Comparison of SentiImgBank with other datasets.

Dataset	Size	Modality	Rosource	Annotation	Emotion Score	Catagory
IAPSa	246	I	IAPS	emotion	✗	✗
Artphoto	806	I	Deviantart.com	emotion	✗	✗
Flickr	301,903	I	Flickr.com	emotion	✗	✗
Multi-ZOL	5,288	T,I	Zol.com	sentiment	✗	✗
MOSEI	23,453	T,I,S	YouTube	sentiment, emotion	✗	✗
IEMOCAP	302	T,I,S	Performance	emotion	✗	✗
MELD	1,433	T,I,S	TV Show	sentiment, emotion	✗	✗
Emotion6	1,980	I	Flickr.com	emotion	✓	✗
SentiImageBank	**648,946**	**T,I**	**Social Media**	**sentiment, emotion**	✓	✓

exploring the rich and diverse aspects of images. The IAPSa dataset includes 246 images selected from the International Affective Picture System (IAPS), it is the first dataset annotated with discrete emotion categories. The Artphoto dataset is obtained by searching with emotion categories as keywords. The true label of each image is determined by the uploader. The Flickr dataset studies the relationship between image emotion and dialogue among friends. And the Emotion6 involves 15 labelers who give discrete emotion distributions for every image, expressing the different and nuanced emotions that the images trigger. However, the existing datasets lack comprehensiveness and diversity, as well as thoroughness and meticulousness in the annotation of emotions and sentiments. In contrast, the SentiImageBank dataset is an innovative dataset that not only annotates the images with both sentiment and emotion, but also adopts fine-grained scores to describe the subtle emotional changes triggered by the images. It allows that each image is annotated with multiple sentiment and emotion scores. Table 1 compares all above-mentioned datasets with their properties.

3   Dataset Construction

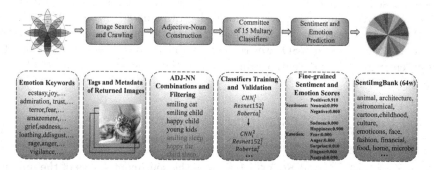

Fig. 2. Overview of the proposed framework for constructing the SentiImgBank.

An overview of the proposed method for SentiImgBank construction is shown in Fig. 2, which draws on the related techniques of MVSO and Sentiwordnet [6]. First, we obtain a set of images and their tags using seed emotion keyword queries. Next, we use a part-of-speech tagger to automatically label each image tag and combine adjective-noun pairs from the words in the tags. Then, the combinations are filtered according to sentiment, frequency and diversity filters (The red combination in Fig. 2 will be filtered out), thus expanding the ANP set in VSO. Afterward, we retrieve the images that meet the requirements from multiple media resources based on category information through all the filtered ANPs, and further filter the images to obtain a preliminary visual resource library. Then we train a set of multi-task multi-class classifiers, each of which can output the sentiment and emotion scores of an image. Since each multi-task multi-class classifier differs from the other in the training set used to train it, thus different sentiment and emotion scores may be obtained for the same image. In order to obtain the final score of an image, we normalize the scores output by each classifier and the ANP scores, and to make full use of the high-level features, we increase the weight of the sentiment scores of the ANP.

3.1   Data Acquisition

We selected the psychological ontology of *Plutchik's wheel of emotions* as seed emotion ontology because it consists of graded intensities of various basic emotions. A set of synonymous keywords for 24 Plutchik emotions was provided by 10 linguistics experts who were recruited. To identify the part-of-speech of each word in Flickr image tags, we used pre-trained Stanford tagger and HunPos tagger, which achieved high accuracy ($¿95\%$) on most languages. We discover ANPs based on the co-occurrence of adjective-noun pairs in image tags, that is, if an adjective-noun pair is relevant to the specific emotion it should appear at least once as that exact pair phrase in the crawled images for that emotion. This method has been proven to be very effective in MVSO.

From these discovered ANPs, we applied several filters to ensure that they satisfy the following criteria: (1) reflect a non-neutral sentiment, (2) frequently used and (3) contain multiple categories. To eliminate neutral candidates adjective-noun pairs, we performed sentiment scoring on ANPs. If an ANP already existed in VSO, we used the sentiment score in VSO directly. If an ANP was not in VSO, we used two publicly available sentiment ontologies: SentiStrength and SentiWordnet, to calculate the sentiment score of ANP $S(anp) \in [-2,+2]$:

$$S(anp) = \begin{cases} S(a) & : \mathrm{sgn}\{S(a)\} \neq \mathrm{sgn}\{S(n)\} \\ S(a) + S(n) & : \text{otherwise} \end{cases} \quad (1)$$

where $S(a) \in [-1,+1]$ and $S(n) \in [-1,+1]$ are the sentiment scores of the individual adjective and noun words respectively, obtained by the arithmetic mean of the SentiStrength and SentiWordnet scores of the word, *sgn* represents the sign of the score. The piecewise condition implies that if the sentiment scores of the adjective and noun exhibit inconsistent signs, we solely consider the sentiment of the adjective. This reflects our belief that the adjective holds greater influence over the emotional meaning in the adjective-noun pair. ANP candidates with a sentiment score of zero were excluded. Following MVSO, we assessed the "frequency" of ANPs based on their appearance as image tags on platforms like Flickr. A higher frequency threshold (30) was set to ensure significant usage of ANPs. To ensure diversity, we examined the category consistency and user diversity of ANPs, striving to balance the number of ANPs in each category (limited to 100) and eliminating those with fewer than three uploaders.

We categorized the ANPs into 24 categories, including animal, architecture, astronomical, cartoon, childhood, culture, emoticons, face, fashion, financial, food, home, microbe, military, object, photo, plant, politics, profession, recreation, scenery, sports, technology, and weather. Based on these categories, we utilized the finalized ANPs to retrieve images from multimedia sources like Twitter, Getty Images, and Flickr, ensuring compliance with the requirements outlined in the introduction section. Finally, we have obtained 5,487 adjective-noun pairs (ANPs), in total of 648,946 images. Table 2 presents the number of ANPs and the corresponding number of images meeting the specified criteria for each category.

Table 2. Statistics of the SentiImgBank construction.

Category	Animal	Architecture	Astronomical	Cartoon	Childhood	Culture	Emoticons	Face	Fashion	Financial	Food	Home
ANPs	235	255	209	220	302	200	130	146	292	254	218	194
Total samples	31718	32661	24933	20214	34880	22448	20712		38381	23264	22984	27159

Microbe	Military	Object	Photo	Plant	Politics	Profession	Recreation	Scenery	Sports	Technology	Weather	Total
162	312	235	200	205	297	195	306	312	198	295	115	5487
13248	25949	28020	21467	34953	31392	20139	31527	38475	21686	30201	22103	648946

3.2 Training Classifiers

Each multi-task classifier is built based on the semi-supervised method described in [7], in where only a small subset $V \subset S_i$ of the training data (denoted as S_i) have been manually labeled by five annotators. Initially, the training data in $U = S_i - V$ are unlabeled, and the process automatically generates labels by taking the labelled subset V as input of our decision committee.

(1) The seed collection V construction. The subset V (namely our seed collection) can be obtained through the manual annotation of approximately 2000 images per category, in total of 48,000 images. We recruit five well-educated volunteers including three undergraduate and two master students to take part in data annotation. They were instructed to annotate 200 images first, in order to strengthen the inter-annotator agreement, which should in principle should reach 90%.

In the annotation phase, five annotators were randomly assigned with annotation tasks. The gold standard labels of each image is determined via majority voting over all annotations. To guarantee high-quality annotations, we calculate the Fleiss' kappa scores. It is a frequently used statistic that is used to measure inter-annotator reliability, where 0.41–0.60 denotes moderate agreement, 0.61–0.80 represents substantial agreement. The agreement scores of the annotation of sentiment are $\kappa = 0.73$, $\kappa = 0.77$, $\kappa = 0.82$, $\kappa = 0.74$, $\kappa = 0.69$. The agreement scores of the annotation of emotion are $\kappa = 0.55$, $\kappa = 0.62$, $\kappa = 0.59$, $\kappa = 0.64$, $\kappa = 0.61$. This shows that five participators have reached substantial agreement on sentiment and emotion annotations.

(2) Decision committee design and training. Next, we design a committee of 15 pre-trained and large language models based classifiers. We need K iterations to generate the final visual sentiment dataset by iteratively expanding the seed collection and training 15 classifiers. Specially, at k^{th} iteration, two sets S_i^{k+1} and V_i^{k+1} are generated, where $S_i^k \supset S_i^{k-1} \supset ... \supset S_i^1 = V$ and $V_i^k \supset V_i^{k-1} \supset ... \supset V_i^1 = V$, and $V_i^k \subset S_i^k$, $V_i^{k-1} \subset S_i^{k-1}$,..., $V_i^1 \subset S_i^1$. The expansion process at k^{th} iteration can be depicted as:

- Three models CNN_i^k, $Resnet152_i^k$, and $RoBERTa_i^k$ are trained using the seed set V_i^k. The initial and most reliable three models (at $1^{st} iteration$), CNN_i^1, $Resnet152_i^1$, and $RoBERTa_i^1$, are trained from our manually annotated seed set V.
- By utilizing the models CNN_i^k, $Resnet152_i^k$, and $RoBERTa_i^k$, a subset of data from U is subjected to prediction while taking into account the sentiment scores of ANP. This process yields labeled data, S_i^{k+1}. We will augment the seed set with additional data that can minimize the disparity between the results of the three classifiers to a threshold value γ, and the seed set is expanded to V_i^{k+1}.

In [7], the author highlighted that selecting a low value of K leads to smaller training sets, which in turn yields classifiers with low recall rates but high precision. On the other hand, increasing K results in larger training sets, improving recall rates but introducing "noise" that reduces precision. Considering these factors, we chose five different values of $K \in [0, 3, 6, 9, 12]$ to define five distinct training sets. With the utilization of three learners, namely the 6-layer CNN, ResNet152, and RoBERTa, this resulted in a total of 15 multi-task multi-class classifiers. The threshold value γ is set to 0.001, and the ANP sentiment weight is set to 1.2. All the images are resized to $224 \times 224 \times 3$ before they are feed into the model.

3.3   Quality Control

Due to the complex and multifaceted nature of constructing the SentiImgBank dataset, it is inevitable that some noise is introduced into the dataset. To ensure the overall quality of the dataset, we have implemented rigorous quality control measures in the following three aspects: (a) ensuring the correctness and diversity of ANPs, (b) ensuring high-quality annotation, and (c) ensuring the quality of the models used.

(1) ANP quality control. To ensure the accuracy and diversity of ANPs, we conducted a validation task through crowdsourcing. Each ANP was evaluated by a minimum of three high-quality independent workers, selected based on their good reputation on the crowdsourcing platform. Detailed definitions of adjective-noun pairs and examples of correct and incorrect ANPs were provided to the workers. The workers were then assigned ANPs and asked to assess whether each ANP met the following criteria: (1) grammatical correctness (adjective + noun), (2) generality, indicating common usage without referring to named entities, (3) classification into one of the 24 predefined categories, and (4) semantic coherence. The crowdsourcing results indicate an average annotator agreement of 89%. To increase the credibility, the Fleiss' kappa score is also computed. The agreement scores of the annotation are $\kappa = 0.83$, $\kappa = 0.87$, $\kappa = 0.76$. This shows a consensus among workers regarding the correctness or incorrectness of ANPs. The quality of ANPs has been proved.

(2) Annotation quality. Since the annotation of 648,946 images is performed by using 15 pre-trained language models, we aim to evaluate the annotation quality. The reason is that manually annotating about 640K images is a very time-consuming task. To balance the evaluation efficiency and time cost, we propose to adopt the random sampling method. We randomly select 1000 images per category from the SentiImgBank, in total of 24,000 images. Then, we recruit three new annotators to annotate these images, and calculate the mean squared error (MSE) results between the annotators' scores and the golden scores. The results for sentiment and emotion are 0.021 and 0.039, respectively. This shows that our designed committee achieves a high level of agreement in annotation.

(3) Models quality. To verify the superiority and reliability of the classifier committee, we evaluated 15 multi-task multi-class models on the MELD dataset. The sentiment and emotion labels are consistent with the labels of our SentiImgBank. We extracted frames from the videos according to the text and conducted experiments on the obtained images and their counterparts. The experimental results are shown in Table 3.

First, we observe that the F1 scores for most sentiment categories gradually increase as the iteration increases, indicating an improvement in the classifier's performance. For the sentiment category, the CNN_i^5 model achieves F1 scores of 53.35, 69.19, and 52.64 for positive, neutral, and negative sentiments, respectively, which are notably higher compared to the initial iteration's CNN_i^1 model. Similarly, for the emotion category, the $RoBERTa_i^5$ model achieves significantly higher F1 scores compared to the initial iteration's $RoBERTa_i^1$ model. This indicates that through iterative training, all the classifiers gradually learn more accurate and expressive features related to emotions and sentiment. Lastly, we can compare this improvement with the baseline model GPT-3. By observing the F1 scores for most sentiment categories, we can conclude that the iteratively trained models consistently outperform the baseline GPT-3 model in sentiment classification. This further supports the notion that iterative training is effective in enhancing the performance of sentiment classification models.

Table 3. The results of our models on MELD.

Dataset	MELD											
F1	Sentiment				Emotion							
Methods	Positive	Neutral	Negative	Average(MF1)	Sadness	Joy	Fear	Anger	Surprise	Disgust	Neutral	Average(MF1)
CNN_t^1	50.24	64.36	46.28	37.43	39.44	45.46	8.99	29.90	29.55	5.26	67.47	39.63
$ResNet152_t^1$	34.04	49.15	44.22	40.40	26.17	39.64	31.34	36.01	29.62	27.35	56.01	42.87
$RoBERTa_t^1$	49.23	72.84	55.64	62.04	29.23	58.41	15.28	40.00	58.62	22.22	74.34	66.39
CNN_t^2	50.59	66.83	47.82	44.22	41.42	47.23	14.16	36.36	32.98	17.31	70.74	42.25
$ResNet152_t^2$	33.86	50.85	42.83	40.55	25.33	37.32	26.42	36.27	31.20	33.99	57.94	45.48
$RoBERTa_t^2$	52.30	75.52	55.72	64.42	29.57	56.55	18.18	45.26	61.29	30.38	76.89	65.74
CNN_t^3	52.56	67.31	50.28	50.56	41.12	48.65	18.35	38.79	35.13	14.81	69.92	44.96
$ResNet152_t^3$	36.56	53.15	43.22	48.36	23.37	40.39	29.09	41.02	33.38	28.39	57.13	47.69
$RoBERTa_t^3$	51.35	73.51	55.97	68.86	33.45	56.97	16.67	41.00	56.29	32.10	75.77	62.10
CNN_t^4	52.41	66.65	51.40	64.52	48.43	52.59	21.21	39.69	31.85	17.70	69.75	52.56
$ResNet152_t^4$	35.05	51.52	44.30	55.76	28.46	39.20	33.86	40.52	30.96	28.40	57.16	49.09
$RoBERTa_t^4$	49.55	73.43	56.87	70.56	31.92	55.25	17.02	41.92	57.22	29.09	75.52	64.53
CNN_t^5	53.35	69.19	52.64	60.00	47.19	52.96	18.46	38.58	38.16	12.50	71.71	50.00
$ResNet152_t^5$	76.65	81.40	63.77	73.33	65.85	41.92	33.88	39.78	62.62	54.12	57.70	57.24
$RoBERTa_t^5$	71.62	84.26	64.02	75.29	64.18	54.91	65.62	52.04	59.87	28.25	76.89	66.67
$GPT-3$	53.62	64.07	52.52	56.73	18.03	57.94	39.22	55.71	51.25	29.03	63.10	44.90

In conclusion, based on the experimental results, we also see that $RoBERTa_i^5$ achieves comparable F1 scores compared with the SOTA model (i.e., SPCL-CL-ERC [8]). This indicates that iterative training is effective in enhancing the performance of sentiment classification models. This also proves the effectiveness of our 15 multi-task classifiers in assigning sentiment and emotion labels.

4   Dataset Analysis

Table 4 presents the distribution of scores in SentiImgBank. The dataset is mostly composed of neutral samples, accounting for over 50% of the dataset. Positive and negative samples are more prevalent within the score range of 0 to 0.375, but decrease significantly beyond 0.5. This indicates a small fraction of data with clear positive or negative sentiment. Neutral samples are concentrated between scores 0.25 and 0.75, constituting 53% of the dataset.

In terms of emotions, neutral samples have the highest proportion (54%), while disgust samples have the lowest (0.1%). The scarcity of disgust samples may be due to a lack of content that elicits disgust. Happy samples have a relatively high proportion, aligning with the positive sentiment. The proportions of surprise, sadness, fear, and anger are 7%, 3%, 5%, and 1% respectively. However, their quantities sharply decline to almost zero beyond a score of 0.375.

To investigate the association between data categories and sentiment, we determined the representative sentiment and emotion category for the images by selecting the category with the highest sentiment and emotion score, excluding data with identical scores. The distribution of sentiment and emotion across categories is depicted in Fig. 3.

Figure 3 (a) illustrates that samples with positive sentiment polarity are primarily found in the fashion, recreation, and scenery categories, aligning with common sense. Neutral samples constitute the largest proportion, concentrated

Table 4. Statistics of the scores distribution

Score	Sentiment			Emotion						
	Positive	Neutral	Negative	Sadness	Happiness	Fear	Anger	Surprise	Disgust	Neutral
0.0	0.42%	0.12%	1.96%	1.44%	0.16%	0.89%	1.29%	0.19%	1.33%	0.03%
0.125	18.42%	8.65%	46.97%	85.00%	30.05%	76.38%	95.36%	64.35%	97.16%	11.28%
0.25	17.35%	12.30%	19.08%	7.92%	23.11%	13.60%	2.20%	20.78%	1.23%	18.27%
0.375	21.22%	17.39%	16.75%	4.29%	18.98%	6.38%	1.04%	10.4%	0.25%	20.96%
0.5	16.78%	19.62%	7.81%	1.14%	12.92%	1.99%	0.07%	3.23%	0.02%	19.70%
0.625	12.25%	17.94%	4.16%	0.15%	7.80%	0.53%	0.02%	1.04%	0.00%	14.46%
0.75	7.65%	12.93%	2.08%	0.02%	4.33%	0.13%	0.01%	0.29%	0.00%	9.10%
0.875	3.82%	6.69%	0.71%	0.02%	1.99%	0.04%	0.01%	0.05%	0.00%	4.30%
1.0	2.07%	4.35%	0.49%	0.02%	0.65%	0.06%	0.00%	0.01%	0.00%	1.88 %
Avg	0.34	0.53	0.13	0.03	0.30	0.05	0.01	0.07	0.00	0.54

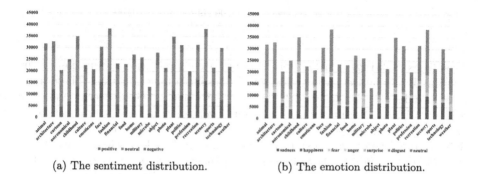

(a) The sentiment distribution.   (b) The emotion distribution.

Fig. 3. Distribution statistics

mainly in the animal, plant, and technology categories. Negative samples are the least frequent and are mainly distributed in the emoticons, face, and military categories, indirectly demonstrating the dataset's robustness.

Figure 3 (b) shows that happiness samples are concentrated in the fashion, childhood, and face categories. Military-related content tends to evoke fear, followed by adverse weather and tragic childhood. The animal, emoticons, and face categories have a higher likelihood of triggering anger emotional polarity. Samples with surprise emotional polarity are prominent in categories such as astronomical, scenery, and weather, which are associated with the unknown. Disgust samples are generally rare across all categories and are not visually represented clearly. Neutral emotional polarity samples are the most common.

5   Experiments

Dataset Split. We randomly split the data of SentiImgBank dataset into train, validation and test subsets by 60%, 20%, 20%. Table 5 shows the detailed statistics for train, validation and test subsets.

Table 5. Dataset statistics.

		SentiImgBank		
		Train	Validation	Test
Sentiment	Positive	133241	44142	44598
	Neutral	205556	68812	68743
	Negative	49526	16506	16127
Emotion	Sadness	12001	3977	3939
	Happiness	119508	39835	39907
	Fear	18678	6196	6232
	Anger	2314	789	711
	Surprise	26562	8944	8719
	Disgust	508	165	181
	Neutral	209796	69884	70100

Baselines Settings. In order to conduct a comprehensive evaluation of the SentiImgBank dataset, we employ several baselines for evaluation. These baselines include: (1) AlexNet model: The AlexNet model is the pioneering model in achieving image classification using multi-layer convolutional neural networks. (2) VGG16 model: The VGG16 model consists of 13 convolutional layers and 3 fully connected layers. (3) Inceptionv3 model: The Inceptionv3 model utilizes the Inception module as its core component. (4) DenseNet121 model: The DenseNet121 network comprises four dense blocks and three transition layers. (5) ResNext model: The ResNeXt-50 network consists of a total of 50 layers organized into 4 stages, each containing multiple residual blocks.

Results and Discussion. By utilizing Mean Squared Error (MSE), Mean Absolute Error (MAE), and Root Mean Squared Error (RMSE) as evaluation metrics, we present the results in Table 6 and make the following observations.

First, the ResNeXt50 network achieved the best performance in both tasks, highlighting the effectiveness of multiple residual blocks and multi-scale feature extraction. In comparison, AlexNet's performance was comparatively unsatisfactory when compared to some of the later-developed network models, underscoring the progress made by subsequent models through continuous innovation and improvement. Second, in the sentiment task with fewer categories, the models exhibited higher RMSE values, whereas in the emotion task with a greater number of categories, the models showcased lower RMSE values. This disparity may be attributed to the independent execution of the two tasks, resulting in some level of randomness in the outcomes. Finally, various models, particularly those refined in later stages, demonstrated low MSE and RMSE values, affirming the credibility and robustness of the SentiImgBank dataset. These findings demonstrate the performance and reliability of the evaluated models and underscore the quality of the SentiImgBank dataset.

Table 6. Dataset statistics.

Model	Sentiment			Emotion		
	MSE	MAE	RMSE	MSE	MAE	RMSE
AlexNet	0.3978	0.4981	0.6307	0.3739	0.4797	0.6115
VGG16	0.1793	0.3353	0.4234	0.1147	0.2637	0.3387
Inceptionv3	0.1918	0.3501	0.4379	0.1531	0.3097	0.3913
DenseNet121	0.1820	0.3385	0.4266	0.1575	0.3144	0.3969
ResNeXt50	0.1100	0.2629	0.3317	0.0858	0.2265	0.2930

6   Conclusions

In summary, this research introduces SentiImgBank, a large-scale visual repository for sentiment analysis. By annotating images with discrete sentiment and emotion scores, the dataset provides a benchmarking visual sentiment lexicon. The study proposes a committee of multi-task multi-class classifiers to automatically generate sentiment and emotion scores, reducing manual annotation efforts. The state-of-the-art baselines demonstrate the credibility and potential of SentiImgBank. Overall, this research contributes to the field of visual sentiment analysis by providing a publicly available resource and benefiting the multi-modal sentiment analysis community.

References

1. Borth, D., Ji, R., Chen, T., Breuel, T., Chang, S. F.: Large-scale visual sentiment ontology and detectors using adjective noun pairs. In: Proceedings of the 21st ACM International Conference on Multimedia, pp. 223–232 (2013)
2. Jou, B., Chen, T., Pappas, N., Redi, M., Topkara, M., Chang, S. F.: Visual affect around the world: a large-scale multilingual visual sentiment ontology. In: Proceedings of the 23rd ACM International Conference on Multimedia, pp. 159–168 (2015)
3. Plutchik, R.: Emotion, a Psychoevolutionary Synthesis (1980)
4. Xu, N., Mao, W., Chen, G.: Multi-interactive memory network for aspect based multimodal sentiment analysis. Proc. AAAI Conf. Artif. Intell. **33**(01), 371–378 (2019)
5. Poria, S., Hazarika, D., Majumder, N., Naik, G., Cambria, E., Mihalcea, R.: Meld: a multimodal multi-party dataset for emotion recognition in conversations. arXiv preprint arXiv: 1810.02508 (2018)
6. Sebastiani, F., Esuli, A.: SentiWordNet: a publicly available lexical resource for opinion mining. In: Proceedings of the 5th International Conference on Language Resources and Evaluation, pp. 417–422. European Language Resources Association (ELRA) Genoa, Italy (2006)
7. Esuli, A., Sebastiani, F.: Determining term subjectivity and term orientation for opinion mining. In: 11th Conference of the European chapter of the Association for Computational Linguistics, pp. 193–200 (2006)
8. Song, X., Huang, L., Xue, H., Hu, S.: Supervised prototypical contrastive learning for emotion recognition in conversation. arXiv preprint arXiv:2210.08713 (2022)

MSED: A Robust Ellipse Detector with Multi-scale Merging and Validation

Zikai Wang and Baojiang Zhong[✉]

School of Computer Science and Technology, Soochow University, Suzhou, China
zkwangsoochow@stu.suda.edu.cn, bjzhong@suda.edu.cn

Abstract. In the task of ellipse detection, the edge pixels that share the similar elliptic convexity are identified and combined to produce a potential ellipse. Due to the discreteness error and various kinds of image noise inherited in real-world images (e.g., the noise caused by compression attack and low-illustration), some of these elliptic edge pixels might deviate from the positions they supposed to be. Consequently, the ellipse detectors could have inferior performance in terms of precision and recall rate. For that, the existing ellipse detection methods always apply a Gaussian smoothing to the input image before the detection procedure. Due to the scale value (i.e., the standard deviation of Gaussian kernel) needs to be determined in advance, the existing single-scale ellipse detectors still suffer from the inferior precision and recall rate issues. To simultaneously solve these issues, a multi-scale strategy is innovatively developed for ellipse detection, based on which a novel *multi-scale ellipse detector* (MSED) is proposed. In our MSED, the elliptic edge pixels at multiple scales are jointly used to produce ellipses with high quality. For that, the ellipses detected at multiple scales are first produced, followed by merging them with a new *multi-scale ellipse merging* approach. In this approach, a *probabilistic model* is developed, based on which the homologous ellipses at multiple scales are merged. Lastly, a *multi-scale ellipse validation* check is further developed to discard those merged ellipses that have low confidence. Extensive experimental results show that our MSED outperforms the current state-of-the-arts and is robust to noise.

Keywords: Multiple scales · Ellipse detection · Multi-scale ellipse merging · Multi-scale ellipse validation

1 Introduction

In recent years, ellipse detection plays an important role in many computer vision applications, such as cell segmentation and detection [4,15], pose estimation [14,

This work was supported in part by the Natural Science Foundation of the Jiangsu Higher Education Institutions of China under Grant 21KJA520007, in part by the Collaborative Innovation Center of Novel Software Technology and Industrialization and in part by the Priority Academic Program Development of Jiangsu Higher Education Institutions.

Q. Liu et al. (Eds.): PRCV 2023, LNCS 14435, pp. 506–517, 2024.
https://doi.org/10.1007/978-981-99-8552-4_40

21], augmented and virtual reality (AR/VR) [3, 22] and so on. Over the past few decades, a great number of ellipse detection methods have been developed.

Most of these methods rely on a *single scale*, where the *scale* refers to the standard deviation of a Gaussian smoothing implemented at the beginning of the detection procedure. The Gaussian smoothing is used to suppress various kinds of image noise in the input image. In these single-scale ellipse detection methods, the standard deviation, denoted as σ, of Gaussian kernel needs to be determined in advance. However, due to the different degrees of quantization error and image noise in real-world images, it is difficult to determine an optimal scale value for every image in advance. On the other hand, an excessively high scale value will suppress some image features (i.e., elliptic edge pixels in ellipse detection). Reversely, the image noise could seriously impact the detection procedure when an excessively low scale value is selected. In our view, how to form precise elliptic edge pixels is highly instrumental to the success of ellipse detection. To pursue this goal, a multi-scale strategy is considered and innovatively exploited in our proposed method, called the *multi-scale ellipse detector* (MSED), for effectively conducting ellipse detection.

In our method, a set of ellipses are first detected at multiple scales. Based on these multi-scale detections, a new multi-scale ellipse merging approach is developed and used to obtain the merged ellipses. To conduct the merging procedure, the homologous ellipses (i.e., the ellipses that could derive from the same ground-truth ellipse) at multiple scales are identified. After that, a probabilistic model is established, based on which the homologous ellipses are merged. Lastly, a new multi-scale validation check is proposed to discard those wrong-merged and insignificant ellipses. With these two approaches, our proposed MSED always has a superior performance in detection precision and recall rate.

The rest of this paper is organized as follows. Section 2 overviews existing ellipse detection methods. Section 3 describes the basic idea of the proposed method and the analysis of the evolution of an ellipse through multiple scales. Our proposed method is detailed in Sect. 4. Experimental results are shown in Sect. 5. Finally, conclusions are given in Sect. 6.

2   Related Work

Existing ellipse detection methods can be roughly classified into *Hough transform*, *edge* and *learning*-based methods. The Hough transform technique is used to detect a particular shape by mapping the sampled points to the corresponding parameter space. For the elliptic shape, due to the high dimension of the parameter space (5D), the HT-based methods, such as the randomized Hough transform (RHT) method [20] and the iterative RHT (IRHT) [10], always have heavy computation burden and excessive consumption of memory.

The edge-based methods are the second well-known family of ellipse detection, which have received great interest in recent years due to its high accuracy and efficiency. The edge pixels that derive from the same ellipse are linked to form the elliptic arcs. Then, these arcs are grouped to produce the potential

Fig. 1. An illustrative example of elliptic edge pixels at multiple scales: (a) the test image, and the details in red box will be displayed; (b)-(d) the edge pixels detected by the Canny [2] operator at multiple scales with $\sigma = 0.6, 1$ and 10, respectively. (Color figure online)

ellipses. Prasad *et al.* [16] proposed a searching method to find the valid groups of elliptic arcs. Fornaciari *et al.* [6] proposed a fast detection method which classified elliptic arcs into four quadrants and the arcs are further combined under a geometric constraint. Following the work of Fornaciari *et al.* [6], Jia *et al.* [7] further improved the detection speed by using the characteristic number to reduce interference from linear structures. Lu *et al.* [9] extracted line segments to approximate elliptic arcs and an arc-support region based method is developed to filter out the line segments derived from linear structures. Meng *et al.* [13] proposed a fast ellipse detector, which conduct an arc adjacency matrix (AAM) to represent the positional relationship between each pair of arc segments, and then the ellipses are produced by decompose the matrix. Shen *et al.* [18] proposed a method, in which arcs and their connectivity are encoded into a sparse directed graph, and then ellipses are generated via a fast access of the adjacency list. Matrorell *et al.* [12] proposed a multi-scale detection strategy, in which the contours detected at multiple scales are merged and classified into the circle, ellipse and line segment. Since the homologous ellipses detected at multiple scales are directly merged by averaging their parameters, the inferior detections at different scales could also impact the merging result.

The learning-based methods always conduct an ellipse model and produce the ellipse by minimizing the loss function between the ellipse model and observation model. Arellano and Dahyot [1] used the Gaussian mixture model (GMM) to conduct an ellipse model and minimizing the $\mathcal{L}_2$ loss between an ellipse model and an observation model. Some deep learning-based models have also been proposed to detect ellipses [5,8,19]. However, subject to the training datasets and the generalizability of the models, these methods can only detect specific elliptic objects, e.g., pupil and iris in [8], face and fruit in [5] and large-size elliptic object in [19].

3   Basic Idea

Due to the quantization error and image noise caused by compression attack, low-light illumination and so on, the elliptic edge pixels could be easily perturbed. An illustrative example is shown in Fig. 1. In Fig. 1(a) the test image is

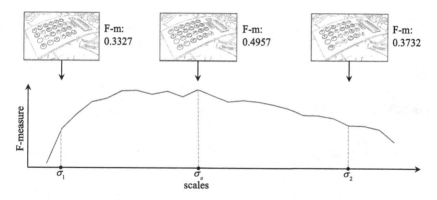

Fig. 2. The objective and subjective performance of a single-scale ellipse detector [9] implemented at different scales.

demonstrated. The elliptic edge pixels detected by the Canny operator [2] at different scales are demonstrated in Figs. 1(b)–1(d), respectively. One can see from Fig. 1(b) that the distribution of elliptic edge pixels are disturbed (discontinue and many noisy pixels) at a common-used scale ($\sigma = 0.6$). An increase of the scale value might benefit to suppress the noise influence. However, some elliptic edge pixels could be suppressed as well, as shown in Fig. 1(c), where the edge pixels extracted at a higher scale ($\sigma = 1$) are depicted. On the other hand, the deviation of an edge pixel becomes more intense as the scale value increases [11]. For demonstration, the edge pixels extracted at a much higher scale ($\sigma = 10$) is shown in Fig. 1(d). It is easy to see that the image noise is suppressed, while some edge pixels are in wrong locations or completely disappeared.

Therefore, the existing single-scale ellipse detectors could have different performance as scale value increases. In Fig. 2, the performance of a single-scale ellipse detector [9] in terms of the *F-measure* (referred to F-m) is demonstrated. As the scale value increases, the image noise is suppressed, and thus the F-m increases. When the scale value is larger than a certain scale σ_o, the F-m gradually decreases. This phenomenon can further prove our analysis of elliptic edge pixels at multiple scales. Thus, in our work, the elliptic edge pixels detected at multiple scales are jointly used to simultaneously address these issues.

In order to conduct our method, the multi-scale representation of the input image $I(x, y)$ is generated by using a set of Gaussian kernels which take the following form

$$g(u, v; \sigma_i) = \frac{1}{2\pi\sigma_i^2} \exp\left(-\frac{u^2 + v^2}{2\sigma_i^2}\right), \tag{1}$$

where (u, v) denotes the coordinate in Gaussian kernel and σ_i is the standard deviation of the Gaussian kernel which is determined as

$$\sigma_i = 0.3\sqrt{2}^{i-1} \tag{2}$$

with $i = 1, 2, \cdots, 5$. Note that the number of used scales determines the trade-off between the efficiency and accuracy. In this work, this number is empirically

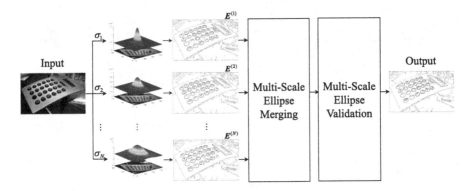

Fig. 3. Outline of our proposed multi-scale ellipse detector.

set to 5. Based on this set of Gaussian kernels, the multi-scale representation of $I(x, y)$ is computed by

$$I^{(i)}(x, y) = I(x, y) * g(u, v; \sigma_i), \tag{3}$$

where the symbol $*$ denotes the convolutional operation. Based on the obtained multi-scale images, our *multi-scale ellipse detection* (MSED) method are proposed to simultaneously address the above-mentioned poor precision and inferior recall rate issues.

4   Methodology

Figure 3 depicts the outline of our proposed *multi-scale ellipse detector* (MSED). To conduct our method, a single-scale ellipse detection method [9] is individually implemented at multiple scales to obtain the multi-scale detection results, due to its efficiency and high quality. Based on these detections, a *multi-scale ellipse merging* approach is developed. In this approach, the optimal scale of the input image is first determined by counting the number of elliptic edge pixels. After that, the homologous ellipses are identified. A probabilistic model is then established to merge these homologous ellipses. Lastly, a multi-scale validation check is developed to discard those ellipses with low confidence.

4.1   Identification of Optimal Scale

Given an input image $I(x, y)$, the multi-scale representation of $I(x, y)$ is obtained by (2) and (3). The set of multi-scale images is denoted as $\boldsymbol{I}_\sigma = \{I^{(i)}(x, y)\}_{i=1}^N$, where N is the number of scales and is set to 5 in this work. For the image at each scale, a single-scale ellipse detection method [9] is individually implemented to produce the ellipse detection at this scale. The set of detected ellipses at the ith scale is denoted as $\boldsymbol{E}^{(i)} = \{E_j^{(i)}\}_{j=1}^{N_i}$ with N_i is the number of ellipses detected at ith scale and $E_j^{(i)} = \{c_x, c_y, a, b, \vartheta\}$ is the jth ellipse detected at the ith scale, where (c_x, c_y) is the ellipse center, a and b are the semi-axes, respectively, and ϑ is the rotational angle with respect to the horizontal axis.

With the ellipses obtained at multiple scales, the optimal scale of $I(x, y)$ is identified, which plays an important role in the follow-up steps. Due to the importance of elliptic edge pixels to the success of ellipse detection, the optimal scale is identified by counting the number of elliptic edge pixels at each scale. For an ellipse $E_j^{(i)}$ detected in $I^{(i)}(x, y)$, the elliptic edge pixels of $E_j^{(i)}$ are those edge pixels (extracted by the Canny operator [2]) that are adjacent to $E_j^{(i)}$ and have similar gradient orientations to the normal orientation of the nearest point on $E_j^{(i)}$ [9,13]. Assuming that there are N_p elliptic edge pixels of $E_j^{(i)}$ are extracted, the pixel set is denoted as $p\left(E_j^{(i)}\right) = \{p_k\}_{k=1}^{N_p}$, where $p_k = (x_k, y_k)$ is the coordinate of an edge pixel. Based on these elliptic edge pixels, the *optimal scale* of $I(x, y)$ is defined as

$$\sigma_o = \arg\max_{\sigma_i} \frac{\sum_{j=1}^{N_i} \left| p\left(E_j^{(i)}\right) \right|}{N_i}. \tag{4}$$

Then, the scale range $[\sigma_{o-1}, \sigma_{o+1}]$ is selected for performing our proposed multi-scale ellipse merging and validation.

4.2 Multi-scale Ellipse Merging

The ellipses detected at each scale are merged progressively via an iterative scheme as follows:

$$E_m^{(k+1)} = \mathcal{M}(E_m^{(k)}, E^{(o-1+k)}) \tag{5}$$

with $k = 1, 2$, $E_m^{(1)}$ is set to $E^{(o-1)}$, $E_m^{(k+1)}$ is the merged result after k iterations and $\mathcal{M}(\cdot, \cdot)$ represents the merging operation that aggregates two input ellipse sets into a merged one which is detailed as follows.

Assuming that we start with the jth ellipse $E_j^{(o-1)}$ in $E^{(o-1)}$. First, the homologous ellipses of $E^{(o-1)}$ in $E^{(o)}$ are identified by comparing the distance between $E_j^{(o-1)}$ and each ellipse in $E^{(o)}$. Given two ellipses E_1 and E_2, the distance between them is calculated as

$$d_e(E_1, E_2) = \sqrt{\sum_{v=1}^{5} (E_{1,v} - E_{2,v})^2}, \tag{6}$$

where $E_{k,v}$, with $k = 1, 2$, is the vth dimension in the parameters of E_i. These two ellipse will be regarded as homologous ellipses if $d_e(E_1, E_2)$ is lower than a pre-determined threshold τ_d. The ellipse $E_j^{(o-1)}$ together with its homologous ellipses are then grouped in a homologous-ellipse set. This set is denoted as $E_h = \{E_j\}_{j=1}^{N_h}$ with N_h ellipses in total.

With this homologous-ellipse set, the merged ellipse is produced by jointly using the elliptic edge pixels of each ellipse in E_h detected at σ_o. To be specific, for each $E_j \in E_h$, it is overlaid on the image at σ_o, and then its elliptic edge pixels $\boldsymbol{p}(E_j) = \{p_k\}_{k=1}^{N_p}$ are extracted from $I^{(o)}(x, y)$. Next, the *most reliable ellipse* in $\boldsymbol{E}_h$ is identified and used to produce the merged ellipse. Here, the 'most reliable ellipse' denotes the ellipse in $\boldsymbol{E}_h$ that is most similar to the *ground-truth ellipse*. However, since the information of the unknown ground-truth ellipse is not available, a probabilistic model is established and used to measure the reliability of each ellipse. To conduct this probabilistic model, the alignment of elliptic edge pixels to the corresponding ellipse is measured as

$$\mathcal{A}(E_j) = \frac{\Sigma_{k=1}^{N_p} \left(\varphi(E_j, p_k) \cdot \mathrm{grad}(p_k) \cdot \psi(E_j, p_k) \right)}{\Sigma_{k=1}^{N_p} \varphi(E_j, p_k) \cdot \Sigma_{k=1}^{N_p} \mathrm{grad}(p_k)}, \tag{7}$$

where $\mathrm{grad}(p_k)$ is the gradient magnitude measured at p_k, $\psi(E_j, p_k)$ and $\varphi(E_j, p_k)$ are two functions used to assess the alignment of each elliptic edge pixels to the current ellipse E_j in terms of *gradient orientation* and *location*, respectively. The two functions are calculated as

$$\psi(E_j, p_k) = \exp\left(-\frac{1}{2\sigma_\psi^2} \left(\theta(p_k) - \nabla E_j(p_k) \right)^2 \right) \tag{8}$$

with $\sigma_\psi = 2$, $\theta(p^{(k)})$ is the gradient orientation of p_k measured at σ_o, $\nabla E_j(p_k)$ is the normal vector of point p_k to E_j, and

$$\varphi(E_j, p_k) = \frac{1}{\sqrt{2\pi}\sigma_\varphi} \exp\left(-\frac{d(p_k, E_j)^2}{2\sigma_\varphi^2} \right) \tag{9}$$

with $\sigma_\varphi = 25$, $d(p_k, E_j)$ is the Rosin approximation distance [17] (i.e., an approximation of the normal distance) from the point p_k to E_j. With the estimated alignment of elliptic edge pixels, the reliability of E_j is computed as

$$\mathcal{R}(E_j) = |\boldsymbol{p}(E_j)| \cdot \mathcal{A}(E_j). \tag{10}$$

The ellipse with the highest $\mathcal{R}(E_j)$ is selected to produce the merged ellipse. That is, the elliptic edge pixels of the most reliable ellipse are extracted across all the scales in $[\sigma_{o-1}, \sigma_{o+1}]$. The union of these pixels are jointly used to fitting a new ellipse as the final merged ellipse E_m.

4.3   Multi-scale Ellipse Validation

A validation approach is developed as follows to measure the quality of each merged ellipse. With this approach, the ellipses with quality below a given threshold will be discarded. In our developed validation check, the *density* and *uniformity* cues of the elliptic edge pixels at multiple scales are analyzed to measure the quality of each merged ellipse, specified as follows.

For a merged ellipse E_m, its elliptic edge pixels extracted at each scale in $[\sigma_{o-1}, \sigma_{o+1}]$ are extracted. Denote the union of these elliptic edge pixels as $\{p_k\}_{k=1}^{S}$ with S pixels. The density is defined as the number of elliptic edge

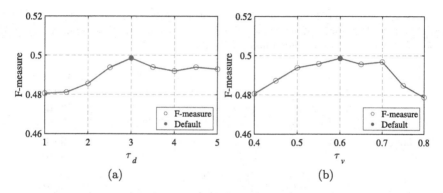

Fig. 4. Parameter tuning of (a) the threshold τ_d used to determine the homologous ellipse; (b) the threshold τ_v used in our validation approach.

pixels divided by the perimeter of the merged ellipse. Then, the quality in terms of the density of elliptic edge pixels is measured as $S/\mathcal{B}(E_m)$, where $\mathcal{B}(E_m)$ is the perimeter of E_m. In general, a higher density of the elliptic edge pixels indicates a better quality of E_m; that is, the E_m is more likely to be a true positive.

The uniformity of the elliptic edge pixels distributed around the merged ellipse is another crucial cue to measure the quality of E_m. An ideally uniform distribution is supposed to produce the best ellipse, while a highly nonuniform one could result in inferior performance. Thus, the uniformity of the elliptic edge pixel distribution is quantized as follows for assessing the quality of merged ellipse. First, E_m is divided into 360 accumulator bins $\{B_k\}$ in the polar coordinate system, with the kth bin B_k corresponding to the angle range of $((k-1)°, k°]$, for $k = 1, 2, \cdots, 360$, respectively. Based on these accumulator bins, the number of the bins that have at least one elliptic edge pixel is incremented as

$$B_k = \sum\nolimits_{i=1}^{S} \delta(\lceil \arctan(\nabla E_m(p_i)) \rceil - k), \tag{11}$$

where $\delta(\cdot)$ is the *Dirac delta* function. Then, the quality of E_m in terms of the uniformity is calculated as

$$U = \sum\nolimits_{k=1}^{360} H(B_k)/360, \tag{12}$$

where $H(\cdot)$ is the *Heaviside step* function.

By integrating the cues from the density and uniformity of elliptic edge pixels, the quality of each merged ellipse is calculated as

$$\mathcal{Q}(E_m) = \alpha S/\mathcal{B}(E_m) + (1-\alpha)U, \tag{13}$$

where $\alpha = 0.6$ is a weighting factor. Finally, each merged ellipse with $\mathcal{Q}(E_m)$ lower than a pre-determined threshold τ_v will be discarded.

Table 1. A comparison of different ellipse detection methods in terms of the MOR (1st entry) and the F-measure (2nd entry). The highest score of the MOR and F-measure achieved on each dataset is highlighted in boldface.

Dataset	YAED [6]	CNED [7]	Lu [9]	AAMED [13]	EllDet [18]	EllSeg [8]	Proposed (Ours)
Prasad [16]	.9138 .3437	.9078 .3586	.9362 .4815	.9333 .4566	.9171 .4303	.8798 .0345	**.9364** **.4986**
Random [6]	.9124 .4163	.9105 .4328	.9357 .5259	.9344 .4894	.9237 .4677	.0000 .0000	**.9367** **.5351**
SmartPhone [6]	.9077 .4003	.9022 .4190	.9214 .6180	.9230 .5266	.9080 .5333	.9008 .0512	**.9241** **.6232**
Satellite [13]	.9110 .3503	.9058 .3735	.9533 .5452	.9440 .5023	.9089 .5672	.8834 .1207	**.9572** **.5712**
Average	.9112 .3777	.9066 .3959	.9367 .5427	.9337 .4937	.9144 .4996	.6667 .0516	**.9386** **.5570**

5   Experiments

5.1   Experimental Setup

The following widely-used datasets are used to evaluate our proposed method, including the Prasad Dataset [16] with 198 real-world images with complex backgrounds; the Random Dataset [6] which has 400 real-world images collected from MIRFlickr and LabelMe repositories [6]; the SmartPhone Dataset [6] which has 629 images about traffic sign and bicycle; the Satellite Dataset [13] including 757 images taken by the NextSat spacecraft in space.

The performance of our method is compared with six state-of-the-art ellipse detectors including the YAED [6], the CNED [7], the Lu *et al.* [9], the AAMED [13], the EllDet [18] and the EllSeg [8]. Codes of these methods are all publicly available online. Note that the EllSeg [8] is a deep-learning method that developed to detect pupil and iris in the input image. The model proposed by the authors is directly used to conduct the experimental results on a NVIDIA Titan Xp GPU. All experimental results produced by other methods are performed with default parameters on the same computer.

There are only two parameters in our proposed method, which are τ_d for identifying homologous ellipses and τ_v for validating merged ellipses. Both of the two parameters are tuned on the Prasad dataset and then applied to the other benchmark datasets. Figure 4 shows the tuning process. In this work τ_d and τ_v are set to 3 and 0.6, respectively.

5.2   Objective Evaluation

The metrics of *F-measure* and *mean overlap ratio* (MOR) measured over all the images from each dataset is utilized to conduct our objective evaluation.

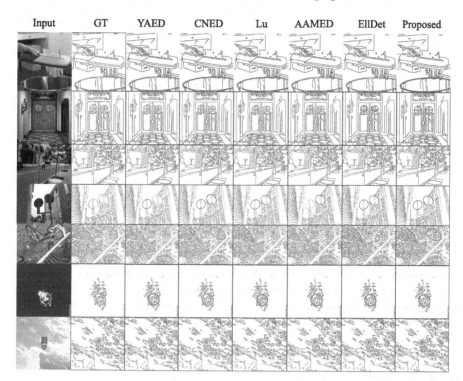

Fig. 5. Ellipse detection results by using different methods on a set of test images taken from the used datasets.

The F-measure is the harmonic mean of the precision and recall rate, and MOR denotes the confidence of a detected ellipse can be regarded as a true positive. A detected ellipse is identified as a true positive if its MOR to the corresponding ground truth is larger than a given threshold D_0, which is set to 0.8 as a common practice in the literature. The detection result of each method is documented in Table 1. One can see that our proposed method outperforms all the state-of-the-art methods in every comparison case.

5.3 Subjective Evaluation

Then, seven images taken from the used datasets are utilized for comparing subjective performance with five single-scale methods. The detection results are depicted in Fig. 5. By comparing with these single-scale methods, one can see that our proposed MSED outperforms these methods in terms of detection precision and recall rate.

5.4 Performance of Robustness to Noise

Two noise categories are employed on the Prasad dataset [16] for comparing with five single-scale methods. First, each images in the Prasad Dataset [16] is

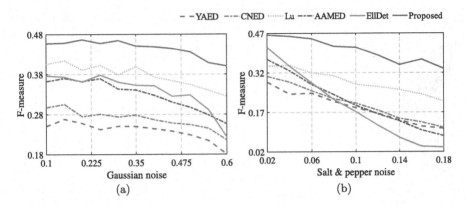

Fig. 6. The performance of compared methods in terms of (a) the Gaussian noise and (b) the salt and pepper noise.

corrupted by the Gaussian noise with increasing variance. The result is shown in Fig. 6(a). Second, each image in the Prasad Dataset [16] is corrupted by the salt and pepper noise with increasing noise density, as demonstrated in Fig. 6(b). By comparing with other method, one can see that our MSED outperforms other methods by a large margin.

6   Conclusion

In this work, a novel multi-scale method is proposed to detect ellipse with superior performance. First, the variation of elliptic edge pixels at multiple scales is analysed. Based on this analysis, our multi-scale ellipse detector (MSED) is developed. Then, a set of ellipses detected at multiple scales is generated from the multi-scale representation of the input image. Based on the generated multi-scale detections, a multi-scale ellipse merging approach is developed and used to obtain the merged ellipses. To conduct the merging procedure, the optimal scale of the input image is obtained by counting the number of elliptic edge pixels at each scale. With the calculated optimal scale, the homologous ellipses are identified, and then merged by using a new probabilistic model. Lastly, for each merged ellipse, a new ellipse validation check is proposed to discard the ellipse with low confidence. In this validation check, the density and uniformity cues of elliptic edge pixels are integrated to assess a merged ellipse. Extensive experimental results show that our MSED can outperform the state-of-the-art methods and be robust to noisy images.

References

1. Arellano, C., Dahyot, R.: Robust ellipse detection with Gaussian mixture models. Pattern Recognit. **58**, 12–26 (2016)
2. Canny, J.: A computational approach to edge detection. IEEE Trans. Pattern Anal. Mach. Intell. **8**(6), 679–698 (1986)

3. Cholewiak, S.A., Love, G.D., Srinivasan, P.P., Ng, R., Banks, M.S.: ChromaBlur: rendering chromatic eye aberration improves accommodation and realism. ACM Trans. Graphics **36**(6), 1–12 (2017)

4. Das, P.K., Meher, S., Panda, R., Abraham, A.: An efficient blood-cell segmentation for the detection of hematological disorders. IEEE Trans. Cybern. **52**(10), 10615–10626 (2021)

5. Dong, W., Roy, P., Peng, C., Isler, V.: Ellipse R-CNN: learning to infer elliptical object from clustering and occlusion. IEEE Trans. Image Process. **30**, 2193–2206 (2021)

6. Fornaciari, M., Prati, A., Cucchiara, R.: A fast and effective ellipse detector for embedded vision applications. Pattern Recognit. **47**(11), 3693–3708 (2014)

7. Jia, Q., Fan, X., Luo, Z., Song, L., Qiu, T.: A fast ellipse detector using projective invariant pruning. IEEE Trans. Image Process. **26**(8), 3665–3679 (2017)

8. Kothari, R.S., Chaudhary, A.K., Bailey, R.J., Pelz, J.B., Diaz, G.J.: EllSeg: an ellipse segmentation framework for robust gaze tracking. IEEE Trans. Visual Comput. Graphics **27**(5), 2757–2767 (2021)

9. Lu, C., Xia, S., Shao, M., Fu, Y.: Arc-support line segments revisited: an efficient high-quality ellipse detection. IEEE Trans. Image Process. **29**, 768–781 (2020)

10. Lu, W., Tan, J.: Detection of incomplete ellipse in images with strong noise by iterative randomized Hough transform (IRHT). Pattern Recognit. **41**(4), 1268–1279 (2008)

11. Lu, Y., Jain, R.C.: Behavior of edges in scale space. IEEE Trans. Pattern Anal. Mach. Intell. **11**(4), 337–356 (1989)

12. Martorell, O., Buades, A., Lisani, J.L.: Multiscale detection of circles, ellipses and line segments, robust to noise and blur. IEEE Access **9**, 25554–25578 (2021)

13. Meng, C., Li, Z., Bai, X., Zhou, F.: Arc adjacency matrix-based fast ellipse detection. IEEE Trans. Image Process. **29**, 4406–4420 (2020)

14. Meng, C., Li, Z., Sun, H., Yuan, D., Bai, X., Zhou, F.: Satellite pose estimation via single perspective circle and line. IEEE Trans. Aerosp. Electron. Syst. **54**(6), 3084–3095 (2018)

15. Panagiotakis, C., Argyros, A.: Region-based fitting of overlapping ellipses and its application to cells segmentation. Image Vision Comput. **93**, 103810 (2020)

16. Prasad, D.K., Leung, M.K., Cho, S.Y.: Edge curvature and convexity based ellipse detection method. Pattern Recognit. **45**(9), 3204–3221 (2012)

17. Rosin, P.L.: Ellipse fitting using orthogonal hyperbolae and stirling's oval. Graph Models Image Process. **60**(3), 209–213 (1998)

18. Shen, Z., Zhao, M., Jia, X., Liang, Y., Fan, L., Yan, D.M.: Combining convex hull and directed graph for fast and accurate ellipse detection. Graph. Models **116**, 101110 (2021)

19. Wang, T., Lu, C., Shao, M., Yuan, X., Xia, S.: ElDet: an anchor-free general ellipse object detector. In: Proceedings of Asian Conference Computer Vision, pp. 2580–2595 (2022)

20. Xu, L., Oja, E., Kultanen, P.: A new curve detection method: randomized Hough transform (RHT). Pattern Recognit. Lett. **11**(5), 331–338 (1990)

21. Zins, M., Simon, G., Berger, M.O.: Object-based visual camera pose estimation from ellipsoidal model and 3D-aware ellipse prediction. Int. J. Comput. Vis. **130**(4), 1107–1126 (2022)

22. Zubizarreta, J., Aguinaga, I., Amundarain, A.: A framework for augmented reality guidance in industry. Int. J. Adv. Manuf. Tech. **102**(9), 4095–4108 (2019)

Author Index

© The Editor(s) (if applicable) and The Author(s), under exclusive license
to Springer Nature Singapore Pte Ltd. 2024
Q. Liu et al. (Eds.): PRCV 2023, LNCS 14435, pp. 519–521, 2024.
https://doi.org/10.1007/978-981-99-8552-4

Printed in the United States
by Baker & Taylor Publisher Services